The **Teacher-Friendly** Guide™ to the Earth Science of the Southwestern US

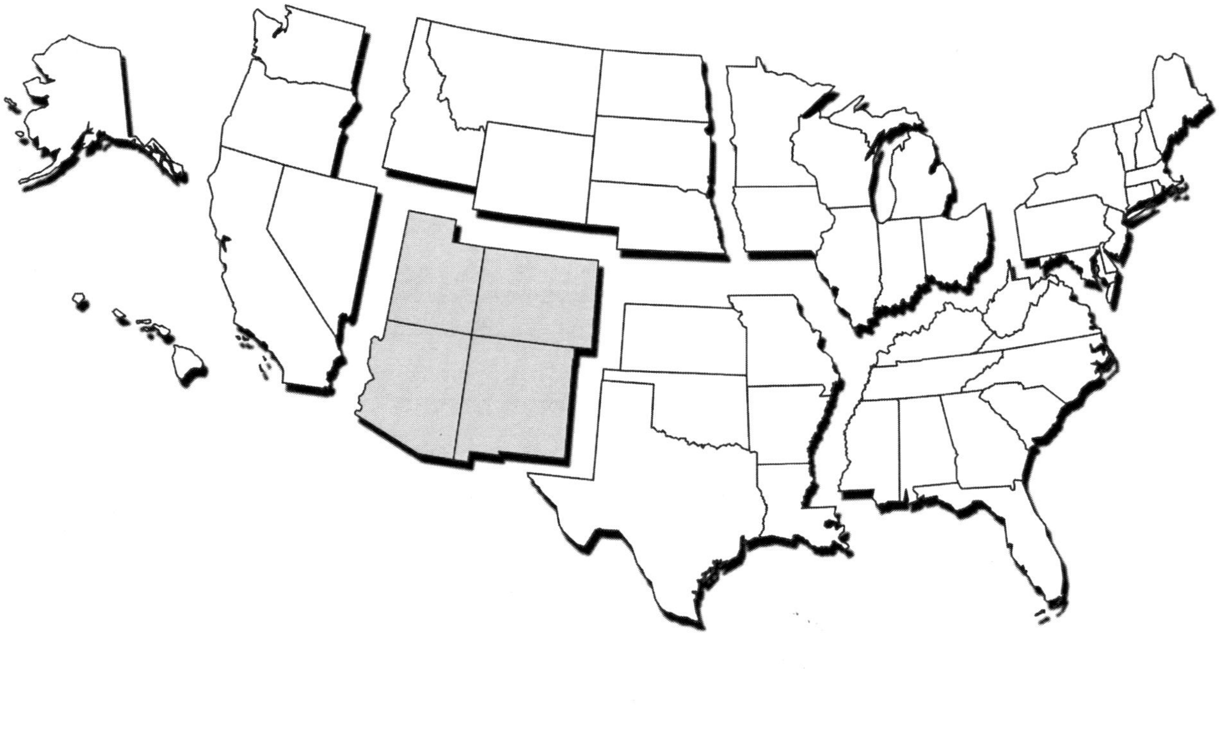

Edited by Andrielle N. Swaby, Mark D. Lucas, & Robert M. Ross

Paleontological Research Institution
2016

ISBN 978-0-87710-514-5
Library of Congress no. 2016952807
PRI Special Publication no. 50

1259 Trumansburg Road
Ithaca, New York 14850 USA
priweb.org

First printing September 2016

This material is based upon work supported by the National Science Foundation under grant DRL-0733303. Any opinions, findings, and conclusions or recommendations are those of the author(s) and do not necessarily reflect the views of the National Science Foundation. The publication also draws from work funded by the Arthur Vining Davis Foundations and The Atlantic Philanthropies.

The interactive online version of this *Teacher-Friendly Guide*™ (including downloadable pdfs) can be found at http://teacherfriendlyguide.org. Web version by Brian Gollands.

The Teacher-Friendly Guide™ series was originally conceived by Robert M. Ross and Warren D. Allmon. Original illustrations in this volume are mostly by Jim Houghton (The Graphic Touch, Ithaca), Wade Greenberg-Brand, and Christi A. Sobel.

Layout and design by Paula M. Mikkelsen, Andrielle Swaby, Elizabeth Stricker, Wade Greenberg-Brand, and Katherine Peck.

Cite this book as:
Swaby, A. N., M. D. Lucas, & R. M. Ross (eds.), 2016, *The Teacher-Friendly Guide to the Earth Science of the Southwestern US.* Paleontological Research Institution (Special Publication 50), Ithaca, New York, 458 pp.

Cite one chapter as (example):
Allmon, W. D., and R. A. Kissel, 2016, Fossils of the Southwestern US. Pages 89–157, in: A. N. Swaby, M. D. Lucas, & R. M. Ross (eds.). *The Teacher-Friendly Guide to the Earth Science of the Southwestern US.* Paleontological Research Institution (Special Publication 50), Ithaca, New York.

On the back cover: Blended geologic and digital elevation map of the Southwestern US. Each color represents the age of the bedrock at the surface. Adapted from Barton, K.E., Howell, D.G., Vigil, J.F., *The North America Tapestry of Time and Terrain*, US Geological Survey Geologic Investigations Series I-2781, http://pubs.usgs.gov/imap/i2781.

Preface

Earth science is an inherently local subject. No two places share exactly the same sequence of events that led to the way they are today. In this sense, Earth science is a subject to be explored in one's own neighborhood, examining the detailed sequence of rocks for the history that has gone on under our feet. What is not possible from only one location is making sense of why this particular sequence of events took place when and where it did, particularly relative to sequences in other places around it.

The distribution of rocks and landforms can be explained by processes that shape areas covering thousands of kilometers, such as the evolution of the Colorado Plateau and the series of alternating uplands and valleys of the Basin and Range. These processes link widely separated sequences in a common history.

Earth science educators at the Paleontological Research Institution, in working with teachers, have noted that no single source for educators exists that attempts to make sense of the disparate local features of the Southwest United States in terms of a basic sequence of historical events and processes. Though nationally distributed textbooks make references to famous geological features of the Southwest and a number of reasonably good resources exist for individual states, these do not take enough geographic scope into account to show why, say, geothermal energy and volcanic deposits are abundant in western Utah, but fossil fuel resources and sedimentary rocks are abundant in eastern Utah and throughout Colorado, and what that has to do with the formation of the Rocky Mountains and the distribution of crystalline igneous and metamorphic rocks at the surface. Further, these resources are not necessarily "teacher-friendly," or written with an eye toward the kind of information and graphics that a secondary school teacher might need in their classroom. This *Teacher-Friendly Guide*™ is intended to fill this need for teachers.

Explaining why (for example, certain kinds of rocks and their mineral resources are found where they are) is the most effective way of providing students with a tool to remember and predict the nature of local Earth science. The Southwestern US (though, like states, an artificial political area) is of the right scale to discuss the evolution of significant portions of sedimentary basins, but also include a variety of igneous rocks. This means most Earth processes are illustrated by rocks present within a day's drive, and that Earth phenomena can be illustrated with examples in areas students and teachers are likely to have been to or at least heard of. Since the rocks and landforms are relatively accessible, regional Earth science is an excellent subject for hands-on, inquiry-based teaching using, for example, real rocks and landforms. A transect across the Southwestern US in several places will reveal most major rock types that students should know and will come into contact with over the course of their lifetimes.

The chapters chosen are by no means an exhaustive list, but reflect especially the historical side of "solid Earth" geosciences. Each chapter starts with a brief review, then (in most chapters) describes the Earth science of four natural physiographic regions of the Southwest. There is a resource list at the end of each chapter. There is a chapter on field work, not only on suggestions for how to do it, but how to integrate the field into a curriculum through "virtual fieldwork experiences." There are chapters on Big Ideas in Earth system science—a few major conceptual ideas that run throughout the subject—and on using real-world regional Earth science in the context of the Next Generation Science Standards (NGSS).

This volume is part of a national series of seven *Teacher-Friendly Guides*™ to regional Earth science, covering all 50 states. We also have two *Teacher-Friendly Guides*™ to evolution, and other Guides in development.

We all hope for our students that, years from now, they will be able to make sense of the place they live and the places they visit, through a comprehension of a few Big Ideas and a basic grasp of the "big picture" story of the geological history of their area. It is our hope that this book might help teachers, and their students, grasp such a coherent understanding of their regional and local Earth system science.

Robert M. Ross, Associate Director for Outreach
Don Duggan-Haas, Director of Teacher Programs
Paleontological Research Institution
September 2016

Table of Contents

Contributors

Warren D. Allmon
Paleontological Research Institution
Ithaca, New York

Carlyn S. Buckler
Cooperstown Graduate Program
Cooperstown, New York

Don Duggan-Haas
Paleontological Research Institution
Ithaca, New York

Lisa R. Fisher
Escalante Mines Inc.
Golden, Colorado

Thomas R. Fisher
Escalante Mines Inc.
Golden, Colorado

Bryan L. Isacks
Cornell University
Ithaca, New York

Richard A. Kissel
Yale University
New Haven, Connecticut

Luke P. McCann
Western Washington University
Bellingham, Washington

Judith T. Parrish
University of Idaho
Moscow, Idaho

Robert M. Ross
Paleontological Research Institution
Ithaca, New York

Andrielle N. Swaby
Paleontological Research Institution
Ithaca, New York

Ingrid H. H. Zabel
Paleontological Research Institution
Ithaca, New York

How to Use this Guide

General philosophy of the Teacher-Friendly Guides™

This Guide is organized by regional geologic history because it helps make sense of local Earth science—*Why does this place look the way it does? Why is this particular set of rocks, soil, landforms, water bodies, and local climate here?* We recommend introducing geologic history into your curriculum early.

The idea of systems also runs through the Guide. Through systems we understand, for example, why geologic history controls where different types of rocks occur, helping us make sense of landforms and water bodies. Landforms and water bodies in turn influence local climate, and all of it influences life. Understanding a few essentials of geologic history and Earth systems allows us to make sense of the world around us.

Please incorporate ideas from the Guide into your existing curriculum. This Guide is a resource rather than a curriculum itself.

Understanding real-world Earth science is a lifelong learning experience. Don't be intimidated by rocks that you don't recognize, fossils with long names, or complicated weather patterns. Enjoy learning alongside your students and show that enjoyment.

A National Series of Guides

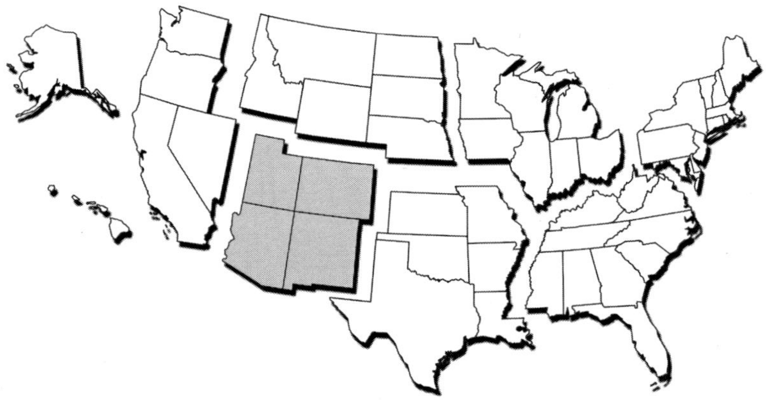

This Guide is one of seven covering the United States. There are also two *Teacher-Friendly Guides*™ to evolution, one focused upon bivalves and another focused on maize genetics. To learn more, visit www.teacherfriendlyguide.org, a website of the Paleontological Research Institution.

For the interactive web version of this Guide, visit www.teacherfriendlyguide.org. To download individual chapters for printing, visit the website for the Southwestern Guide and click "Downloads" on the chapter menu to the left.

To purchase a printed grayscale copy, visit "Publications" at the Paleontological Research Institution website. Images in the printed version, which are in grayscale, are available in color in the digital versions.

Design of the Guides

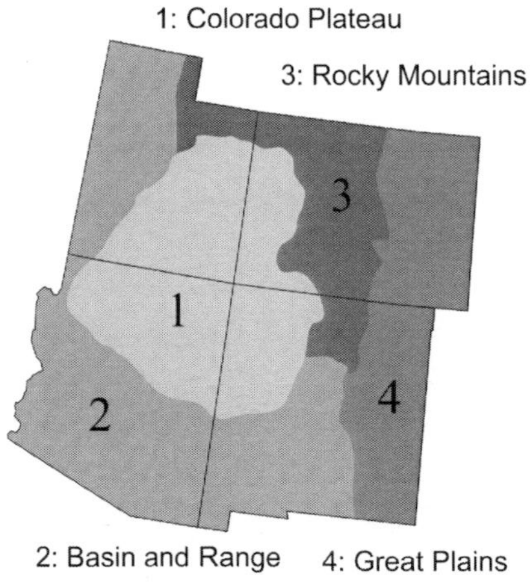

Most chapters in this guide divide the Southwestern US into four broad physiographic regions, each of which has a different geologic history and thus varies in rocks, fossils, topography, mineral resources, soils, and Earth hazards. The Geologic History chapter explains the history of all four regions within the context of Earth history. Chapters on climate and hazards are not divided by region because these tend to be driven by processes at broader geographic scales.

Each Guide begins with five cross-cutting Big Ideas of Earth science. These have applications across the curriculum. Deep study of specific Earth science sites gives context and meaning to these most fundamental ideas, and in turn understanding these ideas facilitates a lifetime of making sense of Earth processes anywhere.

Each Guide ends with a chapter on fieldwork—even from the classroom. You and your students can begin to interpret the Earth science in your area, and bring back photos and data to re-visit your field sites—using "virtual fieldwork"—throughout the year. More information is available at www.virtualfieldwork.org.

Use the color geologic map as a reference tool while you read this Guide. The map is on the back cover of the printed Guides and available as a downloadable graphic on the website.

Cross-referencing

You do not have to read this Guide from front to back! Each chapter is written to stand alone. Main concepts are repeated in more than one chapter. In this way you can read just what you need, in any order, as you approach particular units through the school year.

The chapters are cross-referenced, should you need to find more information about a particular concept or region. Bold-faced words are defined in a separate glossary, with selected words defined in chapter side bars.

For Further Information...

At the end of each chapter are lists of resources specific to that topic. There are lists of national and state-based resources, many of which cover multiple topics, at the end of the Guide.

Earth System Science: The Big Ideas

Like all scientific disciplines, the Earth sciences continually evolve over time. New discoveries fuel new ideas, providing an ever-increasing understanding of the planet. But of the overwhelming number of observations, theories, and principles that form the foundation of Earth **system** science, what is essential for every American to understand? All too often, curricula are too ambitious and, as a result, may fail to cover topics in any substantial depth. An alternative approach is to build one's curriculum upon a foundation of focused, interconnected big ideas. A well-designed set of big ideas can provide an all-encompassing conceptual framework for any discipline, including Earth system science. Developed alongside scientists and Earth science teachers, this coherent set of big ideas illuminates what is fundamental to the Earth sciences:

1. *The Earth is a system of systems.*
2. *The flow of* ***energy*** *drives the cycling of matter.*
3. *Life, including human life, influences and is influenced by the environment.*
4. *Physical and chemical principles are unchanging and drive both gradual and rapid changes in the Earth system.*
5. *To understand (deep) time and the scale of space, models and maps are necessary.*

These ideas are designed to cover the breadth of any Earth science curriculum, but they must be dissected to build deep understanding. Each idea is essentially bottomless; that is, while a meaningful understanding of these ideas is readily attainable, the details contained within are endless. Each of the ideas can be understood, but the depth of understanding can vary greatly.

Introduction of these ideas also invites discussion of the nature of science. As curricula are designed and implemented, the traditional topics of Earth system science should be complemented with ideas on how we have come to know what we know about the natural world. Within our big ideas framework, we draw attention to the nature of science with two overarching questions:

1. *How do we know what we know?*
2. *How does what we know inform our decision making?*

These questions, when addressed in concert with the big ideas, provide a gateway into the nature and utility of the range of scientific ideas.

system • *a set of connected things or parts forming a complex whole.*

energy • *the power derived from the use of physical or chemical resources.*

CHAPTER AUTHORS

Richard A. Kissel
Don Duggan-Haas

Big Ideas

Big Idea 1: The Earth is a system of systems

__plate tectonics__ • the process by which the plates of the Earth's crust move and interact with one another at their boundaries.

__weathering__ • the breakdown of rocks by physical or chemical means.

__erosion__ • the transport of weathered materials.

__uplift__ • upward movement of the crust due to compression, subduction, or mountain building.

__delta__ • a typically wedge-shaped deposit formed as sediment is eroded from mountains and transported by streams across lower elevations.

__plates__ • large, rigid pieces of the Earth's crust and upper mantle, which move and interact with one another at their boundaries.

__rift__ • a break or crack in the crust that can be caused by tensional stress as a landmass breaks apart into separate plates.

The Earth is composed of many systems, which cycle and interact in both space and time. It is also part of a multitude of systems, nested in larger systems such as the solar system and the universe. Systems are composed of an untold number of interacting parts that follow simple rules; they can and do evolve. For example:

> Outlining the geologic history of any region demonstrates the concept of the Earth as a system of systems. **Plate tectonics** drives the formation of mountains. Subsequent **weathering** and **erosion** of the **uplifted** mountains leads to the formation of **deltas** in adjacent shallow seas. And with uplifted continents, shorelines change and the distribution of marine communities are altered.
>
> The planet's systems are intimately connected: the forces of one system affect other systems nested within it. As **plates** collide, systems that drive plate tectonics are obviously linked to the formation of mountains, but they are ultimately linked to and influence much smaller systems and a wide range of landscape types, not just mountains. Much of the terrain in the Southwestern US is rugged and mountainous, reflective of relatively recent tectonic forces at work. Mountains and plateaus rise and fall; **rifts** open; and basins form. These forces have also driven movement of Earth's tectonic plates, such that the Southwestern US sits between 30° and 40° North latitude.
>
> Inland from the Pacific Ocean, the **climate** is largely arid, producing vast desert landscapes. The interplay of climate, rock, and water has shaped every natural landscape on the planet. Humans and other living things build upon (or tear down) the foundations lain down by these other systems, furthering their interplay.

See Chapter 8: Climate to learn how climate has affected the Southwest's life and landscape.

Each of the remaining ideas operates across multiple systems within the larger Earth system.

Big Idea 2:
The flow of energy drives the cycling of matter

The Earth is an open system. Energy flows and cycles through the system; matter cycles within it. This cycling is largely driven by the interaction of the differential distribution of solar radiation and internal heat: the constant flow of solar radiation powers much of Earth's ocean and **atmospheric** processes on the surface of the system, while the flow of heat from **radioactivity** within the Earth drives plate tectonics. For example:

> One of the fundamental processes known to Earth system scientists is the rock cycle. The rock cycle illustrates the steps involved in the formation of one type of rock from another. It is a system that has operated since the Earth's origin, and it continues today. The energy that drives weathering and erosion, melting, or an increase in heat or pressure, drives the continuation of the rock cycle.
>
> The landscape we see today in the Southwest has been shaped by the geologic forces of the past, and these forces are still active today. Evidence throughout the Southwest's terrain tells a story that began billions of years ago with the formation of tectonic plates, and this story continues to evolve. The movement of Earth's plates is driven by plate tectonics, illustrating how the flow of energy drives the cycling of matter—the flow of heat from radioactivity within the Earth drives plate tectonics. Through **geologic time**, the Southwestern US has been shaped by the collision of the North American Plate with the Pacific Plate in a process driven by **convection** within the **mantle**. In addition to tectonic processes, energy flows and cycling matter also shape the landscape through erosion, deposition, sea level change and the direct action of humans.
>
> See Chapter 1: Geologic History for more about the tectonic processes that led to the formation of North America as we know it today.
>
> In the recent geologic past, the Colorado River's water has moved a tremendous mass of sediment from the interior of North America into the Gulf of California. Other Southwestern rivers, including the Rio Grande, Brazos, and Mississippi (through major tributaries like the Arkansas and Red), have moved great quantities of sediment to the Gulf of Mexico. The flow of sediment is, of course, driven by the water cycle, and, especially for the Colorado River, greatly affected by human activity. Like the rock cycle, and plate tectonics, the water cycle is convection driven. Without convection, Earth would be extraordinarily different, if it were here at all.

climate • *a description of the average temperature, range of temperature, humidity, precipitation, and other atmospheric/hydrospheric conditions a region experiences over a period of many years (usually more than 30).*

atmosphere • *a layer of gases surrounding a planet.*

radioactivity • *the emission of radiation by an unstable atom.*

geologic time scale • *a standard timeline used to describe the age of rocks and fossils, and the events that formed them.*

convection • *the rise of buoyant material and the sinking of denser material.*

mantle • *the layer of the Earth between the crust and core.*

Big Ideas

Big Idea 3: Life—including human life—influences and is influenced by the environment

greenhouse gas • *a gas in the atmosphere that absorbs and emits heat.*

global warming • *the current increase in the average temperature worldwide, caused by the buildup of greenhouse gases in the atmosphere.*

hectare • *a metric unit of area defined as 10,000 square meters.*

Across its four-billion-year history, the course of life's evolution has been intimately tied to the Earth's physical environment. Global cooling led to the relatively recent spread of grasslands, which then triggered an evolutionary shift in many herbivorous mammals from browsing to grazing. Conversely, the evolution of life has altered the physical environment. Photosynthetic bacteria released free oxygen into the early oceans and atmosphere, making Earth habitable for later types of organisms. Humans, with their increasing population and expanding technology, have altered the landscape, the flow of rivers, the distribution of flora and fauna, and atmospheric chemistry in ways that affect the climate. Earth system processes also influence where and how humans live. For example:

With human populations increasing the world over, the emission of **greenhouse gases** has also increased dramatically. These gases alter the chemical composition of the atmosphere and directly influence the planet's climate. It is generally agreed that the rapid and immense pouring of carbon dioxide into the atmosphere will lead to **global warming**, which will have incredible impacts throughout the world.

See Chapter 9: Climate to learn more about the effect of greenhouse gases.

Around three million years ago, a land bridge formed between North and South America. For the first time in more than 150 million years, the two continents were linked, and the mammals inhabiting both lands migrated across the bridge. Horses, mastodons, cats, and dogs moved south, while opossums, porcupines, ground sloths, and armadillos moved north (to name a few). Today, half the mammal species in South America are descended from North American migrants.

Throughout the Southwest, water diversion for agriculture and other uses has changed and continues to change remarkably. The Colorado River effectively no longer reaches the Gulf of California due to a series of dams and diversions along its course. Much of its flow is diverted outside of its basin, for use in California and Nevada. Water is also taken to quench the thirst of nearby cities like Denver and Phoenix. The Colorado River serves the needs of 30 million people in seven US states and Mexico, and 70% of its flow is diverted to irrigate 2.2 million **hectares** (5.5 million acres) of land. It also supports production of 4,200 megawatts of electric generating capacity. These changes coupled with long-standing drought in the region—likely enhanced by human-induced climate change—have brought a wide range of changes to the basin. When we ask, "Why does this place look the way it does?" the role of humans must be central to our answer.

Big Idea 4:
Physical and chemical principles are unchanging and drive both gradual and rapid changes in the Earth system

The Earth processes operating today—everything from local erosion to plate tectonics—are the same as those operating since they first arose in Earth's history, and these processes are obedient to the laws of chemistry and physics. While the processes that constantly change the planet are essentially fixed, their rates are not. Tipping points are reached that can result in rapid changes cascading through Earth systems. For example:

> During the **Precambrian**, the evolution of photosynthetic organisms led to significant changes in the planet's atmosphere. Prior to this event, there was little free oxygen in the atmosphere, but with photosynthesis producing oxygen as a waste product, the very existence of these organisms flooded the seas and atmosphere with free oxygen, changing the planet forever. But life's evolution represents just one of the processes working upon Earth systems.
>
> Tectonic processes have been at work in the same way for billions of years, opening and closing oceans and building up and tearing down landscapes. The Grand Canyon offers a large cross-section of Earth history—a window into nearly two billion years of North America's formation. Some 3900 meters (13,000 feet) of **lava** and sediment are exposed there, and though these rocks date back over a huge span of Earth's history, the processes that made them are still at work today. **Schists** still form from other rocks put under heat and pressure; sediments still become **limestones** and **sandstones**, and lavas still cool to form **basalts**. Rocks born from all of these processes are visible in a number of places throughout the Southwestern US.

Precambrian • *a geologic time interval that spans from the formation of Earth (4.6 billion years ago) to the beginning of the Cambrian (541 million years ago).*

lava • *molten rock located on the Earth's surface.*

schist • *a medium grade metamorphic rock with sheet-like crystals flattened in one plane.*

limestone • *a sedimentary rock composed of calcium carbonate ($CaCO_3$).*

sandstone • *sedimentary rock formed by cementing together grains of sand.*

basalt • *an extrusive igneous rock, and the most common rock type on the surface of the Earth.*

Big Idea 5:
To understand (deep) time and the scale of space, models and maps are necessary

The use of models is fundamental to all of the Earth sciences. Maps and models aid in the understanding of aspects of the Earth system that are too big or small for direct observation, or where observation is not possible. They also help make complex systems comprehensible through strategic simplification. When compared to the size and age of the universe, humanity is a speck in space and a blip in time; models assist in the comprehension of time and space at both sub-microscopic and immense scales. For example:

> Much of scientists' understanding of the inner workings of our planet is derived from mathematical modeling. It is not possible to directly measure the movement that occurs below Earth's surface, but modeling of convection currents brings us closer to the true nature of these monumental geologic phenomena.
>
> The observation of natural phenomena today, such as deposition along a riverbed, is critical for interpreting the geologic record. But for processes that operate on much larger, slower scales, modeling within the lab is required. The formation of mountain ranges like the Rocky Mountains and basins like the Gulf of Mexico is better understood by examining the effects of stress and strain in the laboratory.

In Conclusion

Taken individually, these big ideas and overarching questions represent important aspects of Earth system science, but together they are more significant. Keeping these ideas in mind—and considering how they arose through scientific methods and investigation—is invaluable as one proceeds throughout his or her curriculum, and it can provide a conceptual framework upon which to build an enduring understanding of the discipline.

Resources

Books

Donovan, S., & J. Bransford, 2005, *How Students Learn: Science in the Classroom*, National Academies Press, Washington, DC, http://books.nap.edu/catalog.php?record_id=10126.

Wiggins, G. P., & J. McTighe, 2005, *Understanding by Design, 2nd edition*, Association for Supervision and Curriculum Development, Alexandria, VA, 382 pp.

Wiske, M. S., ed., 1998, *Teaching for Understanding: Linking Research with Practice*, Jossey-Bass, San Francisco, CA, 379 pp.

Websites

Exploring Geoscience Methods with Secondary Education Students, by J. Ebert, S. Linneman, & J. Thomas, http://serc.carleton.edu/integrate/teaching_materials/geosci_methods/index.html.

Chapter 1: Geologic History of the Southwestern US: Reconstructing the Geologic Past

Geologic history is the key to this Guide and to understanding the story recorded in the rocks of the Southwestern US. By knowing more about the geologic history of your area, you can better understand the types of rocks that are in your backyard and why they are there. In this chapter, we will look at the history of the Southwest as it unfolded: as a series of major events that created and shaped the area over the past one billion years. These events will act as the framework for the topics in the chapters to follow and will shed light on why your region looks the way it does!

Examination of the available geological record tells us that the shape and position of North America has changed dramatically over the last billion years, and geologic processes continue these changes today. The Earth's outer layer—the **crust**—is dynamic, consisting of constantly moving **plates** that are made of a rigid continental and oceanic **lithosphere** overlying a churning, plastically flowing **asthenosphere**—part of the Earth's **mantle** (*Figure 1.1*). These plates are slowly pulling apart, colliding, or sliding past one another with great force, creating strings of **volcanic islands**, new ocean floor, **earthquakes**, and mountains. The continents are likewise continuously shifting position relative to each other. This not only shapes the land, but also affects the distribution of rocks and **minerals**, natural resources, **climate**, and life.

crust • *the uppermost, rigid outer layer of the Earth.*

plates • *large, rigid pieces of the Earth's crust and upper mantle, which move and interact with one another at their boundaries.*

lithosphere • *the outermost layer of the Earth, comprising a rigid crust and upper mantle broken up into many plates.*

asthenosphere • *a thin semifluid layer of the Earth, below the outer rigid lithosphere, forming much of the upper mantle.*

mantle • *the layer of the Earth between the crust and core.*

volcanic island • *one of a string of islands created when molten rock rises upward through oceanic crust.*

How do we know what the past is like?

Reconstructing the geologic past is a lot like solving a mystery. Geologists use scraps of evidence to piece together events they have not personally observed, but to do so they must contend with two major complications. First, the overwhelming majority of geologic history occurred long before there were any human witnesses. Second, much of the evidence for the older events is highly fragmented. By studying rocks, **fossils**, and other geologic features, however, scientists can still reconstruct a great deal of what the ancient Earth might have looked like.

Rocks and sediments are indicators of past geologic processes and the environments in which those processes took place. In general, **igneous rocks**, created through tectonic activity, reflect the history of molten rock, both below the surface (**plutonism**) and at the surface (**volcanism**).

See Chapter 2: Rocks to learn more about the different types of rocks found in the Southwest.

CHAPTER AUTHORS

Lisa R. Fisher
Richard A. Kissel
Warren D. Allmon

Geologic History

Reconstructing

ultramafic rocks • *igneous rocks with very low silica content (< 45%), which are composed of usually greater than 90% mafic minerals.*

metamorphic rocks • *rocks formed by the recrystallization and realignment of minerals in pre-existing sedimentary, igneous, and metamorphic rocks when exposed to high enough temperature and/or pressure.*

heat • *a form of energy transferred from one body to another as a result of a difference in temperature or a change in phase.*

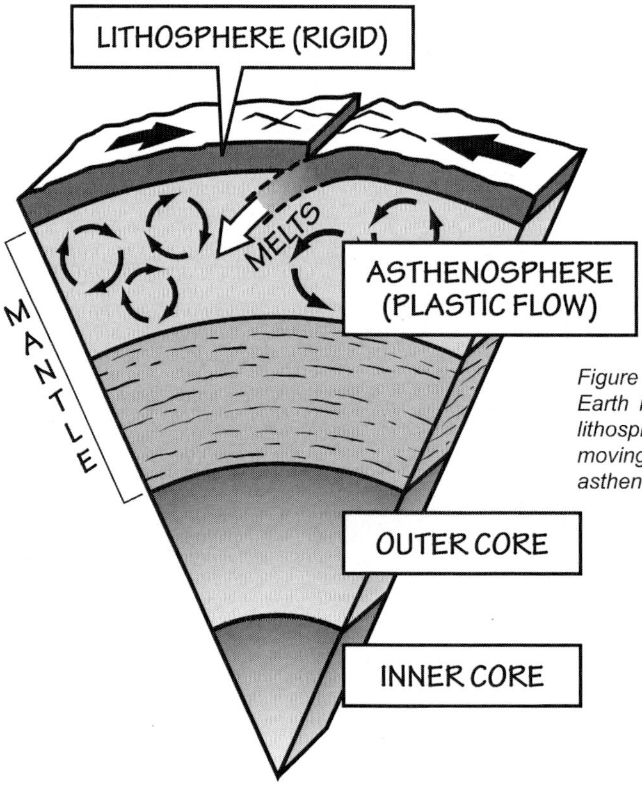

Figure 1.1: The layers of the Earth include the rigid crust of the lithosphere, which is constantly moving over the plastically flowing asthenosphere.

Lithosphere and Asthenosphere: What's the difference?

The lithosphere is the outermost layer of the Earth, a rigid layer of crust and upper mantle broken up into fragments called plates. Although the rock of the asthenosphere would seem very solid if you could observe it in place, under long-term stress it slowly bends and flows, like very thick syrup. The difference between crust and mantle is mainly chemical: the lithosphere's composition typically varies between basalt in oceanic crust and granite in continental crust, while the mantle is composed of homogenous *ultramafic* material. The boundary between rigid lithosphere and flowing asthenosphere is usually found *within* the mantle, and is largely a result of temperature increase with depth beneath the surface. In tectonically active regions of extension such as a mid-ocean ridge, where temperature rises rapidly with depth (compared to more tectonically stable regions), the asthenosphere begins nearly at the base of the crust.

Likewise, **metamorphic rocks** provide important clues about past mountain-building events. Geologists often use these rocks, created when sediment is subjected to intense **heat** and pressure, to map the extent of now-vanished mountain ranges. **Sedimentary rocks** tell perhaps the most comprehensive story of the Earth's history, as they record characteristics of far-away mountain ranges, river **systems** that transported the sediments, and the final environment in which the sediments accumulated and **lithified**. The size and shape of sediments in sedimentary rocks, as well as the presence of fossils and the architecture of sedimentary rock layers (sedimentary structures), can help us infer how the sediments were transported and where they were finally deposited. However, because rocks are often reformed into different rock types, ancient information is lost as the rocks cycle through the igneous, metamorphic, and sedimentary stages.

Fossils indicate both the type of life that once flourished in an area and the kind of climate in which that life existed. Paleontologists use groups of fossils found in the same place to construct pictures of ancient ecosystems. These ecosystems of the past are matched to similar present-day ecosystems, whose climate conditions are then used to infer what sort of climate the fossilized organisms lived in. Unfortunately, few organisms can be easily preserved as fossils, and many environments also do not lend themselves to preserving organisms as fossils. As a result, the clues that fossils give us provide only incomplete glimpses of the ancient world, with many important details missing.

See Chapter 3: Fossils for more information about the Southwest's prehistoric life.

Landscapes and geologic structures are also indicators of past geologic processes and the environments in which they occurred. For instance, the shape of a valley reflects the forces that carved it. Valleys with V-shaped profiles tend to be the products of stream **erosion**, whereas U-shaped valleys are more likely to have been carved by **glaciers**. Layers of intensely folded rock indicate a violent past of tectonic plate collisions and mountain building. Sedimentary structures, such as **ripple marks** or **cross-bedding**, can demonstrate the direction and energy level of the water that transported the sediment. Although landscapes tell us much about the geologic processes that created them, they inevitably change over time, and information from the distant past is overwhelmed by the forces of the more recent past.

See Chapter 4: Topography for more detail about the landscape of the Southwest.

Ultimately, geologists rely upon the preserved clues of ancient geologic processes to understand Earth's history. Because younger environments retain more evidence than older environments do, the Earth's recent history is better known than its ancient past. Although preserved geologic clues are indeed fragmentary, geologists have become increasingly skilled at interpreting them and constructing ever more detailed pictures of the Earth's past.

sedimentary rock • *rock formed through the accumulation and consolidation of grains of broken rock, crystals, skeletal fragments, and organic matter.*

system • *a set of connected things or parts forming a complex whole.*

lithification • *the process of creating sedimentary rock through the compaction or sementation of soft sediment.*

erosion • *the transport of weathered materials.*

glacier • *a body of dense ice on land that does not melt away annually and has sufficient mass to move under its own weight.*

ripple marks • *surface features created when sediment deposits are agitated, typically by water currents or wind.*

cross-bedding • *layering within a bed in a series of rock strata that does not run parallel to the plane of stratification.*

1 Geologic History

Reconstructing

geologic time scale • a standard timeline used to describe the age of rocks and fossils, and the events that formed them.

Precambrian • a geologic time interval that spans from the formation of Earth (4.6 billion years ago) to the beginning of the Cambrian (541 million years ago).

Cambrian • a geologic time period lasting from 541 to 485 million years ago.

Archean • a geologic time interval that extends from 4 billion to 2.5 billion years ago.

Proterozoic • a geologic time interval that extends from 2.5 billion to 541 million years ago.

Sedimentary Structures

Sedimentary rocks often reveal the type of environment in which they formed by the presence of structures within the rock. Sedimentary structures include ripple marks, cross-beds, mud cracks, and even raindrop impressions. Consider the type of environments in which you see these sedimentary structures today in the world around you.

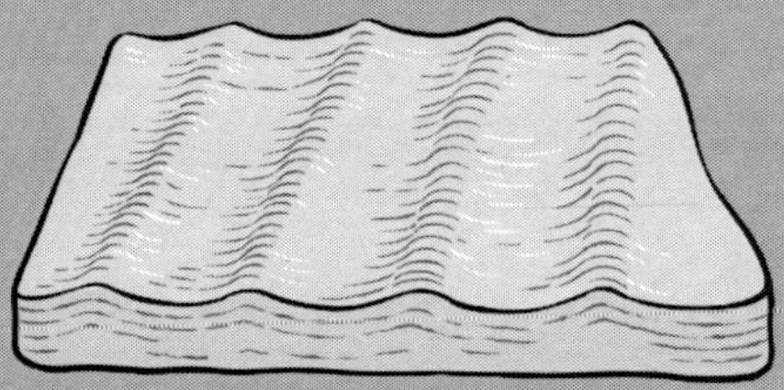

Ripple marks suggest the presence of moving water (though wind can also create ripples and even dunes). Mud cracks indicate that the sediment was wet but exposed to the air so that it dried and cracked.

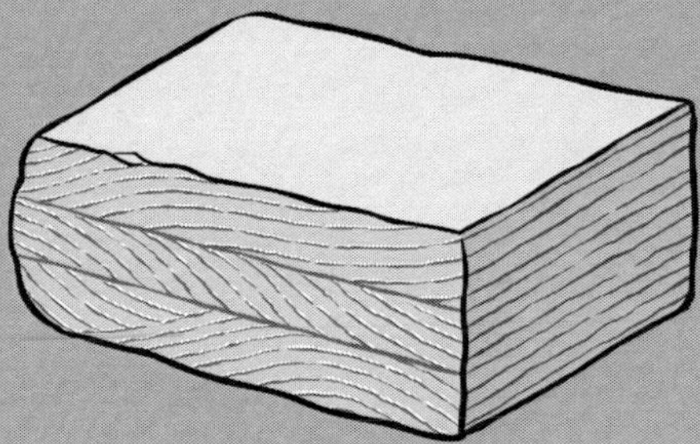

Cross-beds form as flowing water or wind pushes sediment downcurrent, creating thin beds that slope gently in the direction of the flow as migrating ripples. The downstream slope of the ripple may be preserved as a thin layer dipping in the direction of the current, across the natural flat-lying repose of the beds. Another migrating ripple will form an additional layer on top of the previous one.

Earth's Timeline

The **geologic time scale** (*Figure 1.2*) is an important tool used to portray the history of the Earth—a standard timeline used to describe the age of rocks and fossils, and the events that formed them. It spans Earth's entire history and is divided into four principal sections.

The first of these four divisions, the **Precambrian**, extends from the beginning of the Earth, around 4.6 billion years ago, to the beginning of the **Cambrian** period, around 541 million years ago. The Precambrian, in turn, is subdivided into two sections: the **Archean** (before 2.5 billion years ago) and the **Proterozoic** (2.5 billion to 541 million years ago). Less is known about the Earth during the

Geologic History 1

Silurian • a geologic time period spanning from 443 to 419 million years ago.

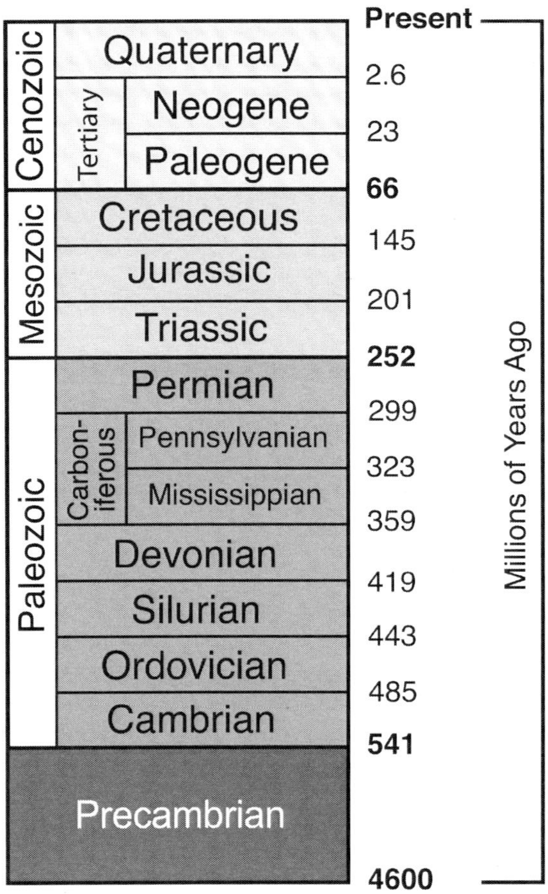

About the Time Scale:
The time scale in The Teacher-Friendly Guides™ *follows that of the International Commission on Stratigraphy (ICS). The Tertiary period, though it was officially phased out in 2008 by the ICS, remains on the scale in the Guides, since "Tertiary" is found extensively in past literature. In contrast, the Carboniferous and Pennsylvanian & Mississippian periods all enjoy official status, with the latter pair being more commonly used in the US.*

Figure 1.2: The Geologic Time Scale (spacing of units not to scale).

Geologic Time

How did geologists come up with the timeline for the history of the Earth? The geologic time scale was developed over the course of many years—beginning in the early 19th century—and through the combined work of many geologists around the world. No single location on Earth contains the complete sequence of rocks from Precambrian to present. Geology as a science grew as geologists studied individual stacks or sections of rock and connected them to each other. Gradually, successions of fossils were discovered that helped geologists determine the relative ages of groups of rocks. These layers could then be correlated with similarly aged rock units from around the world. The names you see for the different periods on the geologic time scale have diverse origins; most are based on geographic areas where rocks of that age were first well studied. Time periods were named after dominant rock types, geography, mountain ranges, and even ancient tribes like the Silures of England and Wales, from which the "*Silurian*" period was derived.

1 Geologic History

Big Picture

Paleozoic • *a geologic time interval that extends from 541 to 252 million years ago.*

Mesozoic • *a geologic time period that spans from 252 to 66 million years ago.*

dinosaur • *a member of a group of terrestrial reptiles with a common ancestor and thus certain anatomical similarities, including long ankle bones and erect limbs.*

Cenozoic • *the geologic time period spanning from 66 million years ago to the present.*

extinction • *the end of species or other taxonomic groups, marked by death of the last living individual.*

Precambrian than during later parts of its history, since relatively few fossils or unaltered rocks have survived. Nevertheless, the evidence that has been preserved and discovered reveals much about the planet's first several billion years, including clear evidence that life first appeared on the planet some 3.9 billion years ago in the form of single-celled organisms.

The second division, the **Paleozoic**, extends from 541 to 252 million years ago. Geological evidence shows that during this time period, continents moved, mountains formed, and life evolved in the oceans and gradually colonized the land.

The third division, the **Mesozoic** (from 252 to 66 million years ago), is also called the "Age of Reptiles" since **dinosaurs** and other reptiles dominated both marine and terrestrial ecosystems. It is also noteworthy that during this time the last of the Earth's major supercontinents, **Pangaea**, formed and later broke up, producing the Earth's current geography.

Pangaea, meaning "all Earth," began to assemble over 300 million years ago and lasted for almost 150 million years. All of the Earth's continents were joined as one to form a giant supercontinent.

The last and current division, the **Cenozoic**, extends from the **extinction** of the dinosaurs, nearly 66 million years ago, to the present. With the demise of the dinosaurs, mammals became much more diverse and abundant. We humans didn't come into the picture until the last two million years. To put this in perspective, if the entire geologic time scale were reduced to 24 hours, we wouldn't come onto the stage until two seconds before midnight!

The Southwestern States:
The Big Picture

The geologic history of the Southwestern United States spans more than two billion years. This history is unusually well exposed throughout the area's extraordinary geological landscape, including spectacular mountains, canyons, and countless other striking features. The Southwest's dynamic geologic history has resulted in the formation, preservation, and exposure of rocks from a wide variety of ages and sources.

Rocks exposed in the Southwest reveal that a large continental mass—a supercontinent—was repeatedly assembled and disassembled over hundreds of millions of years. By around 600 million years ago, the core of what would eventually become most of North America was a recognizable entity. Over the next several hundred million years this continent was mostly tectonically stable and flat, and was repeatedly flooded and exposed by rising and falling sea levels. Around 300 million years ago, episodes of tectonic activity and volcanism added land to the continent along what would become the West Coast. Major mountain building (**orogenesis**) began around 100 million years ago, and reached its peak around 65 million years ago, at the very end of the Mesozoic era. These orogenic episodes formed the modern Rocky Mountains, which have dominated the geology and landscape of western North America ever since. At the same time that the Rockies were rising, globally high sea level caused an enormous shallow sea—the Western Interior Seaway—to form across what is today the Great Plains, from Texas to Alaska. This seaway disappeared in the early Cenozoic era, and was replaced by a changing landscape of forest and grasslands filled with an amazing diversity of life, especially mammals, which replaced dinosaurs in most of the ecological niches for large terrestrial vertebrates. In the late Cenozoic, glaciers shaped the mountain landscape and the creatures that inhabited it.

In this volume, the Southwestern states are divided into four different **physiographic** provinces or regions (*Figure 1.3*): the Colorado Plateau (1), the Basin and Range (2), the Rocky Mountains (3), and the Great Plains (4). Each of these regions has a different geological history and thus varies in terms of rocks, fossils, **topography**, mineral resources, **soils**, and other geological features.

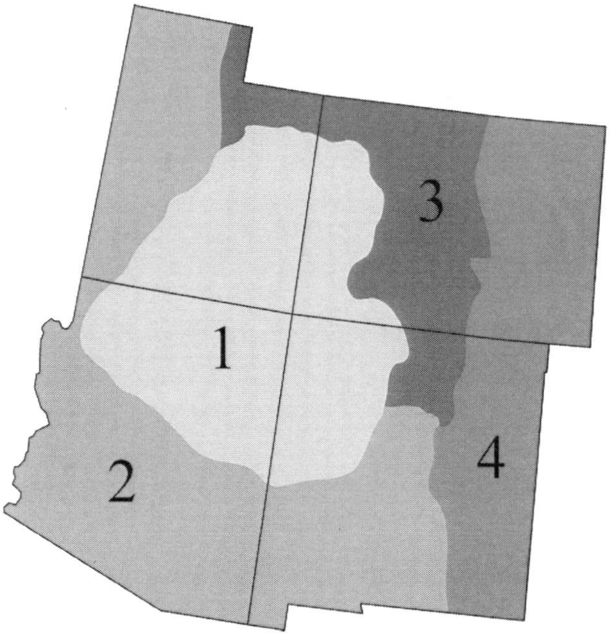

Figure 1.3. Physiographic regions of the Southwest: 1) Colorado Plateau, 2) Basin and Range, 3) Rocky Mountains, and 4) Great Plains.

Big Picture

orogeny • *a mountain-building event generally caused by colliding plates and compression of the edge of the continents.*

physiography • *a subfield of geography that studies the Earth's physical processes and patterns.*

topography • *the landscape of an area, including the presence or absence of hills and the slopes between high and low areas.*

soil • *the collection of natural materials that collect on Earth's surface, above the bedrock.*

1 Geologic History

Precambrian

meteorite • a stony or metallic mass of matter that has fallen to the Earth's surface from outer space.

Canadian Shield • the stable core of the North American continental landmass, containing some of the oldest rocks on Earth.

fault • a fracture in the Earth's crust in which the rock on one side of the fracture moves measurably in relation to the rock on the other side.

craton • the old, underlying portion of a continent that is geologically stable relative to surrounding areas.

magma • molten rock located below the surface of the Earth.

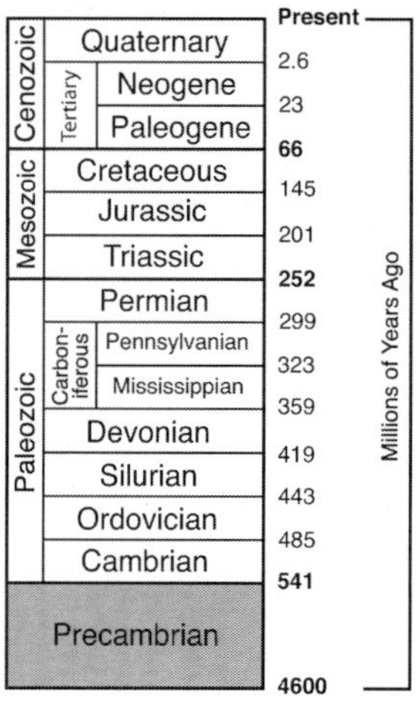

Precambrian Beginnings

Roots of the Southwest

The Earth is estimated to be approximately 4.6 billion years old—an age obtained by dating **meteorites**. Rocks dating to around four billion years old are found on almost every continent, but they are not found at the Earth's surface anywhere in the Southwest. The oldest rocks known on Earth are 4.3-billion-year-old rocks found along the eastern shore of Hudson Bay in northern Quebec. These are part of the **Canadian Shield**, the ancient core of the North American continental landmass, which has experienced very little tectonic activity (**faulting** and folding) for millions of years. Shields, or **cratons**, are the stable cores of all continents and are often covered by layers of younger sediments. They formed and grew during pulses of **magmatic** activity, as bodies of molten rock deep in the Earth's crust contributed to form new crust. In the Southwestern US, the main cratonic elements are referred to as the Mojave Province (western Arizona and Utah), the Yavapi Province (Colorado, central Arizona, northwestern New Mexico, and southeastern Utah), and the Mazatzal Province (southeastern Arizona and New Mexico) (*Figure 1.4*). The composition of many of these rocks, suggests that they formed very deep in the Earth's crust, perhaps 20–25 kilometers (12–15 miles) below the surface. Others originated in island arcs formed by **subduction** and associated volcanism and sedimentation. All were subsequently intruded by magma from deep in the mantle.

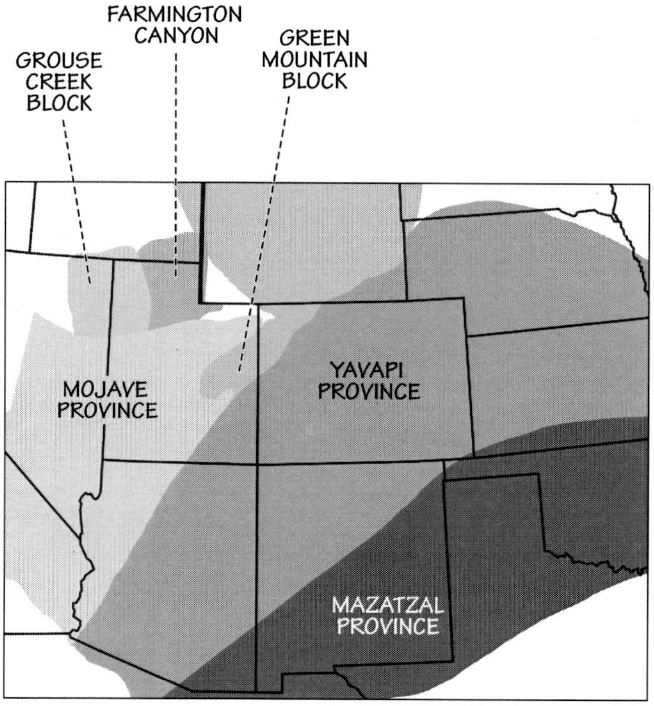

Figure 1.4: Cratonic elements of the Southwestern US.

Geologic History 1

Precambrian

In the Southwestern US, the oldest known rocks found at the surface are late Archean to early Proterozoic metamorphic rocks known as the Grouse Creek Block and the Farmington Canyon Complex, in what is now northwestern Utah (*see Figure 1.4*). These two sets of rocks formed between 2.5 and 2.6 billion years ago. Together with the other cratonic elements (frequently referred to as provinces or **terranes**), they make up the core of what is now the Southwest. As separate volcanic island arcs collided with each other in the early Proterozoic (between 1.8 and 1.6 billion years ago) (*Figure 1.5*), **plate tectonics** and the process of collision eventually formed a supercontinent known as either Nuna or Columbia. The collisional zones between these continental fragments are preserved as long belts of deformed metamorphic rock. Today, they form the subsurface **basement rock** that underlies much of the Colorado Plateau. Most of the Southwest's cratonic rocks are deeply buried, but Yavapai Province rocks are visible at the bottom of the Grand Canyon in northern Arizona.

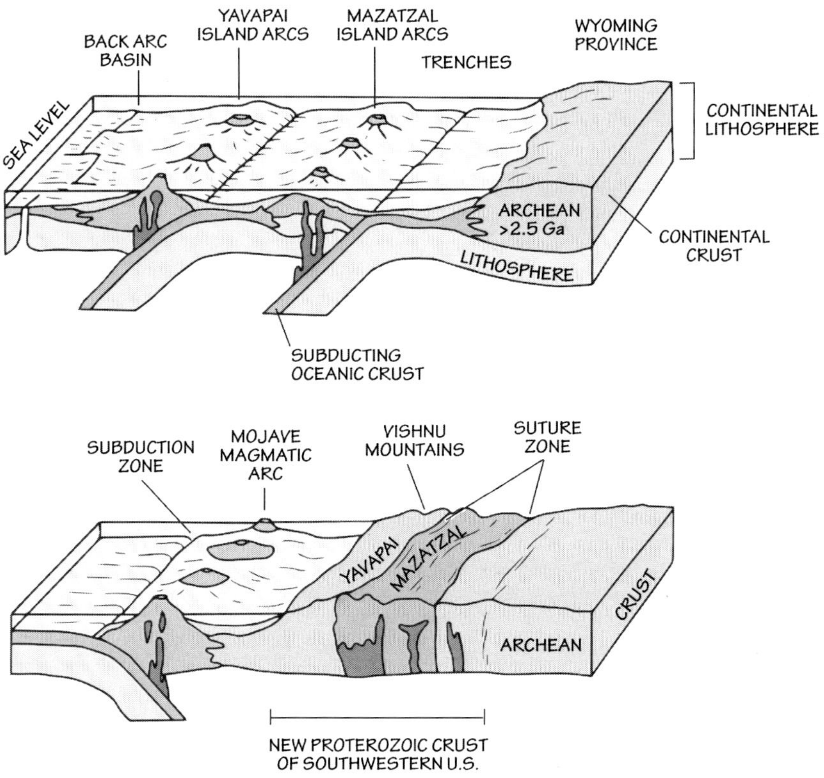

Figure 1.5: Plate tectonic model for the formation of crust by the addition of island arcs to the continent. Some of these terranes are now represented by the oldest rocks at the bottom of the Grand Canyon. The area between the trench and the arc is the fore-arc region, and the area behind the arc (i.e. on the side farthest from the trench) is the back-arc region.

Following its final assembly, the Nuna supercontinent continued to grow by volcanic and magmatic activity along its margins. It began to fragment about 1.6 billion years ago, in a process that lasted for several hundred million years. At this point, the continents began moving back toward each other, and the remainder of the Precambrian period saw the formation of a third supercontinent, called **Rodinia** (*Figure 1.6*). This landmass was fully formed

subduction • *the process by which one plate moves under another, sinking into the mantle.*

terrane • *a piece of crustal material that has broken off from its parent continent and become attached to another plate.*

plate tectonics • *the process by which the plates of the Earth's crust move and interact with one another at their boundaries.*

basement rocks • *the foundation that underlies the surface geology of an area, generally composed of igneous or metamorphic crystalline rock.*

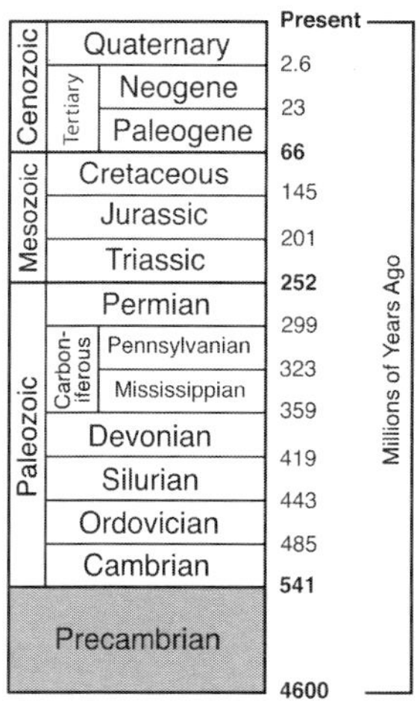

Precambrian

rift • *a break or crack in the crust that can be caused by tensional stress as a landmass breaks apart into separate plates.*

passive margin • *a tectonically quiet continental edge where crustal collision or rifting is not occurring.*

accretion • *the process by which a body of rock increases in size due to the addition of further sedimentary particles or of large chunks of land.*

tillite • *glacial till that has been compacted and lithified into solid rock.*

by about 1.1 billion years ago. Preserved remnants of the continental collisions that formed this supercontinent are found widely across modern North America, including the Pikes Peak Granite in Colorado. Other remnants are found outside the Southwestern US in the Llano province and El Paso region of western Texas, and farther to the east in the Grenville rocks of the Adirondack and Appalachian mountains.

See Chapter 2: Rocks for more on the ancient rocks formed during Precambrian collisions.

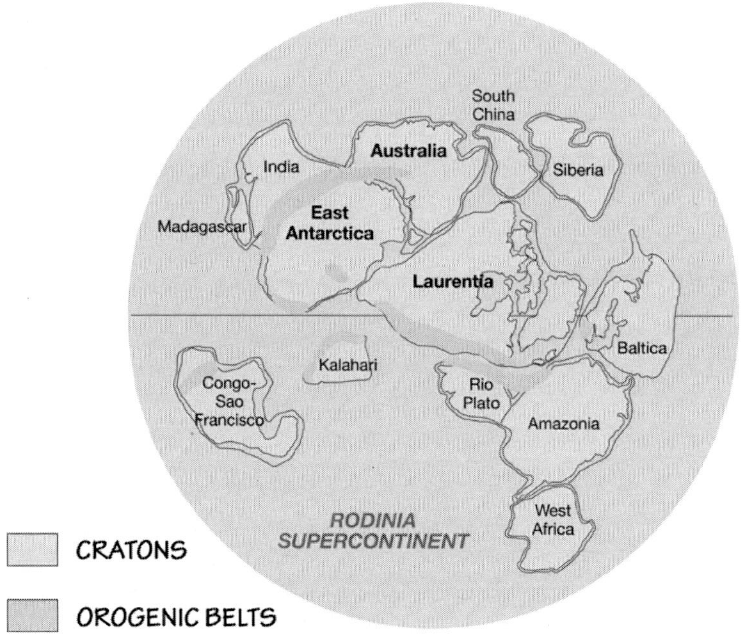

Figure 1.6: The supercontinent Rodinia, circa 1.1 billion years ago. Laurentia represents proto-North America. (See TFG website for full-color version.)

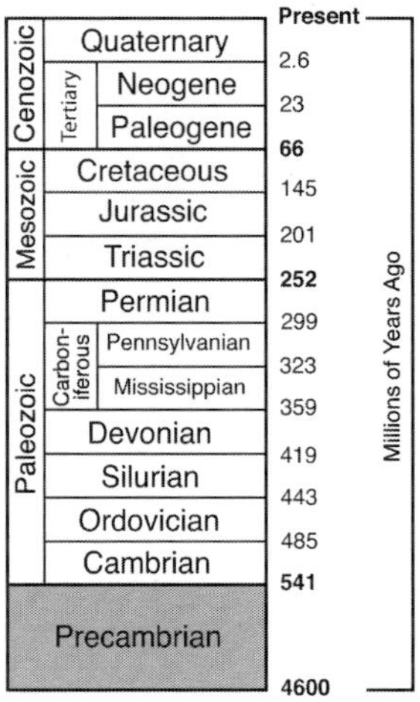

The breakup of Rodinia, beginning nearly 800 million years ago, was associated with the formation of **rifts** throughout North America. Igneous activity occurred in rifted zones and continued slowly and irregularly until about 600 million years ago. North America's rifted edges formed **passive margins**, where sediments were deposited on continental shelves into the early Paleozoic era. The rifted margin of western North America became a shallow continental shelf that extended from Canada through western Utah into southern California. Rivers brought sediment from the continent's interior and deposited it on the shelf. In Arizona, these sediments form a thick sequence of sedimentary layers known as the Grand Canyon Supergroup, which constitutes one of the most complete middle to late Proterozoic geologic records in North America. These rocks record a long history—perhaps 200 million years—between the formation of continental crust by terrane collision and **accretion** to the more familiar events of the Cambrian period and later.

A rift occurs when tectonic plates move away from each other. Magma rises up into the margin, cooling to produce new oceanic crust. The resulting action is similar to two conveyor belts moving away from each other. A failed rift occurs when the existing crust is stretched thin and magma begins to well up, but the plate is never completely broken.

During the late Proterozoic, large areas of the continents were repeatedly covered with glaciers, including some that reached into low latitudes. Many geologists think that this interval may have included the most intense and widespread glacial development in Earth's history. The term "Snowball Earth" is used to describe this proposed state, during which most of the Earth's surface (pole-to-pole) was covered by glaciers. In the Southwest, geological evidence for this glacial expansion is represented by rocks formed from glacially derived sediments (**tillites**) found in western Utah.

See Chapter 8: Climate to learn more about Snowball Earth and other ancient glaciations.

The Paleozoic
Formation of a Continent

The rifting of Rodinia produced a passive margin on the western edge of Laurentia. At the start of the Paleozoic, during the Cambrian, the area that now comprises the states of California, Oregon, Washington, and western Nevada did not exist as part of the North American continent (*Figure 1.7*). The continent's western coastline was located at approximately the Utah-Nevada and Arizona-California state lines, where a broad continental shelf extended westward from the coast. The shore moved eastward over time as sea level rose, covering much of the western US under a shallow (epicontinental) sea, and sheets of **sand** and **carbonate** sediment were deposited on the shelf. During the late Cambrian and early **Ordovician**, the entire shelf was covered by a huge carbonate platform, much like the region around today's Bahama Islands. At other times, such as the late Ordovician, **transgression** paused or reversed (**regression**), and the sea deposited thin sheets of nearly pure **quartz** sand. At the end of the Ordovician, global temperatures and sea level fell abruptly, probably due to glaciation caused by the movement of the large southern supercontinent Gondwana over the South Pole. As a result of these environmental changes, a major **mass extinction** of marine life took place at this time. Sea level continued to rise and fall throughout the first half of the Paleozoic, depositing marine sediments during transgressions and allowing the exposed carbonate rocks to **weather** during regressions.

Paleozoic

sand • *rock material in the form of loose, rounded, or angular grains, and formed as a result of the weathering and decomposition of rocks.*

carbonate rocks • *rocks formed by accumulation of calcium carbonate, often made of the skeletons of aquatic organisms.*

Ordovician • *a geologic time period spanning from 485 to 443 million years ago.*

transgression • *a relative rise in sea level in a particular area, through global sea level rise or subsidence of land.*

regression • *a drop in sea level.*

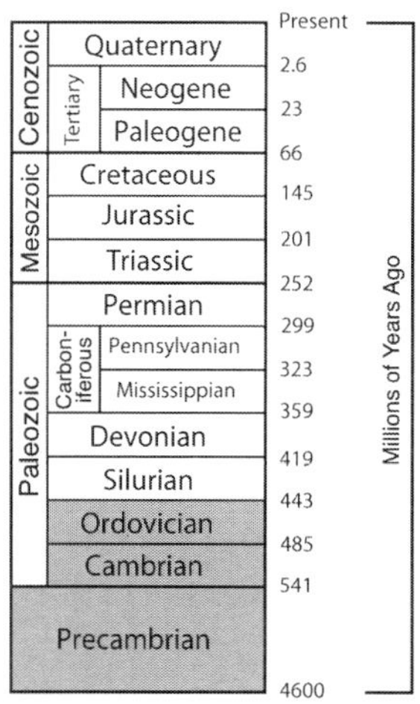

Paleozoic

Baltica • *a late-Proterozoic, early-Paleozoic continent that included ancient Europe.*

Gondwana • *the super-continent of the Southern Hemisphere, composed of Africa, Australia, India, and South America.*

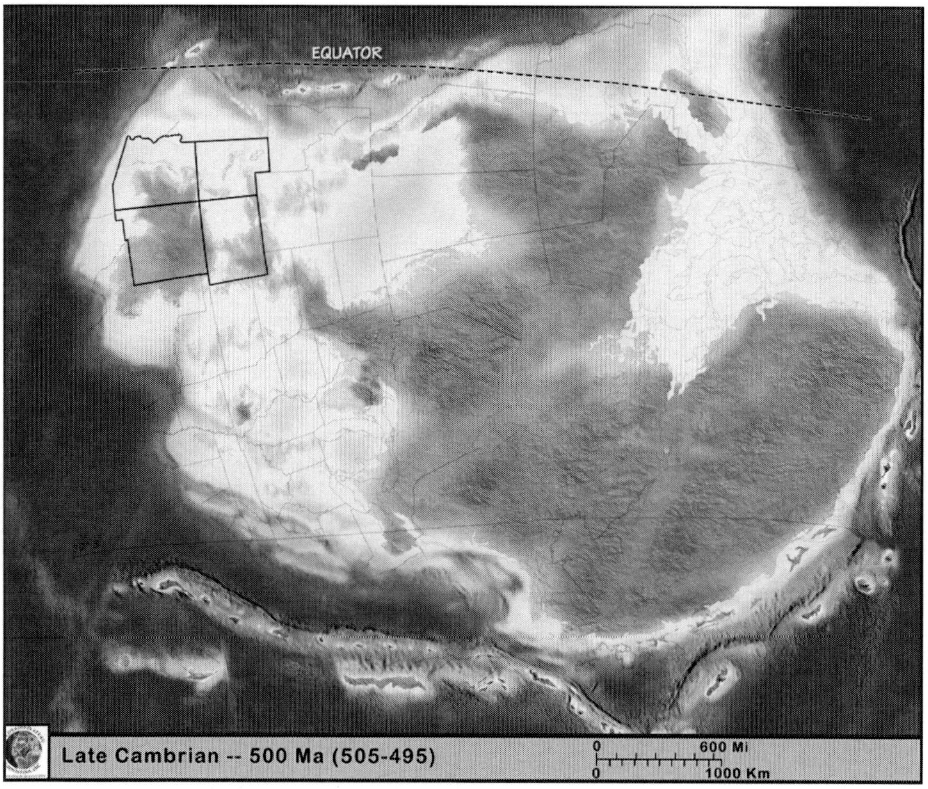

Figure 1.7: The Southwestern US during the late Cambrian, approximately 500 million years ago. The entire region is located in the southern hemisphere—note the position of the equator.

Ancient Continents and Their Names

It has taken hundreds of millions of years for the continents to take on the shapes we see today. Ancient continents looked very different. To simplify descriptions of ancient geography, geologists have given names to earlier "proto-continents" to distinguish them from their modern counterparts. Proto-Europe (northwestern Europe without Ireland and Scotland) in the early Paleozoic is known as *Baltica*; proto-North America is known as Laurentia; and proto-Africa was part of a larger continent known as *Gondwana*, which included what are now Africa, Australia, Antarctica, India, and South America. To simplify descriptions of geological events on these ancient continents, compass directions generally refer to modern, rather than ancient orientations. Thus, "western Laurentia" means the margin of proto-North America that today faces west, but which faced north in the Paleozoic.

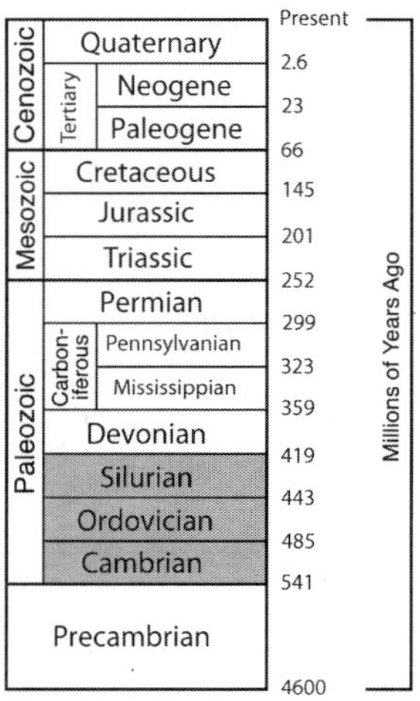

Paleozoic

In the late **Devonian** period (approximately 370 million years ago) a major geological change took place in the Southwest. A portion of the continental shelf adjacent to present-day Idaho and Nevada changed from a quiet passive margin to an active subduction zone, where oceanic crust plunged beneath the continent. Here, as oceanic crust descended deep into the upper mantle, the rock above the descending crust melted to form a line of volcanoes on the surface. Subduction also led to accretion—sediment, sedimentary rock, and even bits of the oceanic crust itself were scraped off the descending crustal plate and pushed onto the overlying plate (*Figure 1.8*). Just as a rug develops folds when pushed from the side, these rocks were wrinkled up into mountains. Volcanic islands carried along by the subducting plate also accreted to the edge of the continent. The landmass began to rotate, moving the North American plate into a more modern orientation (*Figure 1.9*).

Devonian • *a geologic time period spanning from 419 to 359 million years ago.*

density • *a physical property of minerals, describing the mineral's mass per volume.*

Continental and Oceanic Crust

The lithosphere includes two types of crust: continental and oceanic. Continental crust is less *dense* but significantly thicker than oceanic crust. The higher density of the oceanic crust means that when continental crust collides with oceanic crust, the denser oceanic crust (made mostly of dense rocks such as basalt) will be dragged (or subducted) under the buoyant continental crust (made mostly of less dense rocks such as granite). Although mountains are created at these oceanic/continental crust collisions due to the compression of the two plates, much taller ranges are produced by continental/continental collisions. When two buoyant continental crusts collide, there is nowhere for the crust to go but up! The modern Himalayas, at the collision site of the Asian and Indian plates, are a good example of very tall mountains formed by a collision between two continental crusts.

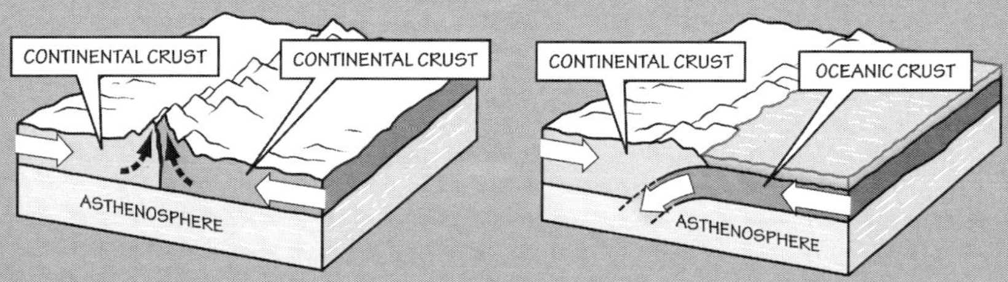

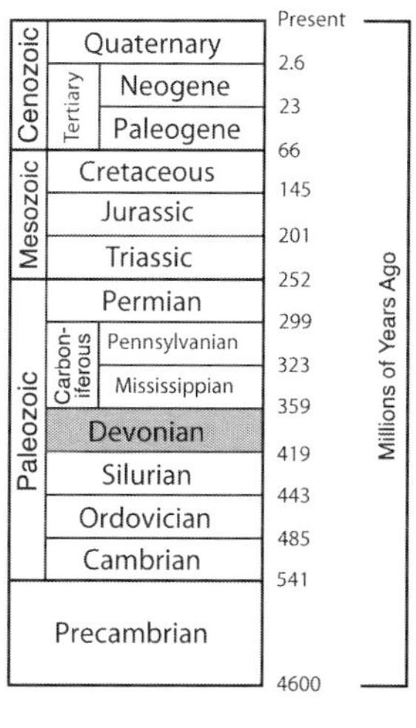

Paleozoic

Carboniferous • a geologic time period that extends from 359 to 299 million years ago.

Mississippian • a subperiod of the Carboniferous, spanning from 359 to 323 million years ago.

Antler Orogeny • a period of mountain building that deformed rocks in a belt extending from the California–Nevada border northward into Idaho.

transform boundary • an active plate boundary in which the lithospheric plates move sideways past one another.

Era		Period	Millions of Years Ago
			Present
Cenozoic		Quaternary	2.6
Cenozoic	Tertiary	Neogene	23
Cenozoic	Tertiary	Paleogene	66
Mesozoic		Cretaceous	145
Mesozoic		Jurassic	201
Mesozoic		Triassic	252
Paleozoic		Permian	299
Paleozoic	Carboniferous	Pennsylvanian	323
Paleozoic	Carboniferous	Mississippian	359
Paleozoic		Devonian	419
Paleozoic		Silurian	443
Paleozoic		Ordovician	485
Paleozoic		Cambrian	541
		Precambrian	4600

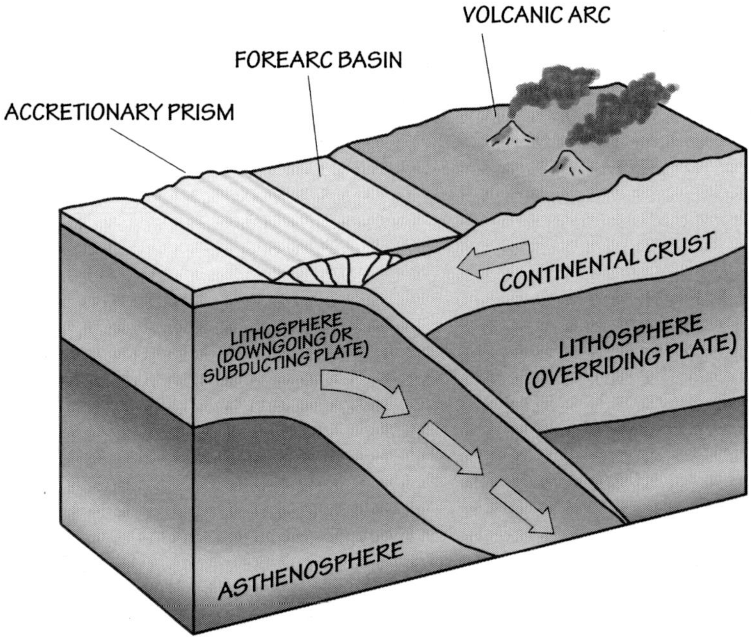

Figure 1.8: Subduction along the western edge of the North American plate. (See TFG website for a full-color version.)

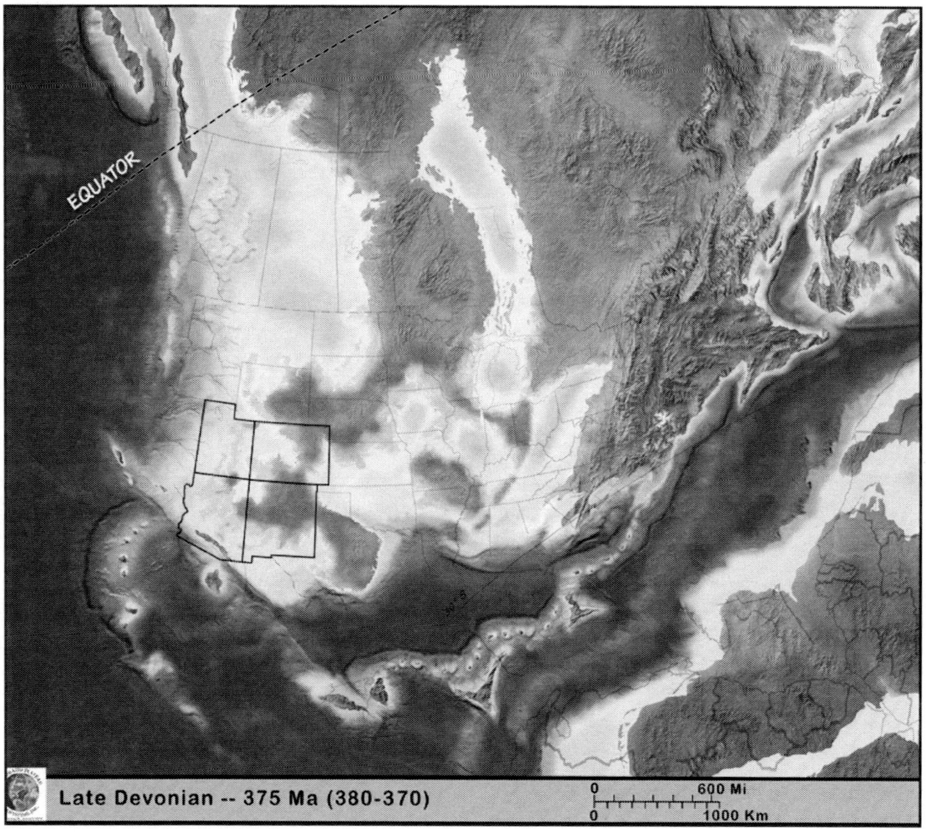

Figure 1.9: The Southwestern US during the late Devonian, approximately 375 million years ago.

Paleozoic

By the **Carboniferous**, most of the West Coast had transformed into a subduction zone. During the **Mississippian** period (around 340 million years ago), an island arc collided with and accreted to that coast, generating a major mountain-building event: the **Antler Orogeny** (*Figure 1.10*). This orogeny was the first in a long series of mountain-building events that affected the margin of western North America. The cycle of sedimentation and collision initiated during the Antler event would be repeated many times and with many variations into the Cenozoic era, when most of the subduction zone along the western margin of North America was altered into a **transform boundary** (the most important component of which is the San Andreas fault system). Much of North America, including most of the Southwest, was gently **uplifted** in the late Mississippian, causing sea level to fall across the continent.

As North America began to collide with Gondwana (composed of present-day South America, Africa, India, Australia, and Antarctica), forces from the collision began to affect the continent's topography. During the **Pennsylvanian** (300 million years ago), **compressional** forces from the collision and tension from coastal subduction combined to deform the continent's interior, buckling the crust and creating deep basins between uplifted blocks. Shallow **inland seas** spread across the interior of the continent, covering parts of North America's Precambrian shield (*Figure 1.11*). Associated uplift led to the expansion of terrestrial environments over areas that had once been marine. Geologists call the resulting landscape the Ancestral Rocky Mountains. Sediments that eroded from this range and other uplifted areas were transported to the inland sea and the continental margins, forming deposits of **conglomerates**, sandstones, **shales**, limestones, and **evaporite** minerals. Although these ranges are long eroded away and the **inland basins** filled with sediment, evidence for their existence is preserved in the patterns of sedimentary rocks remaining throughout the Southwest today.

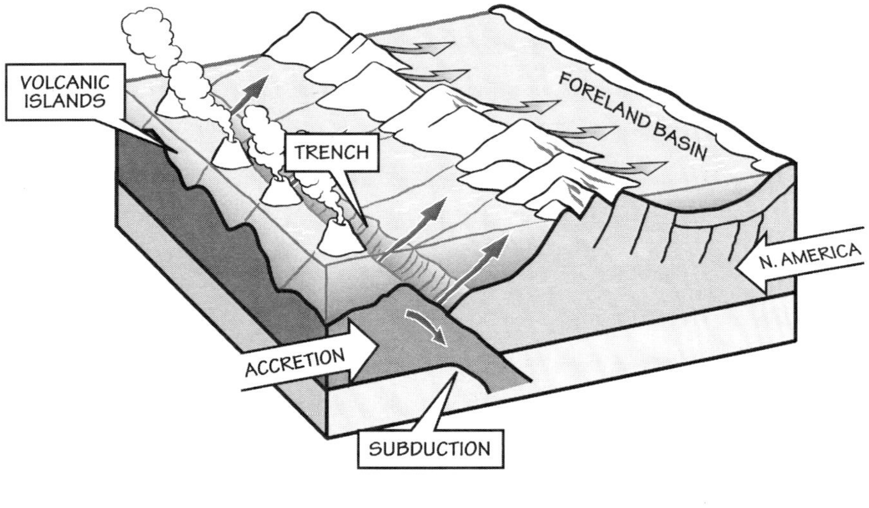

Figure 1.10: Collision of a volcanic island arc with the West Coast during the Antler Orogeny.

uplift • *upward movement of the crust due to compression, subduction, or mountain building.*

Pennsylvanian • *a subperiod of the Carboniferous, spanning from 323 to 299 million years ago.*

compression • *flattening or squeezing as a result of forces acting on an object from all or most directions.*

inland sea • *a shallow sea covering the central area of a continent during periods of high sea level.*

conglomerate • *a sedimentary rock composed of multiple large and rounded fragments that have been cemented together in a fine-grained matrix.*

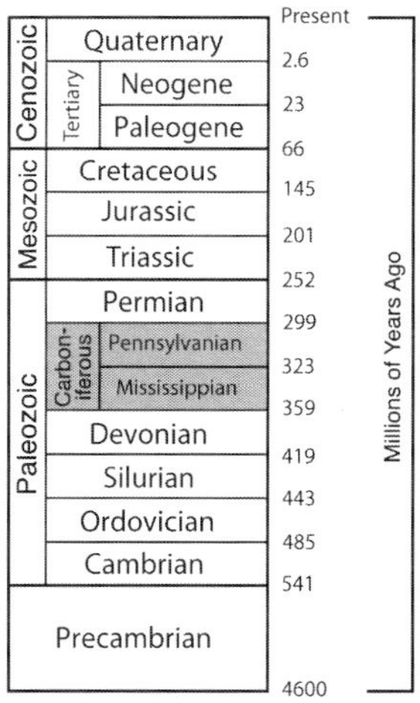

1 Geologic History

Paleozoic

mafic • *igneous rocks that contain a group of dark-colored minerals, with relatively high concentrations of magnesium and iron.*

felsic • *igneous rocks with high silica content and low iron and magnesium content.*

Permian • *the geologic time period lasting from 299 to 252 million years ago.*

iron • *a metallic chemical element (Fe).*

limestone • *a sedimentary rock composed of calcium carbonate ($CaCO_3$).*

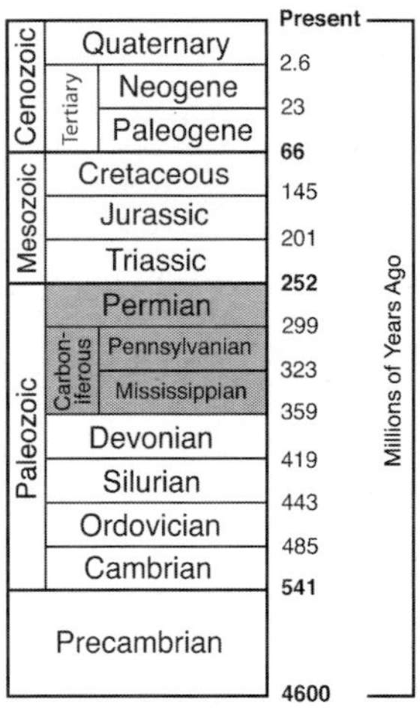

Volcanic Island Arcs

Volcanic islands are common at subduction zones between colliding oceanic plates, where one plate moves (is subducted) beneath the other. They frequently form in curved lines, and are therefore called island arcs. As the plates press together, friction between them generates enough heat and pressure to melt some of the crust. The molten rock rises through the crust and creates volcanoes along the edge of the overlying plate. The Aleutian Islands, Philippine Islands, and Lesser Antilles are all modern examples of volcanic island arcs associated with subduction. Because island arc volcanoes mix the more *mafic* composition of the ocean floor with the more *felsic* composition of overlying sediment derived from continents, they are usually of "intermediate" composition along this spectrum.

Sea level fell again in the late Paleozoic, during the Pennsylvanian and **Permian**, as continental collisions progressed to form the supercontinent Pangaea. As accretion continued over time, the coastline moved farther seaward, running through western Montana, eastern Idaho, central Utah, and western Arizona (*Figure 1.12*). The climate at low latitudes of this supercontinent was hot and dry, and **iron**-rich **limestones**, **sandstones**, and mudstones were **oxidized**. This process generated rocks with a distinctive and characteristic red color, appropriately called "red beds." Permian red beds are very characteristic of the Southwest, particularly in the Colorado Plateau region, where they are the most widespread geologic feature exposed. Many of these red beds represent ancient dunes. Meanwhile, the marine Permian Basin (*Figure 1.13*) of southern New Mexico and west Texas was ringed by **reefs**—the Guadalupe Mountains of southern New Mexico and western Texas contain the largest and best-preserved Paleozoic reef in the world. Regression of the sea near the end of the Permian period spelled the end of the Southwest's magnificent reefs.

See Chapter 2: Rocks to learn more about the Southwest's distinctive red beds and reefs.

Figures 1.11 and 1.12 (AT RIGHT): The Southwestern US during (1.11, top) the Pennsylvanian, approximately 308 million years ago, and (1.12, bottom) the early Permian, approximately 280 million years ago.

Geologic History

1

Paleozoic

sandstone • *sedimentary rock formed by cementing together grains of sand.*

oxidation • *a chemical reaction involving the loss of at least one electron when two substances interact.*

reef • *a feature lying beneath the surface of the water, which is a buildup of sediment or other material built by organisms, and which has positive relief from the sea floor.*

Middle Pennsylvanian -- 308 Ma (311-307)

Early Permian -- 280 Ma (282-275)

Era		Period	Millions of Years Ago
			Present
Cenozoic		Quaternary	2.6
	Tertiary	Neogene	23
	Tertiary	Paleogene	**66**
Mesozoic		Cretaceous	145
		Jurassic	201
		Triassic	**252**
Paleozoic		Permian	299
	Carbon-iferous	Pennsylvanian	323
	Carbon-iferous	Mississippian	359
		Devonian	419
		Silurian	443
		Ordovician	485
		Cambrian	**541**
		Precambrian	**4600**

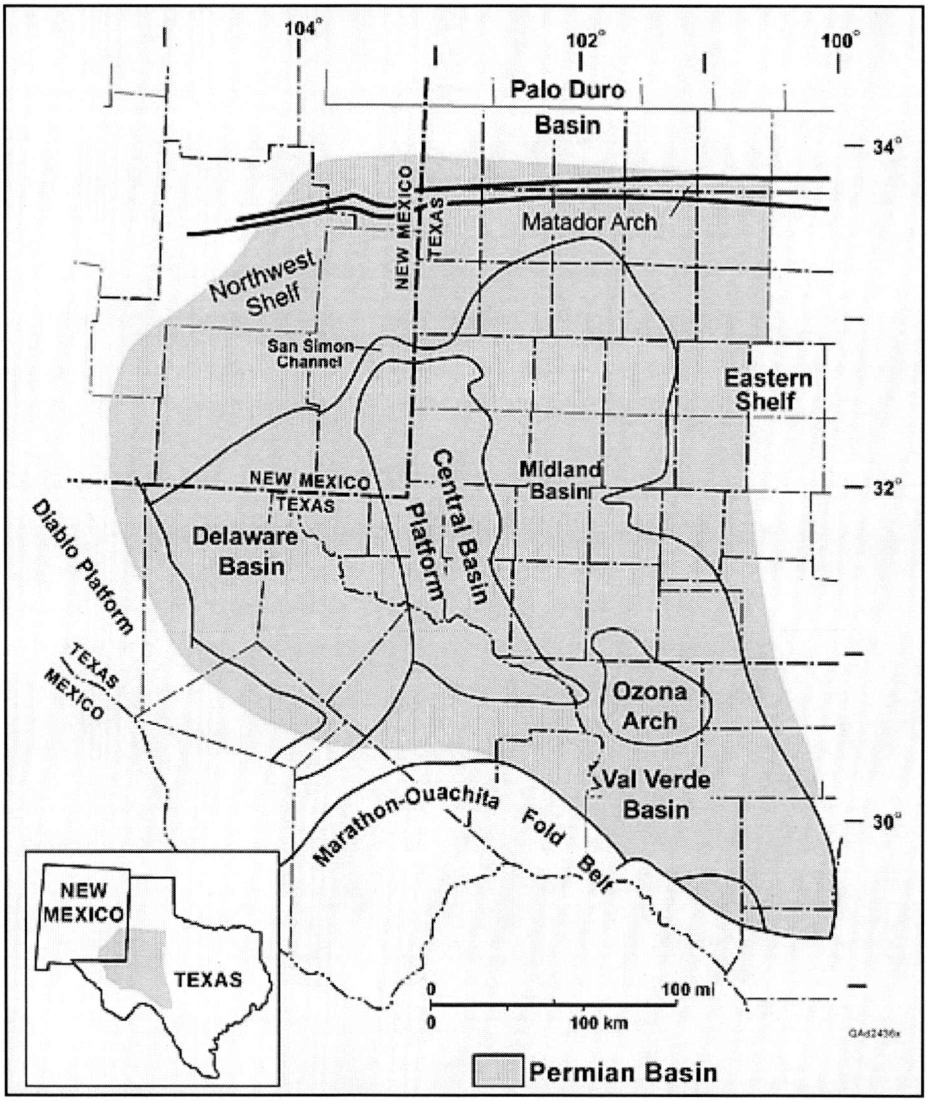

Figure 1.13: The Permian Basin is a large sedimentary basin in western Texas and southeastern New Mexico. It is made up of three main component parts: the eastern Midland Basin, the Central Basin Platform, and the western Delaware Basin. These structures existed from the Carboniferous to the Triassic periods.

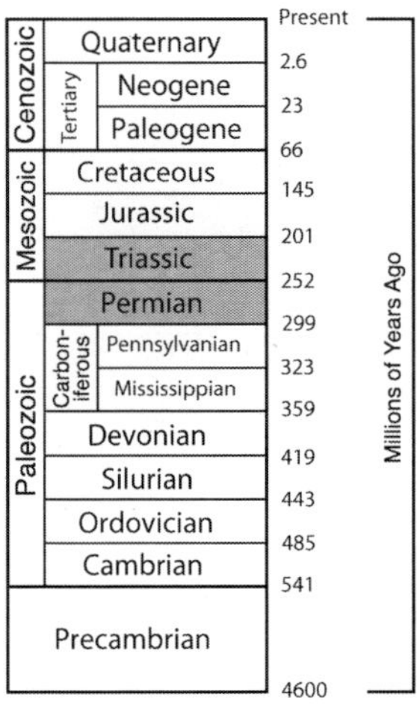

The Mesozoic

A Story of Mountains and Seas

The Mesozoic era is frequently known as the Age of the Dinosaurs or Age of Reptiles, but many other life forms evolved and thrived during this time, including marine invertebrates, flowering plants, birds, and mammals. The Mesozoic was also a time of major geologic change during which great thicknesses of rocks were deposited across the Southwestern US.

Mesozoic

The supercontinent Pangaea was in place by the end of the Permian period, and global sea level was probably at its lowest of any time during the past 600 million years. In the Southwest, the shoreline withdrew from central Utah to western Nevada. During the **Triassic**, Earth's climate was much warmer than today, with an average global temperature of about 25°C. The Southwest was now largely dry land, and rocks from this period record complex and varied deposition in rivers and deserts. Subduction along the western margin of the Americas generated the Cordilleran volcanic arc, which was active throughout much of the Mesozoic. Triassic rocks of the Grand Canyon record an influx of volcanic material from the adjacent arc, as well as sediment from adjacent highlands composed of sedimentary and plutonic rocks. Rocks from the Cordilleran arc are still present today in the Andes, Central America, the Cascades, and the Aleutians.

See Chapter 8: Climate to learn how climate and life have changed throughout geologic history.

Triassic • a geologic time period that spans from 252 to 201 million years ago.

Jurassic • the geologic time period lasting from 201 to 145 million years ago.

Pangaea began to break up during the early **Jurassic**, and sea levels began to rise again. By the middle Jurassic, a shallow arm of the sea reached from Canada south through Montana and parts of Wyoming, Utah, and the Dakotas (*Figure 1.14*), depositing thin layers of limestone, mudstone, and sand. During the Jurassic, mudstone and sandstones were also deposited in lowland areas and river channels throughout the Rockies and Colorado Plateau; these formed the Morrison Formation, which is famous for its abundant dinosaur fossils.

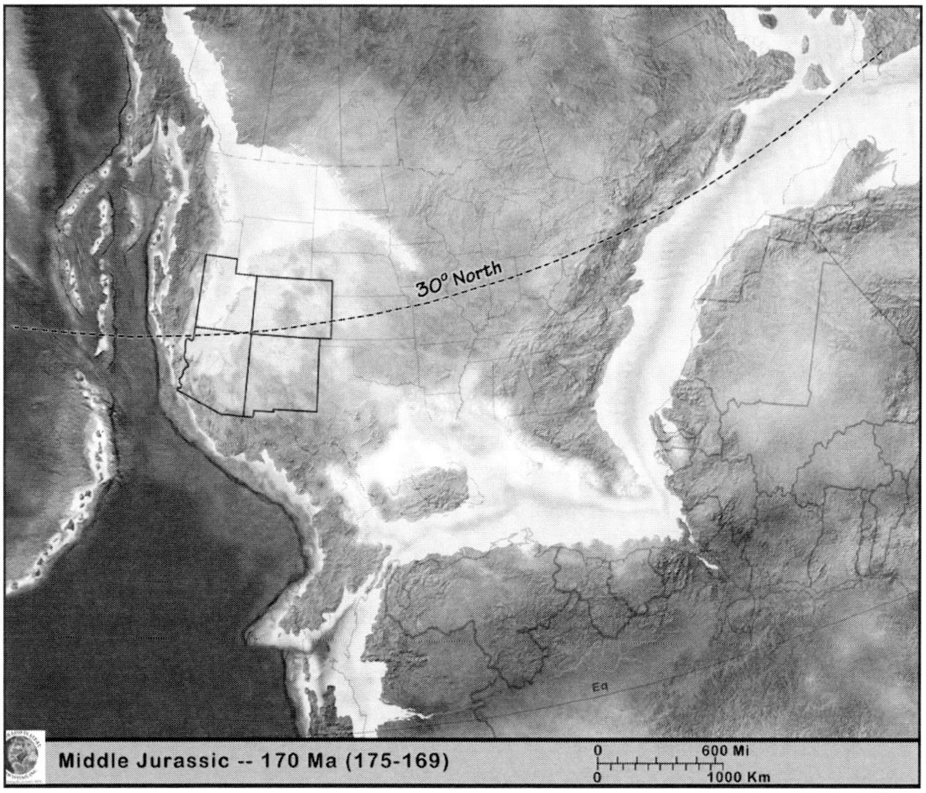

Figure 1.14: The Southwestern US during the Jurassic, approximately 170 million years ago.

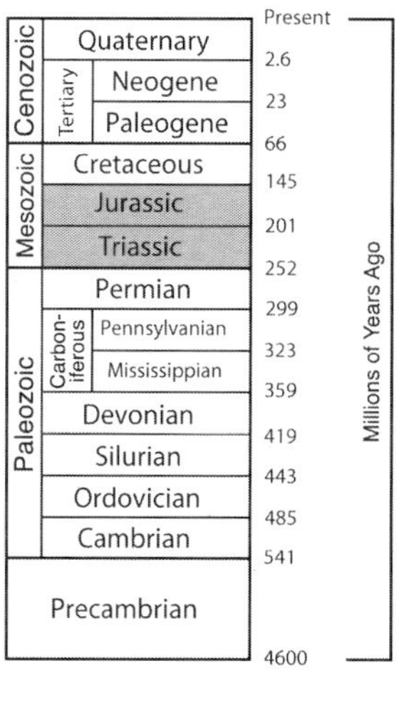

Cretaceous • *a geologic time period spanning from 144 to 66 million years ago.*

downwarp • *a segment of the Earth's crust that is broadly bent downward.*

Sevier Orogeny • *a mountain-building event resulting from subduction along the western edge of North America, occurring mainly during the Cretaceous.*

Laramide Orogeny • *a period of mountain building that began in the late Cretaceous, and is responsible for the formation of the Rocky Mountains.*

During the **Cretaceous**, Pangaea entered its final stages of breakup (*Figure 1.15*). Far to the west, oceanic crust (the Farallon plate) had been subducting under western North America for tens of millions of years, causing a series of volcanic island complexes to collide with and become accreted to that margin of the continent. As oceanic crust was subducted beneath the continent, a new volcanic arc formed the Sierra Nevada of California. The new Atlantic Ocean widened, and sea levels began to rise. The Farallon plate began to subduct at an unusually shallow angle, sliding farther inland beneath western North America before finally sinking into the asthenosphere (*Figure 1.16*). This **downwarped** the center of the continent and created a basin that allowed the waters of the Gulf of Mexico to meet with those in the north, forming the Western Interior Seaway (*Figure 1.17*). This shallow inland sea inundated a 1000-kilometer (620-mile) wide swath from Mexico to Alaska, separating North America into an Appalachian island to the east and a mountainous Cordilleran island to the west. Erosion from these western mountains resulted in deposition of thick layers of sediment throughout the seaway. During the very latest stages of the Cretaceous period, around 70 million years ago, the Western Interior Seaway was displaced by slow uplift of the continent.

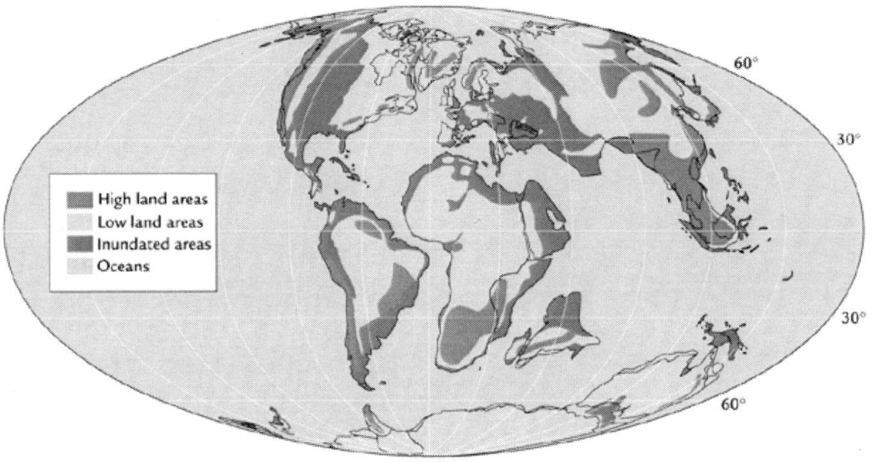

Figure 1.15: Landmasses following the breakup of Pangaea. (See TFG website for a full-color version.)

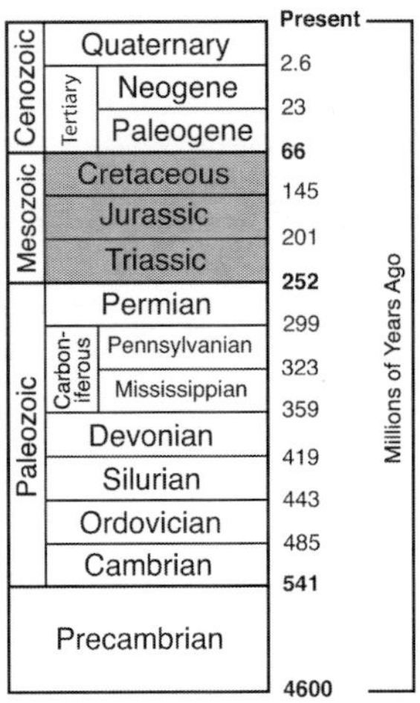

The Farallon plate continued to collide with western North America, thrusting layers of rock up over each other and causing increasing volcanism to the west of the Western Interior Seaway. The compressional forces of subduction faulted the crustal rocks of western North America and uplifted the Rocky Mountains in two major pulses. The **Sevier Orogeny** (100–72 million years ago) formed an extensive belt of and folds, also known as the "Overthrust Belt," extending from the Sierra Nevada in California to the edge of the Colorado Plateau in western Utah. The second event, the **Laramide Orogeny**, peaked around 68–65 million years ago when the angle of the subducting plate became shallower. This formed mountains farther inland than would normally be expected above a subduction zone, uplifting the Rocky Mountains in Wyoming, Colorado, and New Mexico (*see Figure 1.16*). These mountain ranges are bounded by thrust

Evidence for Pangaea

How do we know that Pangaea existed 250 million years ago? Long before the discovery of plate tectonics in the 1960s and early 1970s, fossils and mountain belts provided evidence that the continents had not always been in their current positions. For example, the Permian-aged fossil plant *Glossopteris* had seeds too heavy to be blown across an ocean. Yet *Glossopteris* fossils are found in South America, Africa, Australia, India, and Antarctica! The mountain belts along the margins of North America, Africa, and Europe line up as well and have similar rock types, an indication that the continents at one time were joined as Pangaea. Despite the discovery of *Glossopteris* and other geologic evidence, the theory of continental drift was not accepted for decades, until the mechanisms of continental movement were discovered and reformulated under the modern theory of plate tectonics. The supercontinent Pangaea existed for approximately 100 million years, reaching its largest size during the Triassic period. During the Jurassic, the landmass began to fragment into the modern continents, which slowly moved toward their present-day positions over the following 150 million years.

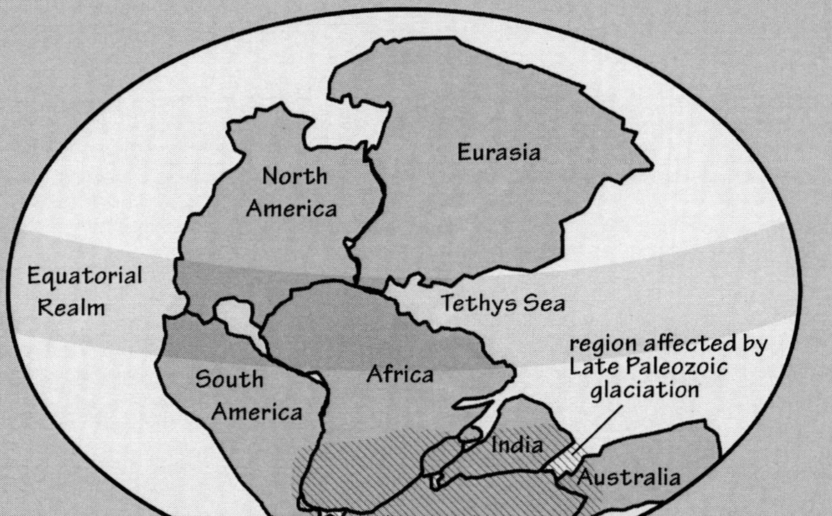

Pangaea during the late Paleozoic era

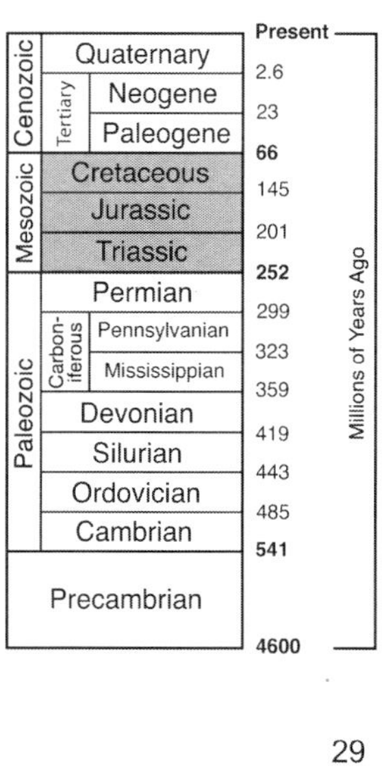

Mesozoic

active plate margin • *the boundary between two plates of the Earth's crust that are colliding, pulling apart, or moving past each other.*

convergent boundary • *an active plate boundary where two tectonic plates are colliding with one another.*

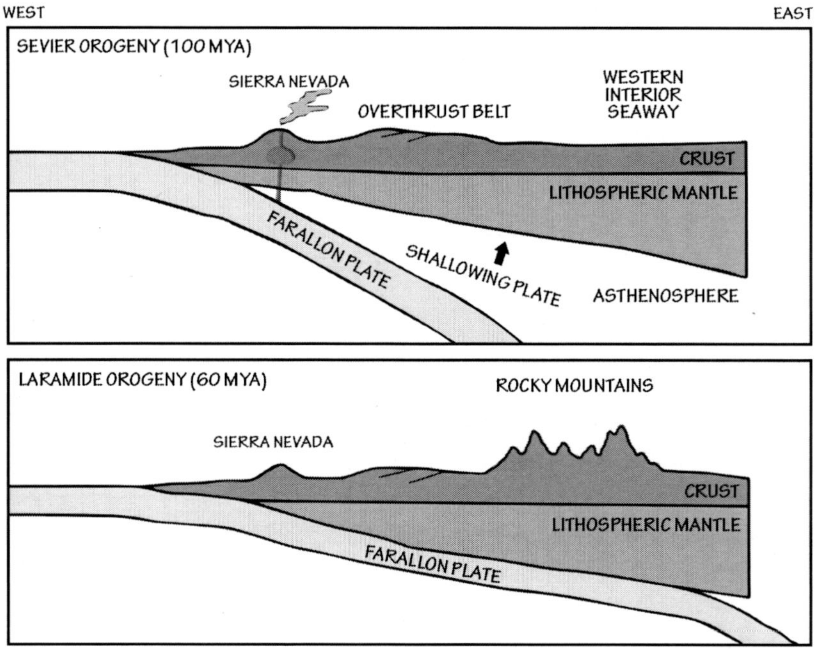

Figure 1.16: The Sevier and Laramide orogenies.

Figure 1.17: The Western Interior Seaway.

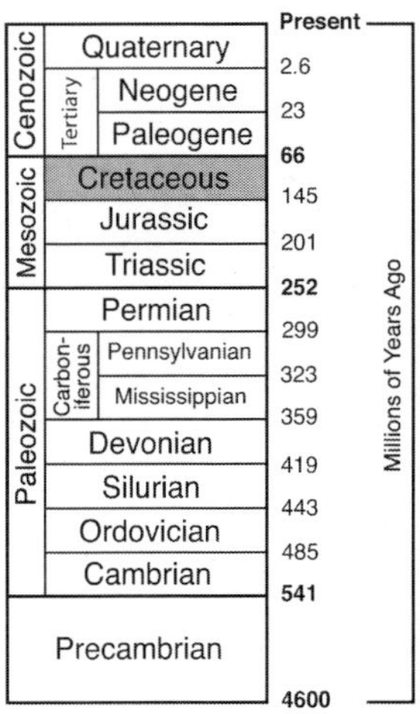

Understanding Plate Boundaries

Active plate margins are the boundaries between two plates of the Earth's crust that are colliding, pulling apart, or moving past each other as they move over the mantle. Some of these plates move as fast as 10 centimeters/year (4 inches/year). The processes of plate movement, spreading, subduction, and mountain building are collectively called plate tectonics.

When one plate slides beneath another, it is called a *convergent boundary* or subduction zone. When two plates pull apart from each other, it is called a *divergent boundary* or rift margin. When the plates slip past each other in opposite directions, it is known as a transform boundary.

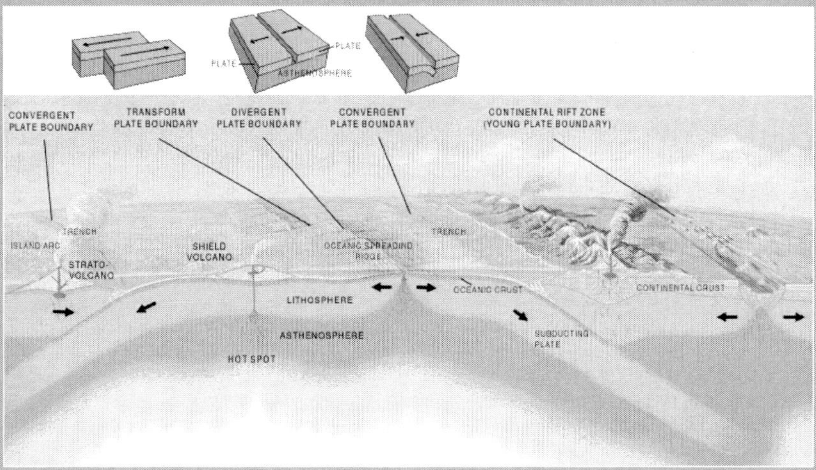

faults. Because the crust flexes or breaks under compression, several inland basins formed between the mountain ranges, and the eroding mountains shed thick layers of sediment into these basins, forming conglomerates, sandstones, and mudstones. The Colorado Plateau remained stable during this time of compression, and persisted during the subsequent episode of extension that followed from the **Paleogene** to the present day.

The Cretaceous-Paleogene (K-Pg) boundary (previously known as the Cretaceous-**Tertiary** [K-T] boundary) marks one of the most significant physical and biological events in Earth history. The boundary marks the contact between the Mesozoic and Cenozoic eras at around 65 million years ago, representing

Mesozoic

divergent boundary • *an active plate boundary where two tectonic plates are pulling apart from one another, causing the mantle to well up at a rift.*

Paleogene • *the geologic time period extending from 66 to 23 million years ago.*

Tertiary • *an unoffical but still commonly used term for the time period spanning from 66 million to 2.5 million years ago, including the Paleogene, Neogene, and part of the Pleistocene.*

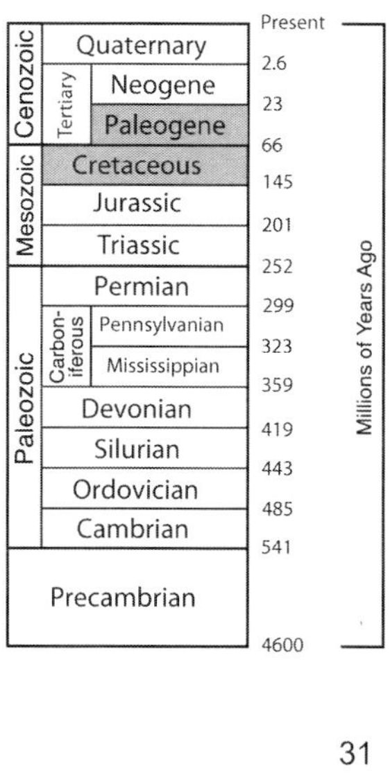

Cenozoic

__basalt__ • an extrusive igneous rock, and the most common rock type on the surface of the Earth.

__lava__ • molten rock located on the Earth's surface.

__Neogene__ • the geologic time period extending from 23 to 2.6 million years ago.

__Quaternary__ • a geologic time period that extends from 2.6 million years ago to the present.

__fluvial__ • see outwash plain: large sandy flats created by sediment-laden water deposited when a glacier melts.

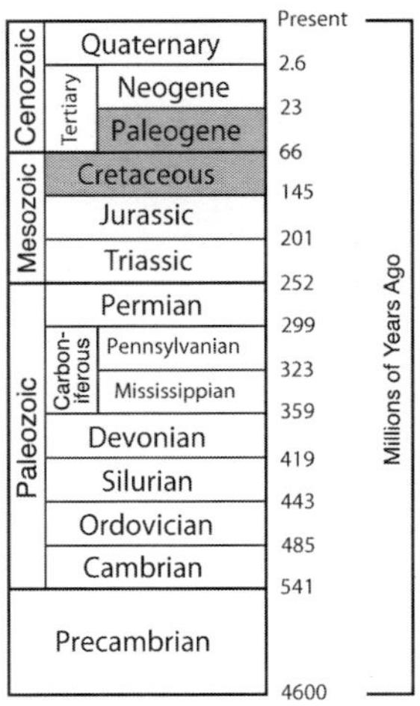

a time during which a large proportion (perhaps 50–70%) of all species of animals and plants (both marine and terrestrial, from microscopic one-celled organisms to massive dinosaurs) abruptly became extinct. Most geologists and paleontologists think these extinctions resulted from the impact of a large comet or asteroid, perhaps associated with an impact crater in the subsurface of Mexico's Yucatan Peninsula. There is also evidence for the occurrence of extensive volcanism at the K-Pg boundary, indicated by large **basaltic lava** flows in India called the Deccan Traps. The end-Cretaceous event greatly altered the history of life on land and in the sea, and these changes are clearly visible in the Southwest's fossil record. The boundary itself is rarely preserved in the geologic record, due to incomplete sedimentation and widespread erosion. However, there are several Southwestern localities, especially in northern New Mexico, that preserve the K-Pg boundary layer.

See Chapter 3: Fossils to learn how the fossil record represents the K-Pg mass extinction.

The Cenozoic
Volcanism and Tectonism

The Cenozoic era (consisting of the Paleogene, **Neogene,** and **Quaternary** periods, 66 million years ago to present) was an age of diversification and evolution of mammals, birds, insects, flowering plants, and coral reefs. The continents continued to spread apart to reach their present-day positions. Sea levels rose and fell, affecting the coastline, but the interior of North America remained relatively high. Sediment deposition, for the most part, occurred as **fluvial** and lake deposits. This was also a time of active volcanism in western North America. The Cenozoic geology of western North America is dominated by three large-scale processes: erosion, subduction and extension, and volcanic activity.

The Paleogene was a time of active volcanism around the world, and North America was no exception. The Paleogene and Neogene periods were warmer than today, with average global temperatures ranging from 17 to 25°C. Erosion of the mountains and highlands that had formed during the Mesozoic produced thick layers of conglomerates, sandstones, and mudstones across stream valleys, **deltas**, and lakes. **Volcanic ash** from nearby eruptions is commonly interlayered with these sediments. Many of these sedimentary layers were deposited by rivers, or in **alluvial** fans coming from the mountain systems. Several such layers are now important **aquifers**, including the enormous Ogallala Aquifer (*Figure 1.18*) which today supplies water for farming and communities on the Great Plains.

Due to crustal deformation during the late Mesozoic and early Cenozoic, numerous basins formed inland lakes or depressions into which sediments were deposited (*Figure 1.19*). Perhaps the best-known example is the Green River Basin of western Wyoming and its equivalents, the Uinta Basin in northeastern Utah and the Piceance Basin in northwestern Colorado. Lakebed shales and

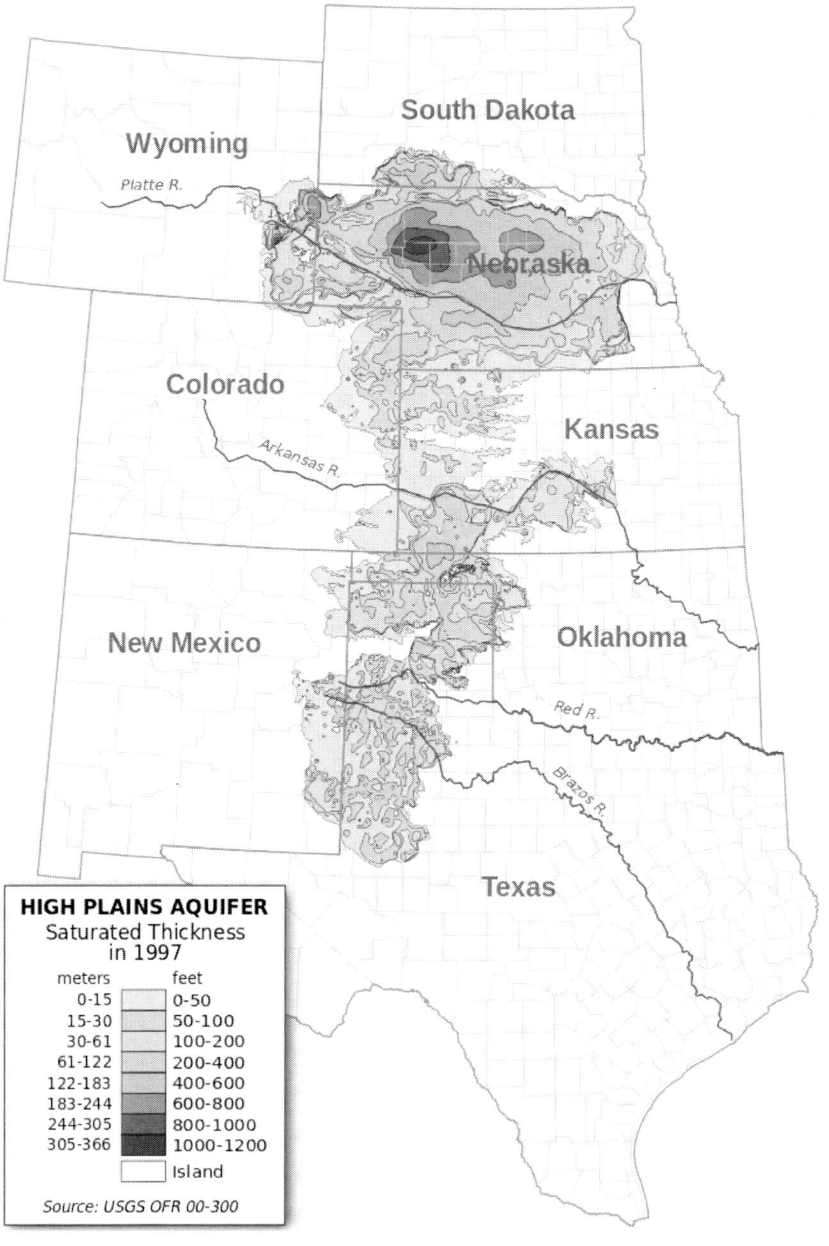

Figure 1.18: Extent and saturated thickness of the Ogalalla Aquifer. (See TFG website for a full-color version.)

mudstones found in all of these basins are famous for their abundant and well-preserved fossils. Other important basins, including the Denver and Raton basins in Colorado and the San Juan and Baca basins in New Mexico and Arizona, produce **coal**, **oil**, and gas, and other industrial minerals.

See Chapter 6: Energy for more on the formation of fossil fuels in the Southwest.

Cenozoic

delta • *a typically wedge-shaped deposit formed as sediment is eroded from mountains and transported by streams across lower elevations.*

volcanic ash • *fine, unconsolidated pyroclastic grains under 2 millimeters (0.08 inches) in diameter.*

alluvial • *a layer of river-deposited sediment.*

aquifer • *a water-bearing formation of gravel, permeable rock, or sand that is capable of providing water, in usable quantities, to springs or wells.*

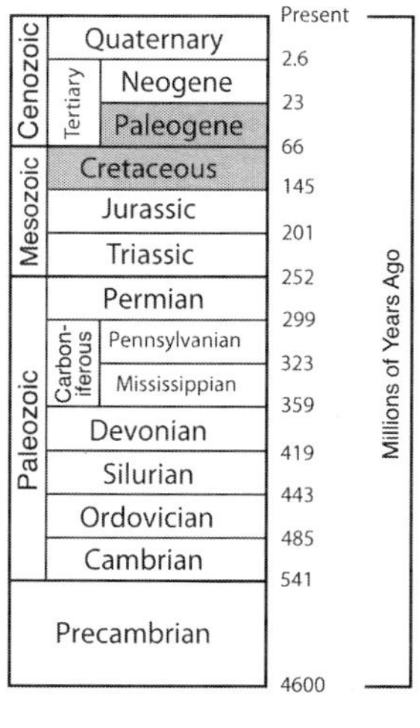

Cenozoic

intrusive rock • *a plutonic igneous rock formed when magma from within the Earth's crust escapes into overlying strata.*

granite • *a common and widely occurring type of igneous rock.*

magnetic • *affected by or capable of producing a magnetic field.*

caldera • *a collapsed, cauldron-like volcanic crater formed by the collapse of land following a volcanic eruption.*

isostasy • *an equilibrium between the weight of the crust and the buoyancy of the mantle.*

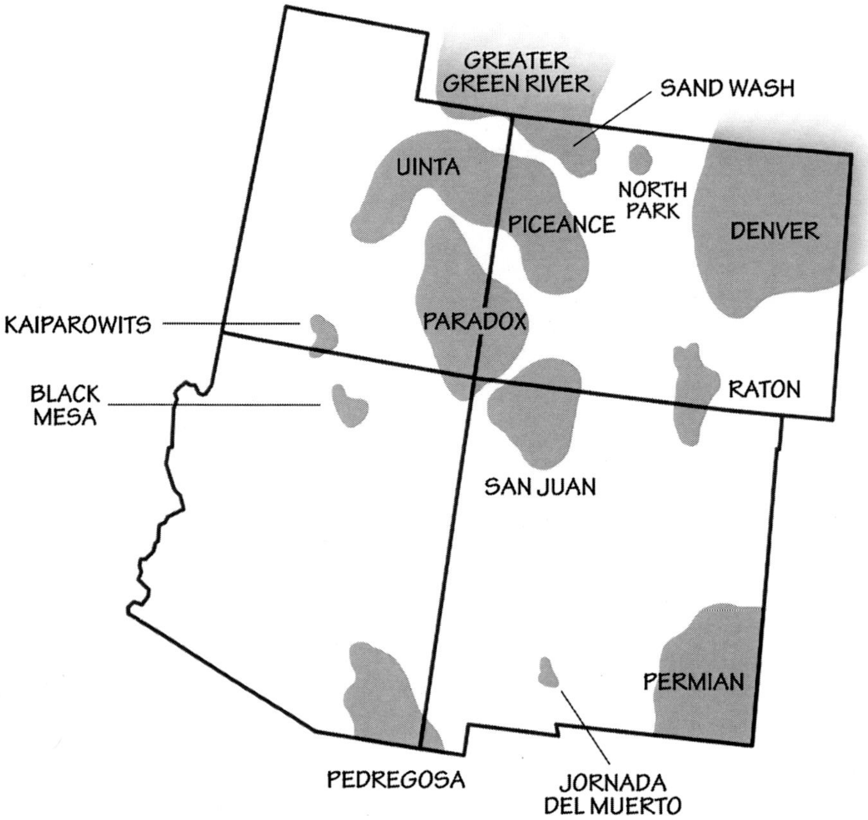

Figure 1.19: Sedimentary basins of the Southwest.

In the late Cenozoic, subduction at the West Coast was progressively replaced to the north and South by the developing San Andreas fault system. Subduction continues today beneath Oregon and Washington (and beneath Mexico south of the Gulf of California). Due to the complex interplay of plate motions, the portion of the subducting plate beneath the Southwestern US overrode hot, upwelling mantle. This, in turn, caused a number of major changes. In the early Paleogene, melting of the lower crust resulted in the **intrusion** of numerous granitic bodies and the formation of large volcanic fields, including the San Juan and Thirtynine Mile in Colorado and the Marysvale in Utah. These **granite** intrusions remain mostly buried beneath the surface, detectable only by **magnetic** surveys. Volcanic lava and ash flows, now deeply eroded, cover the countryside above the buried granite intrusions. Remains of ancient **calderas**, recording the collapse of now-eroded volcanoes, can be detected in the subsurface.

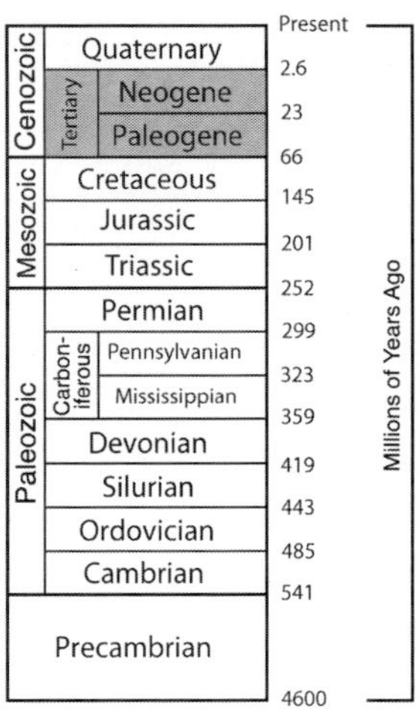

Between 40 and 30 million years ago, a transition from subduction and crustal compression to crustal extension began to occur. By the Neogene, the Farallon plate lay shallowly under the North American plate for hundreds of kilometers eastward of the West Coast. Now situated more fully beneath what are now the South Central, Southwestern, and Northwest Central states, this extra layer of crust caused uplift and extension of the region, as the added thickness of buoyant rock (relative to the mantle) caused the entire area to rise **isostatically**. The Farallon plate was subjected to increasing temperatures as it subducted, causing it to expand. As heat dissipated to the overlying North American plate, that rock expanded as well. Finally, the high temperatures in the upper mantle caused the Farallon plate to melt, and the resulting magma was injected into the North American plate, destabilizing it. These processes, along with the complex crustal movements taking place along the San Andreas fault, caused the surface of the North American plate to pull apart and fault into the mountainous blocks of the huge Basin and Range province that stretches from Idaho, Nevada, and Utah into California, Arizona, New Mexico, and Texas. In the late **Miocene**, around eight million years ago, **epeirogenic** uplift (resulting from upwelling mantle heat pushing the crust upwards) began. The uplift raised not just the mountains but also the base elevation of the entire region to its current level. This process raised the Rocky Mountains and Colorado Plateau to their current "mile-high" elevation, initiating the downcutting of the Grand Canyon in Arizona. The modern direction of the Colorado River's flow toward the Gulf of California was established about six million years ago, at which time the river began to widen and deepen the canyon we know today.

The Quaternary
Mountains of Ice

At the start of the Quaternary period, approximately 2.5 million years ago, continental **ice sheets** began to form in northernmost Canada. Throughout this period, the northern half of North America has been periodically covered by continental glaciers that originated in northern Canada (*Figure 1.20*). The Quaternary period is divided into two epochs: the **Pleistocene** and **Holocene**. During the Pleistocene, ice sheets advanced south and retreated north several dozen times, reaching their last maximum extent 25,000–18,000 years ago. The Holocene epoch is the most recent (and current) period of retreat, and is referred to as an **interglacial** interval. The beginning of the Holocene is considered to be 11,700 years ago, or about 9700 BCE.

The Pleistocene ice sheets did not extend into the Southwest—at their greatest extent in the western US, continental ice sheets reached into Washington, Idaho, Montana, and the Dakotas. However, local alpine glaciers covered the highest parts of the Rocky Mountains as far south as northern New Mexico (*Figure 1.21*). These glaciers carved U-shaped valleys with steep headwalls (**cirques**) and sharp drainage divides (**arêtes**). They also left unsorted deposits of glacial **till** in lateral **moraines** along the sides of valley walls, and in terminal and recessional moraines marking both the farthest extent of the ice and the places where the ice front paused during retreat. Glacial lakes formed in low areas between or in front of glaciers, and also during times between glacial advances. One such lake, Lake Bonneville in Utah, was the precursor of today's Great Salt Lake (*Figure 1.22*)

Quaternary

Miocene • *a geological time unit extending from 23 to 5 million years ago.*

epeirogenic • *large-scale crustal uplift caused by hot or upwelling mantle underlying the surface.*

Pleistocene • *a subset of the Quaternary, lasting from 2.5 million to about 11,700 years ago.*

Holocene • *the most recent portion of the Quaternary, beginning about 11,700 years ago and continuing to the present.*

interglacial • *a period of geologic time between two successive glacial stages.*

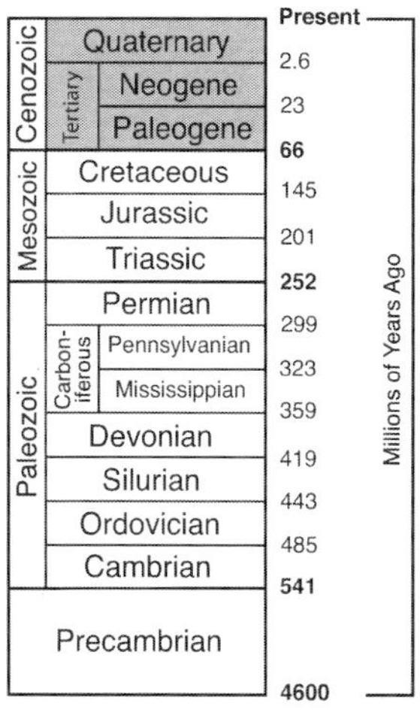

Quaternary

The Quaternary period has been dominated by the advance and retreat of continental ice sheets and alpine glaciers; glacial periods alternate with warm interglacial periods. While the "ice age" continues today, the Earth is in an interglacial stage, since the ice sheets have retreated for now. The glacial-interglacial cycling of ice ages indicates that the world will return to a glacial stage in the future, unless the impacts of human-induced climate change radically shift these natural cycles.

See Chapter 9: Climate to learn more about how climate change affects the environment.

Figure 1.20: Extent of glaciation over North America during the Quaternary.

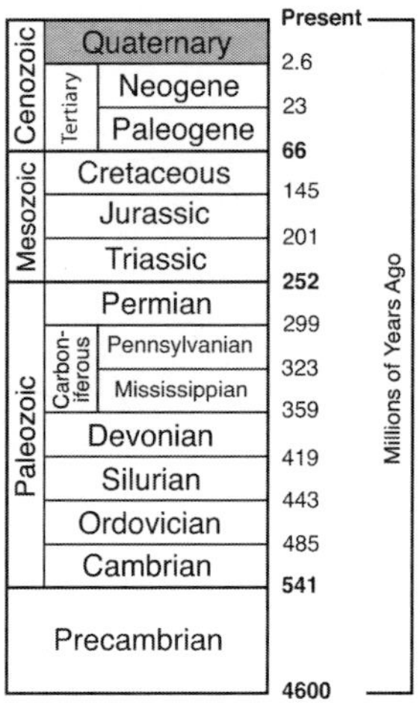

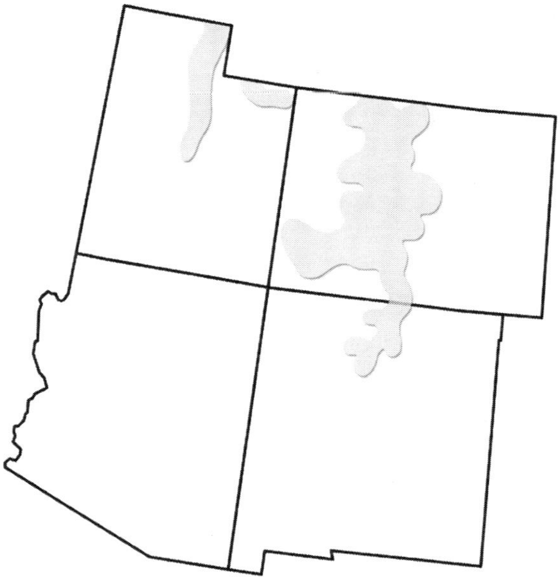

Figure 1.21: Pleistocene alpine glaciers of the Southwest.

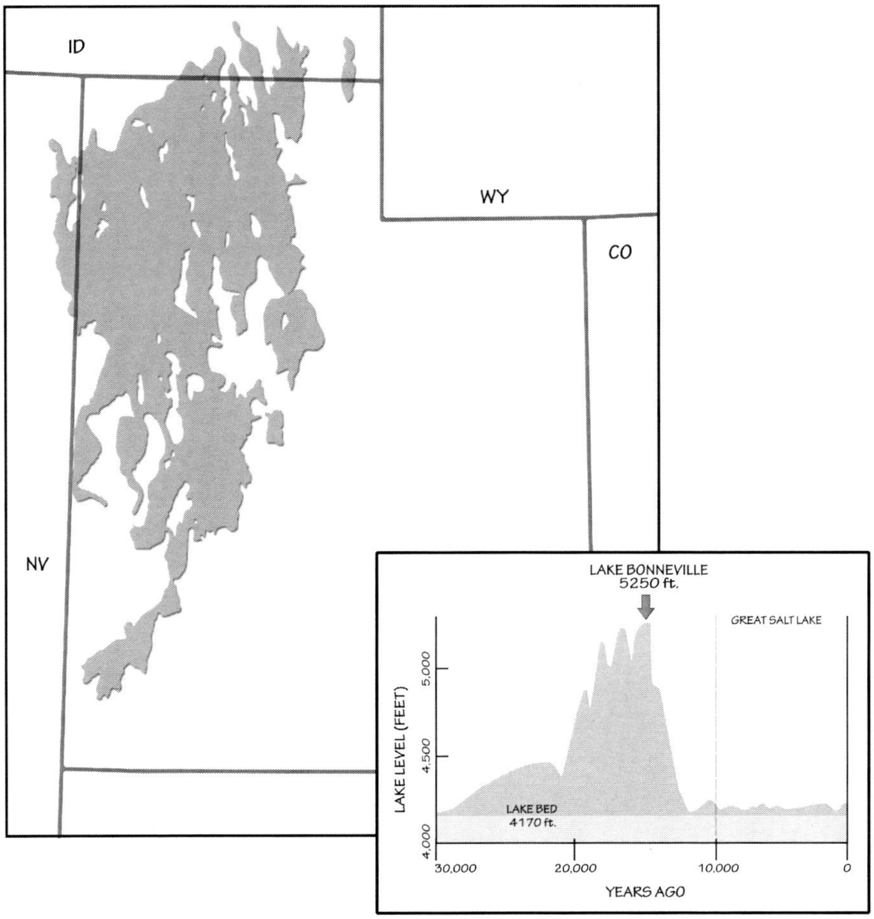

Figure 1.22: Lake Bonneville's maximal extent during the Pleistocene. Inset: Graph of the lake's changing level.

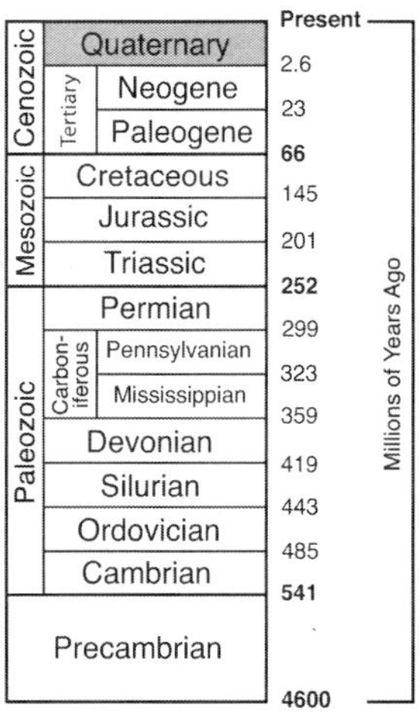

1 Geologic History

Quaternary

Why was there an ice age?

What led to the formation of large continental glaciers in the Northern Hemisphere between 3.5 and 2.5 million years ago? Movement of the Earth's tectonic plates may have been a direct or indirect cause of the glaciation. As plates shifted, continents moved together and apart, changing the size and shape of the ocean basins and altering ocean currents that transported heat from the equator to the poles. Sufficient precipitation in northern Asia and North America also enabled continental glaciers to grow and flow outward. The rise of the Himalayas exposed new rock that trapped carbon dioxide through chemical weathering; in turn, the decreased levels of carbon dioxide led to a global cooling. Finally, and surprisingly, the formation of the Central American Isthmus, which connects North and South America in what is now Panama, likely had a major effect on climate. Ocean currents that had once flowed east to west through the Central American Seaway were now diverted northward into the Gulf of Mexico and ultimately into the Gulf Stream in the western Atlantic. The strengthened Gulf Stream now transported more moisture to high northern latitudes, causing more snow, which eventually formed glaciers.

Era		Period	Millions of Years Ago
			Present
Cenozoic		Quaternary	2.6
	Tertiary	Neogene	23
		Paleogene	**66**
Mesozoic		Cretaceous	145
		Jurassic	201
		Triassic	**252**
Paleozoic		Permian	299
	Carbon-iferous	Pennsylvanian	323
		Mississippian	359
		Devonian	419
		Silurian	443
		Ordovician	485
		Cambrian	**541**
		Precambrian	**4600**

Resources

General Books on Geologic History

Bjornerud, M., 2005, *Reading the Rocks: The Autobiography of the Earth*, Westview Press, Cambridge, MA, 237 pp.

Fortey, R. A., 2004, *The Earth, An Intimate History*, HarperCollins, London, 509 pp.

Hazen, R. M., 2012, *The Story of Earth: The First 4.5 Billion Years, from Stardust to Living Planet*, Viking, New York, 306 pp.

Kious, J., & R. I. Tilling, 1996, *The Dynamic Earth: The Story of Plate Tectonics*, US Geological Survey, Washington, DC, http://pubs.usgs.gov/gip/dynamic/dynamic.html.

Macdougall, J. D., 1996, *A Short History of Planet Earth: Mountains, Mammals, Fire, and Ice*, Wiley, New York, 266 pp.

Morton, J. L., 2004, *Strata: The Remarkable Life Story of William Smith, the Father of English Geology, new edition*, Brocken Spectre, Horsham, UK, 171 pp.

Powell, J., 2001, *Mysteries of Terra Firma: The Age and Evolution of the Earth*, Free Press, New York, 256 pp.

Winchester, S., & S. Vannithone, 2001, *The Map That Changed the World: William Smith and the Birth of Modern Geology*, HarperCollins, New York, 329 pp.

General Websites on Geologic History

Color-coded Continents!, US Geological Survey. [Reconstructions of color-coded continental motions from 620 million years ago through the present; maps from C. Scotese.] http://geomaps.wr.usgs.gov/parks/pltec/ scplseqai.html.

Earth Viewer, by BioInteractive at Howard Hughes Medical Institute. [Free iPad app; an interactive paleogeographic atlas of the world; state and country overlays allows tracking the development of the Western States.] http://www.hhmi.org/ biointeractive/earthviewer.

Geologic Maps of the 50 United States, by Andrew Alden, http://geology.about.com/od/maps/ig/stategeomaps/.

North America During the Last 150,000 Years, compiled by J. Adams, http://www.esd.ornl.gov/projects/qen/nercNORTHAMERICA.html.

The Paleomap Project, C. R. Scotese, http://www.scotese.com.

Paleogeography, R. Blakey. [The older, but free, version of the site.] https://www2.nau.edu/rcb7/RCB.html.

Reconstructing the Ancient Earth, Colorado Plateau Geosystems. [R. Blakey, updated site.] http://cpgeosystems.com/index.html.

Tour of Geologic Time, University of California Museum of Paleontology. [Online interactive geologic calendar exhibit.] http://www.ucmp.berkeley.edu/help/timeform.php.

Books and Articles on Geologic History of the Southwest

Baldridge, W. S., 2004, *Geology of the American Southwest: A Journey through Two Billion Years of Plate-Tectonic History*, Cambridge University Press, 314 pp.

Peal, R. H., 1978, *Colorado Stratigraphic Correlation Chart*, Colorado Geological Survey, http://store.coloradogeologicalsurvey.org/product/colorado-stratigraphic-correlation-chart/.

Rasmussen, J. C., 2012, Geologic History of Arizona, *Rocks & Minerals*, 87(1): 56–63, http://www.janrasmussen.com/pdfs/geol_hist_az_Rasmussen_2012%20preprint.pdf.

Timmons, M. J., & C. A. Manduca, eds., 2012, Grand Canyon geology: Two billion years of Earth's history, *Geological Society of America Special Paper* 489, 156 pp.

For more resources on geologic history, see the section "General geology resources by state" at the end of this volume.

1 Geologic History

Resources

Websites on Geologic History of the Southwest

The Great Unconformity of the Grand Canyon and the Late Proterozoic-Cambrian Time Interval: Part I - Defining It, by J. Share, 2012, http://written-in-stone-seen-through-my-lens.blogspot.com/2012/01/great-unconformity-of-grand-canyon-part.html.

The Great Unconformity and the Late Proterozoic-Cambrian Time Interval: Part II - The Rifting of Rodinia and the "Snowball Earth" Glaciations That Followed, by J. Share, 2012, http://written-in-stone-seen-through-my-lens.blogspot.com/2012_02_01_archive.html.

The Great Unconformity at Baker's Bridge: Part III - Regional Geological and Global Bio-Evolutionary Significance, by J. Share, 2014, http://written-in-stone-seen-through-my-lens.blogspot.com/2014/01/the-great-unconformity-at-bakers-bridge.html.

Sedimentary Rock Formations of the Grand Canyon, In the Playground of Giants, by Ken Bevis (Department of Geology, Hanover College), http://intheplaygroundofgiants.com/geology-of-the-grand-canyon-region/sedimentary-rock-formations-of-the-grand-canyon-region/.

Activities

Okland, L., 1991, Paleogeographic mapping, in: R. H. Macdonald, & S. G. Stover, eds., *Hands-on Geology: K-12 Activities and Resources*, Society for Sedimentary Geology (SEPM), Tula, OK. [Constructing paleogeographic maps for elementary and middle school students.] https://www.beloit.edu/sepm/Fossil_Explorations/Paleogeographic_Mapping.html.

Toilet Paper Analogy for Geologic Time, by J. Wenner, in: Teaching Quantitative Skills in the Geosciences, at Resources for Undergraduate Students and Faculty, SERC. [Demonstration of geological time using a 1000 sheet roll of toilet paper.] http:// serc. carleton.edu/quantskills/activities/TPGeoTime.html.

Understanding Geologic Time, Texas Memorial Museum at the University of Texas at Austin. [Timeline activity for middle school students.] http://www.jsg.utexas.edu/glow/files/Understanding-Geologic-Time-6-8.pdf.

Chapter 2:
Rocks of the Southwestern US

The amazingly diverse rocks in the Southwest span a wide range of ages and types, and they are also well exposed and accessible to the public. For geologists, rockhounds, and people who simply enjoy the outdoors, the Southwestern US is a wonderland filled with an interesting and beautiful array of rocks. Colliding **plates**, rifting, **inland seas**, deposition, **erosion**, igneous and metamorphic activity, and recent **glacial** processes are all part of this story. The Southwest's different rock types influence its **topography** and tell us where to look for certain **fossils** or natural resources. Each type of rock forms in a particular environment under particular conditions (*Figure 2.1*).

Igneous Rocks of the Southwest

rhyolite	granite	basalt
trachyte	tuff	pumice
minette	andesite	breccia
dacite	pyroclastic rocks	

Sediments of the Southwest (not consolidated into rocks)
gravel
sand
silt
clay

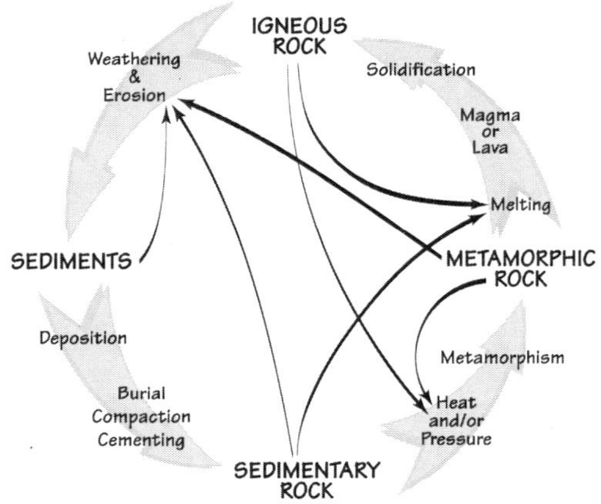

Metamorphic Rocks of the Southwest
gneiss
marble
quartzite
schist
slate

Sedimentary Rocks of the Southwest

arkose	limestone
conglomerate	mudstone
dolomite	sandstone
gypsum	shale
halite (rock salt)	siltstone

Figure 2.1: The rock cycle shows the relationships among the three basic types of rock.

plates • *large, rigid pieces of the Earth's crust and upper mantle, which move and interact with one another at their boundaries.*

inland sea • *a shallow sea covering the central area of a continent during periods of high sea level.*

erosion • *the transport of weathered materials.*

glacier • *a body of dense ice on land that does not melt away annually and has sufficient mass to move under its own weight.*

topography • *the landscape of an area, including the presence or absence of hills and the slopes between high and low areas.*

fossil • *preserved evidence of ancient life.*

CHAPTER AUTHORS

Lisa R. Fisher
Richard A. Kissel

2 Rocks

Rocks

Review

mineral • *a naturally occurring solid with a specific chemical composition and crystalline structure.*

system • *a set of connected things or parts forming a complex whole.*

plate tectonics • *the process by which the plates of the Earth's crust move and interact with one another at their boundaries.*

sedimentary rock • *rock formed through the accumulation and consolidation of grains of broken rock, crystals, skeletal fragments, and organic matter.*

lithification • *the process of creating sedimentary rock through the compaction or sementation of soft sediment.*

weathering • *the breakdown of rocks by physical or chemical means.*

wind • *the movement of air from areas of high pressure to areas of low pressure.*

A rock is a naturally occurring solid substance composed of one or more **minerals**. Broadly speaking, there are three types of rock: sedimentary, igneous, and metamorphic. The rock cycle describes the many processes that produce rocks, while also illustrating differences between the rock types. One type of rock may be transformed into either of the other types, often with the help of other parts of the Earth **system**, such as **plate tectonics**, the water cycle, and biological processes, to name a few.

Sedimentary rock is formed by the **lithification** of sediments (e.g., unconsolidated mineral and organic particles created through the **weathering** of other materials, such as rock and organic matter). Typically, sediments are created in an environment where erosion is a dominant force, and they are transported by **wind**, water, or ice to a depositional environment. For example, a rushing river can wear away the rock it is flowing over, and it also has enough energy to transport the resulting sediment to a lake. The water slows down, losing energy, and deposits the sediment on the bottom of the lake.

Sedimentary Rock Classification

Sedimentary rocks are classified by their sediment size or their mineral content, and each one reveals the story of the depositional environment where its sediments accumulated and were eventually lithified.

Sediment size (decreasing size)	Sedimentary rock	Environment of deposition
gravel	conglomerate	river beds, mountains
sand	sandstone	beaches, river sand bars, sand dunes
sand, silt, clay	graywacke	continental shelf
silt	siltstone	quiet water
clay	shale	very quiet water, lakes, swamps, shallow oceans

Mineral Content	Sedimentary Rock	Environment of Deposition
calcium carbonate skeletons of marine organisms	limestone	tropical reefs, beaches, warm shallow seas
precipitated calcium carbonate	travertine, tufa	hot spings, playas (dry lake beds), drying seas
gypsum	rock gypsum	playas, dryng seas
halite	rock salt	playas, drying seas

Lithification of sediments occurs in several ways. As sediments build up and lower layers are buried more deeply, they may become permeated by water. Minerals dissolved in the water are precipitated, filling the spaces between particles and **cementing** them together. This cementation helps to form many common sedimentary rocks, such as **shale**, **sandstone**, and most **conglomerates**. The evaporation of water may also form sedimentary rocks by leaving behind evaporites (previously dissolved minerals) such as **salt**. Deposits of **calcium carbonate**, usually created through the accumulation of calcium carbonate skeletal material (such as from clams and corals), form the sedimentary rocks **limestone** and **dolostone**.

Igneous rocks form from the cooling of **magma** (molten rock underground) or **lava** (molten rock at the Earth's surface). When magma cools slowly underground, it has time to produce large crystals that are visible to the naked eye. Rocks that form in this manner, such as **granite**, are called **plutonic**. When magma comes to the surface (as lava), it cools quickly so that individual crystals are not visible, resulting in a **volcanic** rock such as **basalt**. In some circumstances, lava may cool so quickly that crystals do not form at all, creating a **glassy rock** such as **obsidian**. Smaller fragmental rocks that cool quickly at the surface form during explosive eruptions; these are called **pyroclastic rocks**, and they are composed of a variety of different volcanic ejecta.

Every rock is capable of being melted, weathered, or changed by **heat** and pressure. Any rock that has been subjected to intense heat and pressure can **recrystallize** into a **metamorphic rock**. This process destroys features in the rock that would have revealed its previous history, transforming it into an entirely

Igneous Rock Classification

Igneous rocks differ not only in their cooling rates and subsequent crystal sizes, but also in their chemical compositions. Rocks found in continental crust, such as granite, have high silica content and low iron and magnesium content. They are light in color and are called *felsic*. Rocks found in oceanic crust, such as basalt, are low in silica and high in iron and magnesium. They are dark in color and are called *mafic*.

Crystal size	Felsic	Intermediate	Mafic	Ultramafic
large (plutonic)	granite	diorite	gabbro	peridotite
small (volcanic)	rhyolite	andesite	basalt	--
none (glassy)	obsidian, tuff, pumice	obsidian	obsidian	--

Review

cementation • *the precipitation of minerals that binds together particles of rock, bones, etc., to form a solid mass of sedimentary rock.*

shale • *a dark, fine-grained, laminated sedimentary rock formed by the compression of successive layers of silt- and clay-rich sediment.*

sandstone • *sedimentary rock formed by cementing together grains of sand.*

conglomerate • *a sedimentary rock composed of multiple large and rounded fragments that have been cemented together in a fine-grained matrix.*

calcium carbonate • *a chemical compound with the formula $CaCO_3$, commonly found in rocks in the mineral forms calcite and aragonite, as well as the shells and skeletons of marine organisms.*

limestone • *a sedimentary rock composed of calcium carbonate ($CaCO_3$).*

dolostone • *a rock primarily composed of dolomite, a carbonate mineral.*

compression *• flattening or squeezing as a result of forces acting on an object from all or most directions.*

protolith *• the original parent rock from which a metamorphosed rock is formed.*

crust *• the uppermost, rigid outer layer of the Earth.*

intrusive rock *• a plutonic igneous rock formed when magma from within the Earth's crust escapes into overlying strata.*

new form as the minerals within realign. The pressure to transform a rock may come from burial by sediment or from **compression** due to plate movements, while the heat may come from very deep burial or from contact with magma.

Metamorphic Rock Classification

Metamorphic rocks are classified differently depending on the *protolith* (parent rock) they are made from. The following chart shows common rocks and the metamorphic rocks that they can become.

Parent rock	Metamorphic rocks
shale	slate, phyllite, schist, gneiss (in order of increasing heat and pressure)
granite	gneiss
sandstone	quartzite
limestone	marble
peridotite	serpentinite

As you read through this chapter, keep in mind that once you understand the geologic events that have affected a given region, you should be able to predict the type of rocks found in that area. For example, when plates collide, compression and friction melt the **crust**. The rising magma forms igneous **intrusions** that crystallize below the surface, producing large-grained igneous rocks such as granite. Rising magma may also break through the surface in the form of volcanoes, creating volcanic rocks such as basalt. Tectonic collision leads to increased heat and pressure, buckling the crust and creating metamorphic rocks. Basins adjacent to mountains fill with transported sediment, producing thick sequences of sedimentary rock. The rocks and sediments exposed at the surface today tell us an important story about the environments in which they were deposited or formed.

Why do we see different kinds of rocks at the surface?

As you walk across the surface of the Earth, you will observe an amazing variety of rock types. If all rocks were flat-lying layers and there was no erosion, then we would only see one type of rock exposed on the surface. Often, however, rocks have been worn away (eroded), and the underlying layers are now exposed at the surface. Layers of rock may also be tilted, folded, or faulted to reveal the underlying rocks at the surface.

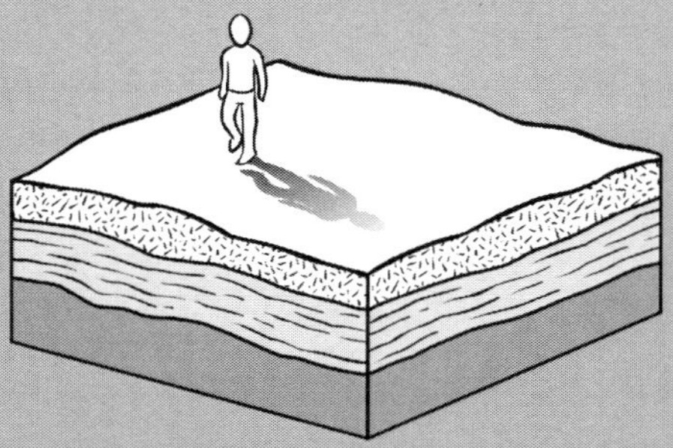

When rocks are flat-lying layers and there is no erosion, folding, or faulting, the person walking across the surface sees only one rock type.

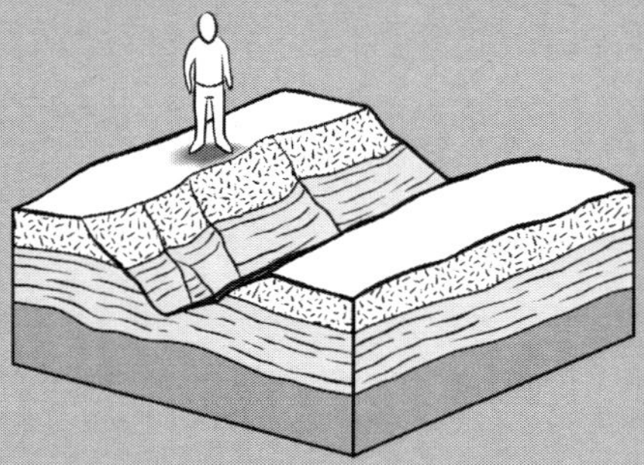

When rocks are worn away (often by streams), the person walking across the surface sees the underlying layers of rock exposed.

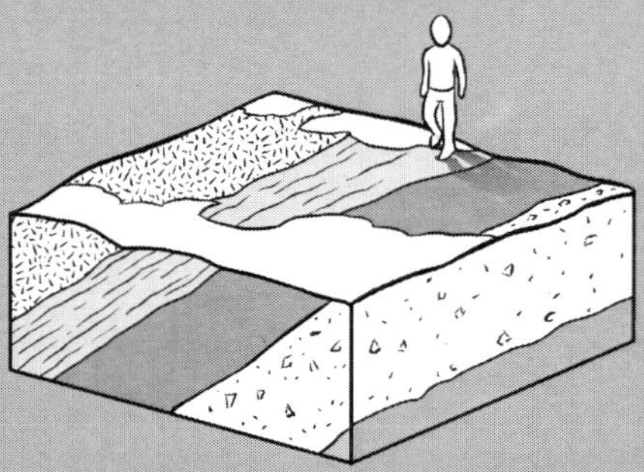

When rocks are folded or tilted, the person walking across the surface sees several layers of rock exposed.

2 Rocks

Region 1

Cenozoic • *the geologic time period spanning from 66 million years ago to the present.*

epeirogenic • *large-scale crustal uplift caused by hot or upwelling mantle underlying the surface.*

uplift • *upward movement of the crust due to compression, subduction, or mountain building.*

Precambrian • *a geologic time interval that spans from the formation of Earth (4.6 billion years ago) to the beginning of the Cambrian (541 million years ago).*

basement rocks • *the foundation that underlies the surface geology of an area, generally composed of igneous or metamorphic crystalline rock.*

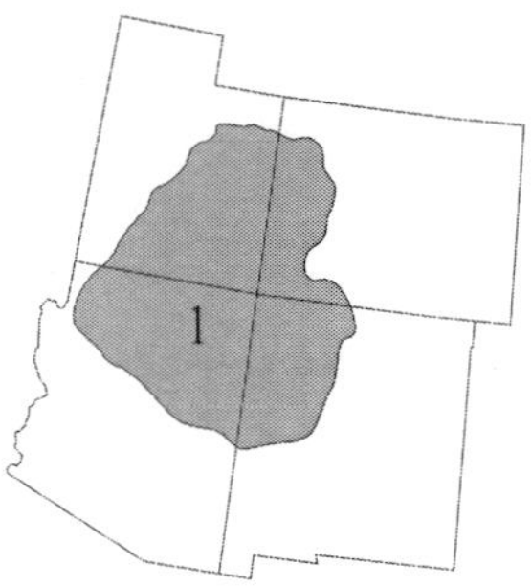

Rocks of the Colorado Plateau
Region 1

The Colorado Plateau is an arid region that developed as a stable crustal block. It was not significantly affected by either the compressional forces that gave rise to the Rockies or by the **Cenozoic** extensional forces that produced the Basin and Range. Between eight and five million years ago, an **epeirogenic** (vertical) **uplift** raised the entire region as a single block to its present elevation of 610–4000 meters (2000–13,000 feet). Despite its name, the Colorado Plateau is not a flat area, but rather a landscape of contrasting features and colors, including dramatic sheer-walled canyons, wind- and water-sculpted formations, towering monoliths, and flat-topped mesas. The region embodies the quintessential Old West, and includes many of the landscape features so often seen in the backdrop of films and books. Interesting rocks and breathtaking vistas can be found almost anywhere on the plateau, which has the highest density of national parks and monuments in the country (*Figure 2.2*). The Colorado Plateau was also home to the ancient Native American culture known as the Anasazi, or Ancestral Puebloan people.

Epeirogenic uplift refers to large-scale crustal uplift caused by hot or upwelling mantle underlying the whole region, which buoys up the overlying crust but does not significantly fold or fault the rocks.

See Chapter 4: Topography for more about the spectacular erosive landscape of the Colorado Plateau.

The Colorado Plateau is underlain by **Precambrian basement rock**, which is exposed in the deep gorge of Arizona's Grand Canyon as well as at the edge of Colorado's San Juan Mountains. The **schists** and **gneisses** found in the Needle Mountains (a subrange of the San Juan Mountains), and at the base of the Grand Canyon, are 1.9 to 1.7 billion years old and comprise metamorphosed volcanic and marine sedimentary rocks. These rocks are the product of plate tectonic activity from the collision and **accretion** of a **volcanic island** arc, the Yavapai **terrane**, with what was then a much smaller North American continent. Igneous rock bodies, such as the Zoroaster Granite,

Terranes are fragments of crustal material that have been broken off from one plate and accreted to a different piece of crust through tectonic forces. Each fragment in a large grouping of accreted terranes shows a distinct geologic history.

See Chapter 1: Geologic History to learn more about the ancient terranes that formed the basement of the Southwest.

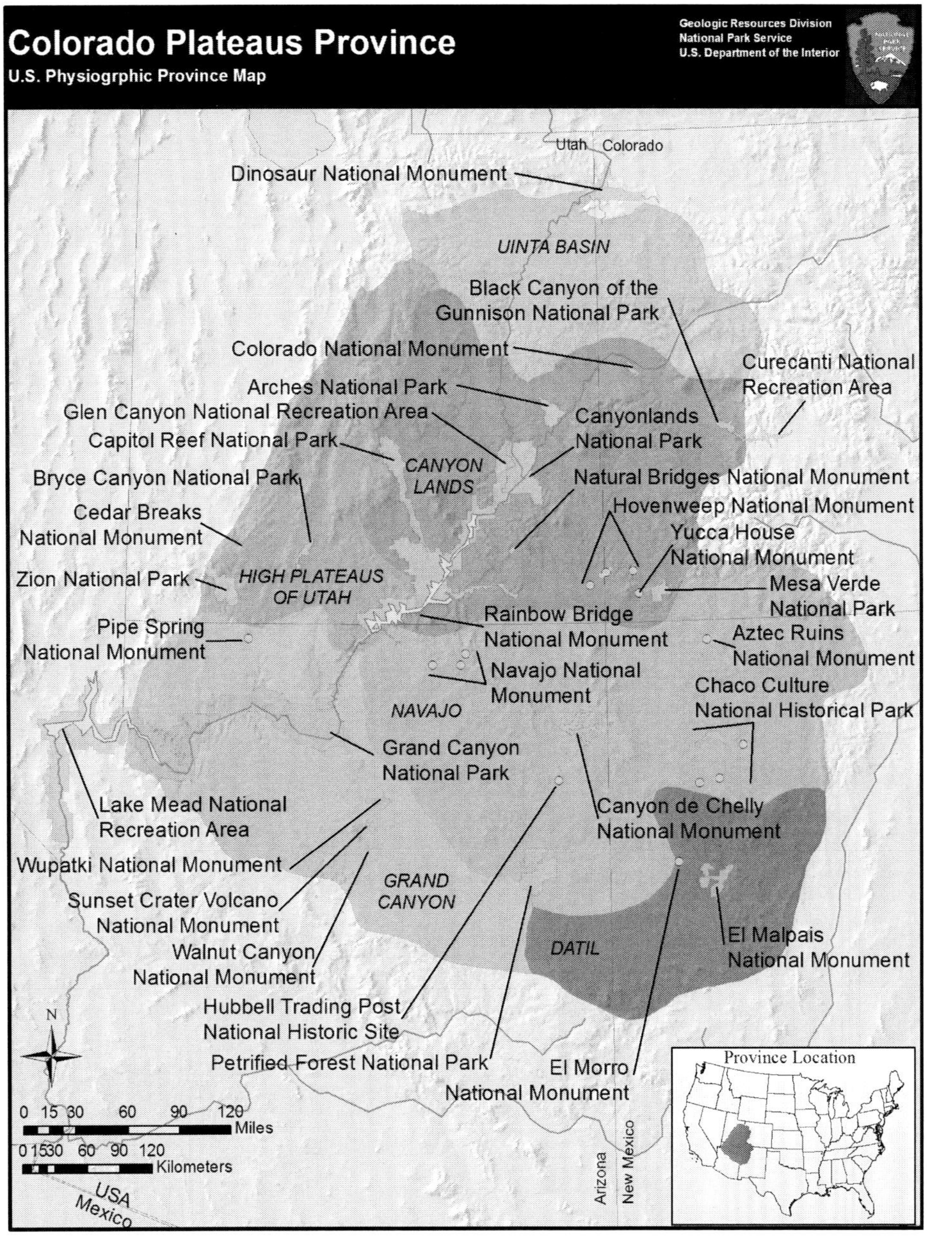

Figure 2.2: Locations of some national parks and monuments in the Colorado Plateau. The region hosts 10 national parks, 27 national monuments, 19 national forests, and 57 state parks. (See TFG website for full-color version.)

were intruded into the plate at approximately the same time. These ancient rocks are overlain by a thick sequence of mildly metamorphosed sedimentary layers known as the **Grand Canyon Supergroup** (*Figure 2.3*), a series of sandstones, shales, and limestones that constitutes one of the most complete middle-to-late **Proterozoic** geologic records in North America. These rocks were generated approximately 1.2 to 1.1 billion years ago during the formation of the supercontinent **Rodinia**, when they were compressed from sedimentary layers originally deposited in **fluvial** and shallow marine environments. Basalt

schist *• a medium grade metamorphic rock with sheet-like crystals flattened in one plane.*

gneiss *• a metamorphic rock that may form from granite or layered sedimentary rock such as sandstone or siltstone.*

accretion *• the process by which a body of rock increases in size due to the addition of further sedimentary particles or of large chunks of land.*

volcanic island *• one of a string of islands created when molten rock rises upward through oceanic crust.*

terrane *• a piece of crustal material that has broken off from its parent continent and become attached to another plate.*

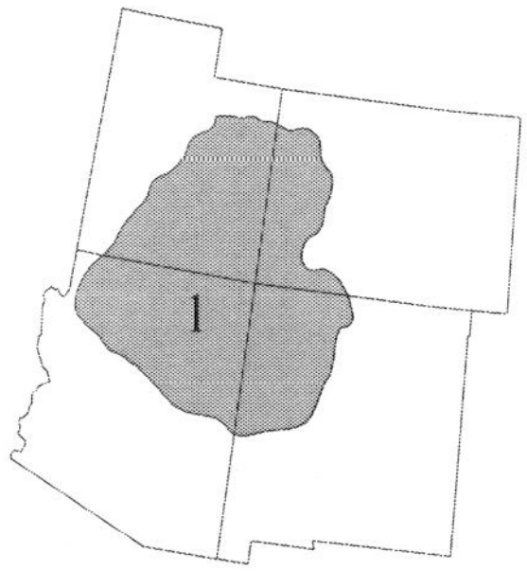

dike • a sheet of intrusive igneous or sedimentary rock that fills a crack cutting across a pre-existing rock body.

Cambrian • a geologic time period lasting from 541 to 485 million years ago.

magnetic • affected by or capable of producing a magnetic field.

earthquake • a sudden release of energy in the Earth's crust that creates seismic waves.

seismic waves • the shock waves or vibrations radiating in all directions from the center of an earthquake or other tectonic event.

sills and **dikes** were intruded throughout the Grand Canyon Supergroup as a result of volcanic activity during the late Proterozoic (*Figure 2.4*). Overall, these layers record a long history—perhaps 200 million years—between the Precambrian formation of continental crust by terrane accretion and the more familiar events of the **Cambrian** period and later.

In the Basement

The Precambrian rocks underlying younger surface rocks in any locality are referred to as the "basement." The basement might be fairly shallow beneath thin layers of young sedimentary rocks or even occasionally exposed at the surface, but it often lies thousands of meters (feet) deep beneath thick sequences of overlying sediment. Basement rocks can be identified in several ways, using both direct and indirect methods.

To investigate the underlying geology directly, holes may be drilled down to the top of the basement. Rock samples are then brought to the surface from the bottom of the hole and analyzed. This procedure is common in government geological surveys or in oil and gas exploration.

Indirectly, sensitive readings of *magnetism* and gravity can be measured from planes and satellites. These give very broad estimates of the characteristics of rocks beneath the surface. Recording of sound waves traveling through, and bouncing off of, layers of subsurface rock can give us an indirect measurement of their properties. This "seismic data" comes both from *earthquakes* (seismology) and from surveys performed by artificially creating *seismic waves* (through pounding the Earth or by creating small explosions) at the surface.

Overlying the uppermost layers of the Grand Canyon Supergroup is one of the most conspicuous features of the rocks in the Grand Canyon, an irregular level called "the Great Unconformity" (*Figure 2.5*). This feature preserves a gap in the geological record where stratified layers have been interrupted or destroyed due to erosion or deformation. The **unconformity** separates rocks of Precambrian age from those of the **Paleozoic** era, and is part of a continent-wide feature that extends across the ancient core of North America. The length of time represented by the Great Unconformity varies along its length—in some

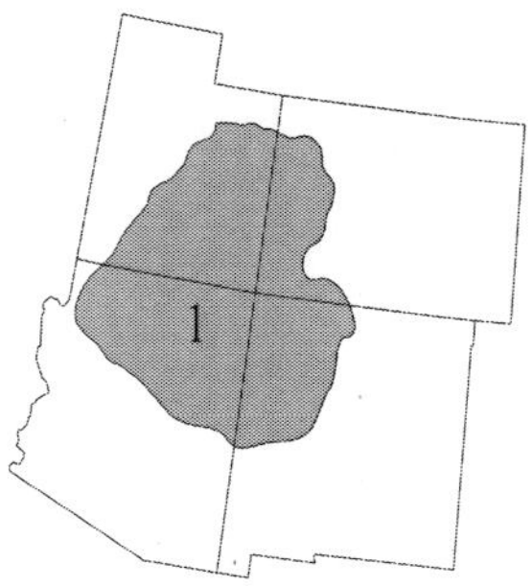

Region 1

unconformity • the relation between adjacent rock strata for which the time of deposition was separated by a period of nondeposition or erosion.

Paleozoic • a geologic time interval that extends from 541 to 252 million years ago.

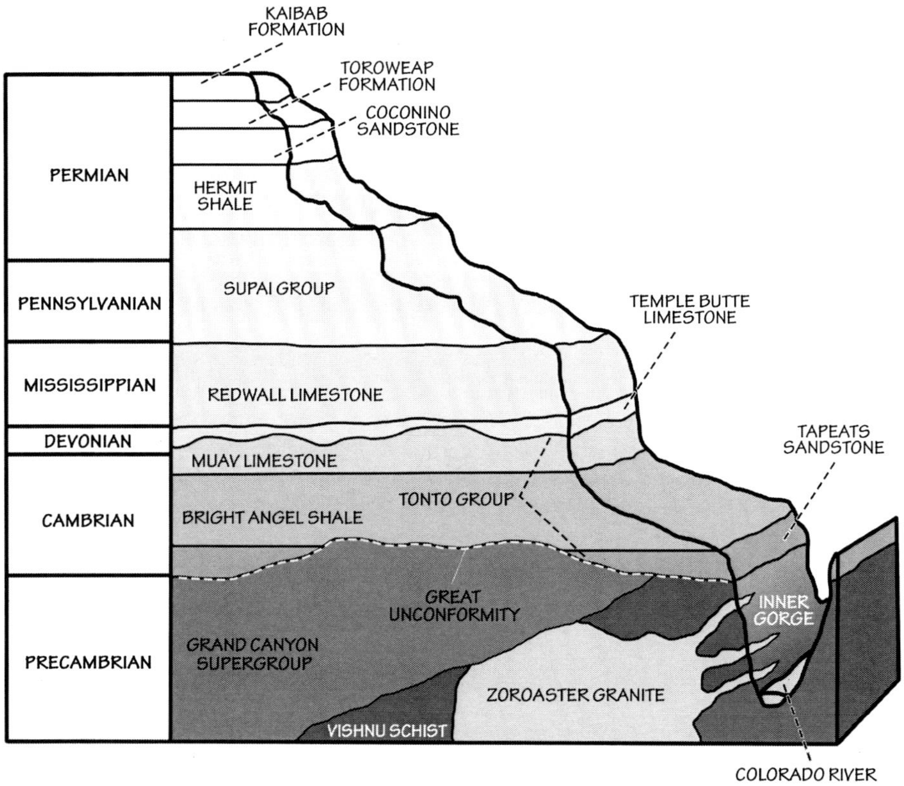

Figure 2.3: Major Paleozoic stratigraphic units of the Grand Canyon and Colorado Plateau.

Figure 2.4: A basalt dike intruded into the orange-red Hakatai Shale, part of the Grand Canyon Supergroup. This exposure is found at Hance Rapid on the Colorado River, Grand Canyon.

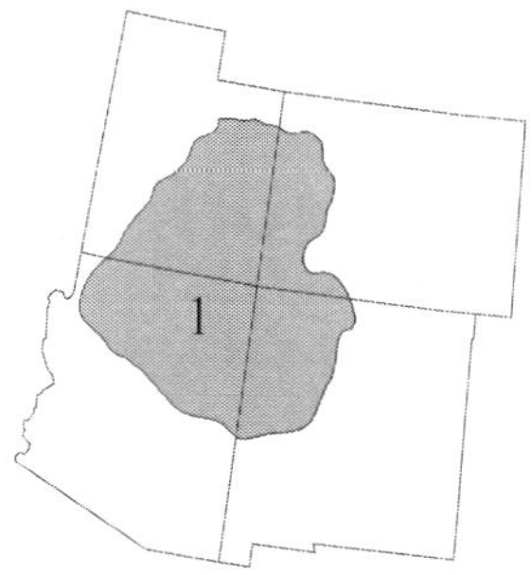

stratigraphy • *the branch of geology specifically concerned with the arrangement and age of rock units.*

craton • *the old, underlying portion of a continent that is geologically stable relative to surrounding areas.*

parts of the Grand Canyon, a period of 175 million years is "missing" between the Cambrian Tonto Group sandstones and the Grand Canyon Supergroup. In other places, there is a gap of over 1.2 billion years where the 550-million-year-old Tapeats Sandstone rests on 1.7-billion-year-old basement rock.

The Great Unconformity is one of the most widely recognized and distinctive **stratigraphic** surfaces in the entire rock record. Geologists are still unsure of its exact origin, but it may have been caused by a major episode of continental uplift following the formation of the North American **craton**. This uplift would have exposed the continent—then completely barren of life—to extensive erosion, degrading the rocks for hundreds of millions of years before they were submerged by a shallow sea in the Cambrian.

Figure 2.5: The Great Unconformity (marked by the line) in the Grand Canyon, Arizona, where the horizontal Tonto Group (Cambrian) overlies the tilted Grand Canyon Supergroup (Proterozoic). (See TFG website for full-color version.)

The Great Unconformity is one of the most prominent sections of "missing time" in North America, but there are other examples of unconformities throughout the US. The absence of rocks deposited during certain time periods does not mean that no rocks were formed during that time. It may mean, however, that very little sediment was deposited, that the sediment was eroded away, or that the rocks are buried beneath the surface. There is no single place on Earth with a complete sequence of rocks from the Precambrian to the Quaternary. Erosion and weathering over time have removed many meters (feet)—and in some cases kilometers (miles)— of rock from the surface of the Southwest.

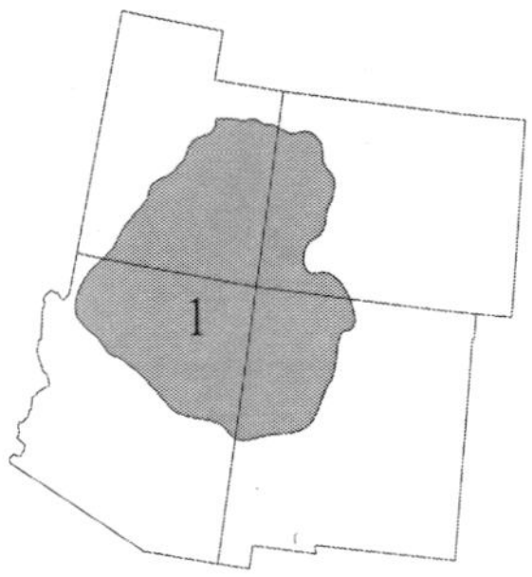

Rocks 2

Region 1

The Colorado Plateau is perhaps the best place to examine a nearly continuous sequence of representative Southwestern sedimentary layers (*see Figure 2.3*). From near the bottom of the Grand Canyon, up through younger layers observed farther east, these rocks tell the story of the entire Paleozoic. Cambrian strata of the Tonto Group comprise a sequence of tan, gray, and brown sandstone, mudstone, and limestone that represent an approaching shoreline (Tapeats Sandstone) and offshore mud layers (Bright Angel Shale, Muav Limestone). Late **Devonian** layers of gray and tan limestone are referred to as the Temple Butte Formation, and they were laid down in an environment similar to that of the modern Yucatan. Sea level rose from the **Ordovician** through **Mississippian** periods, and Mississippian rocks are widespread across much of Arizona and New Mexico, where they typically form prominent cliffs that dominate the landscape. **Orogenic** activity to the northwest (in what is now western Nevada) had little effect on Arizona's clear, warm, shallow marine **carbonate** shelf. Here, the Mississippian Redwall Limestone is one of the most conspicuous rock layers, forming towering vertical cliffs in the Grand Canyon (*Figure 2.6*). Although its name implies that the limestone is red, it is actually gray. The visible portions of the Redwall Limestone have been stained red from the erosion of **iron**-rich sediments in the overlying Supai Group and Hermit Formation, which have washed down across the limestone over time and been absorbed into its surface.

Unless rock layers are overturned, older rocks are found at the bottom and younger rocks are found at the top of a sedimentary sequence. This is known as the *Law of Superposition*. The sedimentary rocks of the Grand Canyon are nearly flat lying, and are a textbook example of superposition.

Figure 2.6: The 340-million-year-old Redwall Limestone forms distinct red cliffs up to 240 meters (800 feet) thick. It is a very hard stone, which often causes it to break at harsh angles, creating pillars.

Devonian • *a geologic time period spanning from 419 to 359 million years ago.*

Ordovician • *a geologic time period spanning from 485 to 443 million years ago.*

Mississippian • *a subperiod of the Carboniferous, spanning from 359 to 323 million years ago.*

orogeny • *a mountain-building event generally caused by colliding plates and compression of the edge of the continents.*

carbonate rocks • *rocks formed by accumulation of calcium carbonate, often made of the skeletons of aquatic organisms.*

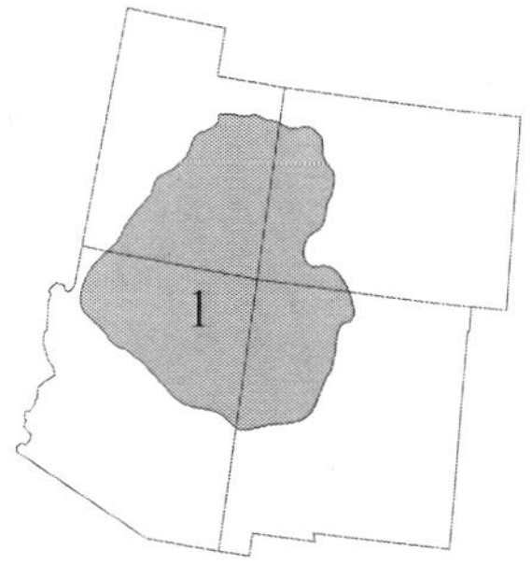

2 Rocks

Region 1

The **Pennsylvanian** and **Permian** periods—the last 70 million years of the Paleozoic era—were marked by uplift and falling sea level across the Southwest, both of which led to the expansion of terrestrial environments. The Redwall Limestone was exposed to subaerial erosion, forming **karst topography**. Streams carved valleys into the limestone, and **estuarine** deposits flooded the valleys as fills during subsequent **transgressions**. Uplifted areas shed large volumes of sediment, forming thick deposits in adjacent basins. The heavy influx of **sand** and mud resulted in preservation of these and other continental sediments as red beds (such as the Hermit Shale), **colored** by the **oxidation** of iron minerals. At this time, the region was near the equator, and an arid **climate** led to the widespread formation of sand dunes. Today, extensive coastal, wind-blown (**aeolian**) sand dune fields are preserved as thick beds of sandstone, with large-scale **cross-bedding**. The late Pennsylvanian Manakacha Formation, in the lower Supai Group, documents the earliest influx of aeolian sand onto what later became the Colorado Plateau. Permian dune sands include the Cedar Mesa Sandstone exposed in Natural Bridges National Monument (*Figure 2.7*) the Needles district of Canyonlands National Park, and the bright pale-yellow Coconino Sandstone visible in Arizona and Utah (*Figure 2.8*).

Along the coasts of shrinking inland seas, salt flats grew along arid coastlines. These newly formed **sabhka** environments became sites of abundant deposition for **evaporite** minerals, including salt, **gypsum**, and anhydrite. Thick sequences of alternating evaporites and shales accumulated in basins across the region, as shallow marine water evaporated and deposited layers of salt. The Paradox Basin in southeastern Utah is best known for its salt deposits; it also contains **oil** reserves and **copper**.

See Chapter 5: Mineral Resources for more information about the plentiful resources found in the Paradox Basin.

Figure 2.7: Sipapu Natural Bridge, formed from Cedar Mesa Sandstone in Natural Bridges National Monument, Utah. Note the cross-bedding at lower right.

Pennsylvanian • *a subperiod of the Carboniferous, spanning from 323 to 299 million years ago.*

Permian • *the geologic time period lasting from 299 to 252 million years ago.*

karst topography • *a kind of landscape defined by bedrock that has been weathered by dissolution in water, forming features like sinkholes, caves, and cliffs.*

estuary • *a place where freshwater and saltwater mix, created when sea level rises to flood a river valley.*

transgression • *a relative rise in sea level in a particular area, through global sea level rise or subsidence of land.*

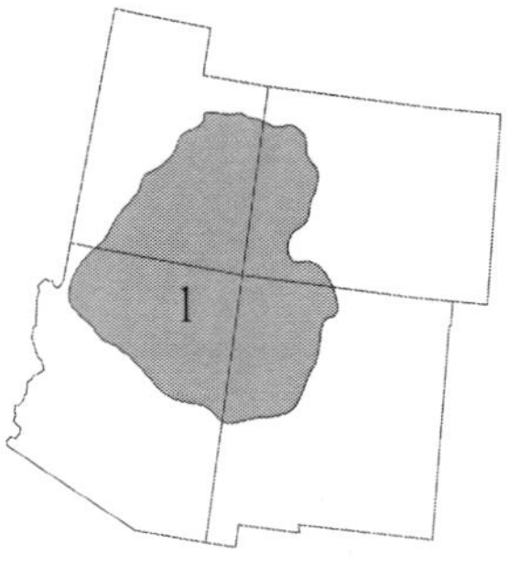

Figure 2.8: Cross-beds in the Coconino Sandstone at Sedona, Arizona.

By the end of the Permian period, the supercontinent **Pangaea** had formed, and the Southwest was largely a terrestrial environment. The **Mesozoic** era generated a succession of near-shore and continental deposits, mostly sandstones and shales, which form the spectacular cliffs and mesas seen throughout much of the Colorado Plateau. Similarly to the way the entire Paleozoic is exposed at the Grand Canyon, the Mesozoic and part of the Cenozoic are beautifully exposed in the Grand Staircase. This immense sedimentary sequence stretches south from Bryce Canyon National Park in Utah to the edge of the Grand Canyon (*Figure 2.9*).

Many Mesozoic sandstones, especially those from the **Triassic**, have a reddish color caused by the oxidation of iron within the rock (*Figure 2.10*). For example, the Triassic Moenkopi Formation consists of reddish-brown, fine-grained sandstone and mudstone. It represents both shallow marine and near-shore terrestrial environments. Terrestrial features of the Moenkopi, such as **ripple marks**, mud cracks, and scour marks, point to an arid environment prone to flash floods. Stream channels, flood plains, fresh or brackish ponds, **playas**, and shallow marine environments can also create similar traces.

Above the Moenkopi, the Chinle Formation consists of diverse sandstones, mudstones, and conglomerates deposited in a somewhat wetter environment—low-**relief** rivers and lakes. The Chinle is famous for its colors, ranging from various shades of reds, blues, and pinks to grays, tans, and browns (*Figure 2.11*). It is also famous for its fossils, which include abundant **trees** (e.g., Petrified Forest National Park) and **dinosaurs**.

See Chapter 3: Fossils to learn more about dinosaurs and tetrapods of the Chinle Formation.

sand • *granular sediment most commonly composed of weathered grains of quartz and feldspar, of grain diameter 1/16 to 2 millimeters.*

color (mineral) • *a physical property determined by the presence and intensity of certain elements within the mineral.*

oxidation • *a chemical reaction involving the loss of at least one electron when two substances interact.*

climate • *a description of the average temperature, range of temperature, humidity, precipitation, and other atmospheric/hydrospheric conditions a region experiences over a period of many years (usually more than 30).*

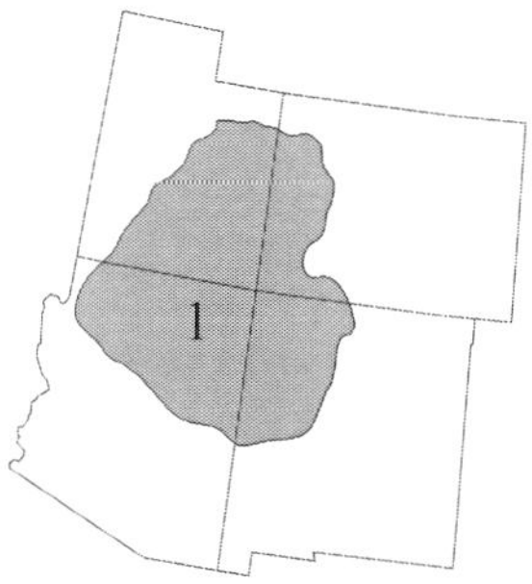

2 Rocks

Region 1

Jurassic *• the geologic time period lasting from 201 to 145 million years ago.*

braided stream *• a stream consisting of multiple, small, shallow channels that divide and recombine numerous times, forming a pattern resembling strands of braided hair.*

erg *• an area of desert, greater than 125 square kilometers (48 square miles), covered by wind-blown sand.*

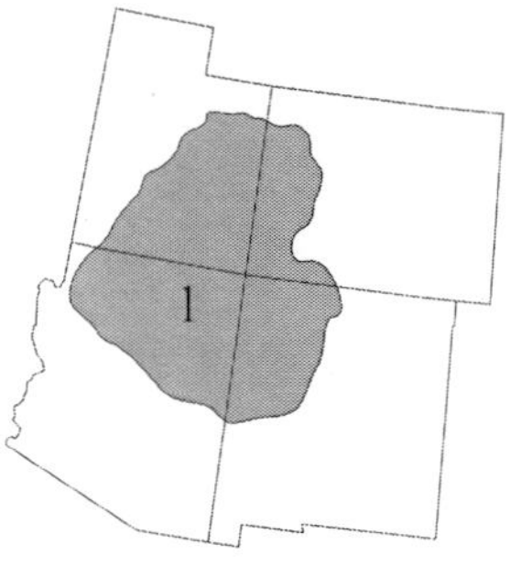

Colors of Sedimentary Rocks
What do they tell us about the environment?

The color of a rock can be an important indicator of the environment in which it formed. The red-brown color so common throughout the Southwest results from oxidized (rusted) iron within the rock. This is most common in sediments deposited in a seasonally hot and dry climate on land, where the iron could be exposed to the air. Red sedimentary rock is also found in the Southwest's Permian rocks, reflecting a time when ocean floor sediments were exposed above water. Red clays may also form in well-oxygenated, deep marine conditions. In some marine environments, however, where iron is reduced rather than oxidized, rocks may take on a greenish hue. Likewise, some greenish sedimentary rocks may indicate the presence of the mineral glauconite, which is found only in marine environments.

In contrast, many shales are gray or black in color, reflecting the abundance of carbon-rich organic material that can accumulate in quiet-water settings. The darker the shale, the more organic material that is preserved within. Shales are most commonly formed in quiet waters where tiny particles have time to settle out onto the sea or lake floor.

Pangaea began to break up during the early **Jurassic**. At this time, vast sand dune fields covered much of the Southwestern US, forming thick cross-bedded layers of sandstone. The early Jurassic sediments of the Colorado Plateau represent some of the world's best geological records from this time interval. These sediments are collectively called the Glen Canyon Group, which includes (from older to younger) the Wingate/Moenave, Kayenta, and Navajo formations. The Moenave and Navajo represent aeolian sediments, while the Kayenta, sandwiched in between them, consists of siltstones and sandstones that were deposited in **braided** and meandering streams.

The Jurassic dune-forming environment (called a "sand sea" or **erg**) is best seen in the Navajo Sandstone, which is exposed widely across northern Arizona and southern Utah (*Figure 2.12*). The Navajo, together with the Nugget and Aztec sandstones to the north and southwest, represents one of the largest ancient sand dune deposits on Earth, covering more than 150,000 square kilometers (58,000 square miles)—an area approximately the size of Illinois (*Figure 2.13*).

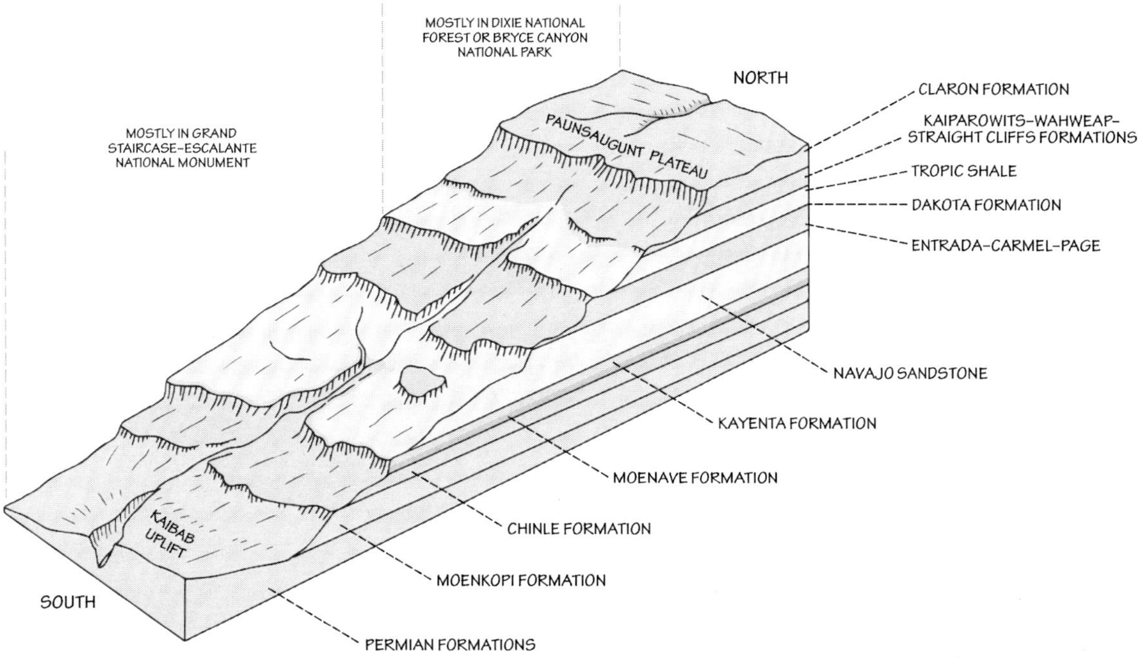

Figure 2.9: Major Mesozoic and Cenozoic stratigraphic units of the Grand Staircase and Colorado Plateau. (See TFG website for full-color version.)

Figure 2.10: Triassic sediments at The Castle, Capitol Reef National Park, Utah. Three strata are visible here: the Wingate Sandstone (top), the Chinle Formation (middle), and the Moenkopi Formation (bottom).

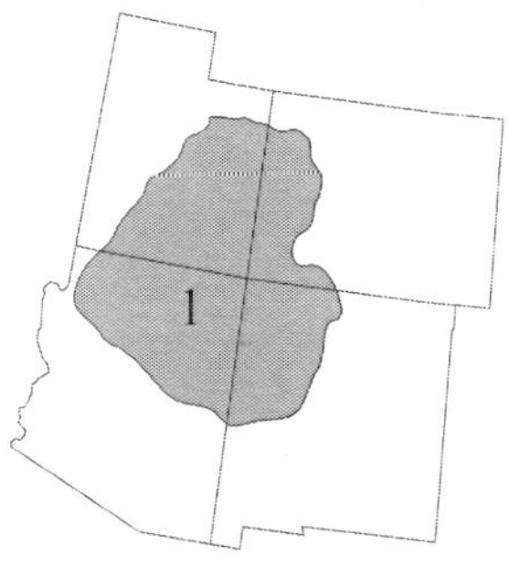

2 Rocks

Region 1

Figure 2.11: The Painted Desert in northern Arizona showcases the Chinle Formation's spectacular colors.

Figure 2.12: The Wave, a series of intersecting U-shaped troughs eroded into Jurassic Navajo Sandstone within the Paria Canyon-Vermilion Cliffs Wilderness, Arizona. The cycling layers in the sandstone represent changes in the direction of prevailing winds as large sand dunes migrated across the desert.

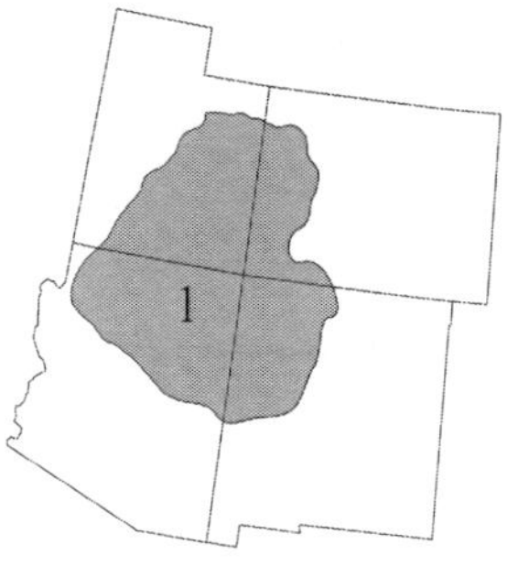

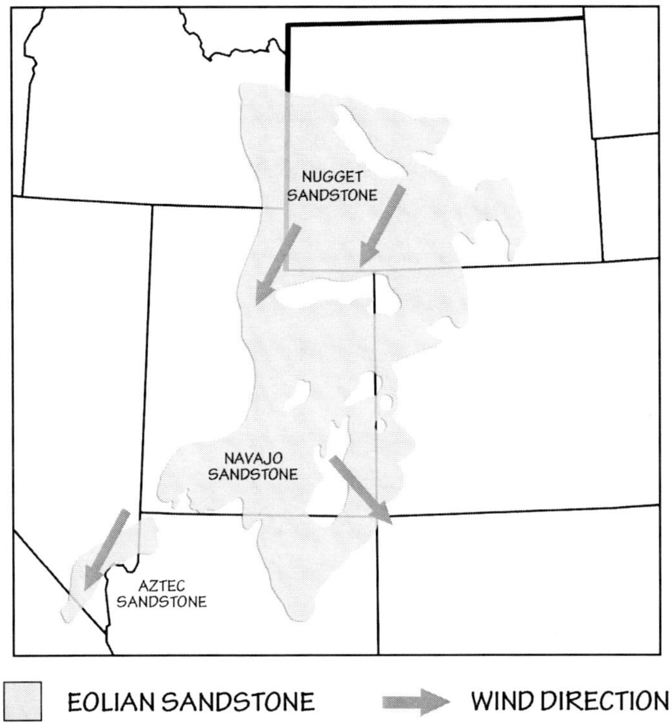

Figure 2.13: Extent of the Navajo, Nugget, and Aztec sandstones, with arrows showing the direction of wind in the dune-forming environment.

Cross-bedded Sand Dunes

Cross-bedded sand dunes form as air movement pushes sediment downwind, creating thin beds that slope gently in the direction of the flow as migrating ripples. The downstream slope of the ripple may be preserved as a thin layer dipping in the direction of the current, across the natural flat-lying repose of the beds. Another migrating ripple will form an additional layer on top of the previous one.

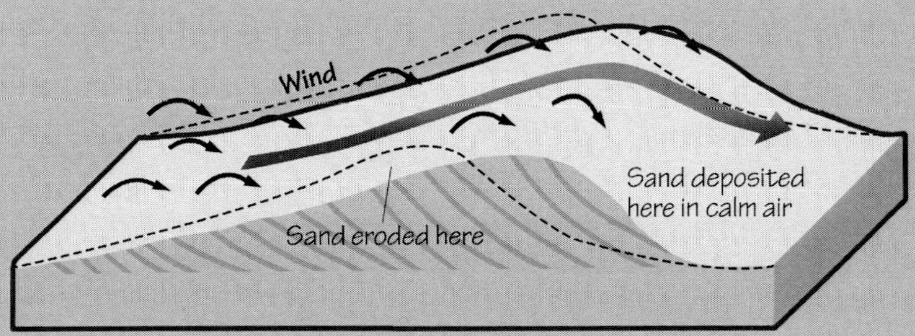

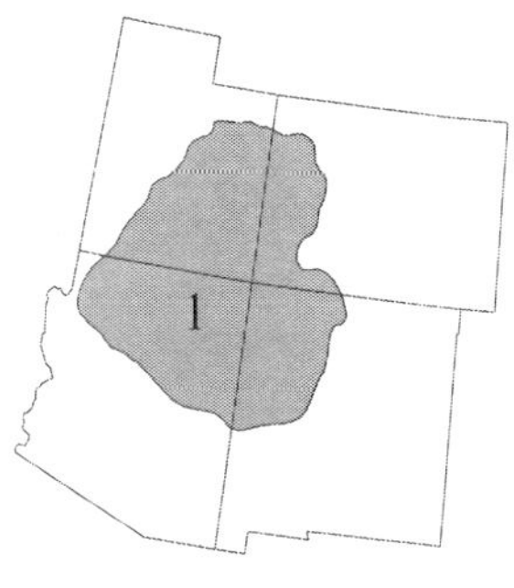

2 Rocks

Region 1

floodplain • *the land around a river that is prone to flooding.*

Cretaceous • *a geologic time period spanning from 144 to 66 million years ago.*

downwarp • *a segment of the Earth's crust that is broadly bent downward.*

lithosphere • *the outermost layer of the Earth, comprising a rigid crust and upper mantle broken up into many plates.*

Laramide Orogeny • *a period of mountain building that began in the Late Cretaceous, and is responsible for the formation of the Rocky Mountains.*

By the late Jurassic, the Colorado Plateau contained a complex environment of coastlines, rivers, lakes, marshes, ponds, **floodplains**, and dunes. The Morrison Formation, a series of variegated mudstones and shales, was deposited in mixed mud flats, river channels, and lakes. Dinosaur bones and trackways are often found in the Morrison's relict stream channels. During the **Cretaceous** period, the interior of North America was **downwarped** by tectonic processes associated with the **subduction** of oceanic **lithosphere** along the western edge of the continent. As the **Laramide** and **Sevier orogenies** occurred to the west, the North American interior was flooded by a particularly vast inland sea called the Western Interior Seaway (*Figure 2.14*). The Dakota Sandstone, formed from beach sand, signals the onset of this sea. Ripple marks created by currents or waves are common in these sediments, indicating wave action at or near a beach (*Figure 2.15*). In Colorado, at Morrison and along the Purgatoire River, spectacular dinosaur **trackways** are preserved in these sandstones.

Inland sea may sound like a contradiction in terms, but there is a very simple, yet important, distinction that differentiates it from other seas: an inland sea is located on continental crust, while other seas are located on oceanic crust. An inland sea may or may not be connected to the ocean. For example, Hudson Bay is on the North American plate and connects to the Atlantic and Arctic oceans, while the Caspian Sea is on the European plate but does not drain into any ocean at all.

As the water deepened, thick sequences of shale (the Mancos Shale) were deposited. When the sea again retreated at the end of the Cretaceous, more sandstones (the Mesaverde Formation) were laid down. During the 12th and 13th centuries, the Ancestral Puebloan peoples of Colorado carved massive, elaborate dwellings and other structures into these sandstones (*Figure 2.16*).

Throughout the Cenozoic era, mountains that had formed during the Sevier and Laramide orogenies experienced significant erosion. Cobbles, sand, and mud were carried by streams and deposited into rivers and lakes, forming conglomerate, sandstone, mudstone, and shale. Large basins— areas of the crust that slowly **subside**, or sink—on the Colorado Plateau received thick layers of sediment that became sources and reservoirs for oil and gas. Swamps and lowlands on basin margins formed widespread **coal** beds. During the **Eocene** period, sediment accumulated in floodplains, shallow lakes, rivers, and **soils** to form the Claron Formation, a set of varied conglomerates, sandstones, mudstones, and carbonates stained by iron to produce a distinctive pinkish hue. These layers alternate

See Chapter 6: Energy for more information about fossil fuel deposits in the Southwest's sedimentary basins.

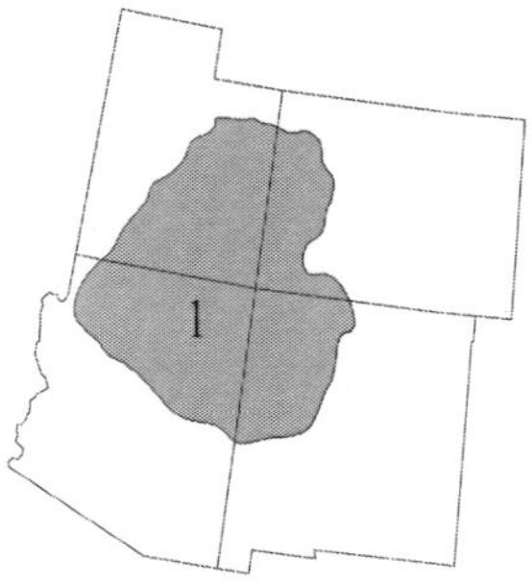

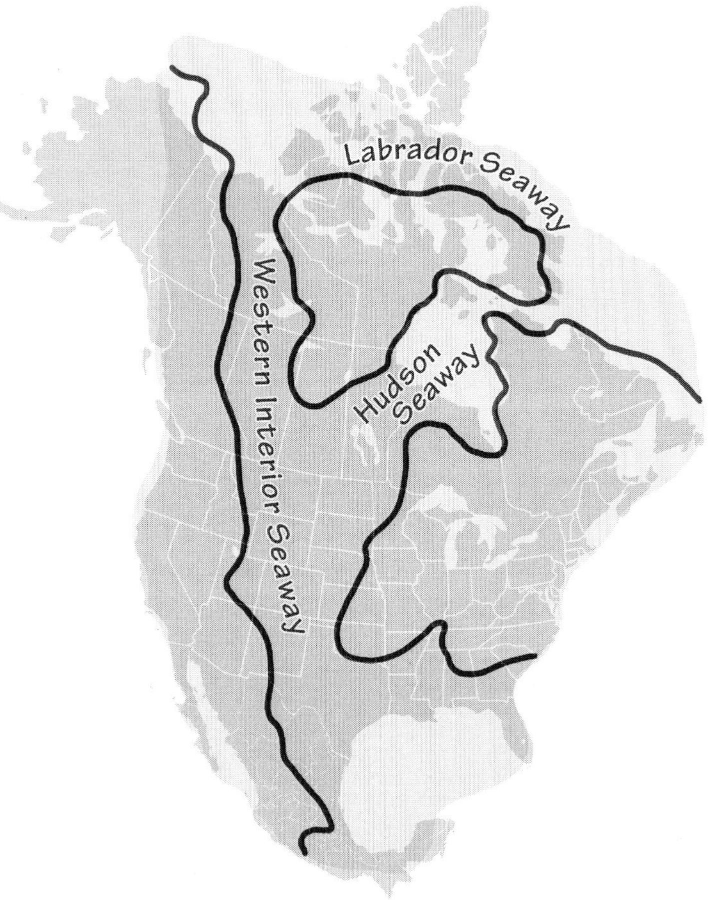

Figure 2.14: The Western Interior Seaway.

Figure 2.15: Ripple marks in the Dakota Sandstone of Dinosaur Ridge, Colorado.

Region 1

Sevier Orogeny • *a mountain-building event resulting from subduction along the western edge of North America, occurring mainly during the Cretaceous.*

subsidence • *the sinking of an area of the land surface.*

trackway • *a set of impressions in soft sediment, usually a set of footprints, left by an animal.*

coal • *a combustible, compact black or dark-brown carbonaceous rock formed by the compaction of layers of partially decomposed vegetation.*

Eocene • *a geologic time period extending from 56 to 33 million years ago.*

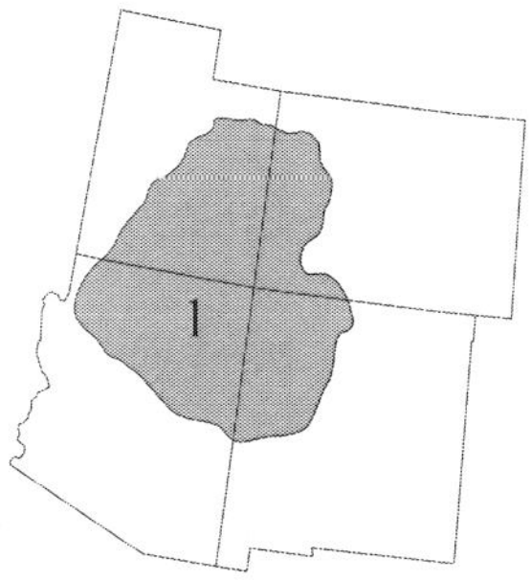

badlands • *a type of eroded topography that forms in semi-arid areas experiencing occasional periods of heavy rainfall.*

capstone • *a harder, more resistant rock type that overlies a softer, less resistant rock.*

Neogene • *the geologic time period extending from 23 to 2.6 million years ago.*

suture • *the area where two continental plates have joined together through continental collision.*

trachyte • *a fine-grained extrusive igneous rock, with a composition high in alkali feldspar.*

tuff • *a pyroclastic rock made of consolidated volcanic ash.*

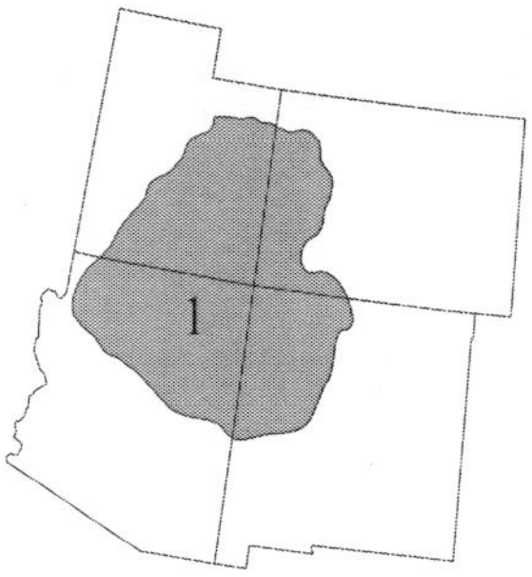

between less and more resistant rocks, such that weathering produces wavy vertical profiles leading to hoodoos and **badlands** topography (*Figure 2.17*). Hoodoos form when weathering erodes a softer material out from underneath a mass of harder **capstone**, leading to "mushroom" formations.

See Chapter 4: Topography for more on hoodoos in the Southwest.

Figure 2.16: Cliff Palace, the largest cliff dwelling in North America, at Mesa Verde National Park, Colorado.

Many of the Colorado Plateau's volcanic rocks occur around the region's edges, and are attributed to Cenozoic volcanism spurred by the formation of the Basin and Range. For example, the southeast margin of the Plateau, in New Mexico, is covered by **Neogene** volcanic rocks that erupted along an ancient **suture** zone called the Jemez Lineament (*Figure 2.18*). The Mt. Taylor volcanic field in New Mexico, part of this igneous zone, contains **trachyte**, **tuff**, **pumice**, **rhyolite**, basalt, and other volcanic flows and deposits. The field is scattered with volcanic necks, the solidifed erosional remnants of volcanoes. The Valles Caldera, a small **supervolcano** also located along the Jemez Lineament, erupted several times between 1.5 million and 60,000 years ago, forming thick layers of solidified **volcanic ash** and tuff. Supervolcanoes—volcanoes capable of producing more than 1000 cubic kilometers (240 cubic miles) of ejecta—can occur when magma rises under the crust from a hot spot, but is unable to break through. Eventually, the crust ruptures when it can no longer contain the built-up pressure. The ashfall from the eruption 1.2 million years ago blanketed an area roughly 800 by 1300 kilometers (500 by 800 miles) in extent (*Figure*

2.19). The Bandelier Tuff is one such layer, formed as extremely hot dust and ash poured out of the Valles Caldera and flowed down its side in thick layers of tuff (*Figure 2.20*). The ancient Ancestral Pueblo peoples built pueblos and dwellings against these cliffs, and also carved rooms from the soft rock.

Figure 2.17: The Claron Formation is prominently exposed in the cliffs and hoodoos (tall skinny spires of rock) of Bryce Canyon, Arizona. Vertical joints through the rocks have become weathered through frost wedging, leading to the formation of hoodoos.

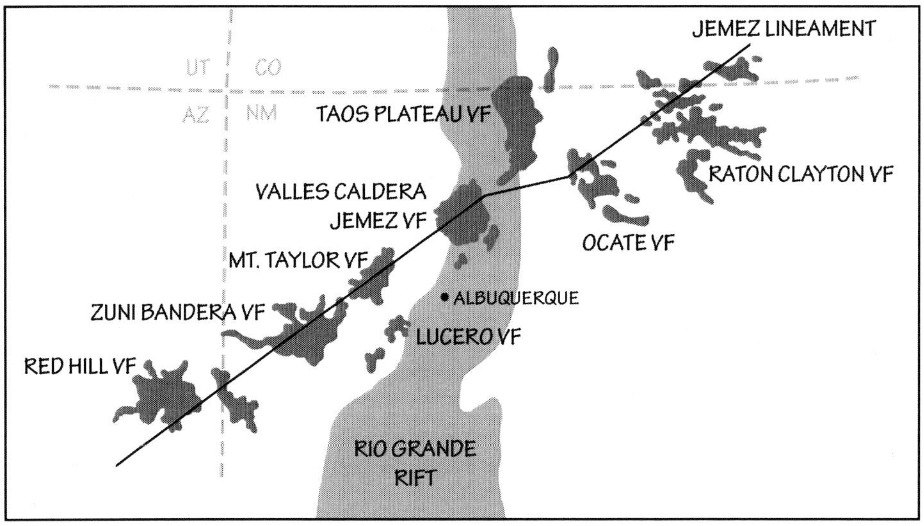

Figure 2.18: The Jemez Lineament and associated volcanic fields.

Region 1

pumice • *a pyroclastic rock that forms as frothing and sputtering magmatic foam cools and solidifies.*

rhyolitic • *a felsic volcanic rock high in abundance of quartz and feldspar.*

supervolcano • *an explosive volcano capable of producing more than 1000 cubic kilometers (240 cubic miles) of ejecta.*

volcanic ash • *fine, unconsolidated pyroclastic grains under 2 millimeters (0.08 inches) in diameter.*

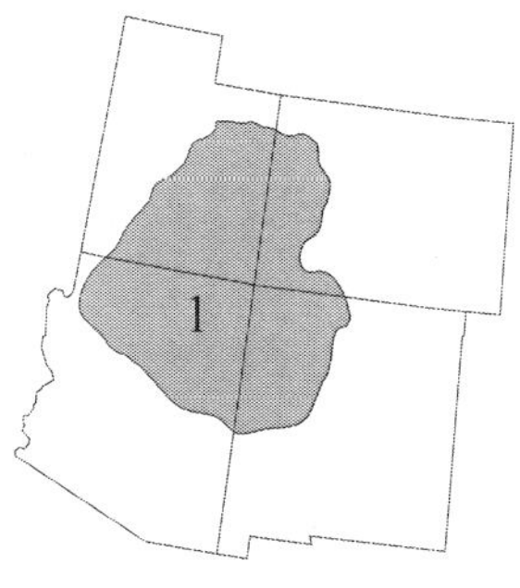

2 Rocks

Region 1

Oligocene • a geologic time interval spanning from about 34 to 23 million years ago.

Miocene • a geological time unit extending from 23 to 5 million years ago.

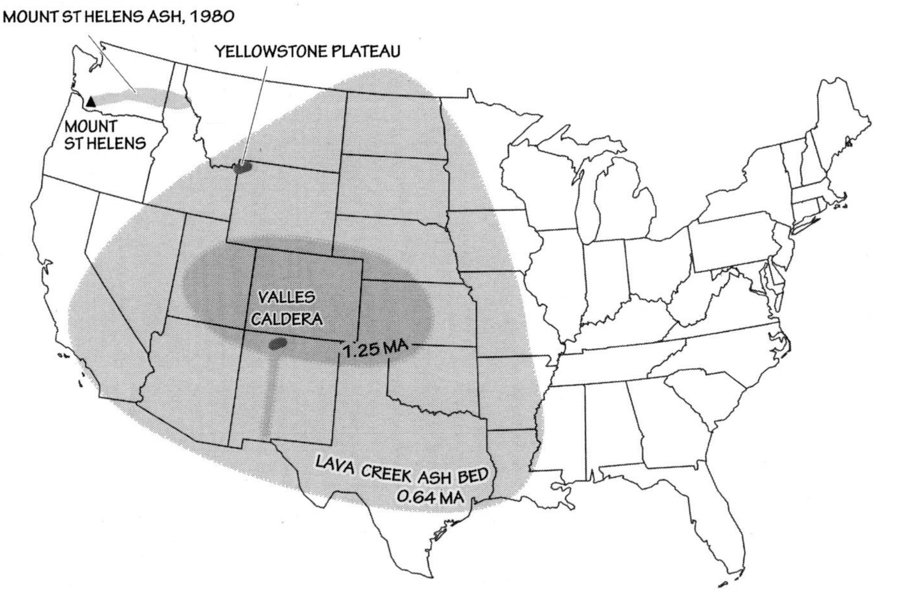

Figure 2.19: Extent of ashfall from the Valles Caldera, as compared to the Lava Creek Ash Bed from Yellowstone and the eruption of Mt. St. Helens. (See TFG website for full-color version.)

Figure 2.20: The Bandelier Tuff, a Pleistocene-aged mixture of volcanic tuff and pumice, at Bandelier National Monument in New Mexico.

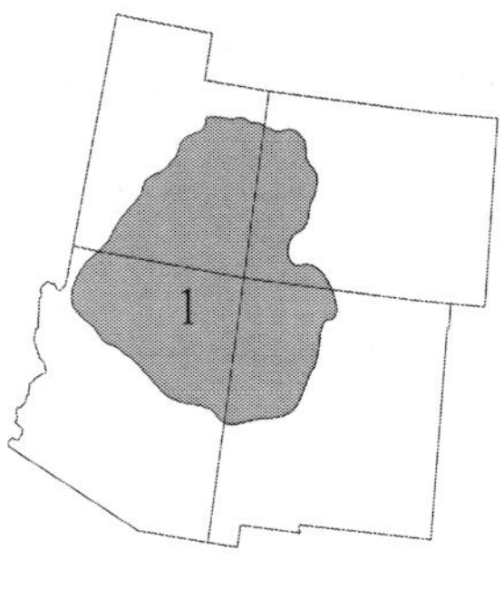

A few notable volcanic deposits are located more centrally within the Colorado Plateau, mainly in the form of volcanic fields. The Navajo Volcanic Field includes over 80 **Oligocene** to **Miocene** volcanoes (approximately 28 to 19 million years old) and associated intrusive igneous rocks that are found in an arc across the region. The rocks in the Navajo Volcanic Field are unusual because they are very potassium rich and highly **mafic** (probably forming from **mantle** magma). Many of the rocks are composed of **minette**, which contains biotite (**mica** with potassium) and orthoclase (potassium fieldspar). **Xenoliths**—rocks from older layers incorporated into the magma—are common, and include rocks from both the crust and upper mantle. Uplift and weathering of the Colorado Plateau exposed these volcanic rocks, including dikes, pipes, necks, sills, and other features formed at the base of volcanoes. Differential weathering produced distinctive landforms such as Shiprock, a volcanic neck made of tuff and **breccia** in northwest New Mexico (*Figure 2.21*). The volcanic crater associated with Shiprock may have been as much as a kilometer (0.6 miles) above the current land surface; the rock itself formed nearly 1000 meters (3000 feet) underground and was eventually exposed after millions of years of erosion. The Uinkaret Volcanic Field, on the north rim of the Grand Canyon, is one of the region's youngest volcanic areas, with igneous rock ranging from 1.2 million years to only 1000 years old. Here, massive basalt lava flows cascaded down into the Grand Canyon, occasionally blocking the Colorado River (*Figure 2.22*). Other volcanic deposits inside the Colorado Plateau include **laccoliths**—dome-shaped igneous intrusions—that core the La Sal, Abajo, and Henry mountains of Utah.

See Chapter 4: Topography to learn more about the La Sal mountains and other igneous intrusions that created Southwestern landforms.

Figure 2.21: Shiprock, a volcanic monadnock in San Juan County, New Mexico, rises roughly 483 meters (1583 feet) above the desert plain.

mafic • *igneous rocks that contain a group of dark-colored minerals, with relatively high concentrations of magnesium and iron.*

mantle • *the layer of the Earth between the crust and core.*

minette • *an unusual, mafic, fine-grained igneous rock made mostly of the potassium-containing minerals biotite mica and orthoclase.*

mica • *a large group of sheetlike silicate minerals.*

xenolith • *a chunk of pre-existing rock incorporated into magma that becomes part of an igneous rock body.*

breccia • *a pyroclastic rock composed of volcanic fragments from an explosive eruption.*

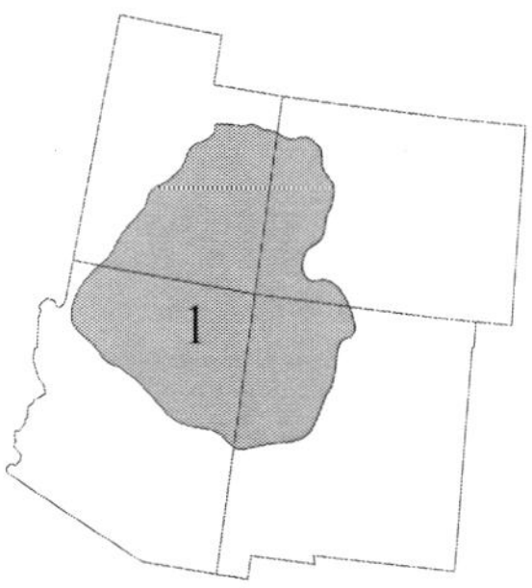

2 Rocks

Region 1

divergent boundary • *an active plate boundary where two tectonic plates are pulling apart from one another, causing the mantle to well up at a rift.*

convergent boundary • *an active plate boundary where two tectonic plates are colliding with one another.*

hot spot • *a volcanic region thought to be fed by underlying mantle that is anomalously hot compared with the mantle elsewhere.*

shield volcano • *a volcano with a low profile and gradual slope, so named for its likeness to the profile of an ancient warrior's shield.*

Quaternary • *a geologic time period that extends from 2.6 million years ago to the present.*

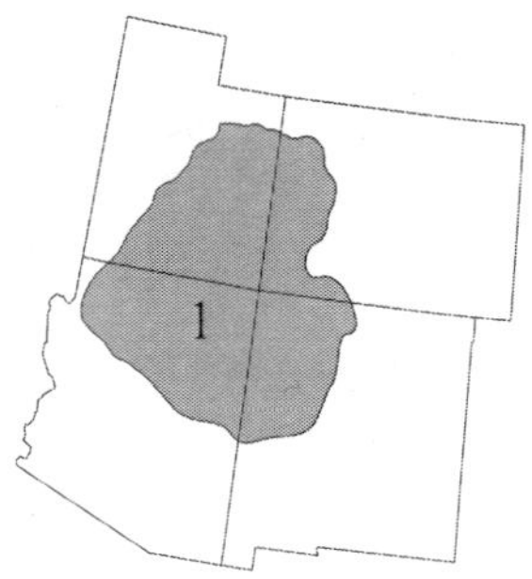

Understanding Volcanism

Most volcanic eruptions occur along tectonic plate boundaries. At *divergent boundaries*, the mantle wells up where two plates pull apart, creating new crust. Mid-ocean ridges are the most common type of divergent boundary and are characterized by the eruption of bulbous, pillow-shaped basalt lavas and hydrothermal fluids. Conversely, *convergent plate boundaries* destroy old lithosphere at subduction zones, where the ocean floor descends into the mantle. Volcanism here results from the subduction of seawater and seafloor sediments that descend into the mantle with the subducting slab, which lowers the melting temperature of mantle rocks enough to generate magma. Explosive eruptions characterize subduction zone volcanism and create arrays of cone-shaped stratovolcanoes that mark the position of the convergent boundary.

Volcanism can also occur at a *hot spot*, where superheated magma plumes well up from a point directly underneath the plate. Large *shield volcanoes* are produced as a direct result. The mechanics of hot spot volcanism are still largely unknown.

Prior to eruption, magma ascends from the mantle to a relatively shallow (1- to 10-kilometer [0.5- to 6-mile] deep) magma chamber. Upward movement reduces the pressure on the magma until it is low enough to permit dissolved gas to exsolve (come out of solution and form bubbles). All eruptions are driven by the exsolution of dissolved gas. As the gas forms bubbles, it expands in volume and forces the magma out of the vent/chamber system onto the surface. The combination of magma viscosity and gas content can produce a range of eruptive styles, from gentle, effusive eruptions to violent explosions.

There are few **Quaternary**-age sedimentary deposits on the Colorado Plateau. Most are related to glacial **outwash** from the alpine glaciers in the Rocky Mountains, as well as modern stream deposits and windblown sand and **silt**.

Figure 2.22: In this aerial view, the black basalt of the Uinkaret Volcanic Field is seen preserved where it cascaded over the side of the Grand Canyon at Lava Falls.

outwash plain • large, sandy flats created by sediment-laden water deposited when a glacier melts.

silt • fine granular sediment most commonly composed of weathered grains of quartz and feldspar crystals, of grain diameter 1/250 to 1/16 millimeters.

physiography • a subfield of geography that studies the Earth's physical processes and patterns.

Phanerozoic • a generalized term used to describe the entirety of geological history after the Precambrian, from 541 million years ago to the present.

Paleogene • the geologic time period extending from 66 to 23 million years ago.

Rocks of the Basin and Range
Region 2

A portion of the Basin and Range—a huge **physiographic** region that extends from southeastern Oregon to west-central Mexico—extends into the Southwest, covering most of western Utah, much of Arizona, and large areas of New Mexico. While the formation of the Basin and Range is a recent event that began only 30 million years ago, the bedrock that makes up the region's up-thrust ranges and down-dropped basins is very old. Here, rocks can be found from nearly all periods of the **Phanerozoic**. This is largely because the region's most recent geologic activity involved crustal extension that has exposed many deeper, older layers. During the **Paleogene**, magma upwelling from the mantle weakened the lithosphere, lowering its **density**. This stimulated uplift, stretching the bedrock in an east-west direction. The crust along the Basin and Range stretched, thinned, and **faulted** into some 400 separate mountain blocks. Movement along the faults led to a series of elongated peaks and down-dropped valleys, also called horst and graben landscapes. In a manner similar to books toppling when a bookend is removed from a shelf, the blocks slid against each other as they filled the increased space (*Figure 2.23*).

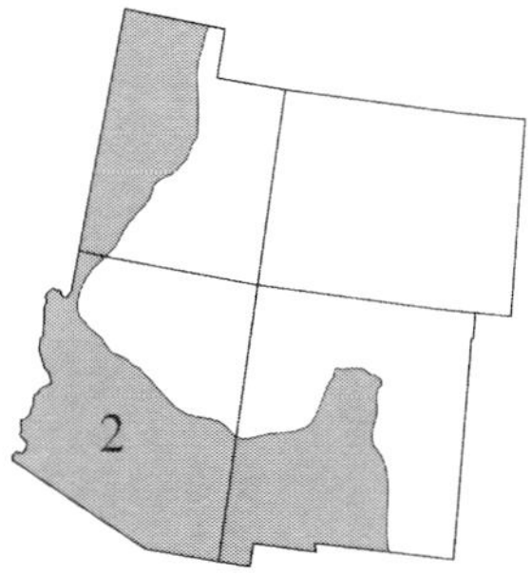

The Basin and Range is characterized by mostly north-south oriented linear mountain ranges of Precambrian, Paleozoic, and Mesozoic-aged rocks. Since the region's formation, the bedrock of the basins has been covered by young deposits, including loose sediment washed down from the mountains and evaporite deposits left behind in dried-out lakes. The ranges, however, expose

dolomite • *a carbonate mineral, consisting of calcium magnesium carbonate* ($CaMg(CO_3)_2$).

tillite • *glacial till that has been compacted and lithified into solid rock.*

phosphate • *an inorganic salt of phosphoric acid, and a nutrient vital to biological life.*

far older materials. The region is underlain by Precambrian basement—granitic gneisses and metamorphosed volcanic and sedimentary rocks ranging from 2.5 to 1 billion years in age. The oldest rocks in the Southwest underlie the northern part of Utah, as well as southwestern Utah and northwestern Arizona. Exposures of these rocks are rare, occuring in the cores of uplifted ranges. Most of the Precambrian rocks exposed in the Basin and Range are related to the Grand Canyon Supergroup, produced approximately 1.2 to 1.1 billion years ago during formation of the supercontinent Rodinia. The mildly metamorphosed conglomerate, sandstone, shale, and **dolomite** layers of the Grand Canyon Supergroup are exposed in the Cricket Mountains of west-central Utah. In addition, ancient **tillites** (glacial deposits) found in Utah represent major glaciation events that occurred during the Proterozoic (*Figure 2.24*).

See Chapter 8: Climate to learn about Snowball Earth and other Proterozoic glacial events.

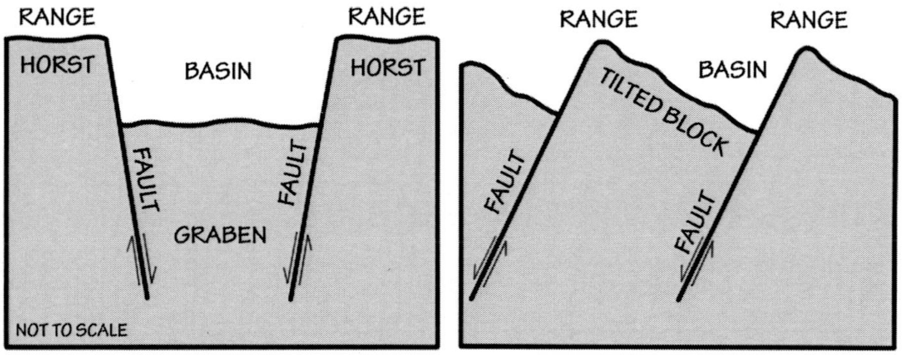

Figure 2.23: Alternating basins and ranges were formed during the past 17 million years by gradual movement along faults. Arrows indicate the relative movement of rocks on either side of a fault.

The Basin and Range's Paleozoic rocks—a succession of sandstones, limestones, and shales—were deposited on the western shore of North America from the Cambrian to the Mississippian. These rocks, like those in the Colorado Plateau, record a story of rising seas and shallow marine environments. Early Ordovician limestones of the El Paso Formation extend from southern Arizona and New Mexico into western Texas; Utah hosts similar limestones, sandstones, and dolomites, with occasional conglomerates formed when tidal currents ripped up newly deposited carbonate layers. This was followed during the Pennsylvanian through the Permian by a transition to shallow and evaporating seas, which deposited sandstones, mudstones, limestones, and **phosphate**-rich rocks. Thick Pennsylvanian sedimentary sequences are present in the Oquirrh Mountains west of Salt Lake City, and in the Sacramento, Robledo, Caballo, and San Andres mountain ranges along the Rio Grande Rift. After the Permian, as receding seas left the continent high and dry, subaerial weathering caused karstic topography to develop in the exposed carbonate rocks. Massive **reefs** became ideal structures for the formation of caves thanks

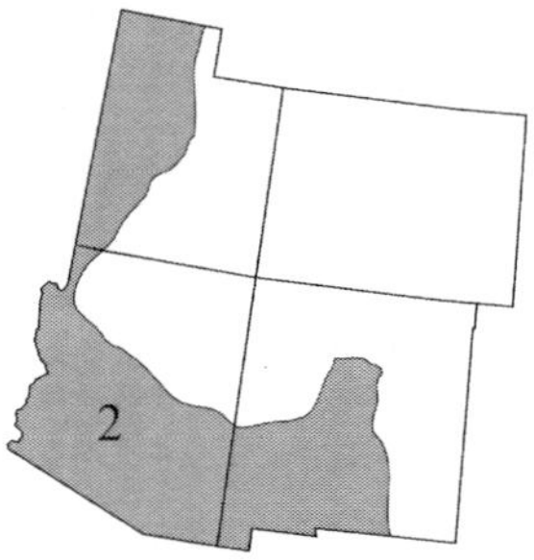

Region 2

reef • a feature lying beneath the surface of the water, which is a buildup of sediment or other material built by organisms, and which has positive relief from the sea floor.

permeability • a capacity for fluids and gas to move through fractures within a rock, or the spaces between its grains.

Figure 2.24: A piece of the Mineral Fork Tillite from Antelope Island, Utah. This rock is thought to have been deposited during the "Snowball Earth" Proterozoic glacial period, and is composed of glacial alluvium as well as embedded aventurine quartz.

to their **permeability** and soluble nature, allowing groundwater to infiltrate, dissolve, and carry away the calcium carbonate material. The Guadalupe Mountains of southeastern New Mexico expose an enormous Paleozoic reef, the Capitan Reef, which was revealed to the elements during the Permian. Carlsbad Caverns, also in the same area, is a massive limestone cave that developed in the region's carbonate bedrock after the mountains were uplifted (*Figure 2.25*).

Mesozoic rocks in the Basin and Range are represented by the red beds, sandstones, mudstones, and limestones of the Moenkopi Formation and the Navajo Sandstone, which formed as sea level dropped relative to the land. These Paleozoic and Mesozoic sediments form the cores of the region's mountain ranges; they were thrusted during the Sevier Orogeny, then involved in the Basin and Range style of extension during the Paleogene. Valleys and basins formed by this extensional faulting were filled with younger rocks from the Cretaceous and the Cenozoic (*Figure 2.26*). These rocks are mainly conglomerates, sand-stones, and mudstones originating from erosion of the nearby uplifts. Neogene sediments from glacial lakes, modern streams, and volcanic activity are also common.

See Chapter 1: Geologic History for more about the tectonic processes, extension, and faulting that shaped the Southwest.

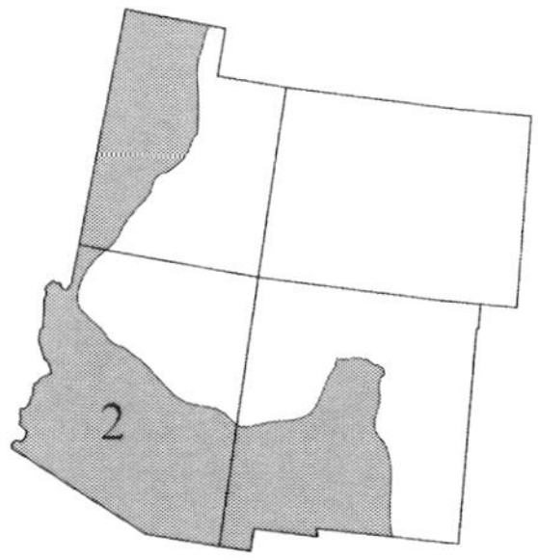

2 Rocks

Region 2

fracture • a physical property of minerals, formed when a mineral crystal breaks.

silica • a chemical compound also known as silicon dioxide (SiO_2).

cinder • a type of pyroclastic particle in the form of gas-rich lava droplets that cool as they fall.

caldera • a collapsed, cauldron-like volcanic crater formed by the collapse of land following a volcanic eruption.

Figure 2.25: Delicate limestone speleothems decorate Carlsbad Caverns, New Mexico, in an array of spectacular formations.

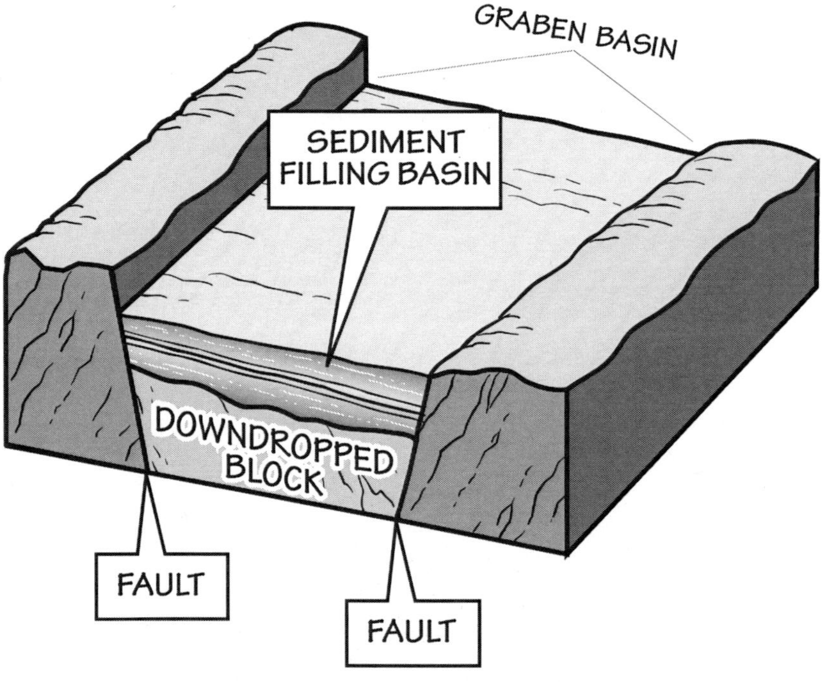

Figure 2.26: Basin fill in the Basin and Range.

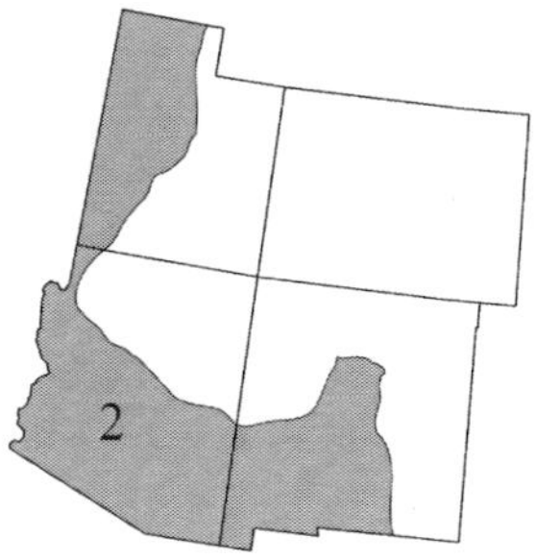

During the Cenozoic, the Southwest experienced significant active volcanism and plutonism. The Paleogene saw magma well up from the mantle from approximately 50 to 45 million years ago, intruding into the existing rock layers. By approximately 35 million years ago, upwelling intensified and the mantle began to uplift the crust, effectively tearing it and causing it to **fracture** and fault at the surface. This process formed a rugged landscape, created numerous igneous intrusions, and fed the region's volcanoes. The partial melting of deeper crust produced large volumes of magma that was both emplaced in the crust and erupted on the surface (for example, plutonic granite and volcanic rhyolite). Many of these rocks host mineral deposits, ranging from precious metals to uranium to industrial materials such as perlite. Most of the Southwest's larger volcanic fields can be found within the Basin and Range, especially around the region's border with the Colorado Plateau.

A pluton is a large body of igneous rock that formed under the Earth's surface through the slow crystallization of magma. The term comes from Pluto, the Roman god of the underworld.

The Marysvale Volcanic Field in western Utah is one of the largest volcanic fields in the western United States. During the Oligocene and Miocene, it and the neighboring Pioche Belt erupted large volumes of **silicic** magma, basalt, and ash, producing a variety of volcanic features that included **cinder** cones, lava domes, and **calderas**. The Boot Heel Volcanic Field, covering an area of more than 24,000 square kilometers (9300 square miles) across New Mexico, Arizona, and Mexico, includes nine calderas, **extrusive** flows (rhyolite, basalt, **andesite**, and tuff), and intrusive granite (*Figure 2.27*). The San Francisco Volcanic Field in Arizona contains the San Francisco Peaks—the eroded remnants of an enormous **stratovolcano**—and cinder cones such as the Sunset Crater, a 340-meter-high (1120-foot-high) cone of **basaltic andesite** that last erupted only 1000 years ago. Other Cenozoic volcanic fields in the Basin and Range include the Ajo (Arizona), the Black Rock Desert (Utah), and the Mogollon-Datil (New Mexico). Igneous rocks in the Basin and Range may also appear outside of volcanic fields, such as the exposures of **jointed** basalt found in Utah's geologically complex Traverse Range (*Figure 2.28*).

Figure 2.27: The Organ Mountains, an exposure of granite and rhyolite in New Mexico's Boot Heel Volcanic Field.

extrusion • *an igneous rock formed by the cooling of lava after magma escapes onto the surface of the Earth through volcanic craters and cracks in the Earth's crust.*

andesite • *a fine-grained, extrusive volcanic rock, with a silica content intermediate between that of basalt and dacite.*

stratovolcano • *a conical volcano made up of many lava flows as well as layers of ash and breccia from explosive eruptions.*

basaltic andesite • *a dark, fine-grained rock that is intermediate between basalt and andesite in silica content.*

joint • *a surface or plane of fracture within a rock.*

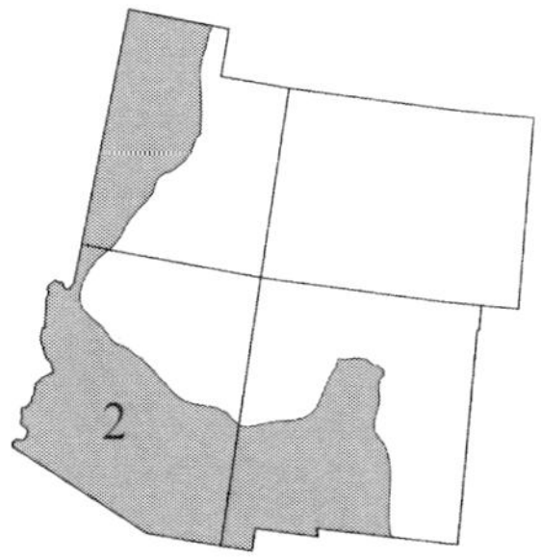

2 Rocks

Region 2

columnar joint • *five- or six-sided columns that form as cooling lava contracts and cracks.*

quartz • *the second most abundant mineral in the Earth's continental crust (after the feldspars), made up of silicon and oxygen (SiO_2).*

aluminum • *a metallic chemical element (Al), and the most abundant metal in the Earth's crust.*

Figure 2.28: Step Mountain, in the Traverse Range, Utah, is composed of overturned basalt that exhibits columnar jointing.

Columnar Jointing

As a lava flow cools, it contracts, and the resulting force may cause the rock to crack. These cracks continue down to the bottom of the flow, resulting in five- or six-sided columns. *Columnar joints* are not restricted to basalt flows and can form in ashflow tuffs as well as in shallow intrusions. The columns are generally vertical, but may also be slightly curved.

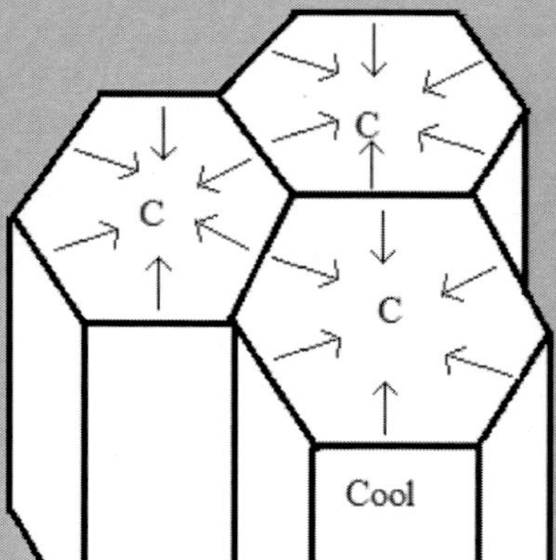

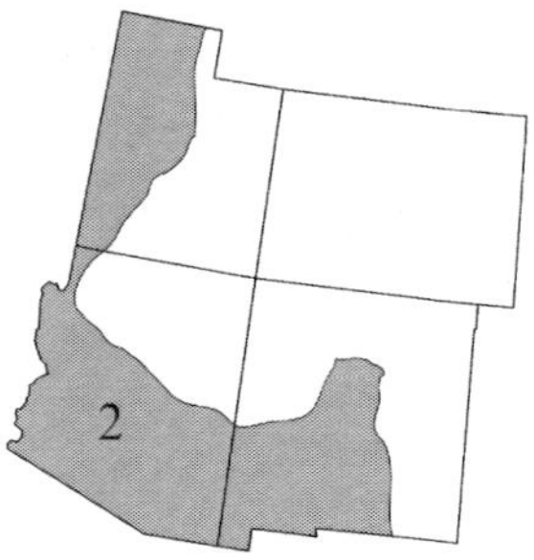

Regions 2-3

Rhyolite and Basalt

Both rhyolite and basalt are lavas, but they behave differently due to their varied densities and melt structures. Rhyolite is composed of felsic minerals including *quartz*, orthoclase, and biotite, and is high in silica and *aluminum*. This composition results in a very viscous magma. The lavas in volcanoes with felsic (rhyolitic) compositions are too viscous to flow easily; pressure builds up beneath them until they erupt explosively. The most explosive volcanoes form calderas, and the ash from such an explosion can travel many kilometers (miles).

Basalt is composed of the mafic minerals plagioclase and *pyroxene*, and may contain *olivine*. These minerals are high in iron and magnesium, and produce a very fluid magma. Volcanoes with mafic or basaltic compositions tend to produce fluid lava flows comparable to those associated with the eruptions presently seen in Hawai'i.

While limited extensional forces continued to act on the Basin and Range up until 10 million years ago, the dominant activity in the region since the beginning of the Neogene has been erosion. The region's valleys are being filled with sediment from the surrounding mountains and mesas, creating thick deposits in northwest-southeast trending bands. Quaternary deposits in the Basin and Range are primarily composed of glacial outwash from rivers, streams, and glacial lakes. These sediments were derived from alpine glaciers in the Rocky Mountains, and were also produced by the **Cordilleran Ice Sheet** during the **ice age**. Windblown sand and dust, both **Pleistocene** and modern, are also common components of the region's Quaternary deposits.

pyroxene • *dark-colored rock-forming silicate minerals containing iron and magnesium.*

olivine • *an iron-magnesium silicate mineral ($(Mg,Fe)_2SiO_4$) that is a common constituent of magnesium-rich, silica-poor igneous rocks.*

Cordilleran Ice Sheet • *one of two continental glaciers that covered Canada and parts of the Western US during the last major Pleistocene ice age.*

ice age • *a period of global cooling of the Earth's surface and atmosphere, resulting in the presence or expansion of ice sheets and alpine glaciers.*

Pleistocene • *a subset of the Quaternary, lasting from 2.5 million to about 11,700 years ago.*

Rocks of the Rocky Mountains

Region 3

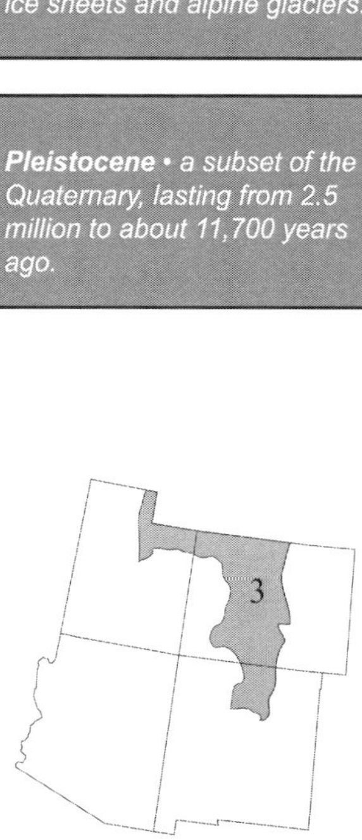

The Rocky Mountains of the Southwest consist of multiple mountain ranges resulting from both the Sevier and Laramide orogenies, which uplifted numerous discrete blocks of terrain along thrust faults that accommodated compressional shortening and thickening of the crust. The overlying sediments were

See Chapter 1: Geologic History to learn more about mountain building during the Laramide and Sevier orogenies.

pillow basalt • *basaltic lava that forms in a characteristic "pillow" shape due to its extrusion underwater.*

quartzite • *a hard metamorphic rock that was originally sandstone.*

slate • *a fine-grained, foliated metamorphic rock derived from a shale composed of volcanic ash or clay.*

subsequently eroded to expose deeper Precambrian rock as well as Paleozoic and Mesozoic sedimentary formations. The thrust-faulted uplift also produced adjacent basins, which subsequently accumulated sediments eroded from the surrounding mountains. In the Southwest, the Rockies are located in central and western Colorado, north-central New Mexico, and northeastern Utah.

Precambrian rocks are well exposed in the Rocky Mountains, where 1.9- to 1.6-billion-year-old metamorphosed volcanic and sedimentary rocks represent the accretion of an ancient terrane to this portion of the North American continent. These ancient rocks are typically metamorphosed and deformed, but occasionally some of the original sedimentary and volcanic textures can still be observed, including bedding, cross-bedding, and **pillow** lavas. Coarse-grained, dark-colored metamorphic rocks are ubiquitous. Many of these Precambrian rock units can be seen in exposures along roads that wind through the mountains. For example, the Uinta Mountains, an east-west segment of the southern Rockies in northeastern Utah, are made of uplifted and metamorphosed sedimentary rocks deposited in a shallow marine basin over 700 million years ago. The rocks at the mountains' peaks include reddish **quartzite** and sandstones, **slate**, and shale. These are part of a thick (4- to 7-kilometer [13,000- to 24,000-foot]) sequence of rocks known as the Uinta Mountain Group that were uplifted during the Laramide Orogeny (*Figure 2.29*).

Figure 2.29: King's Peak, the highest peak in Utah's Uinta Mountains, is composed primarily of Neoproterozoic sandstone, siltstone, and shale.

Colorado's Precambrian rocks are primarily metamorphosed volcanics with intermixed sedimentary units. Schist—metamorphosed sediment—is fairly common in the Rocky Mountains, and can be found throughout the region. It is usually composed of shale and siltstone that have been compressed to

What happens to a rock when it is metamorphosed?

When rocks are subjected to high enough temperatures or pressures, their characteristics begin to change. The weight of overlying rock can cause minerals to realign perpendicularly to the direction of pressure, layering them in a pattern called *foliation*, as exemplified in gneiss and schist. Recrystallization, as seen in marble and quartzite, results as rock is heated to high temperatures. Individual grains reform as interlocking crystals, making the resulting metamorphic rock much harder than its parent rock.

Contact metamorphism describes a metamorphic rock that has been altered by direct contact with magma. Changes that occur due to contact metamorphism are greatest at the point of contact. The farther away the rock is from the point of contact, the less pronounced the change.

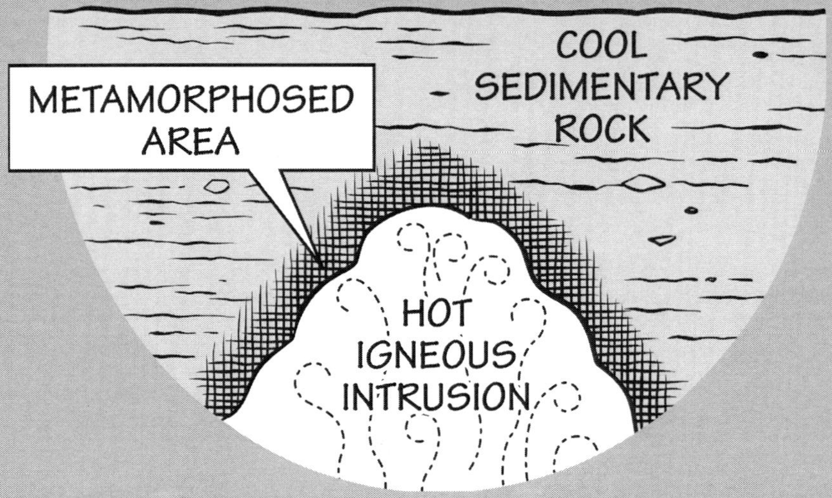

Regional or *dynamic metamorphism* describes a metamorphic rock that has been altered due to deep burial and great pressure. This type of metamorphic rock tends to occur in long belts. Different types of metamorphic rock are created depending on the gradients of heat and pressure applied.

foliation • *the arrangement of the constituents of a rock in leaflike layers.*

regional (dynamic) metamorphism • *a metamorphic rock that has been altered due to deep burial and great pressure.*

form relatively large crystals (especially micas) that are layered into sheet-like structures. Gneisses—metamorphosed rhyolite or granite—are common in the Colorado Rockies (*Figure 2.30*). In contrast, the Precambrian rocks of New Mexico have a higher ratio of sedimentary units, including more quartzite and conglomerate, than is found in Colorado. Quartzite—metamorphosed sandstone—is very resistant to erosion, so it forms steep slopes and is often found at the core of ridges or mountains. This rock ranges in color from white and gray to pink and purple, depending on the amounts of iron and other minor minerals that were deposited with the sand grains. The Picuris Range near Taos in New Mexico contains more Precambrian quartzite than do the mountains of Colorado. (The best quartzite outcrops in Colorado are south of Boulder in El Dorado State Park, and in Coal Creek Canyon, along Highway 72.)

A gneiss is a very highly metamorphosed rock with alternating bands of dark and light minerals. The dark bands are mafic and higher in magnesium and iron, while the lighter bands are felsic and higher in silicates. These bands may form because extreme temperature and pressure cause a *chemical reaction* that forces the different elements into separate layers. Banding may also occur when a set of varied protoliths are subjected to extreme *shearing* and sliding forces, causing them to stretch into stacked sheets.

chemical reaction • *a process that involves changes in the structure and energy content of atoms, molecules, or ions but not their nuclei.*

shearing • *the process by which compressive stress causes the fracturing and faulting of brittle rocks.*

batholith • *a large exposed structure of intrusive igneous rock that solidified at depth, and covers an area of over 100 square kilometers (40 square miles).*

pegmatite • *a very coarse-grained igneous rock that formed below the surface.*

Three major igneous intrusions followed the original deposition of these rocks, forming large granitic **batholiths** approximately 1.8, 1.4, and 1.1 billion years ago. These batholiths are more resistant to erosion than the surrounding rocks are, and form many high mountain peaks including Pike's Peak, Mount Evans, and Long's Peak in the Colorado Front Range. The Pike's Peak granite (1.8 billion years old) is well known for its **pegmatites** (*Figure 2.31*). Although they have essentially the same composition as the surrounding granite, pegmatites are composed of much larger crystals and may also yield rare, potentially valuable, minerals.

See Chapter 5: Mineral Resources to learn about rhodochrosite, diamonds, and other rare Rocky Mountain minerals.

The Rocky Mountains, like other regions of the Southwest, contain a succession of Paleozoic sandstone, limestone, and shale. Between the Cambrian and Mississippian, these rocks were deposited in shallow marine environments on what was then the western shore of North America. Early Paleozoic rock units can be found in the canyons of Utah's Wasatch Mountains, at the edge of the Sawatch Range in Colorado, and along the deep gorge of the Colorado River, where thick layers of Cambrian and Ordovician sediments overlie the Precambrian basement (*Figure 2.32*).

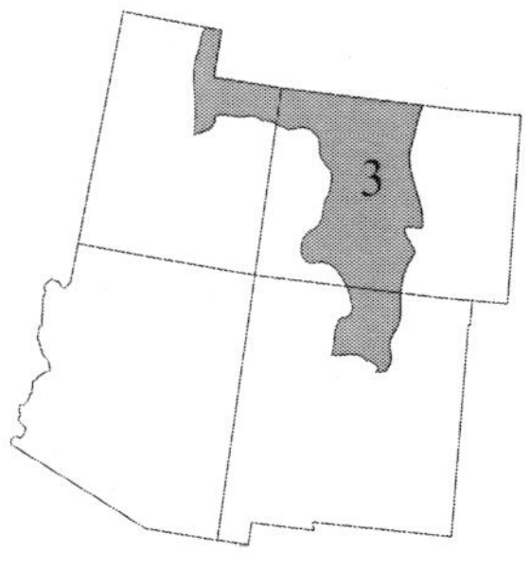

Figure 2.30: Ypsilon Mountain in the Mummy Range of northern Colorado is principally composed of 1.7-billion-year-old biotite schists and gneisses.

Figure 2.31: Coarsely crystalline granite (approximately 1.8 billion years old) from the Pike's Peak batholith in central Colorado. This pegmatite is dominated by feldspar, quartz, and mica.

Carboniferous • *a geologic time period that extends from 359 to 299 million years ago.*

alluvial • *a layer of river-deposited sediment.*

arkose • *sandstone that is > 25% feldspar, reflecting its relative lack of transport and weathering.*

Erosion during the **Carboniferous** deposited sediments on both the east and west sides of the mountains. This resulted in the formation of **alluvial** fans along the mountain flanks, and bedded sediments in lower areas. Mississippian rocks are generally gray limestone and dolomite; Cave of the Winds near Manitou Springs, Colorado is a cavern system developed in Mississippian limestone. The Pennsylvanian Fountain Formation along the eastern flank of the Colorado Front Range was deposited as a series of alluvial fans—wedge-shaped deposits of sediment formed when sediment was deposited at the mouths of ancient streams coming out of the mountains. It is composed of **arkose** covered by conglomerate and thin layers of mud and siltstone (*Figure 2.33*). By the Permian, erosion had leveled the ranges and the area was flooded by shallow and evaporating seas, which deposited marine sandstone, mudstone, limestone, and phosphate-rich rocks (in the deeper zones). Today, these sedimentary layers are best seen around the edges of the Rocky Mountains, where beds

have been tilted and pushed upward by mountain building processes. Fossils of marine invertebrates are common in these Paleozoic rocks.

Figure 2.32: Cambrian and Ordovician strata are exposed in Glenwood Canyon, Colorado, where Interstate Highway I-70 follows the Colorado River.

Why are there different sedimentary rocks in different environments?

Most sedimentary rock deposited in underwater settings originated from material eroded on land and washed down streams or rivers before settling to the bottom of a body of water. Intuitively, the faster the water is moving, the larger the sediments it may carry. As the water slows down, the size of sediments it can carry decreases. Furthermore, the farther the grains of sediment are carried, the more rounded they become as they are tumbled against each other. In this way, rivers emptying into a sea are effectively able to sort sediment. Near the mouth of the river, the water is still relatively high energy, dropping only the largest pieces; farther from the shore, the dropped particles get smaller. Therefore, conglomerates and sandstones are interpreted to have been deposited on or near the shore, siltstone farther from the shore, and shale in deep water quite far from shore where currents are slow enough that even very tiny particles may settle out.

Why are there different sedimentary rocks in different environments? (continued)

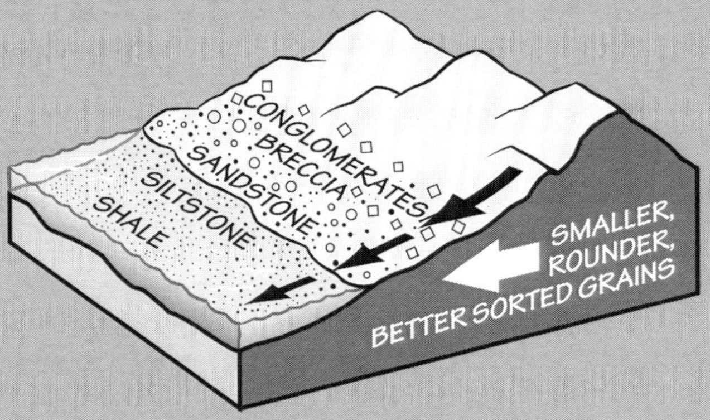

Increased distance from shore and water depth can also reduce the presence of oxygen in the water, causing organic material to decompose less completely. This causes darker, carbon-rich rocks (including some that contain exploitable *fossil fuels*) to form in these areas. Limestone is made primarily of calcium carbonate, the components of which are dissolved in the water. Living creatures, like coral and *foraminifera*, take those components out of the water to make calcium carbonate shells, which, after the creatures die, accumulate to become limestone. These shelled creatures tend to fare better in clear water, so limestone usually forms far from other sources of sediment. While this process happens over much of the seafloor, if more than 50% of the sediment being deposited is from another source, the rock that forms is, by definition, not limestone.

fossil fuels • *fuel for human use that is made from the remains of ancient biomass.*

foraminifera • *a class of aquatic protists that possess a calcareous or siliceous exoskeleton.*

Mesozoic rock units in the Rocky Mountains contain abundant fossils, especially those of dinosaurs and other vertebrates. The region's Triassic and Jurassic rocks were deposited in terrestrial settings, and include red beds, sandstone, mudstone, and limestone. Triassic red beds are composed of iron-rich sandstone or siltstone, while the Jurassic Morrison Formation is a striking variegated siltstone and shale colored in red, purple, gray, tan, and green. The Morrison includes lenses of sandstone, which often contain dinosaur bones that were deposited in streambeds. During the Cretaceous, marine shale and sandstone formed when the epicontinental Western Interior Seaway flooded the area, and coal formed along coasts and swamps. The Cretaceous Dakota Group is a

weather • *the measure of short-term conditions of the atmosphere such as temperature, wind speed, and humidity.*

Pliocene • *a geologic time interval extending from roughly 5 to 2.5 million years ago.*

felsic • *igneous rocks with high silica content and low iron and magnesium content.*

vesicular • *porous or pitted with vesicles (cavities).*

rift • *a break or crack in the crust that can be caused by tensional stress as a landmass breaks apart into separate plates.*

yellow sandstone, often bearing ripple marks formed in beach or tidal zones (*see Figure 2.15*). Dinosaur bones and trackways are present in several places. Cretaceous shale units have thick layers of black, gray, or tan, and often weather into badlands topography. The Dakota Hogback (*Figure 2.34*) is located along the edge of the Front Range just west of Denver, Colorado. This area preserves an important stratigraphic section that includes the Pennsylvanian Fountain Formation (*see Figure 2.33*), Permian Lyons Formation, Permo-Triassic Lykins Formation, Jurassic Ralston Creek Formation and dinosaur fossil-bearing Morrison Formation, Cretaceous Dakota Group, and various Cenozoic formations that are exposed at the edge of the Denver Basin. The modern Rocky Mountains rose near the end of the Cretaceous—earlier in eastern Utah during the Sevier Orogeny, and later in Colorado and New Mexico during the Laramide Orogeny.

See Chapter 3: Fossils for more about the vast array of dinosurs found in the Morrison Formation.

Cenozoic rocks in the Rocky Mountains are mainly sandstones, shales, conglomerates, and mixed sediments that formed when eroding sediment from the uplifted Rocky Mountains was deposited by rivers onto alluvial fans, and into lakes, basins, and swamps. Where they are adjacent to volcanic areas, some stream sediments carry cobbles of volcanic rocks. Thick wedges of Cenozoic sediments were deposited on the flanks of uplifted ranges, and sedimentary basins and lakes were centers for the deposition of thick layers of shale and mudstone. Cenozoic deposits differ from one basin to another, reflecting local differences in **weather** and erosion.

The San Juan Volcanic Field is in the San Juan Mountains of southwestern Colorado and consists mostly of Oligocene lavas and breccias, and Miocene and **Pliocene** basalts. These rocks represent stratovolcano deposits, and include numerous calderas and ash flows, which contain rhyolite tuffs (**felsic**) and andesite (intermediate in composition). The field's largest known caldera, La Garita, is associated with its largest known ash flow deposit. During the Oligocene, this caldera deposited thick layers of coarse volcanic tuff. Ash-flow tuffs are the result of pyroclastic flows—explosions that contain pulverized rock and superheated gases, which can reach temperatures of up to 1000°C (1830°F). The violent expansion of hot gas shreds the erupting magma into tiny particles that cool in the air to form dense clouds of volcanic ash. The tremendous explosions that are necessary to create ash-flow tuffs are caused by rhyolitic magma, which is felsic in nature. High silica content makes the magma quite viscous, preventing gas bubbles from easily escaping, thus leading to pressure build-ups that are released by explosive eruptions. The ash flows from these violent explosions tend to hug the ground, eventually solidifying into tuffs. Tuffs and other pyroclastic materials are **vesicular** (porous) due to gases expanding within the material as it cools. The Wheeler Geologic Area in the La Garita Mountains in southern Colorado is recognized for the unusual jagged shape of its terrain due to the erosion of these tuffs, forming striking hoodoos (*Figure 2.35*). By the Miocene, volcanism in the area had become more mafic, representing a tectonic change to **rifting** during crustal extension.

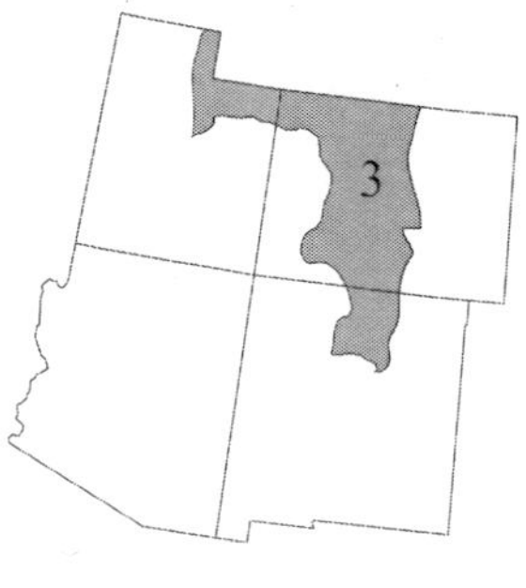

Figure 2.33: Red Rocks Park is nestled within the towering Fountain Formation in Morrison, Colorado. The formation is composed of a sedimentary rock called arkose, which is red in color due to the presence of oxidized iron and a large number of pink feldspar and quartz grains.

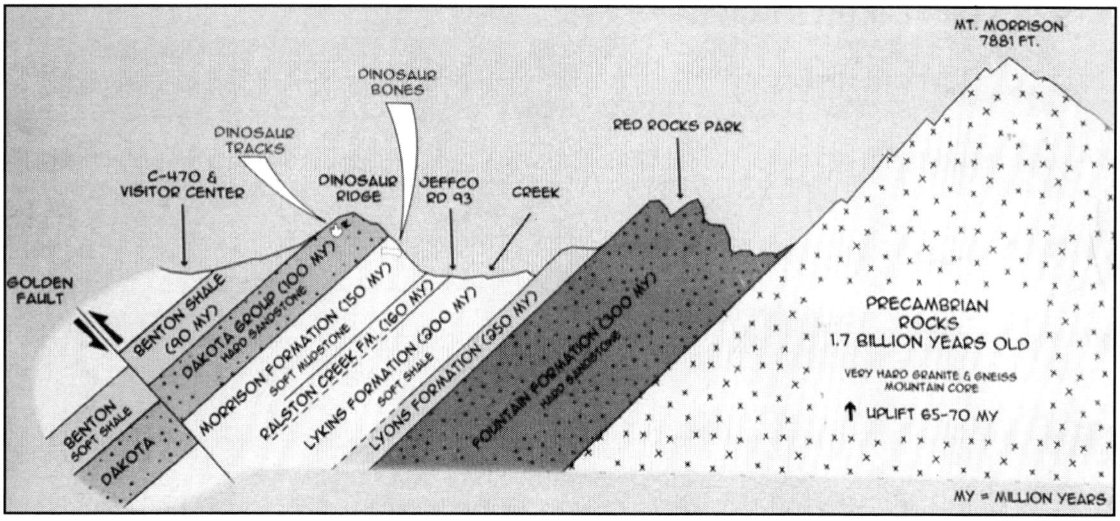

Figure 2.34: Stratigraphic units of the Dakota Hogback, a north-south trending ridge of eastward-dipping Mesozoic sedimentary rocks in north-central Colorado.

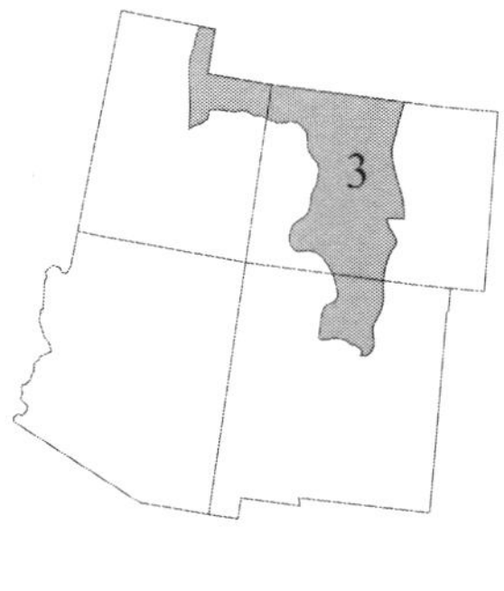

Rocks

Region 3

Paleocene • a geologic time interval spanning from about 66 to 56 million years ago.

gold • a soft, yellow, corrosion-resistant element (Au), which is the most malleable and ductile metal on Earth.

silver • a metallic chemical element (Ag).

hydrothermal solution • hot, mineral-rich water moving through rocks.

contact metamorphism • the process by which a metamorphic rock is formed through direct contact with magma.

calcite • a carbonate mineral, consisting of calcium carbonate ($CaCO_3$).

Figure 2.35: The Wheeler Geologic Area in Mineral County, southern Colorado, is a highly eroded outcropping of Oligocene volcanic ash.

Other than the volcanic fields at the edge of the Colorado Plateau and Great Plains, Phanerozoic volcanic and plutonic rocks are minor in the Rocky Mountain region. There are, however, a few notable exceptions. The Colorado Mineral Belt is a structural weakness in the Precambrian crust through which multiple igneous rock bodies were emplaced or erupted during the **Paleocene** (approximately 65 to 60 million years ago). The accompanying fluids carried **gold**, **silver**, other metals, and uranium, many of which have been mined for profit. (This was the primary source of gold during the Gold Rush of 1859.) In addition to deposits emplaced by **hydrothermal solutions**, **contact metamorphism** from the heat of Paleogene igneous activity altered the surrounding rock to form deposits such as the Yule Marble. This metamorphosed Mississippian limestone was transformed into a distinctive, smooth, white stone that is 99.5% pure **calcite**. Thanks to its smooth texture and pure white appearance, the Yule Marble has been used to cover the exterior of a variety of buildings and monuments including the Tomb of the Unknown Soldier and the Lincoln Memorial (*Figure 2.36*)

See Chapter 5: Mineral Resources for more detail about mineral extraction in the Colorado Mineral Belt.

Other Cenozoic-aged volcanic and intrusive activity can be found at Cripple Creek, Colorado (host of another large gold deposit); Questa, New Mexico

Figure 2.36: The Lincoln Memorial's brilliant white exterior is faced with Yule Marble.

molybdenum *• a metallic chemical element (Mo) which has the sixth-highest melting point of any element.*

ice sheet *• a mass of glacial ice that covers part of a continent and has an area greater than 50,000 square kilometers (19,000 square miles).*

cirque *• a large bowl-shaped depression carved by glacial erosion and located in mountainous regions.*

moraine *• an accumulation of unconsolidated glacial debris (soil and rock) that can occur in currently glaciated and formerly glaciated regions.*

(host of a **molybdenum** deposit); Thirty-Nine-Mile Volcanic Field southwest of Denver; and in various volcanic rocks near Steamboat Springs and North Park, Colorado. Near Gunnison, Colorado, there is an unusual igneous rock known as a carbonatite, a carbonate intrusive rock that does not possess the usual silicate composition. Some of the youngest volcanic activity in the region includes a basaltic cone and lava flow at Dotsero, Colorado, dated at approximately 2200 BCE. Interstate Highway I-70 actually crosses the foot of the flow.

Pleistocene glaciation produced glacial **till** and outwash material in the mountains and basins. Alpine glaciers, rather than continental **ice sheets**, carved **cirques** and deposited **moraines** in mountain valleys. Modern stream sediments are still deposited today.

Rocks of the Great Plains

Region 4

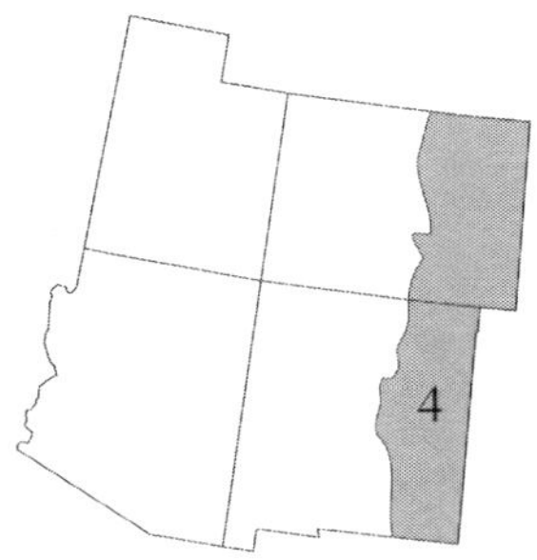

Only the western margin of the Great Plains lies within the Southwestern US, along the eastern border of Colorado and New Mexico. The region's geologic history is chiefly one of sedimentary deposition driven first by marine environments and, more recently, by terrestrial deposition from the erosion of the Rocky Mountains to the west. In general, rocks found at the surface of the Great Plains are young sediments (predominantly Cenozoic with some Cretaceous). Erosional processes along the Platte and Arkansas rivers in

clay • *the common name for a number of very fine-grained, earthy materials that become plastic (flow or change shape) when wet.*

rare earth elements • *a set of 17 heavy, lustrous elements with similar properties, some of which have technological applications.*

kaolinite • *a silicate clay mineral, also known as china clay.*

meteorite • *a stony or metallic mass of matter that has fallen to the Earth's surface from outer space.*

mass extinction • *the extinction of a large percentage of the Earth's species over a relatively short span of geologic time.*

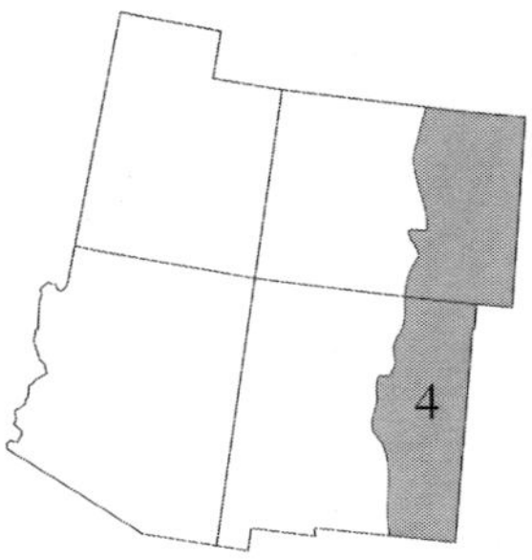

Colorado, and the Canadian and Pecos rivers in New Mexico, have cut into the gently eastward-sloping land surface. Outcrops on the plains are usually exposed by stream erosion, in dissected terrain, or in quarries.

Throughout much of the Paleozoic era until the early stages of the Carboniferous, the Great Plains region was submerged in a shallow sea. During the Paleozoic and Mesozoic, thick sequences of rocks were deposited across the region in environments that varied from marine and coastal to lake, stream, and alluvial. Layers of limestone and shale were deposited when shallow seas repeatedly flooded the area, while sandstones accumulated from sandy beaches were left behind as the seas retreated. With the rise of the Rocky Mountains to the west, erosion and Cenozoic-era volcanism produced sediment that was transported and deposited throughout the Great Plains. The resulting rocks are sandstones, shales, limestones, mudstones, conglomerates, and some evaporites, deposited in beds buried beneath the surface of the plains. Although they are not exposed at the surface, many of these buried rocks are well studied from the cores of numerous oil and gas drillholes throughout the region.

The Great Plains overlie the Raton, Denver, and Permian basins, which are downwarped areas that formed due to tectonic compression in the late Paleozoic. During the Permian period, these basins were carbonate marine environments where reefs grew in the shallows. Deposition of marine sandstone, shale, limestone, and then evaporites (salts and related minerals) eventually ceased as sea level dropped, the basins closed, and the remaining seawater evaporated. The Raton Basin exposes a well-preserved sequence of rock spanning the Cretaceous-Paleogene (K-Pg) boundary. A common characteristic of the K-Pg boundary is the presence of a thin, millimeter-scale layer of **clay** containing a number of **rare earth elements**, including iridium. This layer is represented in the basin by a sheet of compact **kaolinite** clay (*Figure 2.37*). The iridium is thought by most geologists to have come from the impact of a large comet or **bolide**, which was likely a primary cause of the **mass extinction** that marks

Figure 2.37: The Cretaceous-Paleogene (K-Pg) boundary layer is clearly visible in this roadcut along Long Canyon Road near Trinidad Lake, Colorado.

the end of the Cretaceous period. The Denver Basin contains a 55-million-year-old formation of coarse sandstone and conglomerate called the Dawson Arkose, a collection of fluvial and alluvial sediments eroded from the Rockies and deposited at their base during the late Cretaceous and early Paleogene. The Dawson Arkose is beautifully exposed at the Paint Mines Interpretive Park near Colorado Springs, where pastel hoodoos and badlands topography are eroded into the landscape (*Figure 2.38*).

Gravel, sand, and mud dominate the surface of the Great Plains, with progressively younger sediment located farther from the western mountains. Eroded material eventually filled stream valleys and covered hills, creating a massive, gently sloping plain that was in place by five million years ago. Except in river valleys where Mesozoic rocks are exposed, the surface is blanketed by thick soils, **caliche**, windblown sand and silt, and playa lakes. These sediments overlie the Paleogene sediments of the Ogallala Formation, a unit of unconsolidated sands, gravels, and clays that eroded from the Rockies. The Ogalalla Formation is extremely **porous**, and, as a result, it acts as an important **aquifer** for much of the Great Plains.

See Chapter 9: Earth Hazards to find out more about the effect of drought on the Ogallala Aquifer.

Volcanic features are less common in the Great Plains, and are generally found along or near the region's western edge. Castlewood Canyon State Park, near Castle Rock, Colorado, provides evidence for an explosive eruption of rhyolitic ash, which created an ash-flow tuff. This welded rhyolite tuff was formed during

Figure 2.38: Ancient cultures and early settlers mined the brightly colored clays seen at the Paint Mines Interpretive Park, Colorado for use as pigment.

gravel • *unconsolidated, semi-rounded rock fragments larger than 2 millimeters (0.08 inches) and smaller than 75 millimeters (3 inches).*

caliche • *a zone of cemented material within soil, formed when water infiltrates the soil, dissolves soluble materials, and evaporates, leaving behind precipitates in the pore space between soil grains.*

porosity • *the percentage of openings in a body of rock such as pores, joints, channels, and other cavities, in which gases or liquids may be trapped or migrate through.*

aquifer • *a water-bearing formation of gravel, permeable rock, or sand that is capable of providing water, in usable quantities, to springs or wells.*

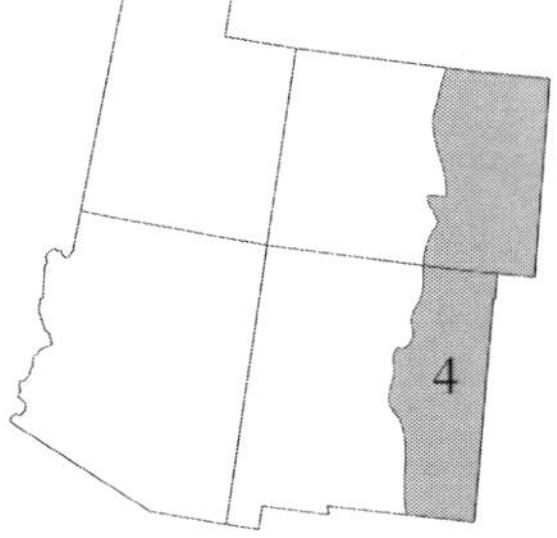

granodiorite • a coarse-grained, plutonic rock rich in the elements sodium and calcium, and in the minerals potassium feldspar and quartz.

syenite • a durable, coarse-grained, intrusive igneous rock, which is similar to granite but contains less quartz.

dacite • a fine-grained, extrusive igneous rock, with a silica content intermediate between that of andesite and rhyolite.

loess • very fine-grained, wind-blown sediment, usually rock flour left behind by the grinding action of flowing glaciers.

a tremendous volcanic eruption during the Eocene, approximately 36 million years ago, which filled the air with superheated ash and rock that solidified as it hit the ground. The resulting attractive pink rock has been mined near Castle Rock for use as a decorative stone for over a century; it has been used as a building stone in several Denver buildings, including the historic Trinity Presbyterian Church in downtown Denver, parts of the Capitol Building, and buildings on the University of Denver campus.

The Spanish Peaks, a 27- to 14-million-year-old igneous intrusion, are located in south-central Colorado near the towns of Walsenburg and Trinidad. The peaks are formed from a varied set of rocks, including granite, **granodiorite**, **syenite**, monzonite, and lamprophyre. In northeastern New Mexico, the 20,700-square-kilometer (8000-square-mile) Raton-Clayton Volcanic Field contains over 100 recognizable volcanoes, which erupted during the Neogene (likely as a result of crustal extension). For example, Capulin Volcano, an extinct cinder cone that rises steeply from the surrounding grasslands, blanketed the surrounding area with more than 39 square kilometers (15 square miles) of lava. Loose cinders, ash, and rock debris piled up to form a conical mountain, while lava erupted from fissures and vents closer to the volcano's base. Lava-capped mesas throughout the volcanic field, including the Barella, Johnson, and Raton mesas, bear witness to these powerful eruptions. Lava and rocks in the area have a wide range of compositions, including felsic rhyolite, intermediate **dacite** and andesite (*Figure 2.39*), and mafic basalt.

Many Quaternary deposits on the Great Plains of eastern Colorado and northeastern New Mexico are partly related to glacial processes. In the Southwest, glacial activity during the ice age was dominated by alpine glaciers in the mountains west of the Plains. Glacial erosion in the mountains resulted

Figure 2.39: A piece of vesicular andesite from the flank of the Sierra Grande shield volcano, the largest peak in New Mexico's Raton-Clayton Volcanic Field. Specimen is 7.3 centimeters (2.9 inches) wide.

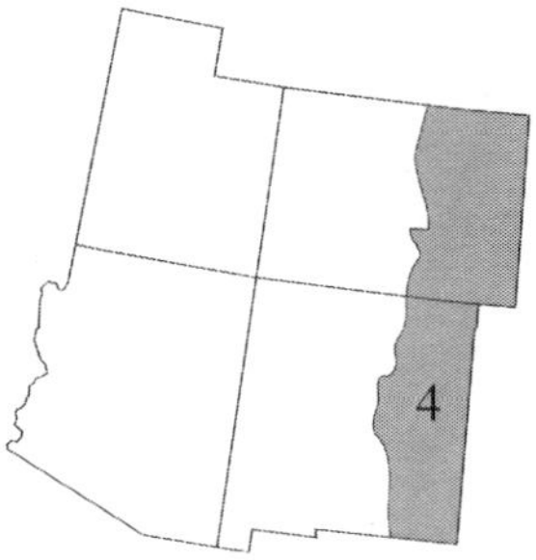

in vast amounts of fine- to medium-sized sediment being carried onto the Great Plains by rivers such as the Platte in Colorado. Finer sediment known as glacial flour, the product of glaciers grinding over rock, was picked up by the wind and deposited across large swaths of the region in layers called **loess**.

State Rocks, Minerals, and Gems

Arizona

Arizona has no state rock.

State mineral: copper
Copper mining has been a major industry in Arizona since the 19th century. The state has six of the 10 largest copper mines in the US, including the Morenci mine in Greenlee County, Arizona, which is the largest in the country. In 2007, Arizona produced 60% of the country's copper.

State **gem**: turquoise
Arizona is a major producer of turquoise, an opaque, blue-green mineral that has been popular in jewelry since Native American times. It contains copper and is often found in abundance near copper deposits. Arizona is currently the leading US state for turquoise production.

Colorado

State rock: Yule Marble
Yule Marble is found in the Yule Creek Valley in the West Elk Mountains of Colorado. This white, crystalline **marble** is composed of 99.5% calcite. It has been used in the construction of a number of prominent buildings and monuments.

State mineral: rhodochrosite
Rhodochrosite is a deep red to rose pink carbonate mineral found in association with Colorado's gold, silver, **lead**, **zinc**, and molybdenum **ores**. The world's largest rhodochrosite crystal is on display at the Denver Museum of Nature and Science. Because rhodochrosite is so specifically associated with Colorado, it was chosen as the state mineral over other common minerals such as gold and silver.

State gem: aquamarine
Aquamarine is a blue-colored variety of the mineral **beryl**. It was first discovered in 1881 on Mount Antero, in the Rockies of central Colorado. The state continues to be a major producer of gem-quality aquamarine.

gem *• a mineral that has aesthetic value and is often cut and polished for use as an ornament.*

marble *• a metamorphic rock composed of recrystallized carbonate minerals, most commonly calcite or dolomite.*

lead *• a metallic chemical element (Pb).*

zinc *• a metallic chemical element (Zn, atomic number 30).*

ore *• a type of rock that contains minerals with valuable elements, including metals, that are economically viable to extract.*

beryl *• a white, blue, yellow, green, or pink mineral, found in coarse granites and igneous rocks.*

State Rocks

bituminous coal *• a relatively soft coal containing a tarlike substance called bitumen, which is usually formed as a result of high pressure on lignite.*

hardness *• a physical property of minerals, specifying how hard the mineral is, and its resistance to scratching.*

opal *• a silicate gemstone lacking a rigid crystalline structure.*

New Mexico

New Mexico has no state rock or mineral.

State gem: turquoise

Large turquoise deposits in New Mexico are present in the metamorphosed volcanics of the Los Cerillos Mountains, where they have been mined since prehistoric times. Due to the processes under which it formed, this turquoise developed in over 75 different colors. New Mexico was the largest US producer of turquoise until the 1920s; Arizona and Nevada have since become larger producers.

Utah

State rock: coal

Coal is found in 17 of Utah's 29 counties, but coal mining is primarily concentrated in Emery and Carbon counties, where coals formed in the Uinta Basin. Most coal mined in Utah is **bituminous**.

State mineral: copper

The Kennecott's Bingham Canyon mine in the Oquirrh Mountains is the world's largest open-pit copper mine. Copper is a versatile metal widely valued for its capacity to conduct heat and electricity, and is used in electronics, transportation, plumbing, and alloys, among many other areas.

State gem: topaz

Topaz is a **hard,** semiprecious gem that is found in Beaver, Juab, and Tooele counties in Utah. At Topaz Mountain, which is in the Thomas Mountain Range in Juab County, topaz and other minerals such as beryl and **opal** are found in relatively high abundance in the cavities of Neogene-aged rhyolites.

Resources

Rock and Mineral Field Guides

Bonewitz, R. L., 2008, *Rock and Gem*, Dorling Kindersley, NY, 360 pp.

Chesterman, C. W., 1979, *National Audubon Society Field Guide to North American Rocks and Minerals*, Knopf, New York, 850 pp.

Dixon, D., & R. L. Bernor, 1992, *The Practical Geologist: The Introductory Guide to the Basics of Geology and to Collecting and Identifying Rocks*, Simon and Schuster, New York, 160 pp.

Hausel, W.D., 2014, *A Guide to Finding Gemstones, Gold, Minerals, and Rocks*, Gemhunter Publications: Gilbert, AZ, 370 pp.

Mitchell, J., 2008, *The Rockhound's Handbook, revised edition*, Gem Guides Book Company, Baldwin Park, CA, 299 pp.

Pellant, C., 2002, *Rocks & Minerals*, Dorling Kindersley (Smithsonian Handbooks), New York, 256 pp.

Prinz, M., G. Harlow, & J. Peters, eds., 1978, *Simon & Schuster's Guide to Rocks & Minerals*, Simon and Schuster, New York, 607 pp.

For additional resources on minerals, see the resources section in Chapter 5: Mineral Resources.

General Books and Websites on Rocks

Atlas of Igneous and Metamorphic Rocks, Minerals and Textures, University of North Carolina Geology Department, http://leggeo.unc.edu/Petunia/IgMetAtlas/mainmenu.html.

Raymond, L., 2007, *Petrology: The Study of Igneous, Sedimentary and Metamorphic Rocks*, Waveland Press, Long Grove, IL, 768 pp. [College textbook.]

Rocks, Kentucky Geological Survey and University of Kentucky, https://www.uky.edu/KGS/rocksmn/rocks.htm.

Rock Types, by S. Sepp, Sandatlas, http://www.sandatlas.org/rock-types.

Vernon, R. H., 2000, *Beneath Our Feet: The Rocks of Planet Earth*, Cambridge University Press, Cambridge, UK, 216 pp.

Books, Articles, and Websites on Rocks of Specific Areas of the Southwest

Multistate areas

Romaine, G., 2014, *Rocks, Gems, and Minerals of the Southwest*, FalconGuides, an imprint of Globe Pequot Press: Guilford, CT, 160 pp.

Rygle, K. & S. Pedersen, 2016, *Southwest Treasure Hunter's Gem and Mineral Guide: Where and How to Dig, Pan and Mine Your Own Gems and Minerals, 6th edition*, GemStone Press: Woodstock, VT, 224 pp.

Arizona

Anthony, J., 1995, *Mineralogy of Arizona, 3rd edition*, University of Arizona Press, Tucson, AZ, 508 pp.

Bearce, N., 2006, *Minerals, Fossils, and Fluorescents of Arizona: a Field Guide for Collectors*, Arizona Desert Ice Press, Tempe, AZ, 401 pp.

Blair, G., 2008, *Rockhounding Arizona: a Guide to 75 of the State's Best Rockhounding Sites*, Falcon Guides, Guilford, CT, 232 pp.

Kappele, W. & G. Warren, 2014, *Rockhounding Utah: a guide to the State's Best Rockhounding Sites*, FalconGuides, Guilford, CT, 208 pp.

Lynch, B. & D. Lynch, 2010, *Arizona Rocks & Minerals: a Field Guide to the Grand Canyon State*, Adventure Publications, Cambridge, MN, 252 pp.

2 Rocks

Resources

Colorado

Eckel, E., R., Cobban, & S. Mosburg, S., 1997, *Minerals of Colorado*, Fulcrum Publishers, Golden, CO, 676 pp.

Eckel, E.B., & Mendenhall, C., 2015, *Minerals of Colorado: A 100-Year Record*, CreateSpace Independent Publishing Platform, North Charleston, SC, 484 pp.

Kappele, W. & C. Kappele, 2004, *Rockhounding Colorado*, Falcon Press, Helena, MT, 256 pp.

Mitchell, J., 2007, *Gem Trails of Colorado*, Gem Guides Book Company, Baldwin Park, CA, 224 pp.

Romaine, G., 2014, *Rocks, Gems, and Minerals of the Rocky Mountains*, FalconGuides, an imprint of Globe Pequot Press, Guilford, CT, 176 pp.

New Mexico

Lynch, D. & Lynch, B., 2014, *New Mexico Rocks & Minerals: a Field Guide to the Land of Enchantment*, Adventure Publications, Cambridge, MN, 320 pp.

Mitchell, J., 2010, *Gem Trails of New Mexico*, Gem Guides Book Company, Baldwin Park, CA, 280 pp.

Rockhounding Guide to New Mexico, New Mexico Bureau of Geology and Mineral Resources, 111 pp., http://geoinfo.nmt.edu/publications/guides/rockhound/RockhoundGuide.pdf.

Vaskys, R. & M. Freed, 2008, *Rockhounding New Mexico: a Guide to 140 of the State's Best Rockhounding Sites*, Globe Pequot Press, Guilford, CT, 272 pp.

Voynick, S., 1997, *New Mexico Rockhounding: a Guide to Minerals, Gemstones, and Fossils*, Mountain Press Publishing Company, Missoula, MT, 309 pp.

Utah

Kappele, W. & G. Warren, 2014, *Rockhounding Utah: a Guide to the State's Best Rockhounding Sites*, FalconGuides., Guilford, CT, 208 pp.

Mitchell, J., 2006, *Gem Trails of Utah*, Gem Guides Book, Baldwin Park, CA, 232 pp.

Wilson, J. R., 1995, *A Collector's Guide to Rock, Mineral, & Fossil Localities of Utah*, Utah Geological Survey Miscellaneous Publication, 148 pp.

Chapter 3: Fossils of the Southwestern US

Fossils (from the Latin word *fossilis*, meaning "dug up") are the remains or traces of organisms that lived in the geologic past (older than the last 10,000 years), now preserved in the Earth's **crust**. Most organisms never become fossils, but instead decompose after death, and any hard parts are broken into tiny fragments. In order to fossilize, an organism must be buried quickly before it is destroyed by **weathering**, decomposed, or eaten by other organisms. This is why fossils are found almost exclusively in sediment and **sedimentary rocks**. **Igneous rocks**, which form from cooling **magma** or **lava**, and **metamorphic rocks**, which have been altered by **heat** and pressure, are unlikely to contain fossils (but may, under special circumstances). Different fossils are found in different regions because of the presence of rocks deposited at different times and in a variety of environments. Since rapid burial in sediment is important for the formation of fossils, many fossils are from marine environments, where sediments are more likely to accumulate.

Fossils come in many types. Those that consist of an actual part of an organism, such as a bone, shell, or leaf, are known as **body fossils**; those that record the actions of organisms, such as footprints and burrows, are called **trace fossils**. Body fossils may be preserved in a number of ways. These include preservation of the original **mineral** skeleton of an organism, mineral **replacement** (chemical replacement of the material making up a shell by a more stable mineral), **recrystallization** (replacement by a different **crystal form** of

Lagerstätten

The "soft" tissues of an organism, such as skin, muscles, and internal organs, are typically not preserved as fossils. Exceptions to this rule occur when conditions favor rapid burial and mineralization or very slow decay. The absence of oxygen and limited disruption of the sediment by burrowing are both important for limiting decay in those deposits where soft tissues are preserved. The Southwestern states contain several examples of such exceptional preservation, also called Lagerstätten, including trilobites and other marine animals preserved in Utah's Cambrian shales, and the Cenozoic insects and plants preserved at Florissant, Colorado.

crust • *the uppermost, rigid outer layer of the Earth.*

weathering • *the breakdown of rocks by physical or chemical means.*

sedimentary rock • *rock formed through the accumulation and consolidation of grains of broken rock, crystals, skeletal fragments, and organic matter.*

igneous rocks • *rocks derived from the cooling of magma underground or molten lava on the Earth's surface.*

magma • *molten rock located below the surface of the Earth.*

lava • *molten rock located on the Earth's surface.*

mineral • *a naturally occurring solid with a specific chemical composition and crystalline structure.*

CHAPTER AUTHORS

Warren D. Allmon
Richard A. Kissel

3 Fossils

Overview

clay • *the common name for a number of very fine-grained, earthy materials that become plastic (flow or change shape) when wet.*

filter feeder • *an animal that feeds by passing water through a filtering structure that traps food.*

ripple marks • *surface features created when sediment deposits are agitated, typically by water currents or wind.*

intertidal • *areas that are above water during low tide and below water during high tide.*

permineralization • *a fossilization method in which empty spaces (such as in a bone or shell) are filled by minerals.*

extinction • *the end of species or other taxonomic groups, marked by death of the last living individual.*

stratigraphy • *the branch of geology specifically concerned with the arrangement and age of rock units.*

Discovering Ancient Environments

The kinds of animals and plants living in a particular place depend on the local environment. The fossil record preserves not only fossil organisms, but also evidence of what their environments were like. By studying the geological and biological information recorded in a rock that contains a fossil, scientists can determine some aspects of the paleoenvironment.

Grain size and composition of the rock can tell us what type of sediment surface the animal lived on, what the water flow was like, and whether the sediment was transported in a current. Grain size also tells us about the clarity of the water. Fine-grained rocks such as shales are made of tiny particles of silt or *clay* that easily remain suspended in water. Thus, a fossil found in shale might have lived in muddy or very quiet water. *Filter-feeding* organisms, such as clams or corals, are not usually found in muddy water because the suspended sediment can clog their filters.

Sedimentary structures, such as asymmetrical *ripple marks* and cross-beds, can indicate that the organism lived in moving water. Mud cracks or symmetrical ripples are characteristic of shoreline or *intertidal* environments.

Broken shells or *concentrated layers of shells* may indicate transportation and accumulation by waves or currents.

Color of the rock may indicate the amount of oxygen in the water. If there is not enough oxygen in the water, organic material (carbon) in sediments will not decompose, and the rock formed will be dark gray or black in color.

the same chemical compound), **permineralization** (filling of empty spaces in a bone or shell by minerals), and molds and casts, which show impressions of the exterior or interior of a shell. **Chemical fossils** are chemicals produced by an organism that leave behind an identifiable trace in the geologic record, and it is these fossils that provide some of the oldest evidence for life on Earth.

Fossils 3

Paleontologists use fossils as a record of the history of life. They tell us that an incredible multitude of organisms lived prior to the species that we see on Earth today; that most species that ever lived have become extinct; and that living things have changed through evolution over time, from one species into another, and adapted to changing environments. Fossilized organisms are also extremely useful in understanding the ancient environment that existed when they were alive. The study of the relationships of fossil organisms to one another and their environment is called **paleoecology**.

Fossils provide the most important tool for dating the rocks in which they are preserved. Because species only exist for a certain amount of time before going **extinct**, their fossils only occur in rocks of a certain age. The relative age of such fossils is determined by their order in the stacks of layered rocks that make up the **stratigraphic** record (older rocks are on the bottom and younger rocks on the top—a principle called **superposition**). Such fossils are known as **index fossils.** The most useful index fossils are abundant, widely distributed, easy to recognize, and occur only during a narrow time span. This use of fossils to determine relative age in geology is called **biostratigraphy**. The **geologic time scale** is in part based on sequences of fossils correlated from around the world.

Some of the most useful index fossils are hard-shelled organisms that were once part of the marine plankton.

Ancient Biodiversity

Since life began on Earth more than 3.7 billion years ago, it has continuously grown more abundant and diverse. It wasn't until the beginning of the **Cambrian** period, around 541 million years ago, that *complex life*—living things with cells that are differentiated for different tasks—became predominant. This event at the beginning of the Cambrian, called the Cambrian Explosion, resulted in the emergence of most major animal phyla. The diversity of life has generally increased through time since then. Measurements of the number of different kinds of organisms—for example, estimating the number of species alive at a given time—are used when attempting to describe Earth's **biodiversity**. With a few exceptions, the rate at which new species evolve is significantly greater than the rate of extinction.

Most species have a lifespan of several million years; rarely do species exist longer than 10 million years. The extinction of a species is a normal event in the history of life. There are, however, intervals of time during which extinction rates are unusually high, in some cases at a rate of 10 or 100 times the normal pace. These intervals are known as **mass extinctions**. There were five particularly devastating mass extinctions in geologic history (*Figure 3.1*), and these specific events have helped to shape life through time. Unfortunately, this is not just a phenomenon of the past—it is estimated that the extinction rate on Earth right now may be as much as 1000 times higher than normal, due mostly to human activity, and that we are currently experiencing a sixth mass extinction event.

Overview

Law of Superposition • *the geologic principle that states that unless rock layers have been overturned or intruded, older rocks are found at the bottom and younger rocks are found at the top of a sedimentary sequence.*

geologic time scale • *a standard timeline used to describe the age of rocks and fossils, and the events that formed them.*

Cambrian • *a geologic time period lasting from 541 to 485 million years ago.*

biodiversity • *the number of kinds of organisms at any given time and place.*

mass extinction • *the extinction of a large percentage of the Earth's species over a relatively short span of geologic time.*

3 Fossils

Overview

erosion • *the transport of weathered materials.*

stromatolite • *regularly banded accumulations of sediment created by the trapping and cementation of sediment grains in bacterial mats.*

carbonate rocks • *rocks formed by accumulation of calcium carbonate, often made of the skeletons of aquatic organisms.*

cyanobacteria • *a group of bacteria, also called "blue-green algae," that obtain their energy through photosynthesis.*

Proterozoic • *a geologic time interval that extends from 2.5 billion to 541 million years ago.*

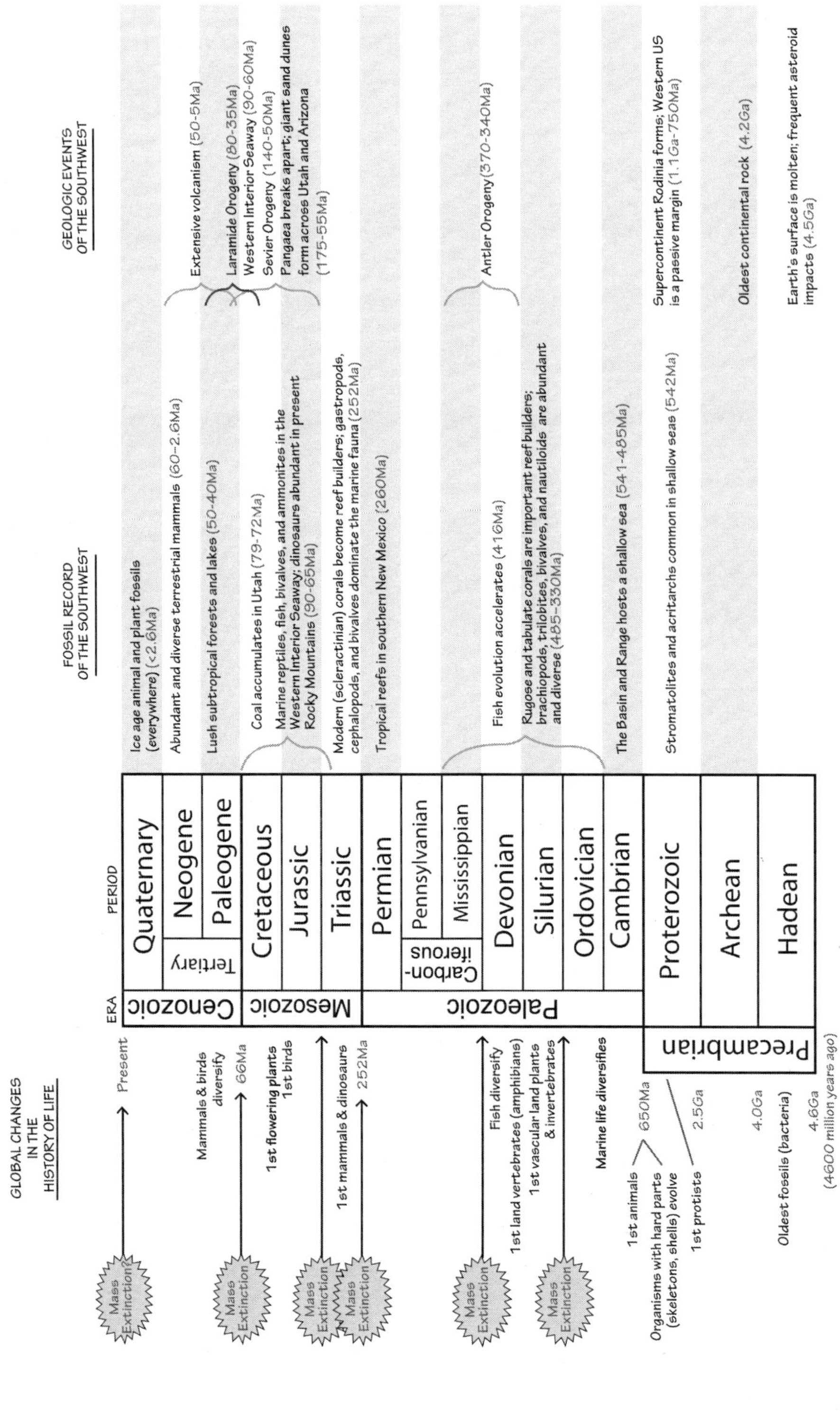

Figure 3.1. The history of life in relation to global and regional geological events and the fossil record of the Southwestern US. (Time scale is not to scale).

Fossils 3

Region 1

acritarch • a microscopic organic-walled fossil, often approximately spherical in shape, found in rocks from over 2 billion to about 500 million years ago.

Phanerozoic • a generalized term used to describe the entirety of geological history after the Precambrian, from 541 million years ago to the present.

trilobite • an extinct marine invertebrate animal characterized by a three-part body and a chitinous exoskeleton divided longitudinally into three lobes.

brachiopod • a marine invertebrate animal characterized by upper and lower calcareous shell valves joined by a hinge, and a crown of tentacles (lophophore) used for feeding and respiration.

Fossils of the Southwestern US

The rocks of the Southwest showcase an excellent fossil record, as the area's sparse vegetation affords extensive rock exposures. A large proportion of these exposed rocks are sedimentary and fossil-bearing, preserving a tremendous variety of fossils and providing glimpses of what life was like over the last 1.2 billion years. Fossils can be found in nearly every part of the Southwest, representing most major categories of organisms and most periods of geologic time. The history of life in the Southwest has been pieced together from the fossil record and different layers of rock in many different areas. Particular fossil organisms lived only in certain environments, and these changing environments did not exist continuously through time, nor were they all necessarily preserved in the rock record. Nowhere in the Southwest (or anywhere) is a complete record of rock from every period preserved. Not all sediment ends up as rock, and likewise, rock that formed long ago may have been **eroded** away. Not all organisms are preserved as fossils, and rocks that contain fossils may have been destroyed by erosion, or they may still be buried well below the current surface of the Earth, out of sight from paleontologists.

In the remainder of this chapter, we will highlight the major types of fossils present in most of the geologic periods represented by rocks in each state. The references at the end of the chapter should be consulted for details, especially for identifying particular fossils you might find.

Fossils of the Colorado Plateau
Region 1

The rocks of the Colorado Plateau represent a diversity of environmental conditions over hundreds of millions of years. These rocks are well exposed in the many canyons that occur across this region, most spectacularly in the Grand Canyon in northern Arizona (*Figure 3.2*). The oldest fossils known from the Colorado Plateau region are **stromatolites**—layered domes of **carbonate** sediment formed by mats of bacteria known as **cyanobacteria**. These are found in a layer called the Bass Limestone, exposed near the bottom of the Grand Canyon (*Figure 3.3*). These stromatolites date to the mid-**Proterozoic**, around 1.2 billion years ago. Late Proterozoic rocks known as the Chuar Group (around 740 million years old) contain a variety of microfossils. Most are referred to as **acritarchs**, which are spherical objects 1–5 millimeters in diameter, thought by many paleontologists to be a resting state of single-celled algae (*Figure 3.4*). Several levels within the Chuar Group also contain stromatolites.

See Chapter 2: Rocks to learn more about Precambrian rocks of the Colorado Plateau.

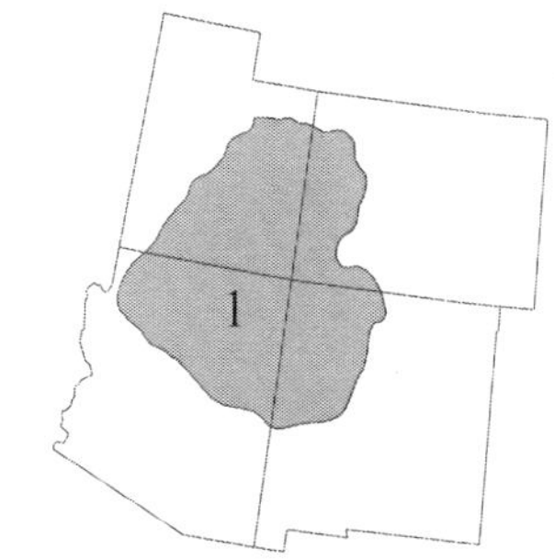

The earliest **Phanerozoic** sediments recorded in the Grand Canyon area are known as the Tonto Group. The fossils in these rocks are usually not very well preserved, but they come in great variety, dominated by **trilobites** (*see Figure 3.43*), of which more than 50 species have been reported. **Brachiopods** are

3 Fossils

Region 1

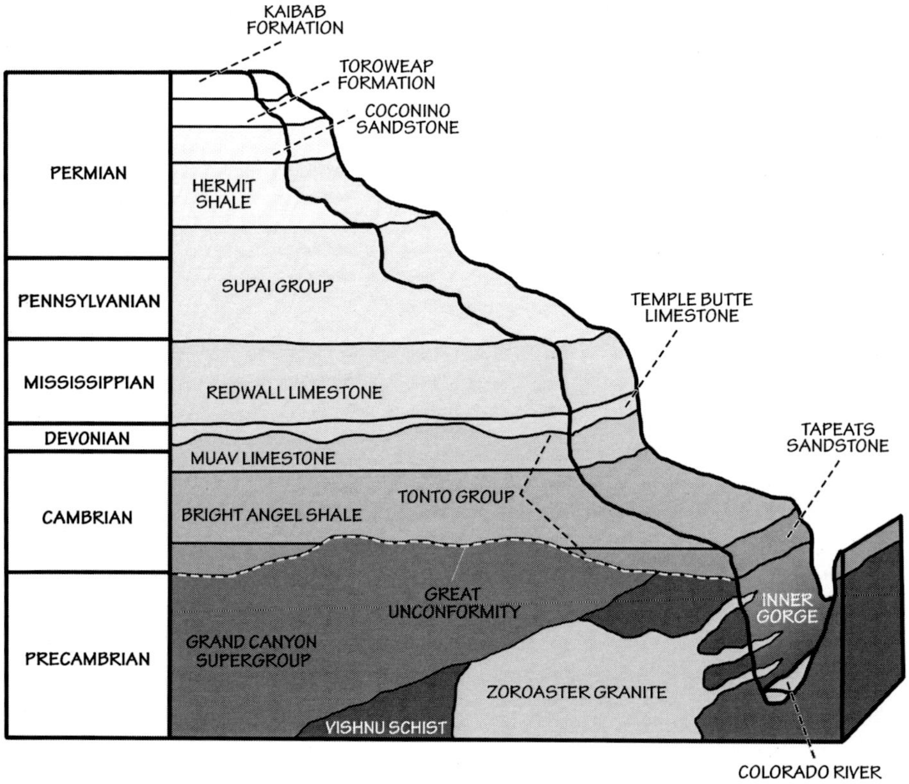

Figure 3.2: Major Proterozoic and Paleozoic stratigraphic units of the Grand Canyon and Colorado Plateau.

Figure 3.3: Stromatolites from the mid-Proterozoic Bass Limestone of the Grand Canyon.

Stromatolites

Stromatolites are regularly banded accumulations of sediment created by the trapping and cementation of sediment grains in bacterial mats (especially photosynthetic cyanobacteria). Cyanobacteria secrete a sticky mucus that binds settling sediment; their photosynthesis creates a chemical environment that leads to the precipitation of calcium carbonate. The calcium carbonate then hardens the underlying layers of bacterial mats, while the living bacteria move upward so that they are not buried. Over time, this cycle of growth combined with sediment capture creates a rounded structure filled with banded layers.

Stromatolites peaked in abundance around 1.25 billion years ago, and likely declined due to predation by grazing organisms. Today, stromatolites exist in only a few locations worldwide, such as Shark Bay, Australia. Modern stromatolites form thick layers only in stressful environments, such as very salty water, that exclude animal grazers.

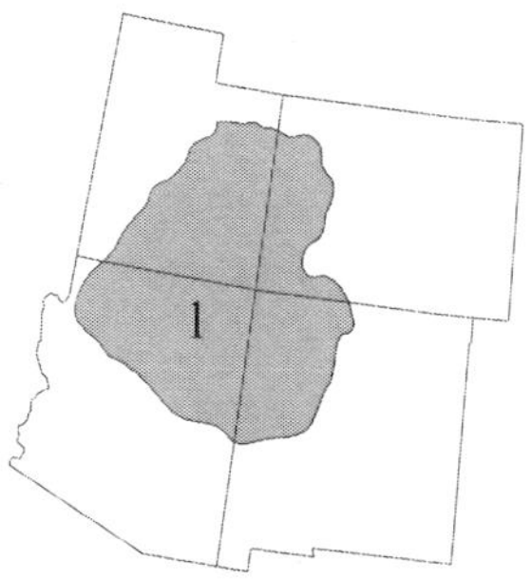

3 Fossils

Region 1

sponge • a marine invertebrate belonging to the Phylum Porifera, and characterized by a soft shape with many pores and channels for water flow.

echinoderm • a member of the Phylum Echinodermata, which includes starfish, sea urchins, and crinoids.

Ordovician • a geologic time period spanning from 485 to 443 million years ago.

Silurian • a geologic time period spanning from 443 to 419 million years ago.

Devonian • a geologic time period spanning from 419 to 359 million years ago.

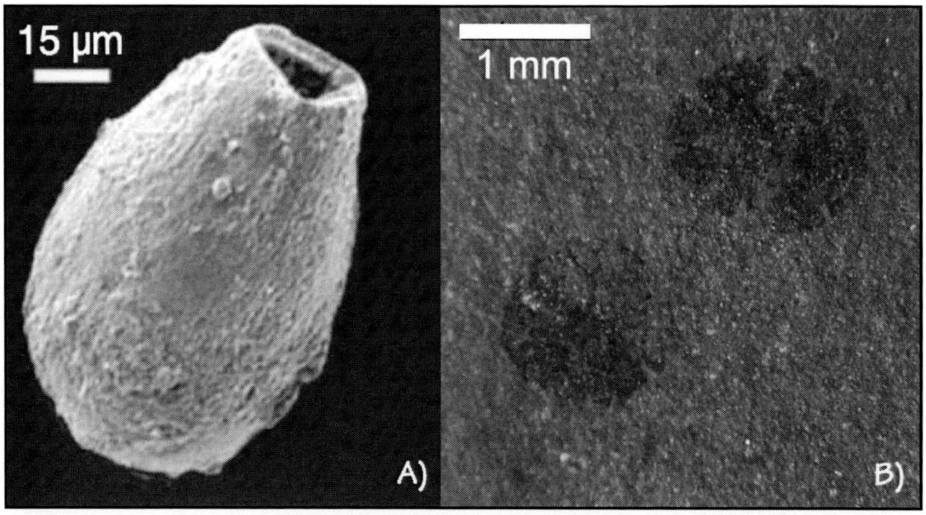

Figure 3.4: Microfossils from the late Proterozoic Chuar Group. A) A "vase-shaped microfossil" (VSM), thought to be the shell of an amoeba-like, single-celled organism (protist). B) Chuaria, *a common acritarch fossil of uncertain affinities, most likely the resting stage of a single-celled eukaryote (protist).*

also frequently found in these rocks (*Figure 3.5*), and, more rarely, **sponges** and **echinoderms** similar to those found in the Cambrian of western Utah (*see Figure 3.44*).

Ordovician and **Silurian** rocks are absent in the southern part of the Colorado Plateau. **Devonian** rocks are present in the area, but are only moderately fossiliferous. The Devonian Temple Butte Formation, exposed in the Grand Canyon, contains poorly preserved brachiopods, corals, **crinoids**, and also occasionally the remains of **placoderms**—an extinct group of fishes that dominated the waters of the Devonian. Their name means "plated skin," and they were characterized by bony armor covering the front part of the body and pectoral fins. Fragments of the bottom-dwelling placoderm *Bothriolepis* (*Figure 3.6*) have been recovered from Devonian rocks in northern and central Arizona, and also in central Colorado, northwest of Denver (Eagle and Garfield counties). Placoderms became extinct during a mass extinction event near the end of the Devonian period. **Conodonts**—small, primitive, eel-like vertebrates—are also known from Devonian and later rocks throughout the region.

Mississippian rocks exposed in the Grand Canyon, including the Redwall Limestone and Surprise Canyon Formation, contain abundant brachiopods, corals, **bryozoans**, **gastropods**, bivalves, nautiloid cephalopods, crinoids, and shark teeth (*Figures 3.7 and 3.8; see also Figure 3.50*). In at least one Redwall locality, straight nautiloids more than 60 centimeters (24 inches) long have been found. Some **Pennsylvanian** layers in the Colorado Plateau also accumulated in terrestrial environments, including swamps, rivers, lakes, and floodplains. Some of these sediments include significant **coal** deposits.

See Chapter 6: Energy to learn more about coal resources in the Southwest.

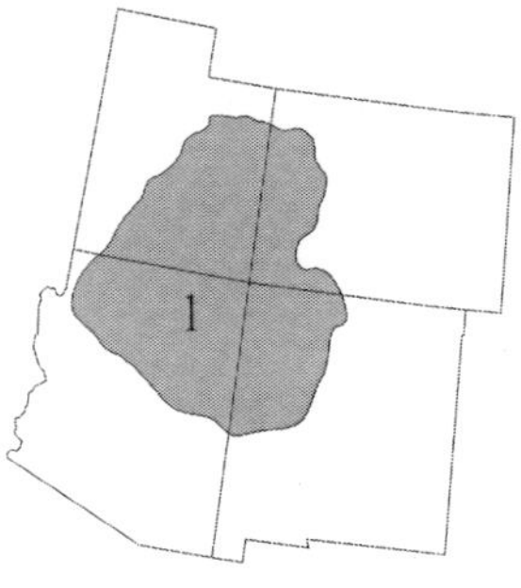

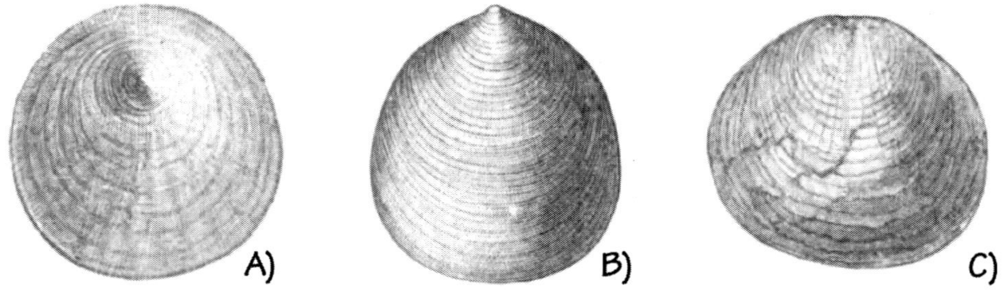

Figure 3.5. Cambrian brachiopods from the Colorado Plateau, each approximately 1 centimeter (0.4 inches) across. A) Acrothele subsidua. *B)* Obolella chromatica. *C)* Lingulella ella.

Conodonts

Conodonts are tiny, tooth-shaped microfossils (0.2–5 millimeters long), found in Cambrian- through Triassic-aged marine rocks. They have long been among the most important index fossils in these rocks, allowing the latter to be dated through biostratigraphy. For many years, paleontologists variously thought they were fragments of arthropods, fish teeth, mollusks, or even plants. In 1983 the discovery of a whole conodont animal in Scotland revealed that they belonged to small, fish-like animals that were distant relatives of bony fish.

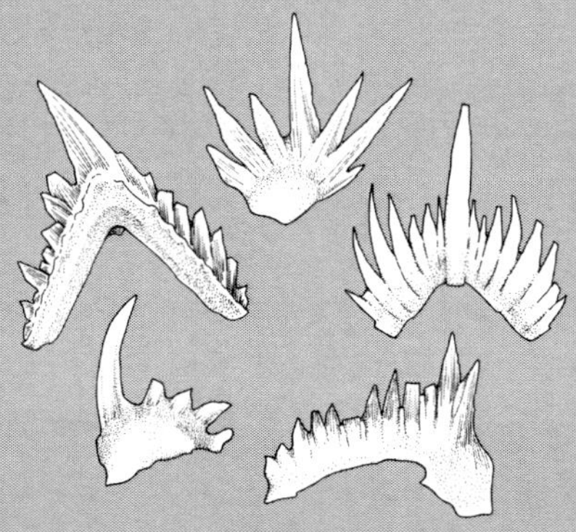

Isolated conodont elements (Silurian).

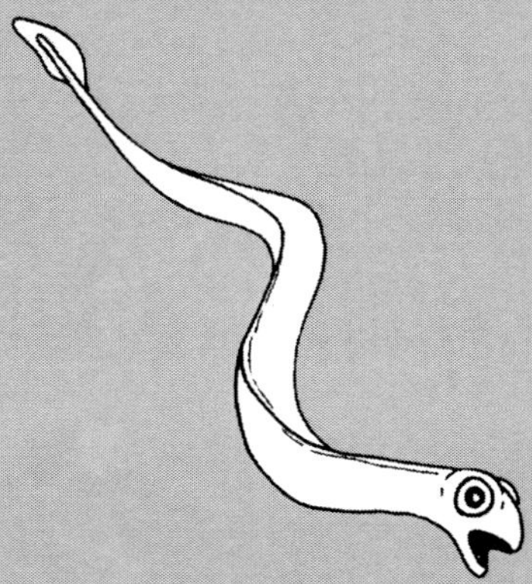

Restoration of a live conodont animal. Length 2–4 centimeters (1–2 inches).

Region 1

crinoid • *a marine invertebrate animal characterized by a head (calyx) with a mouth surrounded by feeding arms.*

placoderms • *an extinct class of heavily armored fishes.*

Mississippian • *a subperiod of the Carboniferous, spanning from 359 to 323 million years ago.*

bryozoan • *a marine or freshwater colonial invertebrate character-ized by an encrusting or branching calcareous skeleton from which multiple individuals (zooids) extend from small pores to filter-feed.*

gastropod • *a marine, freshwater, or terrestrial invertebrate animal characterized by a single, coiled, calcareous shell, a muscular foot for gliding, and internal asymmetry caused by torsion.*

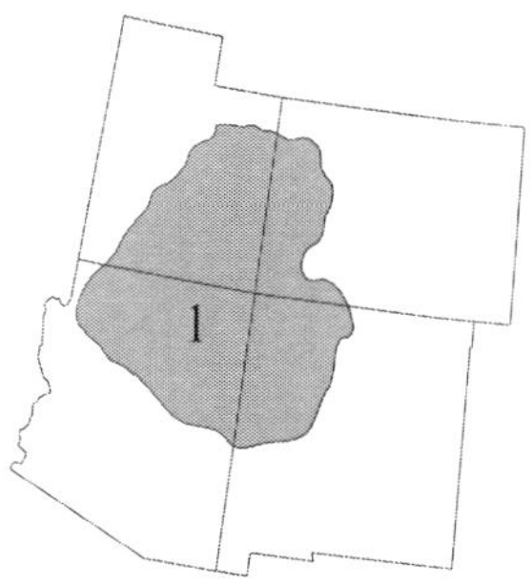

3 Fossils

Region 1

lycopod • *an extinct, terrestrial tree characterized by a tall, thick trunk covered with a pattern of diamond-shaped leaf scars, and a crown of branches with simple leaves.*

seed fern • *an extinct terrestrial plant characterized by a fern-like appearance, but bearing seeds instead of spores.*

horsetail • *see sphenopsid: a terrestrial plant belonging to the family Equisetaceae, characterized by hollow, jointed stems with reduced, unbranched leaves at the nodes.*

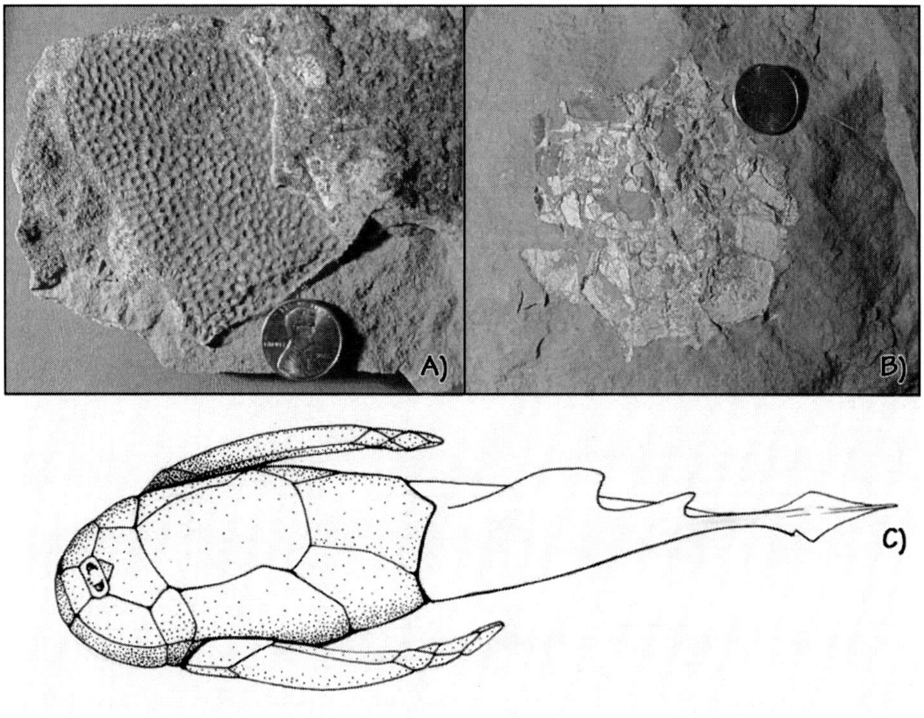

Figure 3.6: Placoderm fish from Devonian rocks near Payson, in northern Arizona. A) Textured dermal plates. B) Ventral median dermal plate. C) Life restoration of a placoderm, Bothriolepis, *approximately 30 centimeters (12 inches) long.*

Figure 3.7: Mississippian marine fossils from the Redwall and Surprise Canyon formations. A) Solitary rugose coral, Zaphrentis, *about 2.5 centimeters (1 inch) tall. B) Colonial rugose coral,* Michelinia, *about 4 centimeters (1.6 inches) wide. C) Brachiopod,* Schizophoria, *about 2.5 centimeters (1 inch) wide. D) Brachiopod,* Punctospirifer, *about 4 centimeters (1.6 inches) wide. E) Gastropod,* Straparolus, *about 1 centimeter (0.4 inches) wide. F) Brachiopod,* Buxtonia, *about 4 centimeters (2 inches) wide.*

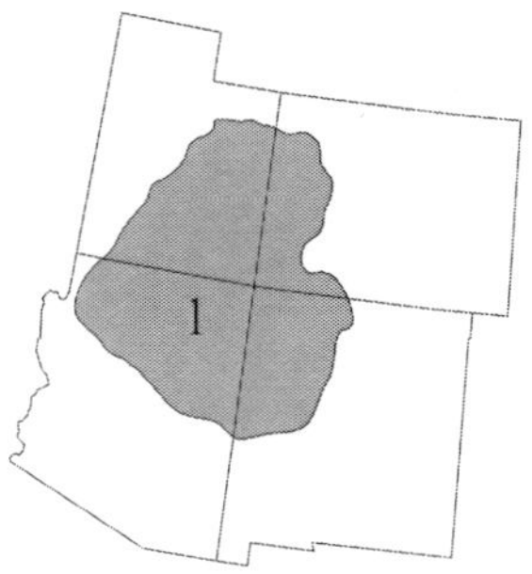

Figure 3.8: Fenestellid bryozoan from the Redwall Limestone, Grand Canyon; 5 centimeters (2 inches) wide.

Permian • *the geologic time period lasting from 299 to 252 million years ago.*

trackway • *a set of impressions in soft sediment, usually a set of footprints, left by an animal.*

foraminifera • *a class of aquatic protists that possess a calcareous or siliceous exoskeleton.*

Bryozoans

Bryozoans are colonial invertebrates, many of which build elaborate skeletons of calcium carbonate. Bryozoans are common in today's oceans, where they are frequently found encrusting rocks or shells. During the Paleozoic era, however, bryozoans commonly grew off of the sea floor as erect structures. After they died, their skeletons accumulated into thick beds of limestone. Although they do not appear to be, bryozoans are actually closely related to brachiopods—both groups have the same distinctive feeding and respiratory structure, the lophophore.

The Surprise Canyon (which extends into the early Pennsylvanian) also contains layers with terrestrial plant fossils, including **lycopods**, **seed ferns**, and **horsetails** (*Figure 3.9*).

Pennsylvanian to **Permian** marine rocks in the southern Colorado Plateau include the Supai Group in northern Arizona and the Honaker Trail Formation in southern Utah. In the Grand Canyon and the surrounding area, the Supai Group contains trace fossils (burrows, **trackways**, and other marks) in almost all layers. A number of beds also contain abundant and diverse shelly fossils, including brachiopods (*Figure 3.10*) and **foraminifera**.

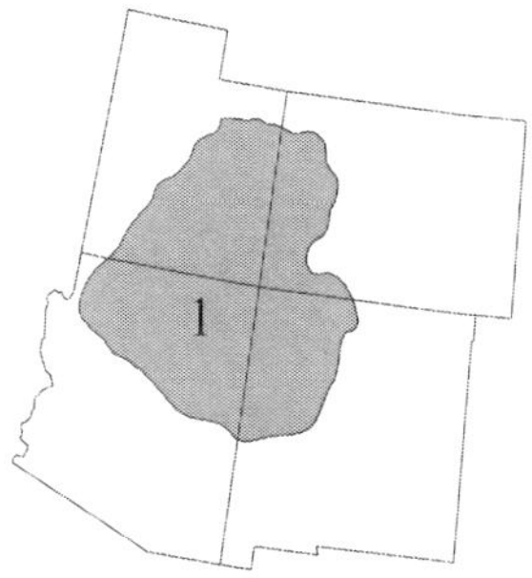

3 Fossils

Region 1

cross-bedding • *layering within a bed in a series of rock strata that does not run parallel to the plane of stratification.*

sand • *rock material in the form of loose, rounded, or angular grains, and formed as a result of the weathering and decomposition of rocks.*

arthropod • *an invertebrate animal, belonging to the Phylum Arthropoda, and posessing an external skeleton (exoskeleton), body segments, and jointed appendages.*

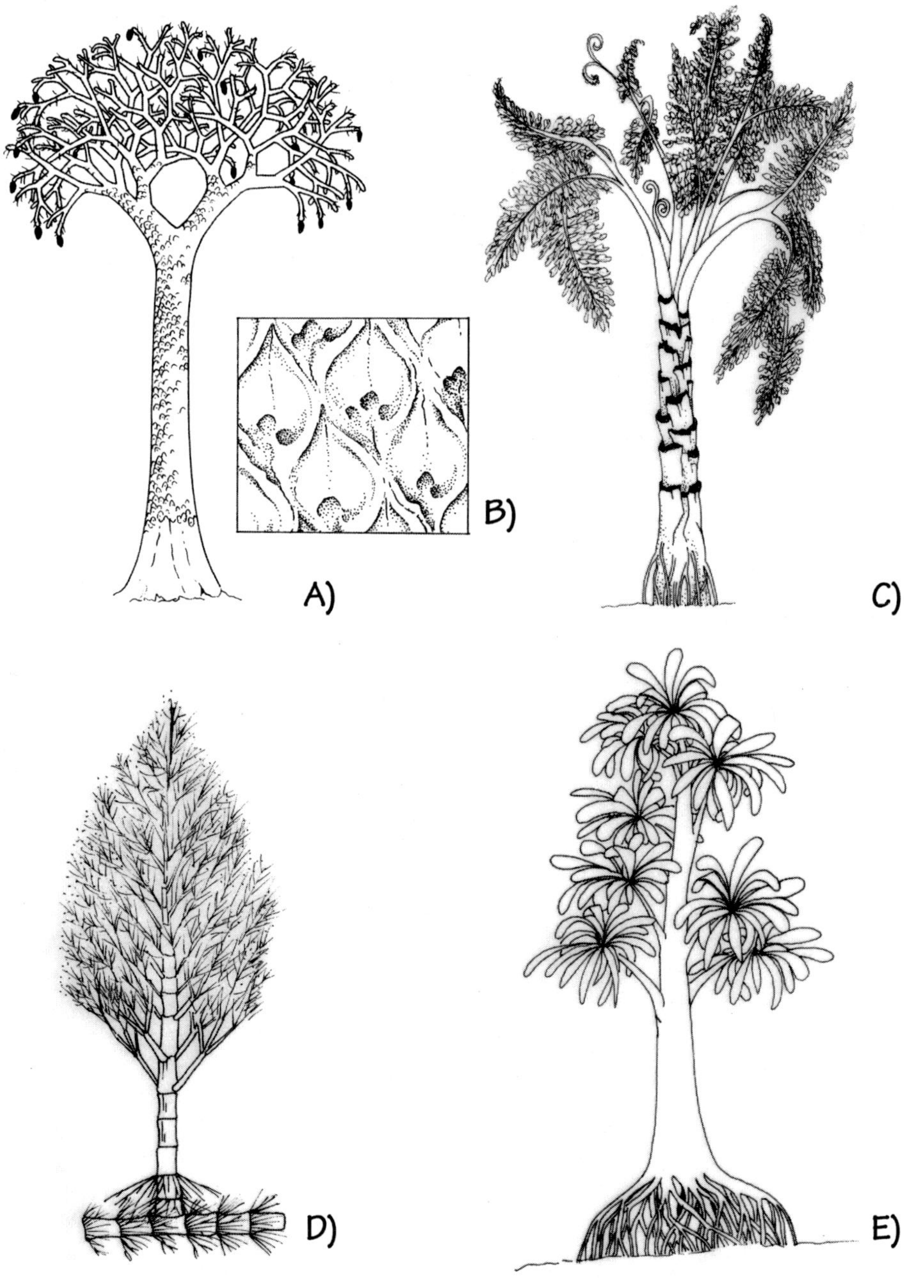

Figure 3.9: Restorations of Pennsylvanian coal swamp plants. A) Lepidodendron, *a lycopod (club moss), reached 30 meters (100 feet) tall. B) Close-ups of leaf scars on a* Lepidodendron *trunk. C)* Medullosa, *a tree fern, reached 10 meters (35 feet) tall. D)* Calamites, *a sphenopsid (horsetail), reached 20 meters (65 feet) tall. E)* Cordaites, *a gymnosperm seed plant, reached 10 meters (35 feet) tall.*

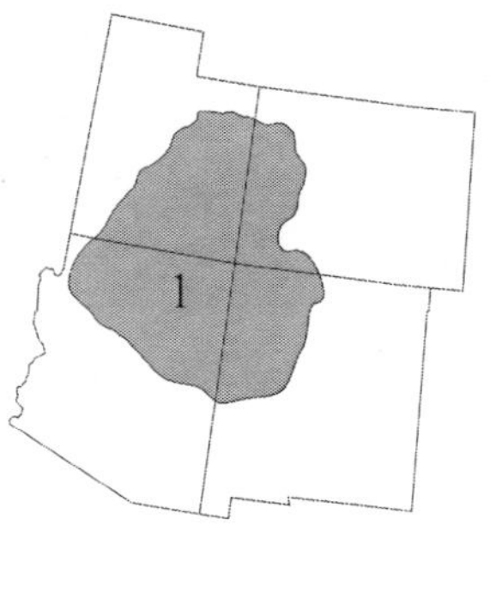

The Permian-aged Coconino Sandstone is a 90-meter-thick (300-foot-thick) terrestrial deposit, with **cross-bedded** layers that are characteristic of **sand** dunes. More than 20 types of vertebrate footprints have been identified within

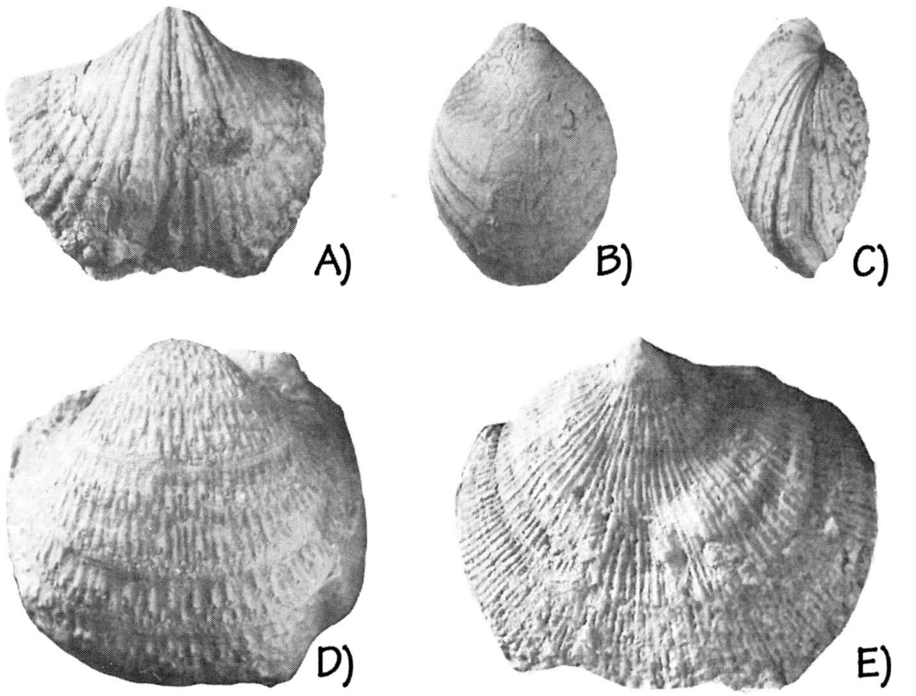

Figure 3.10: Pennsylvanian brachiopods from northern Arizona. A) Anthracospirifer, *2 centimeters (0.8 inches) wide. B) and C)* Composita, *1.5 centimeters (0.7 inches) long. D)* Pulchratia, *2 centimeters (0.8 inches) wide. E)* Orthotetes, *4 centimeters (1.6 inches) wide.*

this unit (*Figure 3.11*), made by a variety of amphibians and reptiles that trekked through damp sand some 280 million years ago. A variety of other tracks are attributed to **arthropods**, including spiders, scorpions, beetles, and millipedes. One of the most important deposits of early Permian vertebrate fossils in North America is known from the Arroyo del Agua area, west of Abiquiu, Rio Arriba County, in northern New Mexico. This site has produced the skeletal remains of many large terrestrial amphibians and reptiles, including *Sphenacodon, Eryops, Ophiacodon,* and *Diadectes* (*Figure 3.12*). *Eryops* and the sail-backed predator *Dimetrodon* are also abundant in the Permian rocks of southern Utah, including Monument Valley and the Valley of the Gods. The small reptile *Seymouria* is found in Permian rocks south of Moab, Utah.

See Chapter 2: Rocks to learn how cross-bedded layers form and where these layers can be found in the Southwest.

Above the Coconino Sandstone is another Permian marine unit, the Kaibab Limestone, which contains abundant fossils from marine vertebrates and invertebrates, including brachiopods, sponges, corals, crinoids, echinoids, gastropods, **bivalves**, nautiloid and **ammonoid cephalopods**, conodonts, and **shark** teeth (*Figure 3.13*). Ammonoid cephalopods are especially important for biostratigraphy in this and other Permian marine layers.

bivalve *• a marine or fresh-water invertebrate animal characterized by right and left calcareous shells (valves) joined by a hinge.*

ammonoid *• a member of a group of extinct cephalopods belonging to the Phylum Mollusca, and posessing a spiraling, tightly coiled shell characterized by ridges, or septa.*

cephalopod *• a marine invertebrate animal characterized by a prominent head, arms, tentacles with suckers, and jet propulsion.*

shark *• a large fish characterized by a cartilaginous skeleton and five to seven gill slits on the side of the head.*

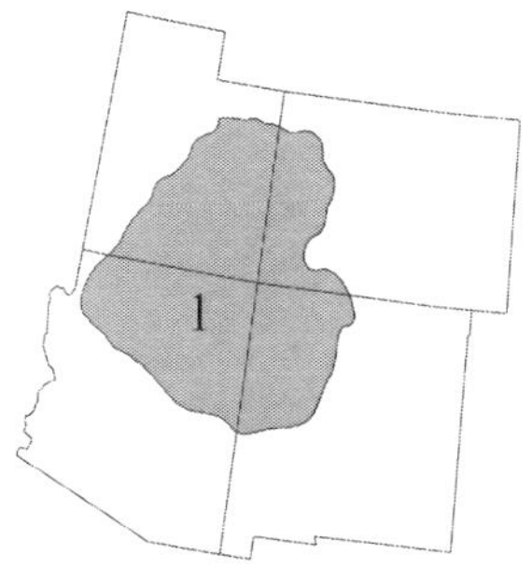

3 Fossils

Region 1

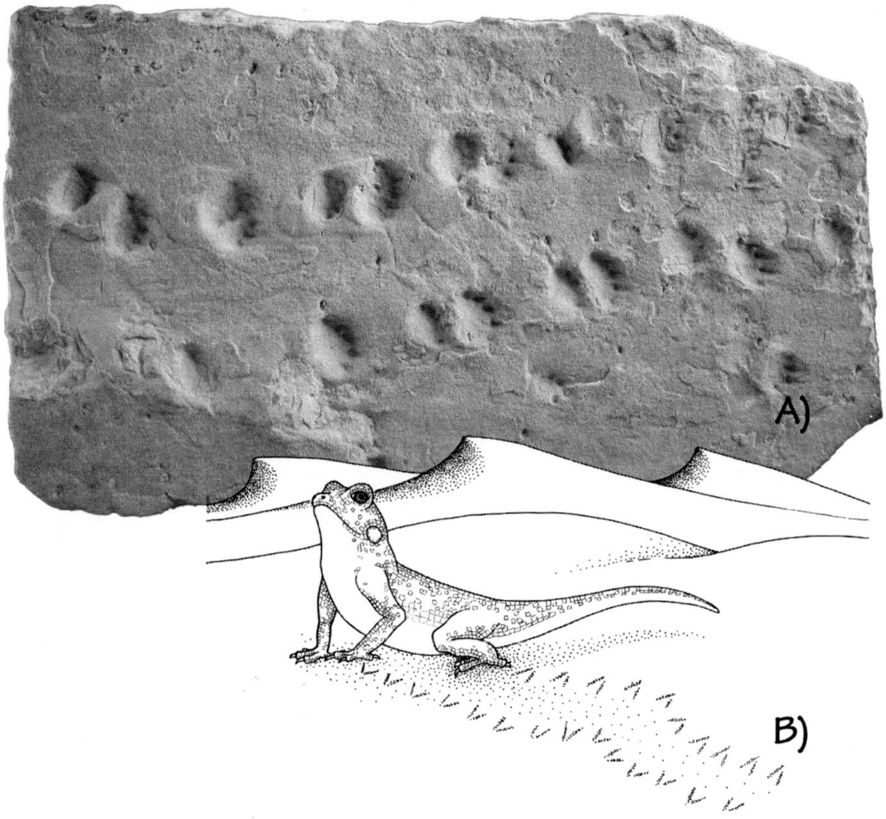

Figure 3.11: Permian dune deposits. A) Tetrapod tracks from the Coconino Sandstone. B) Life restoration of the track-maker, a small reptile.

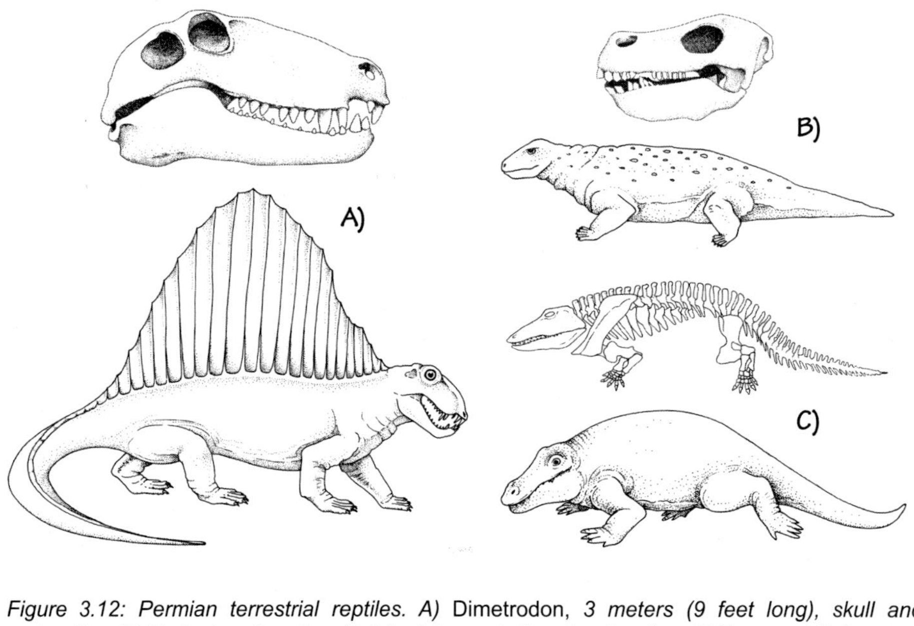

Figure 3.12: Permian terrestrial reptiles. A) Dimetrodon, *3 meters (9 feet long), skull and restoration. B)* Diadectes, *2 meters (6.5 feet) long, skull and restoration. C)* Eryops, *1.5–2 meters (5–6.5 feet) long, skeleton and restoration.*

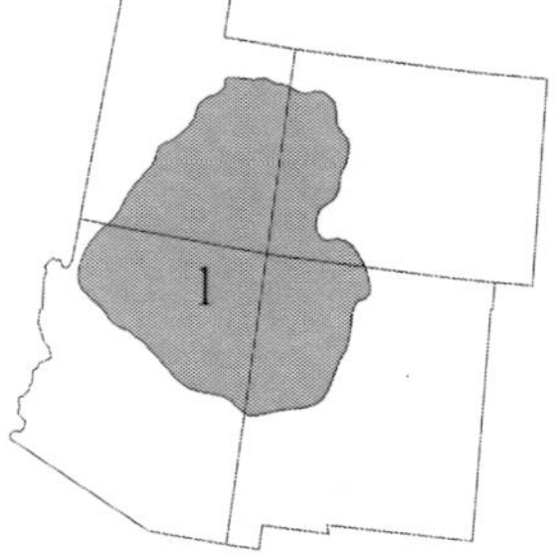

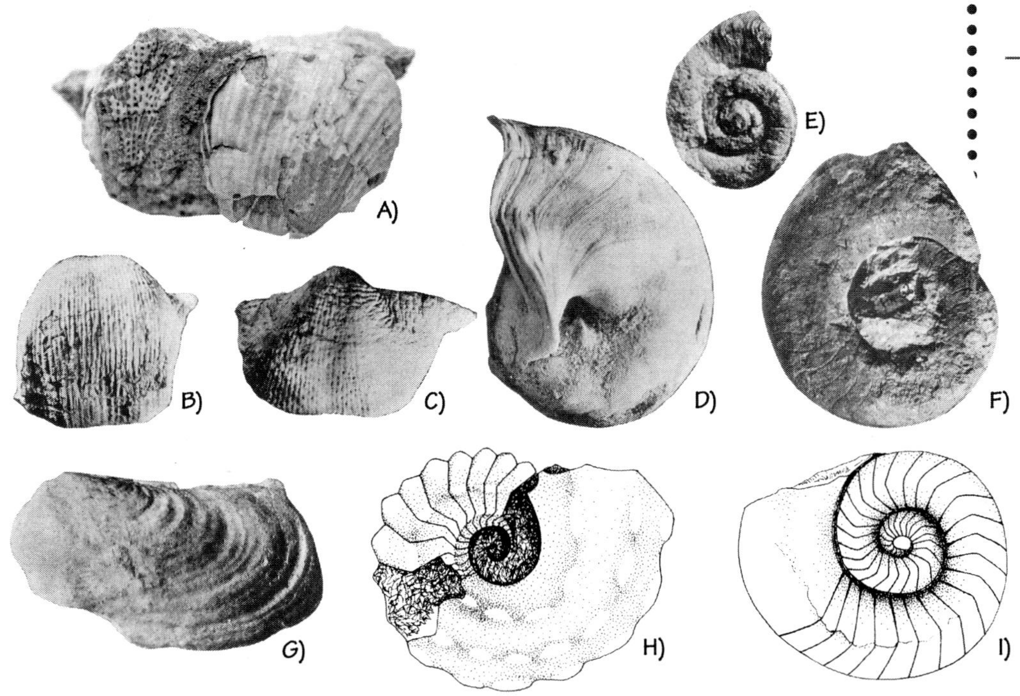

Figure 3.13: Permian marine invertebrates from the Permian Kaibab Limestone. A) Brachiopod, Dictyoclostus, *approximately 7.5 centimeters (3 inches) wide, with encrusting bryozoan. B) Brachiopod,* Productus ivesi, *approximately 5 centimeters (2 inches) wide. C) Brachiopod,* Productus bassi, *approximately 5 centimeters (2 inches) wide. D) Snail-like mollusk,* Bellerophon, *approximately 3.5 centimeters (1.5 inches) wide. E) Gastropod,* Euomphalus, *approximately 1.5 centimeters (0.6 inches) wide. F) Coiled nautiloid, approximately 10 centimeters (4 inches) wide. G) Bivalve,* Allorisma, *approximately 10 centimeters (4 inches) wide. H) Nautiloid,* Tainoceras duttoni, *approximately 10 centimeters (4 inches) wide. I) Nautiloid,* Stearoceras sanandresense, *approximately 10 centimeters (3 inches) wide.*

Gastropods

Commonly known as snails, gastropod mollusks encompass terrestrial, freshwater, and marine species, and include varieties with and without shells (e.g., slugs). Gastropods are among the most diverse groups of organisms—only insects have more named species. The soft parts of gastropods are generally similar to those of bivalves, but the former typically have coiled shells and are usually much more active. Gastropods are present in Paleozoic and Mesozoic rocks, but are especially abundant and diverse in Cretaceous and Cenozoic sediments of the Coastal Plain.

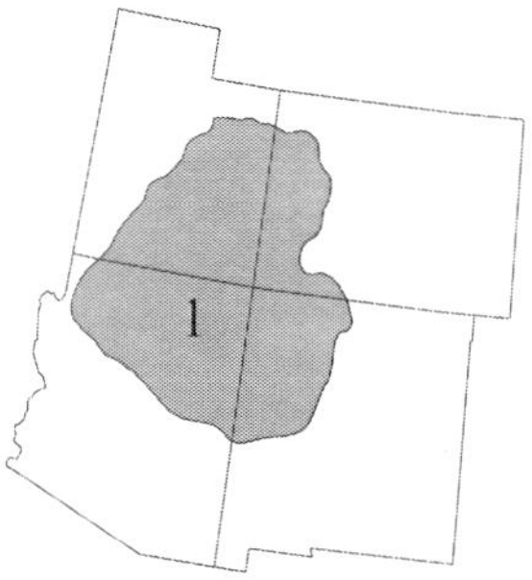

3 Fossils

Region 1

Early and middle **Mesozoic** rocks (**Triassic** and **Jurassic** periods) of the Colorado Plateau are just as fossil rich as **Paleozoic**-aged rocks. The Triassic rocks of northern Arizona are famous for their fossils of terrestrial (land-dwelling) and freshwater vertebrates. The middle Triassic Moenkopi Formation contains one of the best vertebrate fossil assemblages of this age anywhere in the world, including abundant bones and trackways of both large amphibians and reptiles. Skeletal remains in the Moenkopi include freshwater sharks, coelacanths, and lungfish, as well as amphibians, rhynchosaur reptiles, and the non-dinosaur **archosaur** *Arizonasaurus* (*Figure 3.14*).

Mesozoic *• a geologic time period that spans from 252 to 66 million years ago.*

Figure 3.14: Tetrapod fossils of the Moenkopi Formation. A) The trace fossil Cheirotherium barthii, *likely an archosaur footprint. B)* Arizonasaurus, *restoration, approximately 3 meters (10 feet) long.*

Triassic *• a geologic time period that spans from 252 to 201 million years ago.*

Jurassic *• the geologic time period lasting from 201 to 145 million years ago.*

The late Triassic Chinle Formation (or Group) is exposed across much of the Southwest (*Figure 3.15*). This formation accumulated in a shifting set of habitats, from forests to rivers and lakes. Fossils of freshwater fish, clams, and crustacean burrows are common in parts of the Chinle, suggesting abundant water, but other evidence indicates that rainfall was probably highly seasonal, and the environment was frequently dry (*Figure 3.16*). Fossil fishes found here include the coelacanth *Chinlea*, the shark *Xenacanthus*, the lungfish *Ceratodus*, and *Australosomus*—a relative of today's goldfish and tuna (*Figure 3.17*). Land-living reptiles include a parade of forms (*Figures 3.18 and 3.19*), including the bizarre armored group known as aetosaurs (e.g., *Desmatosuchus, Stagnolepis*), the dicynodont reptile *Placerias*, crocodile-like **phytosaurs** (the most common vertebrate in the formation), and the small theropod dinosaur *Coelophysis*. In the 1940s, hundreds of *Coelophysis* skeletons were discovered at Ghost Ranch in Rio Arriba County, north-central New Mexico. This fossil treasure trove may have formed when the dinosaurs died around a shrinking water source during the dry season, and then were swept up and deposited by a flash flood. The study of fossils from this deposit helped to make *Coelophysis* one of the world's best-known

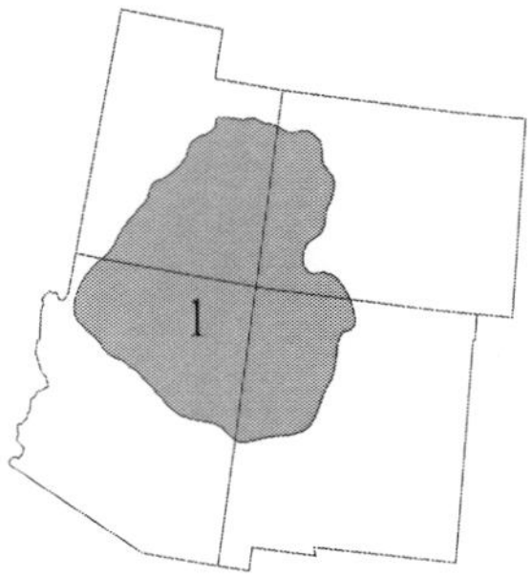

See Chapter 2: Rocks to learn more about the Grand Staircase, which exposes the Chinle and Moenkopi formations.

Region 1

Paleozoic • a geologic time interval that extends from 541 to 252 million years ago.

archosaur • a member of an evolutionary group of reptiles that includes non-avian dinosaurs and birds, pterosaurs, and crocodiles, among others.

phytosaur • an extinct, semi-aquatic reptile from the late Triassic period.

Figure 3.15: Location of Chinle Formation outcrops across the Southwestern US.

Figure 3.16: Reconstruction of a late Triassic depositional environment, similar to that of the Chinle Formation.

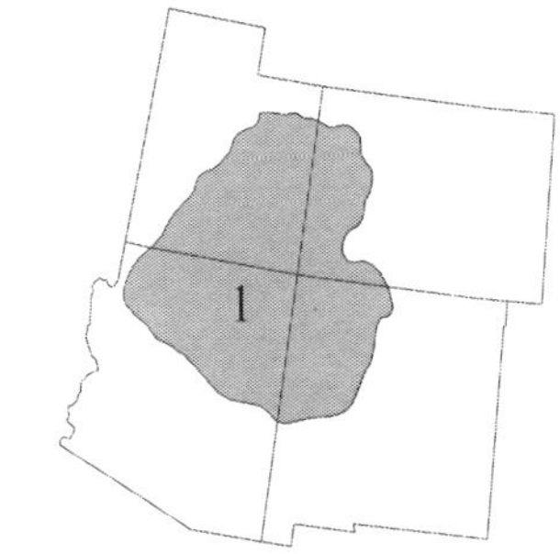

dinosaurs. The Chinle is also famous for its fossil **trees**, spectacularly exposed in Arizona's Petrified Forest National Park (*Figure 3.20*). Forests of these **conifers**, most belonging to the species *Araucarioxylon arizonicum*, were also filled with ferns, **cycads**, horsetails, and **ginkgoes**.

The Jurassic Kayenta Formation overlies the Chinle and is exposed in northern Arizona (Painted Desert and vicinity). It contains fossils of early mammal relatives as well as dinosaurs, including the larger theropod *Dilophosaurus* (*Figures 3.21 and 3.22*). Although it was once desert dune sand, the overlying

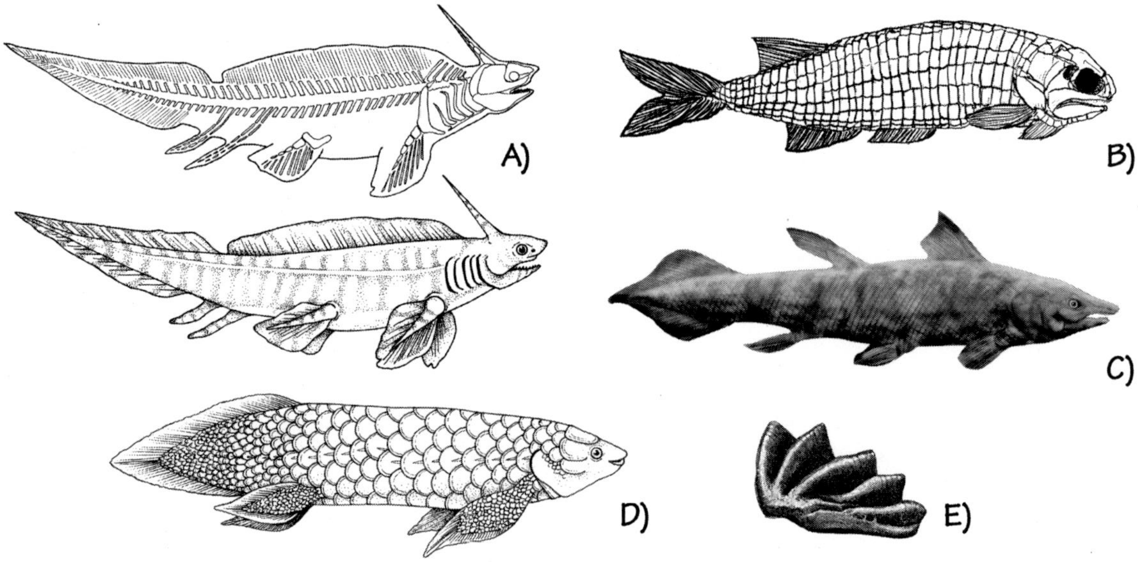

Figure 3.17: Fish of the Chinle Formation. A) Skeleton (top) and restoration (bottom) of the freshwater shark Xenacanthus*; approximately 30 centimeters (1 foot) long. B) Restoration of the ray-finned fish* Australosomus*; approximately 6 centimeters (1.5 inches) long. C) Restoration of the coelacanth* Chinlea*; approximately 1.5 meters (5 feet) long. D) Restoration of the lungfish* Ceratodus*; approximately 1.2 meters (48 inches) long. E)* Ceratodus *tooth; approximately 5 centimeters (2 inches) wide.*

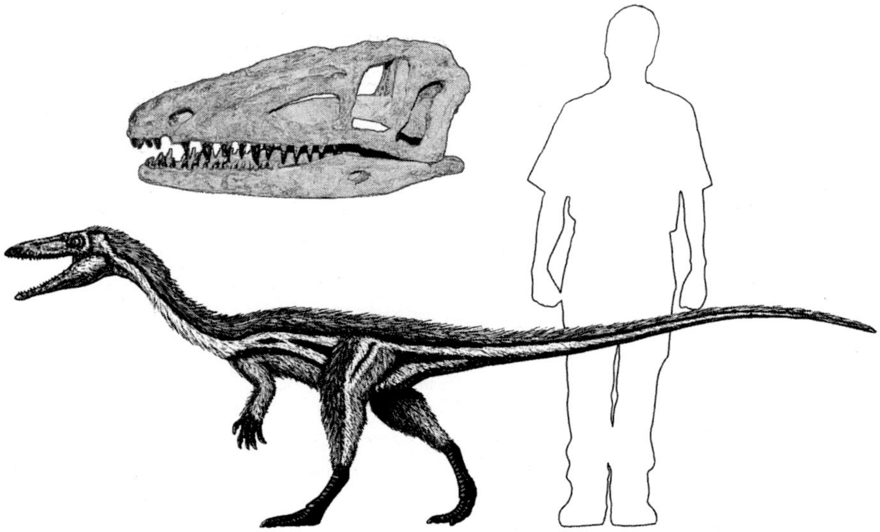

Figure 3.18: Skull and restoration of the Triassic theropod dinosaur Coelophysis*, approximately 2.5 meters (8 feet) long.*

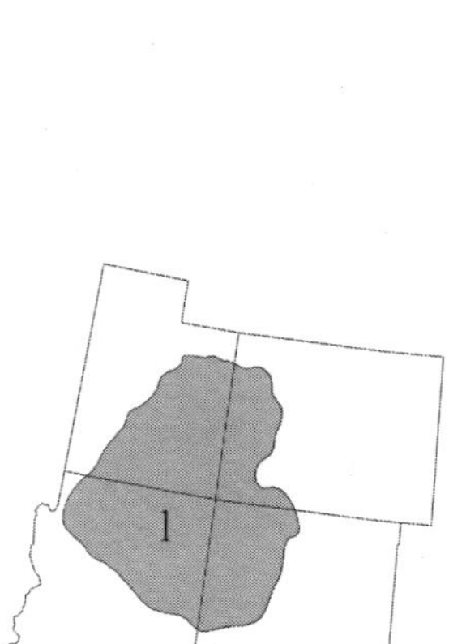

Figure 3.19: Non-dinosaur tetrapods from the Chinle Formation. A) Skeleton and restoration of the armored aetosaur reptile Desmatosuchus, *approximately 4.5 meters (15 feet) long. B) Skeleton and restoration of the synapsid reptile* Placerias, *approximately 3.5 meters (11.5 feet) long. C) Skull and restoration of the carnivorous reptile* Postosuchus, *approximately 4–5 meters (13–16 feet) long. D) Tooth and restoration of the phytosaur* Smilosuchus, *approximately 4 meters (13 feet) long.*

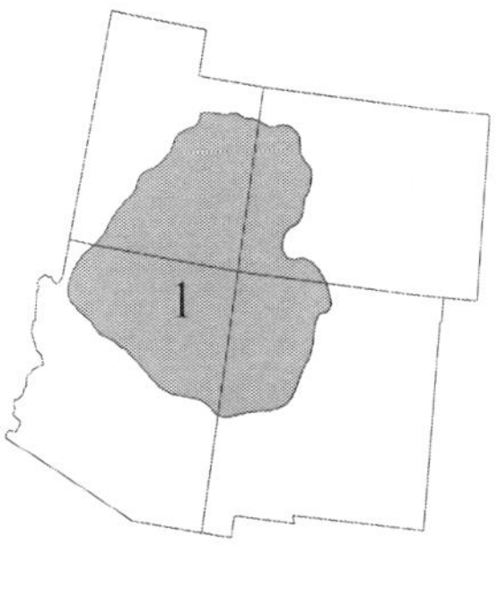

Figure 3.20: Fossil log from Petrified Forest National Park, northern Arizona.

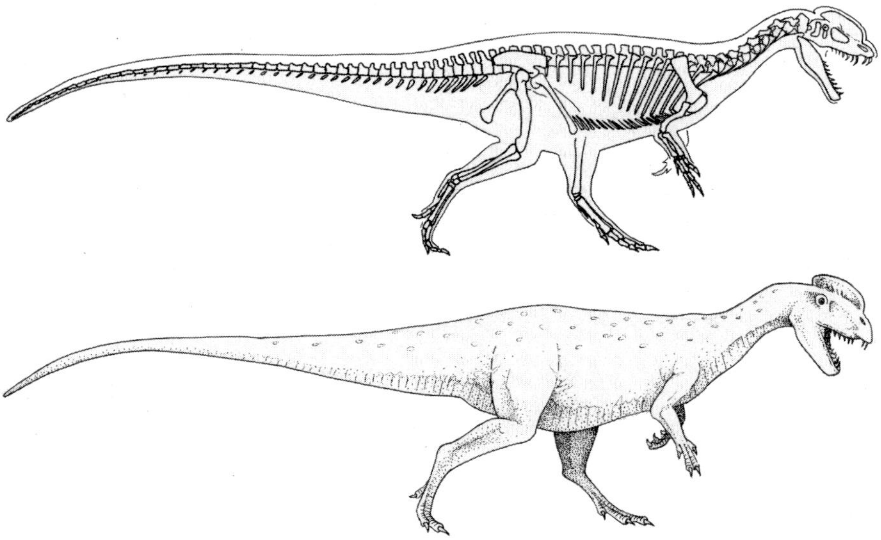

Figure 3.21: Skeleton and restoration of the theropod dinosaur Dilophosaurus, *approximately 6 meters (20 feet) long.*

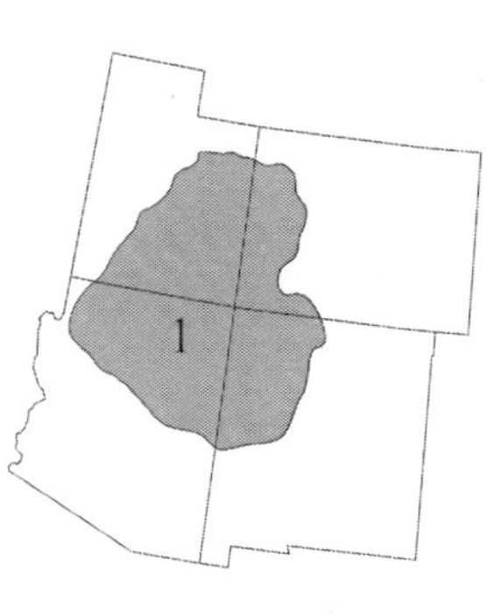

synapsid • *a group of tetrapod vertebrates posessing one opening in the skull behind each orbit (eye hole), and a bony arch beneath.*

Figure 3.22. Partial lower jaw of a small early relative of mammals, *Dinnetherium nezorum*, from the early Jurassic Kayenta Formation, near Tuba City, Arizona.

Early Mammals

Mammals evolved from a branch of reptiles (called the *synapsids*) some time in the early Triassic, around the same time that the first dinosaurs evolved from a different branch of reptiles. The earliest known mammals were very small (frequently only 2–5 centimeters [1–2 inches] long), and their fossils are almost exclusively limited to teeth and bits of lower jaw. Some of the oldest mammal fossils in the world come from the Southwestern region of the US, including the Kayenta Formation of Arizona's Painted Desert. The early Cretaceous rocks of southern Utah also contain abundant tooth fossils from many kinds of small mammals. Mesozoic mammals included many groups that left few or no descendants in the Cenozoic after the extinction of the dinosaurs.

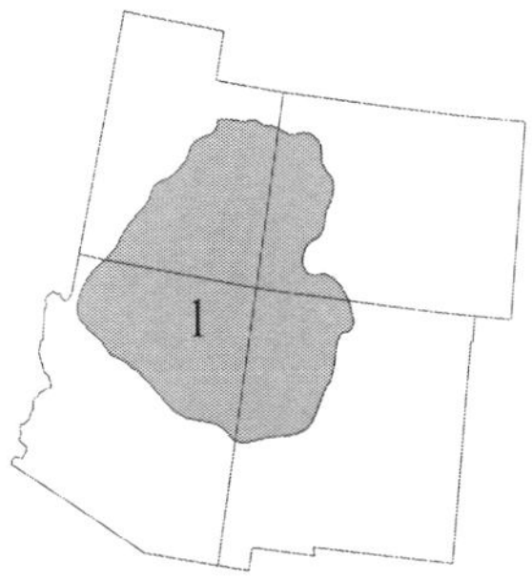

3 Fossils

Region 1

Fossil Mammals: It's (almost) all about the teeth

Mammals have evolved into an amazing variety of shapes and sizes, and much of this diversity and success is due to their teeth! Mammals are "warm-blooded," meaning they can regulate their own body temperature. This requires a high metabolism, the derivation of energy from food. Mammals meet their heavy food requirements with the help of a distinctive chewing system, starting with their teeth. Unlike reptiles, most mammals—including humans—have several different kinds of teeth in their mouths. Also unlike reptiles, some of these teeth are highly complex, with many bumps and grooves on the chewing surfaces. This range of tooth forms allows mammals to efficiently eat many different kinds of food. It also allows different kinds of mammals to eat different foods. This means that different mammals usually have very different teeth, and that you can often identify a mammal species using only its teeth. This is extremely important for studying fossils, because mammal teeth are frequently found as fossils. Mammal paleontology is therefore largely the study of fossil teeth.

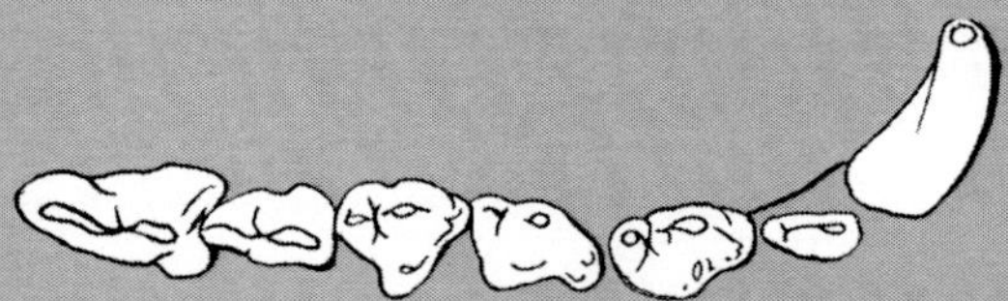

Upper right-side dentition of Hyanodon, *a dog-like carnivore.*

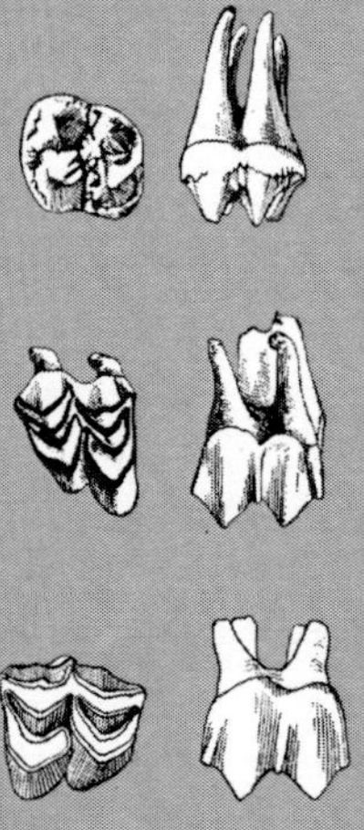

Upper molar of peccary (Tagassu), *deer* (Odocoileus), *and camel* (Poebrotherium).

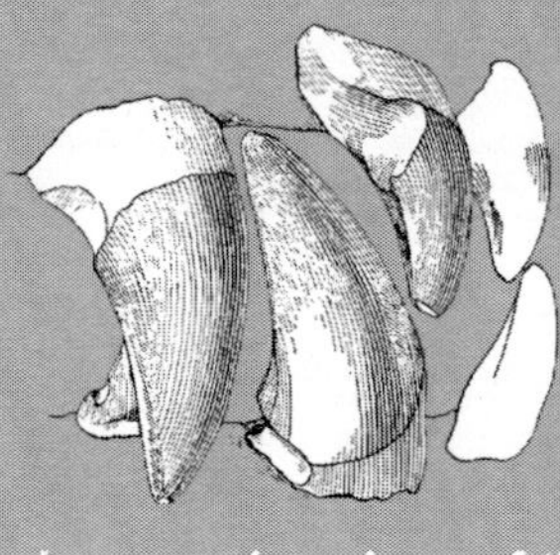

Incisors and canines of the entelodont Archaeotherium.

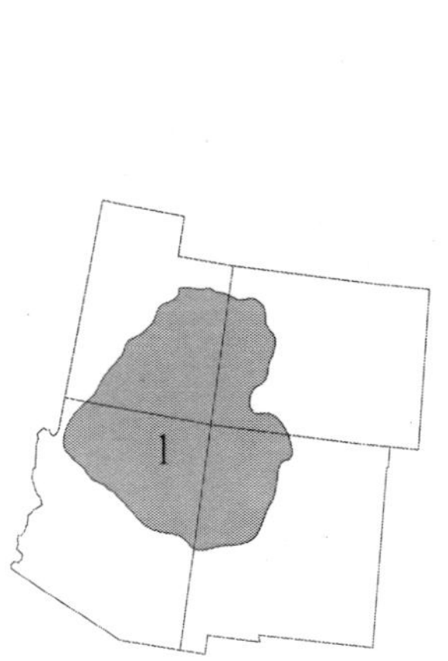

Navajo Sandstone also contains **dinosaurs** (*Figure 3.23*). As the Jurassic continued, dinosaurs continued to expand in diversity and size. Their dominance is abundantly demonstrated in the Jurassic Morrison Formation, an extensive layer of rock that crops out in all four states of the Southwest, as well as in the Dakotas, Montana, Nebraska, Kansas, Oklahoma, Texas, and Idaho (*Figure 3.24*). Although the Morrison Formation contains more than 90 known species of fossil vertebrates, including fish, turtles, amphibians, and early mammals, it is most famous for its abundance and diversity of dinosaurs (*Figure 3.25*), including carnivorous **theropods** such as *Ceratosaurus* and *Allosaurus*, herbivores such as *Dryosaurus, Stegosaurus,* and *Camptosaurus*, and gigantic long-necked **sauropods**, such as *Barosaurus*, *Diplodocus*, *Apatosaurus*, *Brachiosaurus*, and *Camarasaurus*. One giant sauropod from the Morrison, originally named *Seismosaurus*, was initially thought to be the world's largest dinosaur. That honor may instead go to another Morrison titan, *Amphicoelias,* which may have reached more than 50 meters (150 feet) long and weighed more than 120,000 kilograms (160 tons). Two spectacular places to see these dinosaur fossils still in place are Dinosaur National Monument, on the border between Utah and Colorado, and the Cleveland-Lloyd Dinosaur Quarry in Utah.

See Chapter 2: Rocks to learn more about the Navajo Sandstone and other Jurassic dune environments.

The dinosaur fossil beds in Dinosaur National Monument were discovered in 1909 by a paleontologist working for the Carnegie Museum of Natural History in Pittsburgh. Over several years, crews from the Museum excavated thousands of fossil bones and shipped them back to Pittsburgh for study and display. The site was declared a national monument in 1915, but new discoveries continue to this day—in late 2015, one of the oldest known **pterosaurs** was found in the monument's Triassic rocks. The famous "Wall of Bones" (*Figure 3.26*), located within the monument's main building, consists of a steeply tilted rock layer containing hundreds of dinosaur bones. The layer was tilted by the **Laramide Orogeny** millions of years after the dinosaurs lived, died, and were buried.

See Chapter 1: Geologic History for more about the mountain building events that shaped the Southwest.

The Cleveland-Lloyd Dinosaur Quarry, located in northern Emery County, Utah, contains the densest concentration of Jurassic-aged dinosaur bones ever found. Over 10,000 bones (belonging to at least 74 individual dinosaurs) have been excavated at the quarry. Curiously, more than 70% of these bones come from carnivores, primarily *Allosaurus fragilis*. With more than 46 individual specimens of *Allosaurus*, scientists have been able to both deduce how *Allosaurus* aged and compare individuals to better understand variation within the species.

dinosaur *• a member of a group of terrestrial reptiles with a common ancestor and thus certain anatomical similarities, including long ankle bones and erect limbs.*

theropod *• an evolutionary group of bipedal dinosaurs, including all of the carnivorous dinosaurs.*

sauropod *• a member of an evolutionary group of herbivorous dinosaurs characterized by very large body size, and long necks and tails.*

pterosaurs *• extinct flying reptiles with wingspans of up to 15 meters.*

Laramide Orogeny *• a period of mountain building that began in the late Cretaceous, and is responsible for the formation of the Rocky Mountains.*

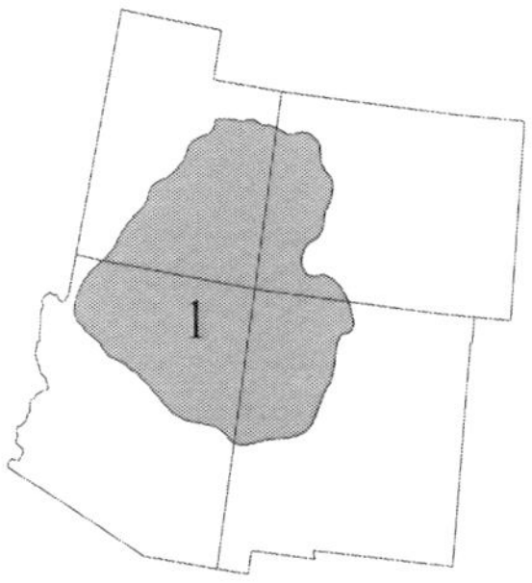

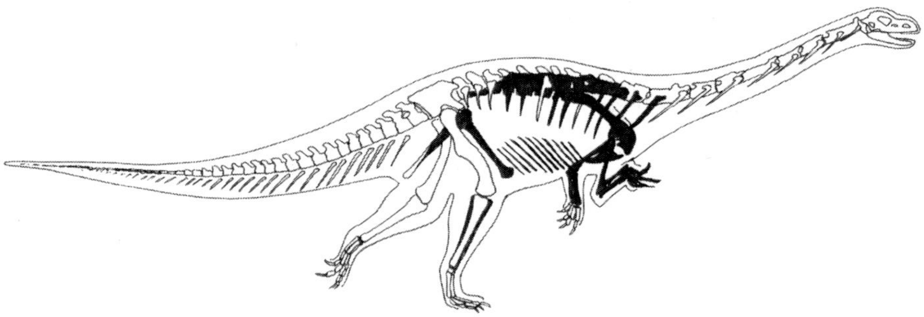

Figure 3.23: The early sauropod dinosaur Seitaad ruessi *from the Navajo Sandstone of Utah, approximately 4 meters (12 feet) long. This skeleton is reconstructed, and only the black bones are known from fossils.*

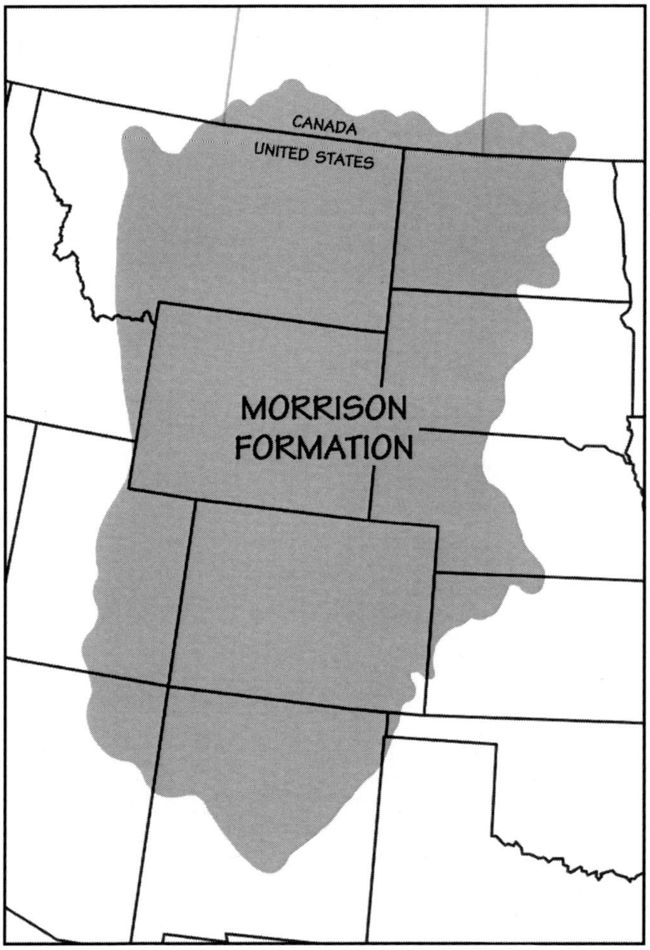

Figure 3.24: Geographic extent of the late Jurassic Morrison Formation.

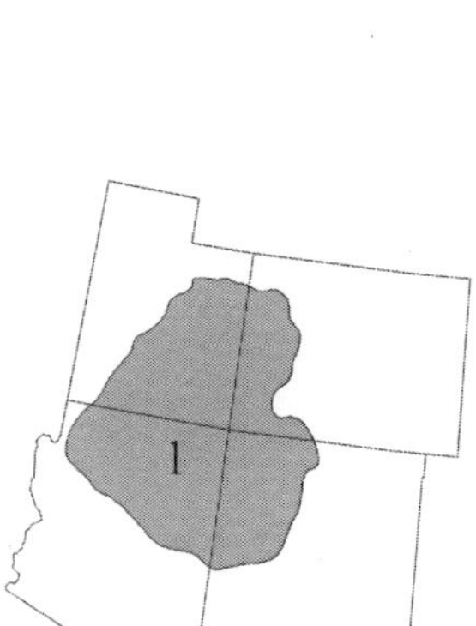

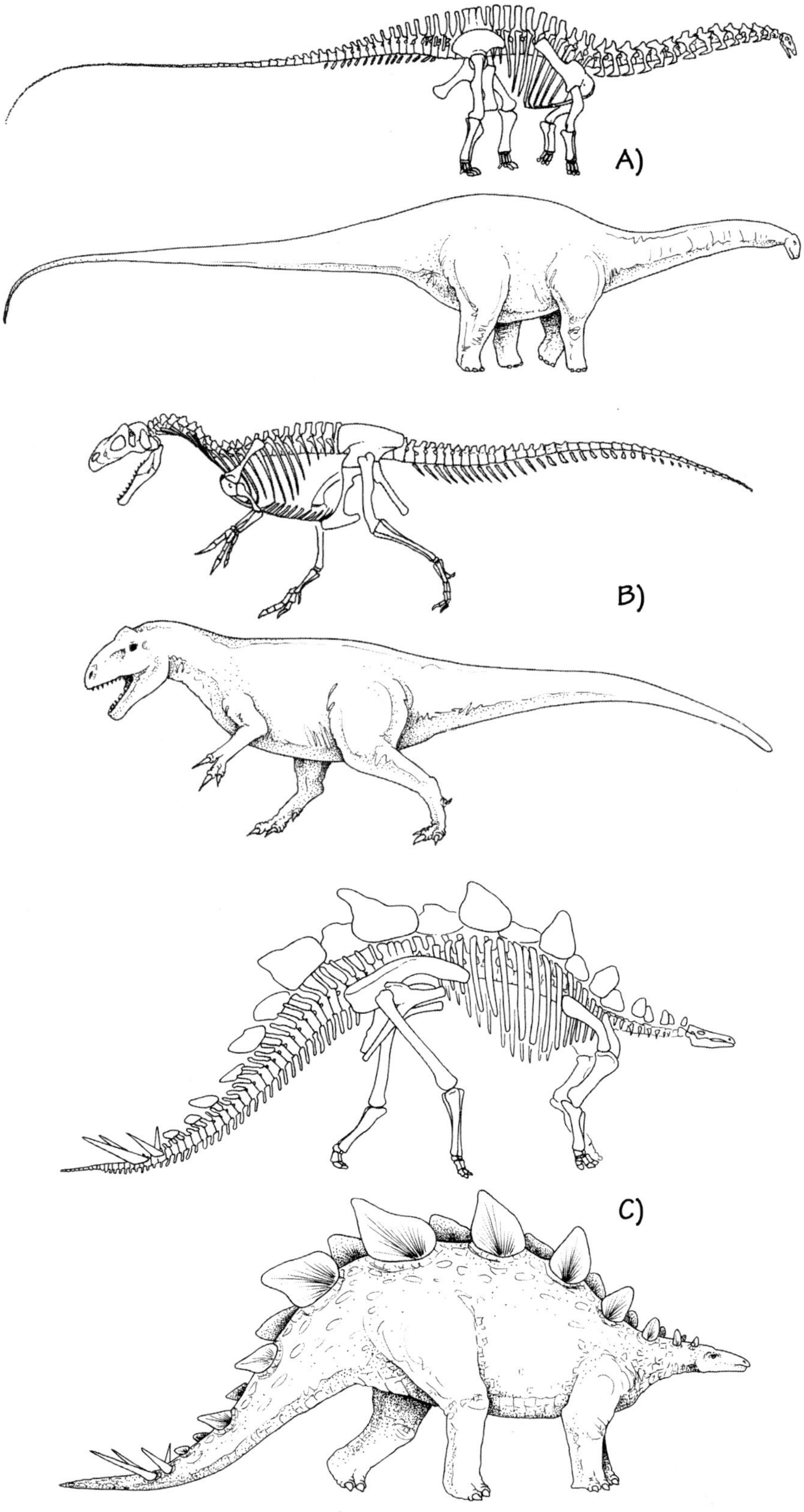

Figure 3.25: Some common and familiar dinosaurs from the Morrison Formation.
A) Apatosaurus *(approximately 23 meters [75 feet] long), skeleton and restoration;*
B) Allosaurus *(approximately 8.5 meters [28 feet] long), skeleton and restoration;*
C) Stegosaurus *(approximately 9 meters [30 feet] long), skeleton and restoration.*

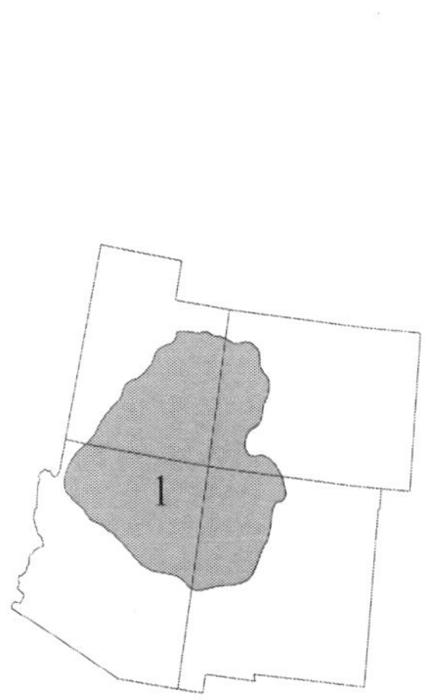

Fossils

Region 1

floodplain • *the land around a river that is prone to flooding.*

alluvial • *a layer of river-deposited sediment.*

Cretaceous • *a geologic time period spanning from 144 to 66 million years ago.*

Figure 3.26: The Wall of Bones at Dinosaur National Monument, Colorado and Utah.

Dinosaurs are also well represented across much of the Southwest through fossil footprints (*Figure 3.27*). The Purgatoire River tracksite, also called the Picketwire Canyonlands tracksite, is one of the largest dinosaur tracksites in North America. The site is located in the Comanche National Grassland along the Purgatoire River south of La Junta, in Otero County, Colorado.

The Morrison Formation also contains abundant plant fossils, including conifers and ferns—flowering plants had not yet evolved (*Figures 3.28 and 3.29*). These fossils suggest that the Morrison ecosystem was a mosaic of river, lake, and **floodplain** environments developed on an enormous **alluvial** plain covered by sediment eroding from the ancestral Rocky Mountains.

Cretaceous rocks are widely exposed across southern Utah, and contain a great abundance and diversity of dinosaurs. Herbivores are represented by hadrosaurs, ankylosaurs, and several kinds of horned dinosaurs (ceratopsians) (*Figures 3.30 and 3.31*). Carnivorous dinosaurs include a variety of theropods, from the large and ferocious-looking *Utahraptor* to the even larger tyrannosaurs (*Figures 3.32 and 3.33*). Theropods are well represented in the southern Colorado Plateau, but not until recently have paleontologists been able to determine exactly which species are represented by the fossils found there. Most large teeth and bones found in the area were originally thought to be from close relatives of *T. rex*, such as the *Albertosaurus*, which are well known farther north in Wyoming and Montana. In 2010, however, some of these fossils were recognized as belonging to a previously unknown theropod, named *Bistahieversor*. This predator is estimated to have been around 9 meters (30 feet) long, weighing at least a ton. Although the Cretaceous was also a golden age for flying reptiles (pterosaurs), they are only rarely found in the Colorado Plateau's Mesozoic rocks.

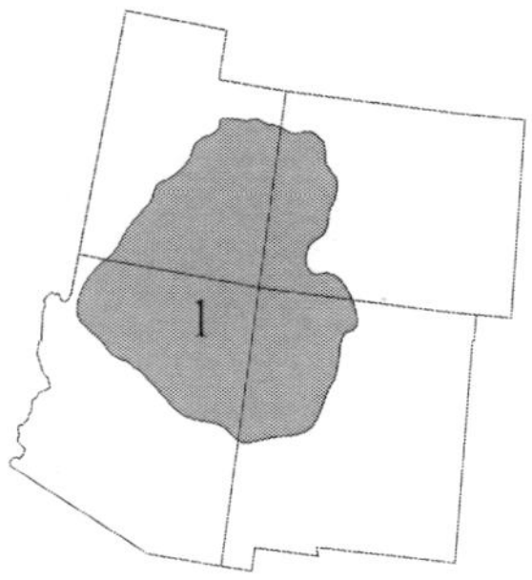

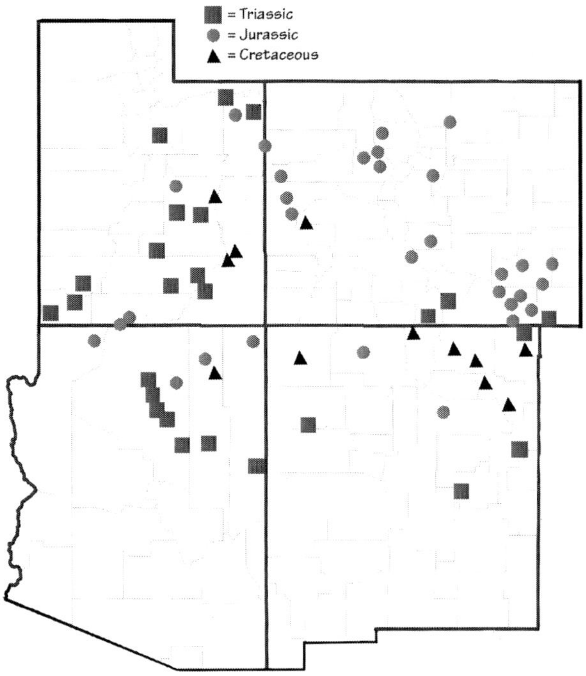

Figure 3.27: Major localities in which Mesozoic footprint assemblages (mostly dinosaurs) have been found.

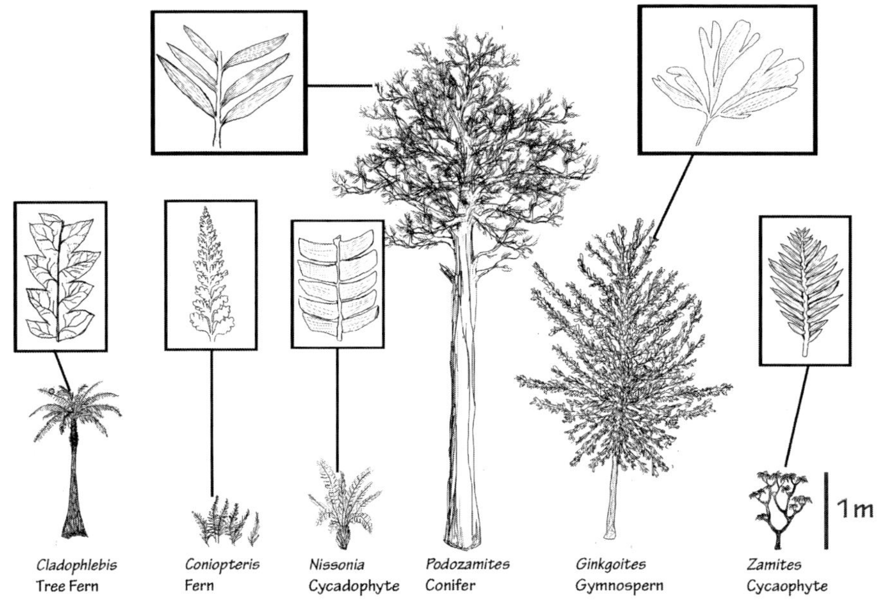

Figure 3.28: Some of the most common plants from the late Jurassic Morrison Formation.

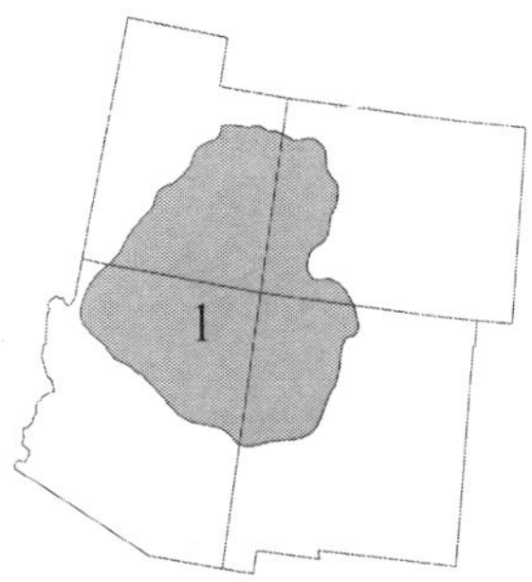

3 Fossils

Region 1

Figure 3.29: Reconstruction of the biological community represented by fossils in the Morrison Formation.

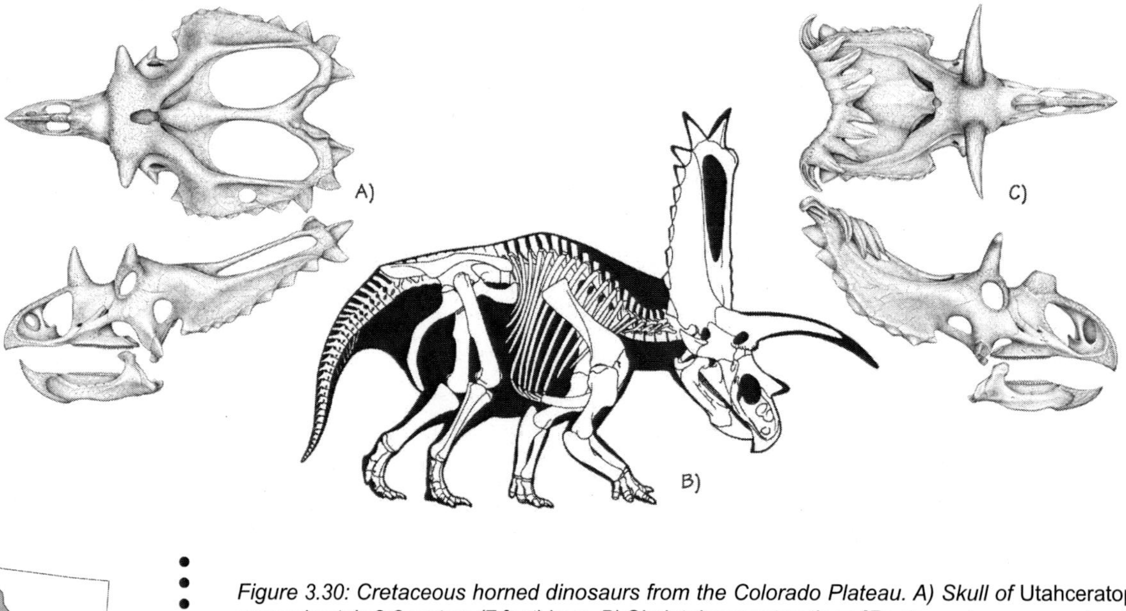

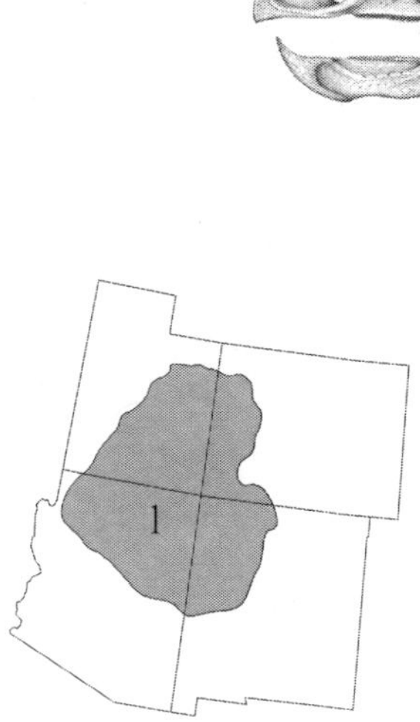

Figure 3.30: Cretaceous horned dinosaurs from the Colorado Plateau. A) Skull of Utahceratops, *approximately 2.3 meters (7 feet) long. B) Skeletal reconstruction of* Pentaceratops, *approximately 6 meters (20 feet) long. C) Skull of* Kosmoceratops, *approximately 2 meters (6 feet) long.*

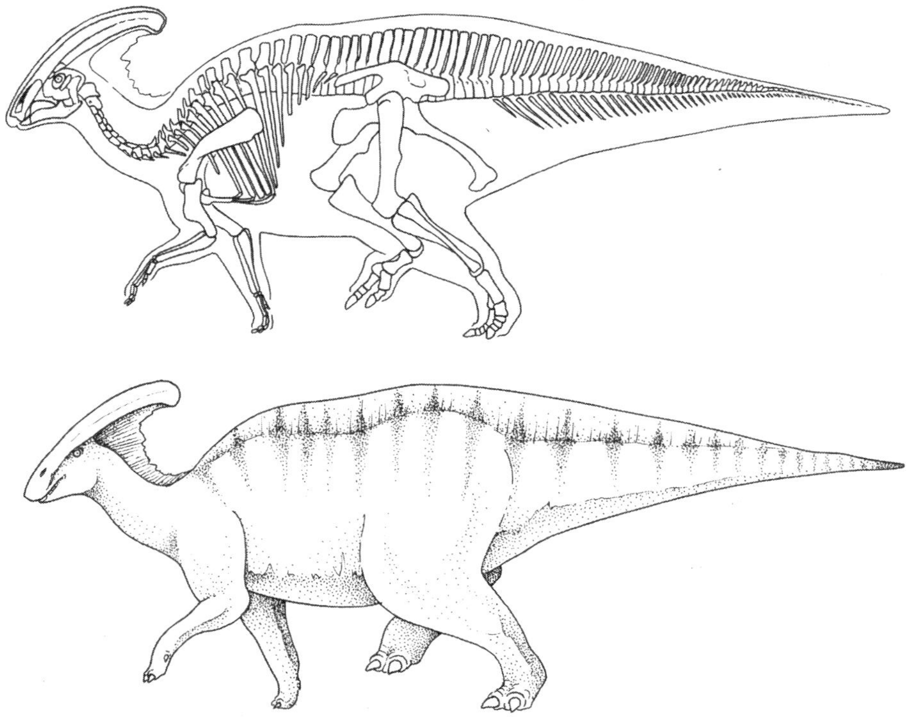

Figure 3.31: *The hadrosaur* Parasaurolophus*; approximately 10 meters (33 feet) long, skeleton and reconstruction.*

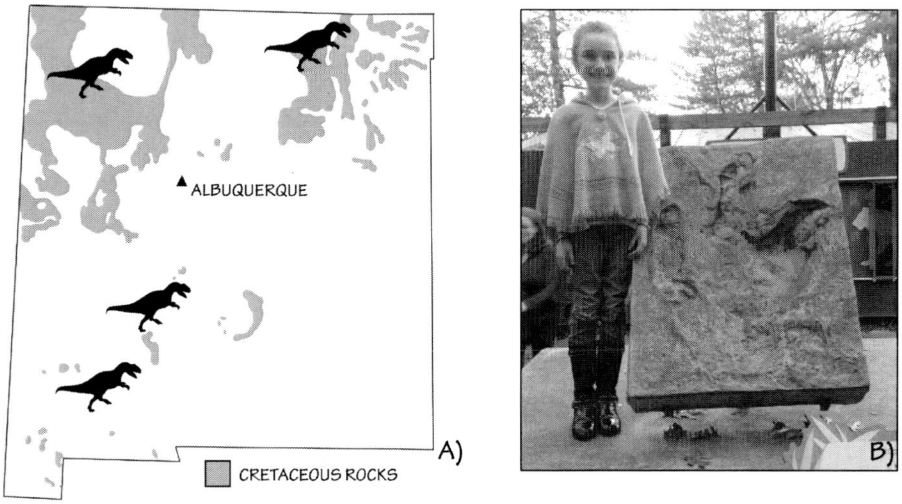

Figure 3.32: Tyrannosaurs in New Mexico. A) Locations where tyrannosaur fossils have been found. B) Cast of the only footprint ever confidently assigned to Tyrannosaurus rex*, found near the town of Cimarron, Colfax County, New Mexico. The print measures 86 centimeters (34 inches) long. This cast is on exhibit at the Smithsonian Institution's National Zoo in Washington, DC.*

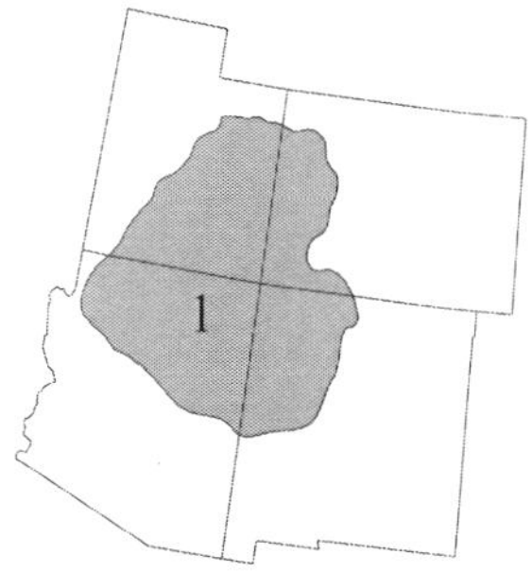

3

Fossils

Region 1

angiosperms • *flowering plants, including taxa from deciduous trees to shrubs and grasses.*

ichthyosaurs • *extinct Mesozoic marine reptiles that were probably similar in size and habitat to the toothed whales, dolphins, and large sharks of today.*

mosasaur • *an extinct, carnivorous, marine vertebrate reptile characterized by a streamlined body for swimming, a powerful fluked tail, and reduced, paddle-like limbs.*

plesiosaur • *a member of a group of extinct, long-necked, Mesozoic marine reptiles.*

cementation • *the precipitation of minerals that binds together particles of rock, bones, etc., to form a solid mass of sedimentary rock.*

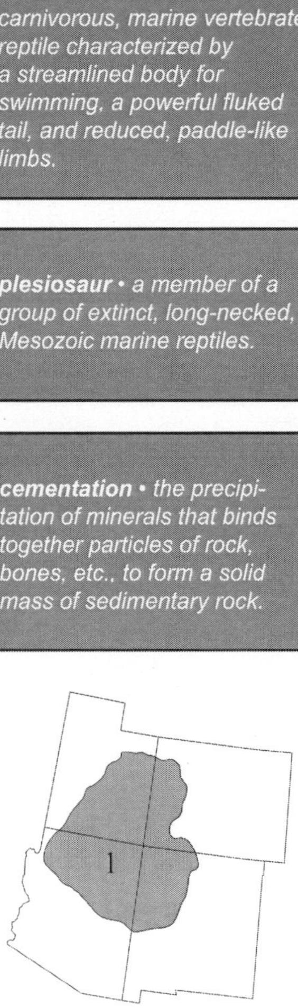

Figure 3.32: Skeleton and reconstruction of the theropod dinosaur Utahraptor*, approximately 6.5 meters (20 feet) long.*

Fossils of mammals—mostly very small teeth—are also found in the Cretaceous rocks of southern Utah (*Figure 3.34*). These fossils represent marsupials, placentals, and several extinct groups that left no modern descendants.

Cretaceous plant fossils are abundant in Utah, especially in Emery and Carbon counties, where large quantities of land plants accumulated in coastal swamps and eventually formed significant deposits of coal (*Figure 3.35*). In addition to conifers and tree ferns (*Figure 3.36*), which were major contributors to these coal deposits, the region's Cretaceous land plants include **angiosperms** (flowering plants) such as palms and magnolias. Angiosperms probably originated in the late Jurassic, but had diversified and taken over many terrestrial ecosystems by the mid-Cretaceous.

As the Western Interior Seaway flooded North America during the late Cretaceous, portions of the Colorado Plateau were submerged and covered in a series of marine layers (*Figure 3.37*). These rocks frequently contain abundant vertebrate and invertebrate fossils. Marine vertebrates include bony fish, sharks, **ichthyosaurs**, **mosasaurs**, and **plesiosaurs** (*Figure 3.38*), while invertebrates encompass a great diversity of mollusks, barnacles, and echinoderms. Relatives of living oysters were diverse and abundant during the Cretaceous; they could **cement** themselves to surfaces, and varied widely in shape and ornamentation. *Inoceramus* was a large, flat bivalve that could reach diameters of up to 1.2 meters (4 feet)! These mollusks were also important hard substrates for other organisms to attach to in soft sea-floor sediments. The Mancos Shale preserves the abundant remains of marine snails, sharks, mosasaurs, crinoids, bivalves, and ammonoids (*Figure 3.39*). The Dakota Group includes abundant ammonites and bivalves, as well as the last known North American lungfish from just east of Arches National Park in southeastern Utah. The Tropic Shale and its equivalents are some of the most fossil-rich marine units in North America; ammonoids found here include especially valuable index fossils such as *Sciponoceras* and *Collignoniceras*.

See Chapter 4: Topography to learn more about Arches National Park and other parks of the Colorado Plateau.

Figure 3.34: Restoration of Alphadon, a tiny marsupial mammal from the Cretaceous of southern Utah; up to 30 centimeters (12 inches) long.

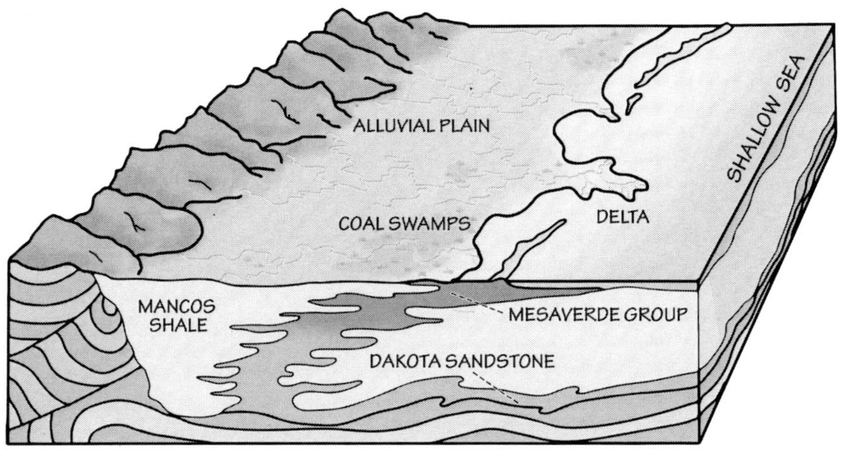

Figure 3.35: Reconstruction of the paleoenvironment in which Utah's Cretaceous coal deposits formed. When trees and other vegetation fall into the stagnant water of a coastal swamp, deterioration of the organic material is delayed, and a thick layer of peat is formed. Over time, the peat is compressed into coal.

The Cretaceous-**Paleogene** (K-Pg) boundary is visible in several areas of the Southwest, especially in the San Juan Basin of New Mexico. This boundary marks the mass extinction that wiped out non-avian dinosaurs, flying reptiles, swimming reptiles, and many other forms of life across the globe (*see Figure 3.1*). Rocks above the K-Pg boundary in the San Juan Basin are rich in fossil land mammals. These include a number of major groups, many of which became extinct within a few million years *(Figure 3.40)*. **Multituberculates** were a group of mostly small, rodent-like mammals that originated in the Cretaceous and disappeared in the early **Oligocene**—their name refers to their distinctive multicusped teeth. Other mammals abundantly represented by fossils in the Basin are commonly referred to as "archaic," meaning that their features are primitive compared to those of later mammals. These archaic mammals include **condylarths**, an informal term for a number of early mammals that are not closely related to each other. Some of the earliest known primates were also present at this time. Late Cretaceous and early **Cenozoic** rocks in the San Juan Basin also contain abundant flowering plants (*Figure 3.41*).

Region 1

Paleogene *• the geologic time period extending from 66 to 23 million years ago.*

multituberculate *• a member of a group of rodent-like animals with a long evolutionary history, from the Jurassic period until at least the Oligocene epoch.*

Oligocene *• a geologic time interval spanning from about 34 to 23 million years ago.*

condylarth *• a member of an informal group of ungulates (a large group of mammals that include most of the relatively large herbivorous mammals).*

Cenozoic *• the geologic time period spanning from 66 million years ago to the present.*

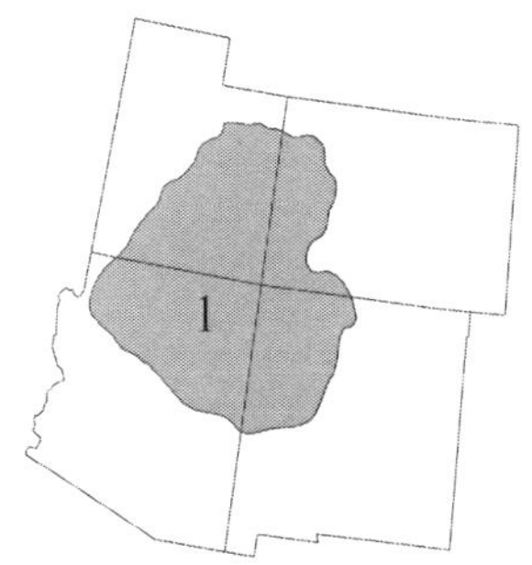

Region 1

compression *• flattening or squeezing as a result of forces acting on an object from all or most directions.*

coalification *• the process by which coal is formed from plant materials through compression and heating over long periods of time.*

coal balls *• masses of calcium carbonate that crystallize inside coals from minerals dissolved in groundwater.*

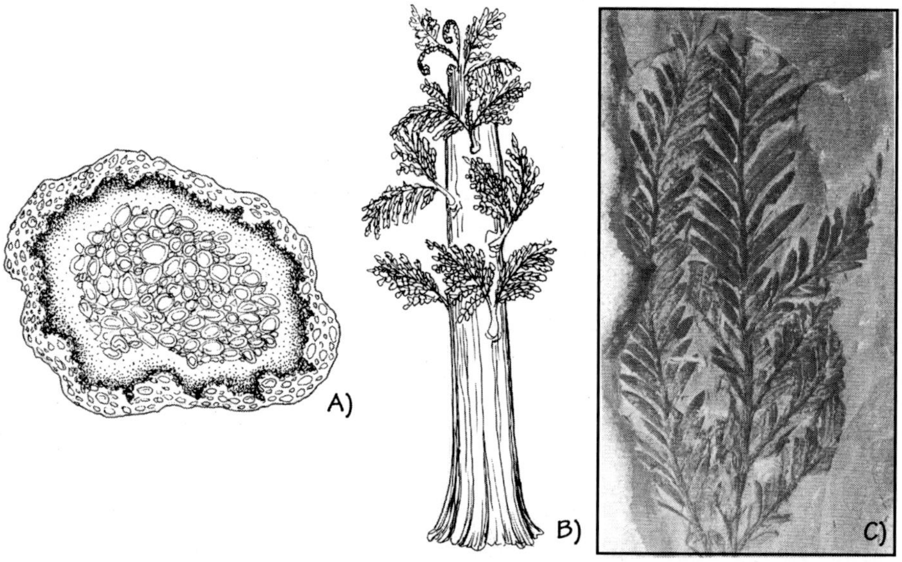

Figure 3.36: Cretaceous land plants of Utah. A) Cross-section of the tree fern Tempskya, *approximately 30 centimeters (1 foot) wide. B) Reconstruction of the tree fern* Tempskya. *C) Fossil leaf of a redwood,* Sequoia coneata, *approximately 10 centimeters (4 inches) long, from eastern Utah. Sequoias were major contributors to Utah's coal beds.*

Fossils and Coal

Coal is technically a metamorphic rock formed of highly *compressed* and altered *peat*. As is the case in most metamorphic rocks, this alteration (sometimes called *coalification*) means that coal itself does not usually contain well-preserved plant fossils. Instead, we learn about the plants that make up coal from two kinds of fossils: impressions and compressed plant parts left in shales deposited above or below coal seams, and *coal balls*, which are masses of calcium carbonate that crystalize inside coals from minerals dissolved in groundwater, protecting the plants they contain from alteration. Coal balls are usually studied by slicing them with a saw, polishing the sliced surfaces, and then making peels of the surface using sheets of acetate. These coal-ball peels are then examined under a microscope. See Chapter 6: Energy for more information on coal in the Southwest.

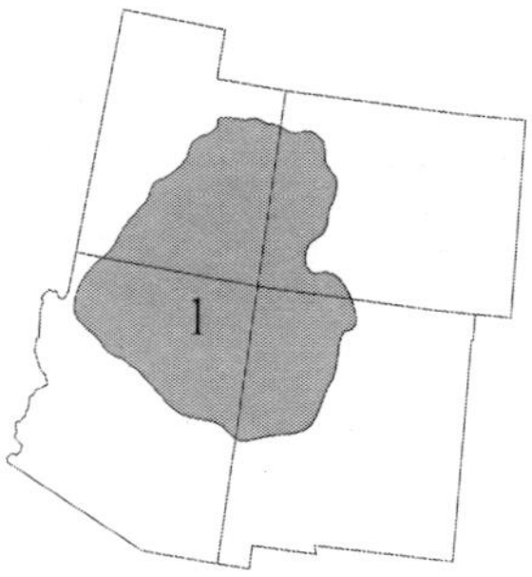

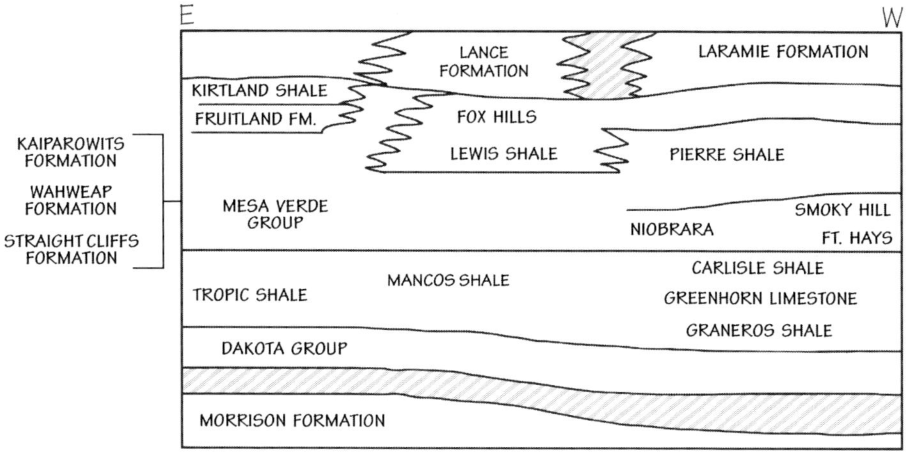

Figure 3.37: Cretaceous stratigraphy of the Colorado Plateau.

Figure 3.38: Cretaceous marine vertebrates of the Southwest. A) Restoration of Elasmosaurus, *a large plesiosaur from the Niobrara Chalk of Kansas and eastern Colorado, approximately 14 meters (46 feet) long. B) Mosasaur tooth, approximately 5 centimeters (2 inches) long. C) Restoration of the late Cretaceous mosasaur* Tylosaurus, *approximately 15 meters (50 feet) long.*

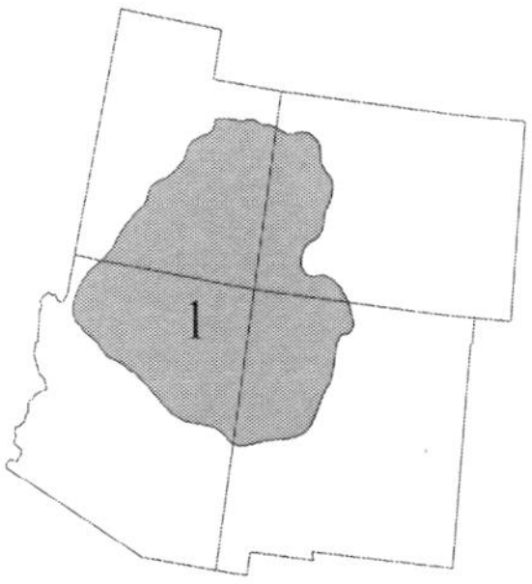

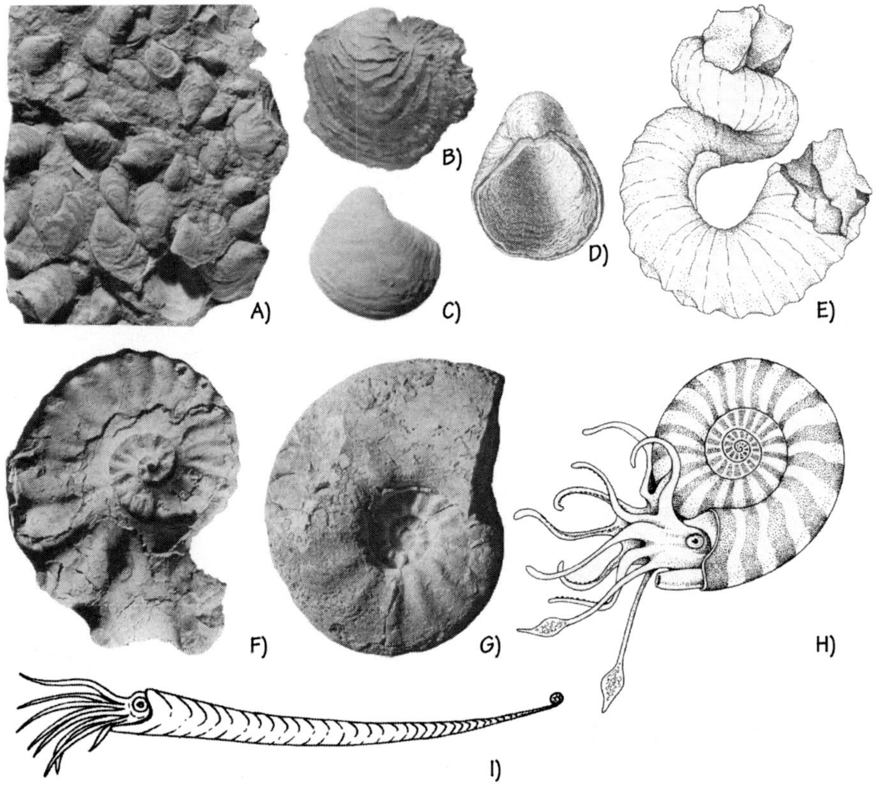

Figure 3.39: Cretaceous marine mollusks of the Mancos Shale and Dakota Group. A) Bivalve, Inoceramus *sp., 10 centimeters (4 inches) wide. B) Bivalve,* Exogyra trigeri, *8 centimeters (3 inches) wide. C) Bivalve,* Exogyra laevis, *4 centimeters (1.5 inches) wide. D) Bivalve,* Gryphaea newberryi, *4 centimeters (1.5 inches) wide. E) Heteromorph ammonite,* Didymoceras, *15 centimeters (6 inches) wide. F) Ammonite,* Plesiacanthoceras, *15 centimeters (6 inches) in diameter. G) Ammonite,* Paracompsoceras, *20 centimeters (8 inches) in diameter. H) Restoration of a living ammonite. I) Restoration of an orthocone cephalopod,* Baculites, *usually 3–4 centimeters (2 inches) in diameter and up to 60 centimeters (2 feet) long.*

Cephalopods

Cephalopods, such as squid, octopods, nautiloids, ammonoids, and belemnites, are mollusks with tentacles and beak-shaped mouths for catching prey. Some cephalopods such as belemnites and living cuttlefish have internal shells, while others have straight or coiled shells, such as those of ammonoids or nautiloids. Still other cephalopods, such as the octopus, have no shells. The mass extinction at the end of the Cretaceous, famous for eliminating the dinosaurs, also eliminated belemnites and ammonoids, which had been extremely diverse during the Mesozoic. Ammonoids are useful index fossils, especially in Mesozoic rocks.

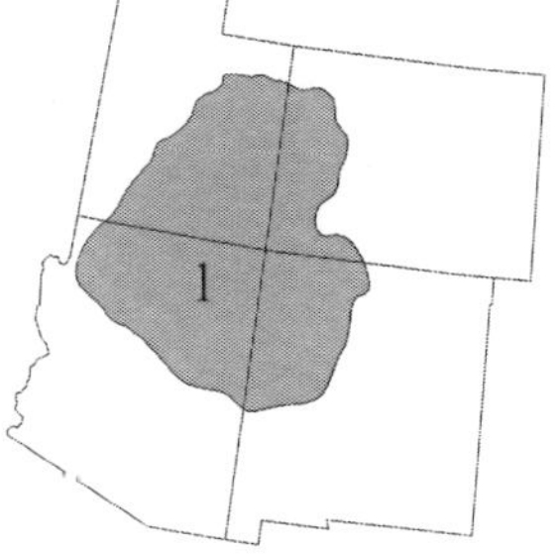

Pleistocene • a subset of the Quaternary, lasting from 2.5 million to about 11,700 years ago.

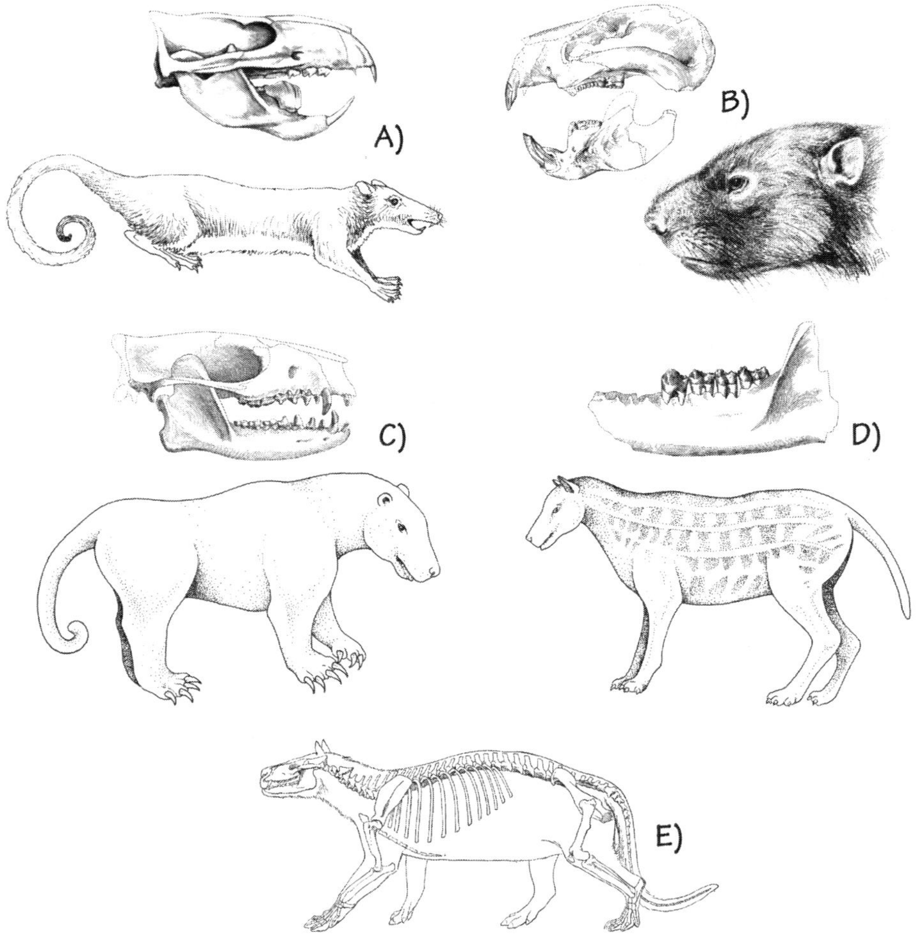

Figure 3.40: San Juan Basin Paleocene mammals. A) Skull and restoration of the multituberculate Ptilodus mediaevus*; body 30–50 centimeters (12–20 inches) long, skull approximately 6 centimeters (2.4 inches) long. B) Skull and restoration of* Taeniolabis*, a member of a condylarth group known as taeniodonts, which had front teeth developed into tusks; body approximately 100 centimeters (39 inches) long, skull approximately 15 centimeters (6 inches) long. C) Skull and restoration of another taeniodont,* Onychodectes*; body 60 centimeters (24 inches) long, skull 10 centimeters (4 inches) long. D) Jaw and restoration of the condylarth* Periptychus*; jaw 10 centimeters (4 inches) long. E) Skeleton and reconstruction of* Tetraclaenodon*, a member of a condylarth group known as phenacodonts, which may have been closely related to the ancestors of the modern perissodactyls (including horses and rhinos); body 75 centimeters (30 inches) long, skull 11.5 centimeters (4.5 inches) long.*

Between 40,000 and 10,000 years ago during the **Pleistocene** epoch, the Shasta ground sloth (*Nothrotherium shastense*) inhabited Rampart Cave in the Grand Canyon (*Figure 3.42*). Dung samples from this and other caves are rich in well-preserved pollen and other plant material, which allows the diet of these extinct animals to be reconstructed. The preserved dung also contains sloth DNA. Like many other species of large mammals, these sloths became extinct abruptly at the end of the Pleistocene, probably due at least in part to human hunting. Until 1976, Rampart Cave contained the thickest and least disturbed deposit of stratified Shasta ground sloth dung known to science.

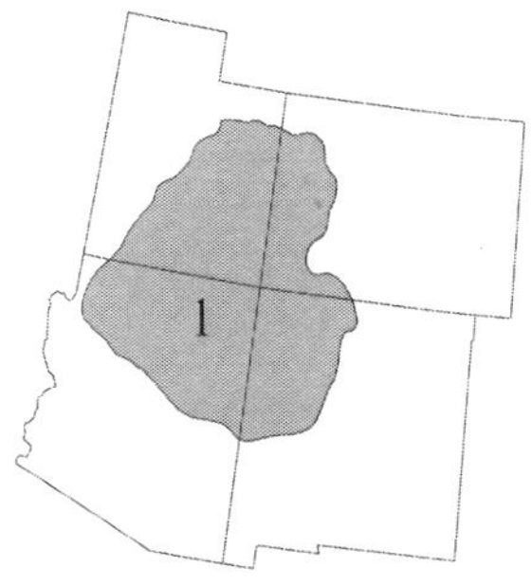

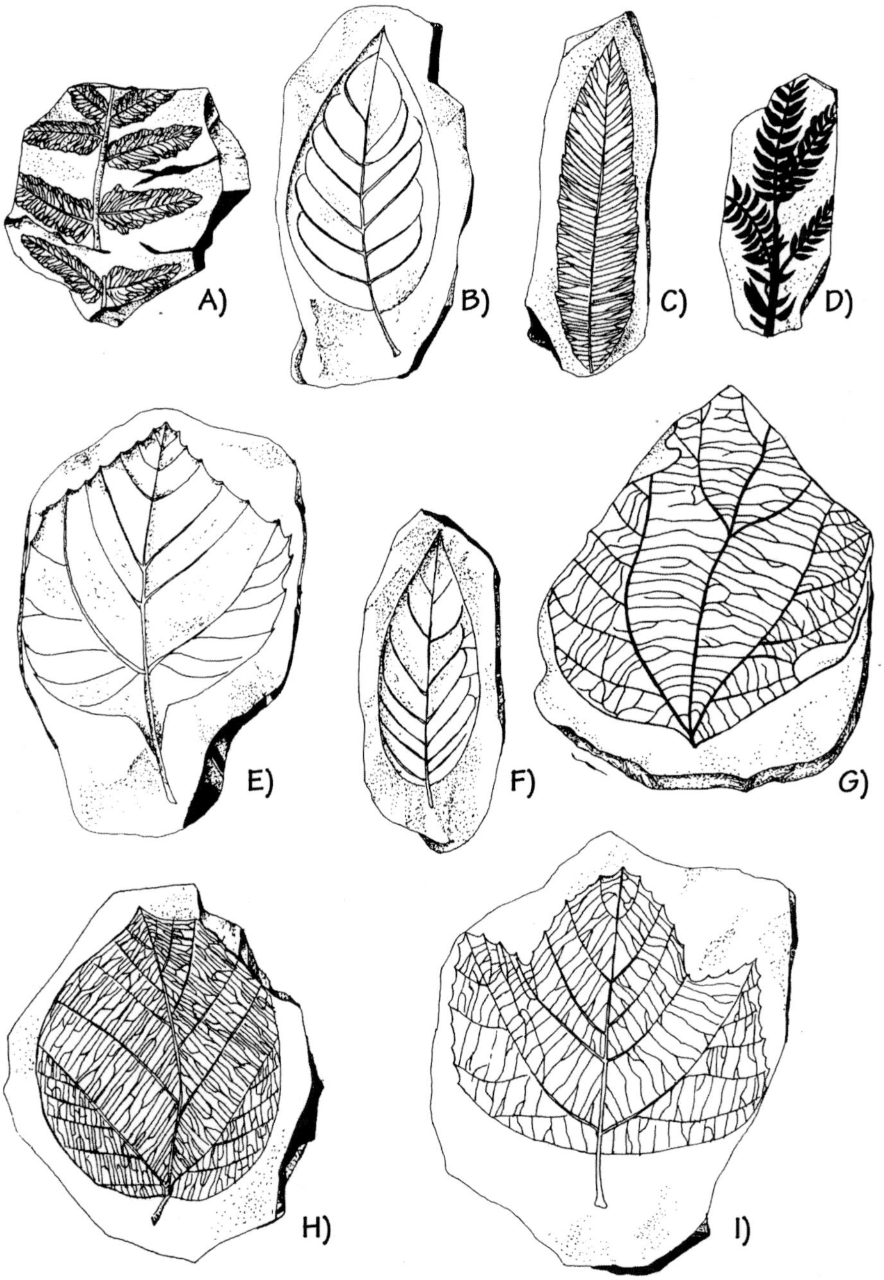

Figure 3.41: Idealized illustrations of fossil angiosperm plants from the late Cretaceous and early Paleogene of the San Juan Basin of northern New Mexico. A). Osmunda, *leaf size 20–150 centimeters (8–59 inches). B)* Carya, *leaflets 20–36 centimeters (8–14 inches). C)* Allantodiopsis, *leaf size 5–17 centimeters (6–7 inches). D)* Sequoia, *individual leaves approximately 0.65 centimeters (0.25 inches) long. E)* Platanus, *20 centimeters (8 inches). F)* Magnolia, *15–17 centimeters (6–7 inches). G and H)* Ficus, *approximately 7.5 centimeters (3 inches). I)* Cissus, *20 centimeters (8 inches).*

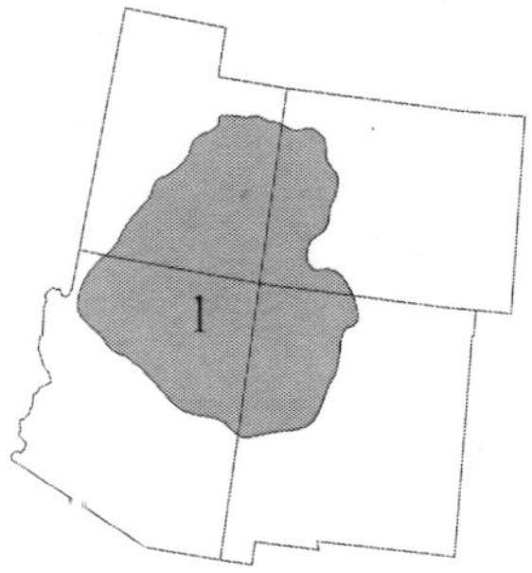

Region 2

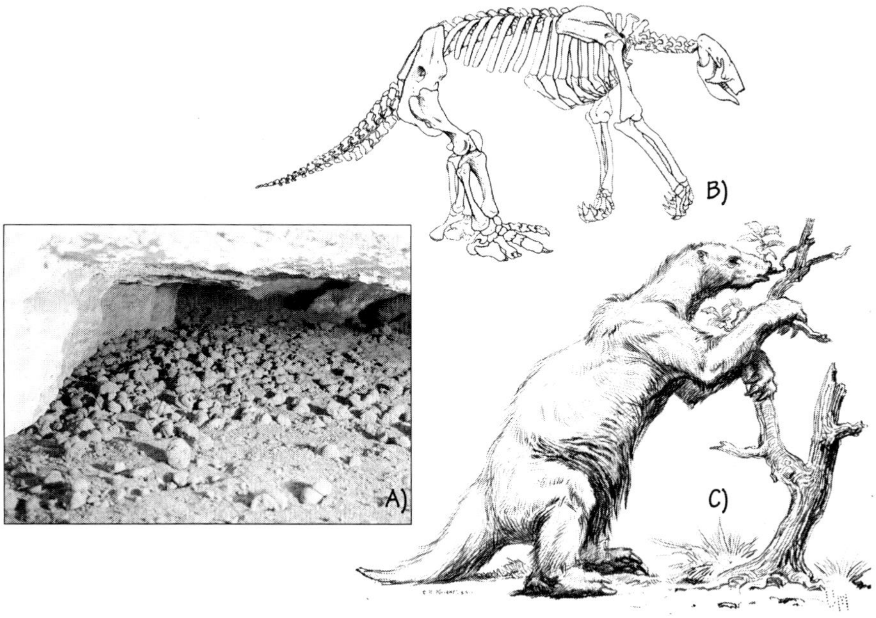

Figure 3.42: The Shasta ground sloth, Nothrotherium shastense. *These sloths were 2.75 meters (9 feet) long as adults and weighed approximately 250 kilograms (550 pounds). A) Preserved sloth dung inside Rampart Cave, Arizona. B) Skeleton. C) Restoration.*

Fossils of the Basin and Range
Region 2

graptolite • an extinct colonial invertebrate animal characterized by individuals housed within a tubular or cup-like structure.

Trilobites are extremely abundant and diverse in west-central Utah, in the middle Cambrian rocks exposed in the House Range of Millard County (*Figure 3.43*). One of these trilobite species is especially familiar: *Elrathia kingii* is one of the most abundant trilobite species found in North America, and is commonly sold commercially. Utah's middle Cambrian rocks (especially the Wheeler Formation) also contain fossils of soft-bodied organisms, including arthropods, sponges, brachiopods, and echinoderms, similar to those of the famous Burgess Shale in British Columbia, Canada (*Figure 3.44*). Such exceptional preservation was possible in part because these rocks formed from marine sediments deposited in anoxic (very low-oxygen) conditions. Cambrian and Ordovician fossils are also common in the rocks of southern Arizona (especially Cochise County) and in southwestern and south-central New Mexico. In New Mexico, the Bliss Formation contains at least 19 species of trilobites as well as numerous brachiopods and conodonts.

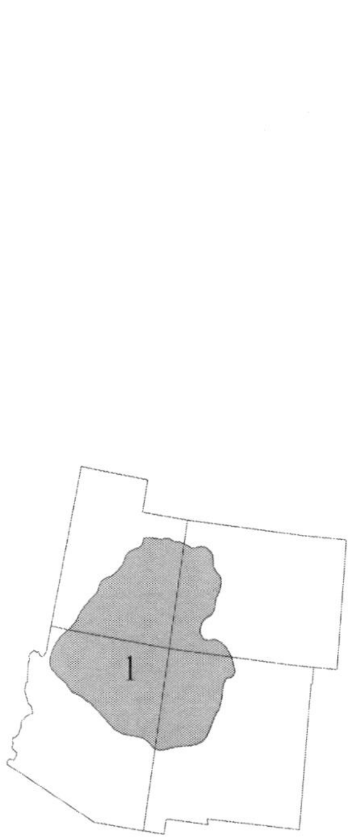

The early Ordovician El Paso Formation and overlying Montoya Formation extend from southern Arizona and New Mexico into western Texas. More than 400 species of marine invertebrates have been reported from these two units, including trilobites, corals, brachiopods, bryozoans, nautiloids, gastropods, **graptolites**, and the oldest known crinoids in the world (*Figures 3.45 and*

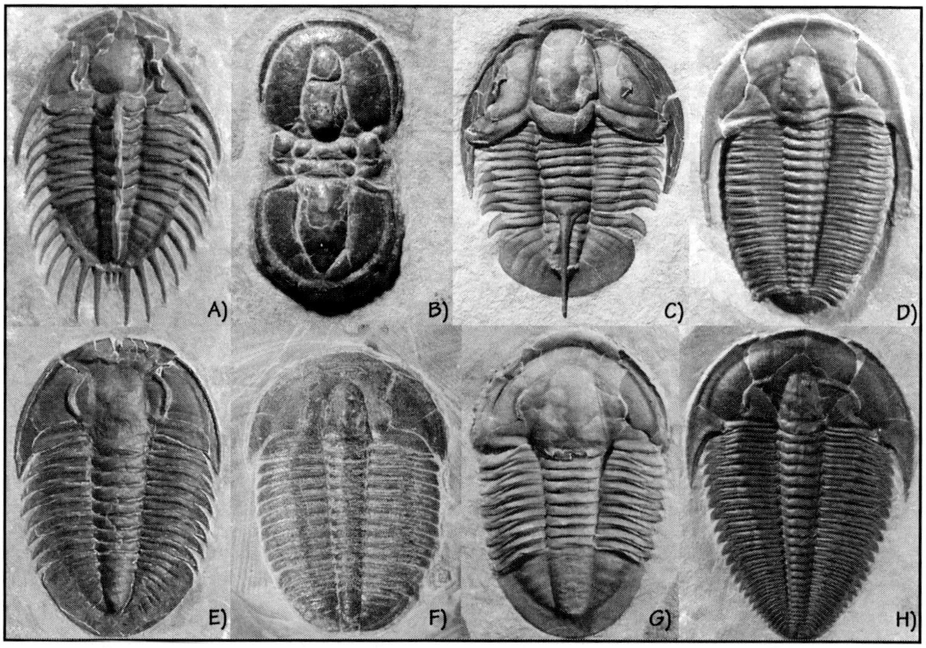

Figure 3.43: Trilobites from the middle Cambrian of western Utah. A) Kootenia spencei, approximately 6 centimeters (2 inches) long. B) Itagnostus interstictus, approximately 0.6 centimeters (0.2 inches) long. C) Genevievella granulata, approximately 3 centimeters (1.3 inches) long. D) Chancia ebdome, approximately 4 centimeters (1.6 inches) long. E) Athabaskia bithus, approximately 5 centimeters (2 inches) long. F) Elrathia kingii, approximately 2.5 centimeters (1 inch) long. G) Asaphiscus wheeleri, approximately 6 centimeters (2.4 inches) long. H) Amecephalus idahoense, approximately 5 centimeters (2 inches) long.

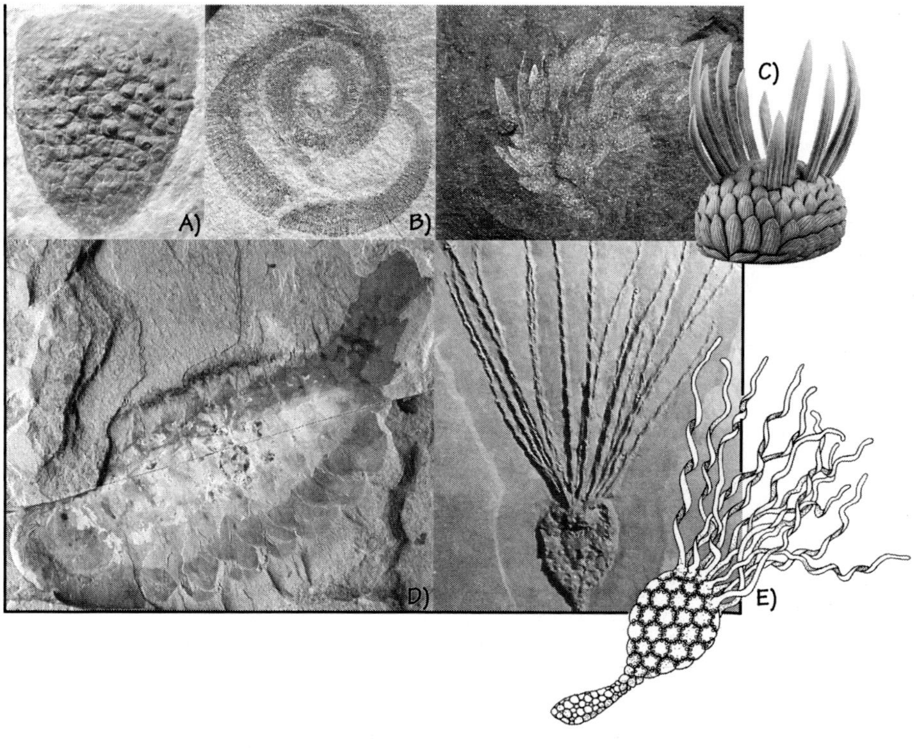

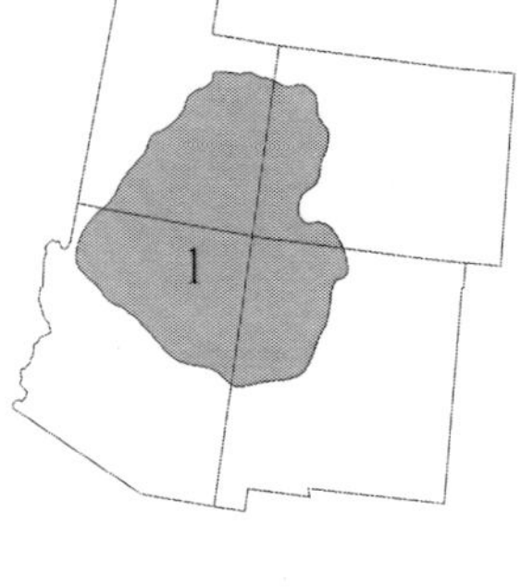

Trilobites

Trilobites are iconic Paleozoic fossils; they were more common in the Cambrian and Ordovician than in later periods, and became extinct at the end of the Permian. They were marine arthropods, and had well-defined head, tail, and thoracic (leg-bearing) segments. Most had large compound eyes, often with lenses that are visible to the naked eye. In life, they had antennae like many other arthropods, but since these were not mineralized, they only fossilize under exceptional circumstances. Many could roll up for protection, and several species also had large spines.

Sponges

Sponges (Phylum Porifera) are the simplest major group of animals; their earliest fossils appear in the late Precambrian. Most modern sponges live in the ocean and usually have basket-shaped bodies. They live by filtering food and oxygen out of water pumped in through openings in their body walls and out through a larger opening at the top. The familiar bath sponge has no mineralized skeleton, but many other kinds of sponges have skeletons composed of tiny structures called spicules, which are made of calcium carbonate ($CaCO_3$) or silica (SiO_2). It is these skeletonized sponges that have the greatest likelihood of becoming fossils. Over their long history, such sponges have frequently been important contributors to reefs and reef-like mounds.

Figure 3.44 (AT LEFT): Exceptionally preserved Cambrian fossils from western Utah Lagerstätten. A) Sponge, Valospongia, *approximately 17 centimeters (7 inches) tall. B) Worm,* Wronascolex ratcliffei, *approximately 12.5 centimeters (5 inches) long. C)* Wiwaxia, *fossil and reconstruction, approximately 3.6 centimeters (1.4 inches) long.* Wiwaxia*'s classification is subject to debate, and it is currently classified as either an annelid or mollusk. D) Arthropod,* Nettapezoura basilica, *approximately 13.5 centimeters (5 inches) long. E) Echinoderm,* Gogia, *fossil and reconstruction, approximately 5 centimeters (2 inches) long.* Gogia *belongs to a primitive group of echinoderms known as eocrinoids, which existed from the Cambrian to Silurian periods.*

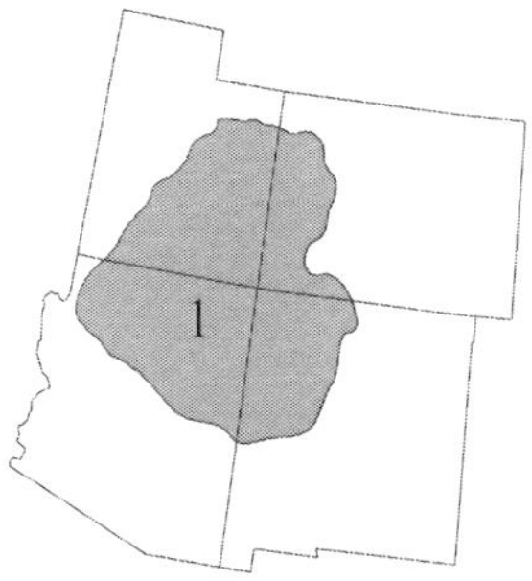

3 Fossils

Region 2

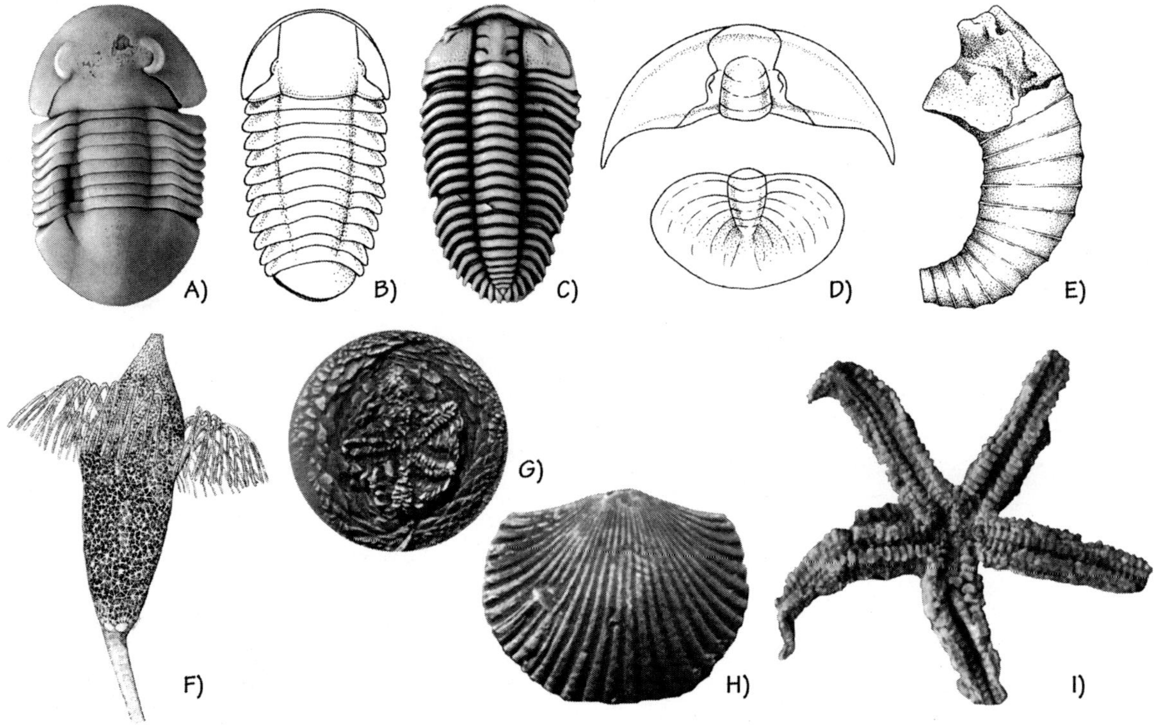

Figure 3.45: Ordovician marine invertebrates from the Basin and Range region (Bliss, El Paso, Montoya formations). A) Trilobite, Presbynileus, *approximately 3.7 centimeters (1.5 inches) long. B) Trilobite,* Illaenurus, *approximately 4 centimeters (1.6 inches) long. C) Trilobite,* Hintzeia, *approximately 2.5 centimeters (1 inch) long. D) Trilobite,* Briscoia *(head and tail), approximately 2 centimeters (0.75 inches) wide. E) Nautiloid,* Bisonoceras, *approximately 10 centimeters (4 inches) long. F) Restoration of the world's oldest known crinoid,* Titanocrinus sumralli, *from the Ordovician of Utah, approximately 20 centimeters (8 inches) tall. G) Edrioasteroid (extinct group of encrusting echinoderms), approximately 7 millimeters (0.3 inches) wide. H) Brachiopod,* Shoshonorthis, *approximately 2 centimeters (0.75 inches) long. I) Sea star, 4 centimeters (1.6 inches) wide.*

Carboniferous • *a geologic time period that extends from 359 to 299 million years ago.*

uplift • *upward movement of the crust due to compression, subduction, or mountain building.*

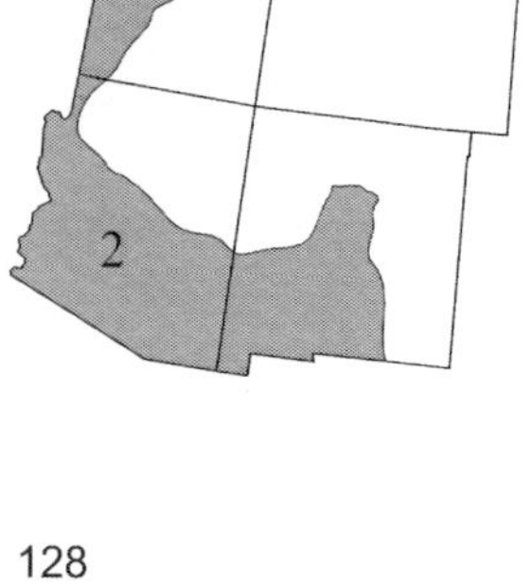

3.46). Ordovician rocks also occur in scattered outcrops across west-central and northwest Utah, and contain some of the most diverse and best preserved marine assemblages of this age anywhere in the world. Silurian-aged rocks are much less abundant in the Basin and Range than are older or younger rocks. In Millard County, Utah, Silurian rocks contain corals and brachiopods. In southern New Mexico and central Arizona, middle and late Devonian deposits contain abundant and diverse marine fossils, especially brachiopods, corals, and bryozoans (*Figures 3.47–3.49*).

Carboniferous rocks in the Basin and Range represent an archipelago of warm, shallow seaways and **uplifted** islands. The Mississippian Escabrosa Limestone in south-central Arizona is similar to the Grand Canyon's Redwall Limestone. Although many fossils in the Escabrosa are not as well preserved as those in the Redwall, the unit still contains abundant crinoids, mollusks,

calyx • *the head of a crinoid.*

Crinoids

Crinoids are echinoderms, related to sea urchins and sea stars. These invertebrate animals feed by using their arms to filter food out of the water. Most are attached to the sediment by a stalk that ends in a root-like structure called the holdfast—some forms, however, are free floating. Crinoid fossils are most commonly found as "columnals," pieces of the stalk that hold the head (*calyx*) above the surface. The calyx and the holdfast are only occasionally preserved as fossils. Crinoids are still around today; those in shallow water are mostly stalkless, while those with stalks are restricted to deep water.

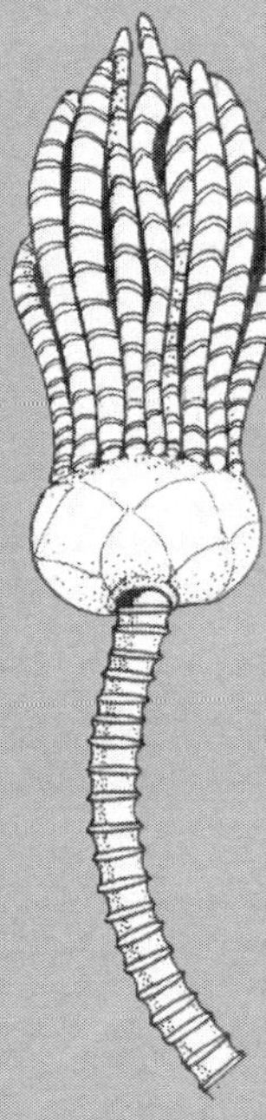

Crown and stem, about 15 centimeters (6 inches) long.

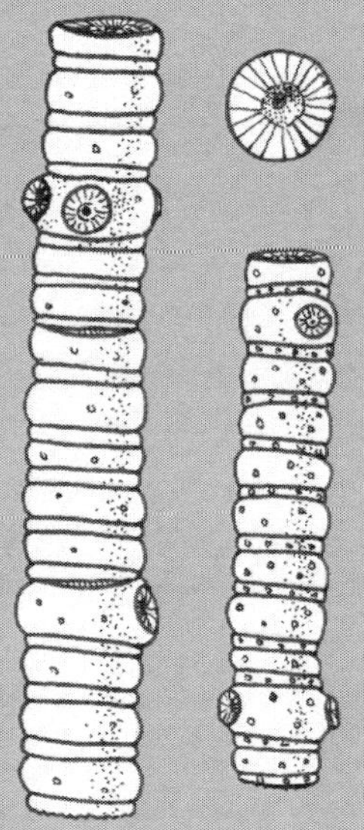

Stem fragments.

bivalves, brachiopods, corals, bryozoans, and foraminifera (*Figure 3.50*). Mississippian marine invertebrates are also abundant in Utah, and there is a strong Pennsylvanian fossil record in New Mexico. Fusulinid foraminifera (*Figure 3.51*) are the most important group for biostratigraphy in the Carboniferous and Permian rocks of the Basin and Range. More than 200 species of these creatures have been described from Pennsylvanian and Permian rocks in the

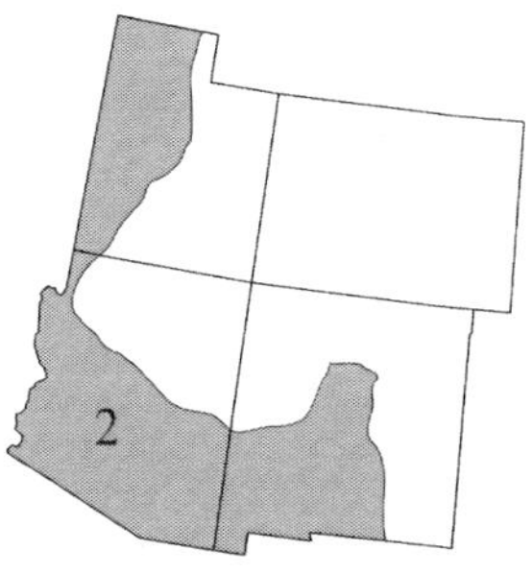

Figure 3.46: Graptolites from the Basin and Range region. A) Didymograptus*, scale in centimeters. B)* Phyllograptus*, approximately 3.5 centimeters (1.4 inches) long. C)* Clonograptus*, scale in centimeters.*

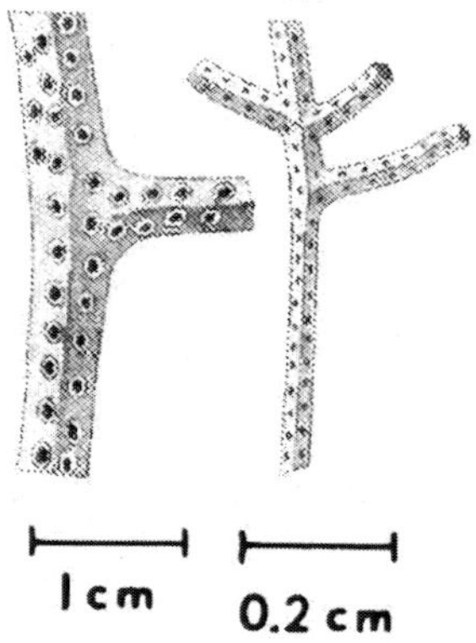

Figure 3.47: Thamnopora*, a Devonian bryozoan from Utah.*

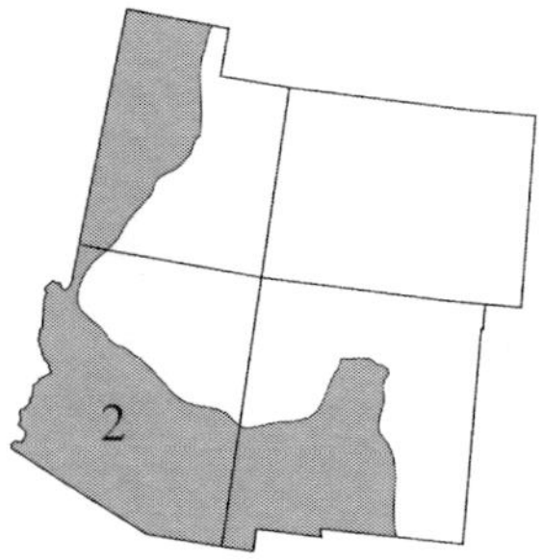

Graptolites

Graptolites (meaning "rock writing") are an extinct group of colonial, free-floating organisms. They lived from the Cambrian to the Carboniferous, and were relatives of modern hemi*chordates* such as acorn worms. Graptolites are frequently preserved as thin, black, sawblade-like streaks across black shale; tiny cups along these structures held individual animals. Graptolites are often useful as index fossils.

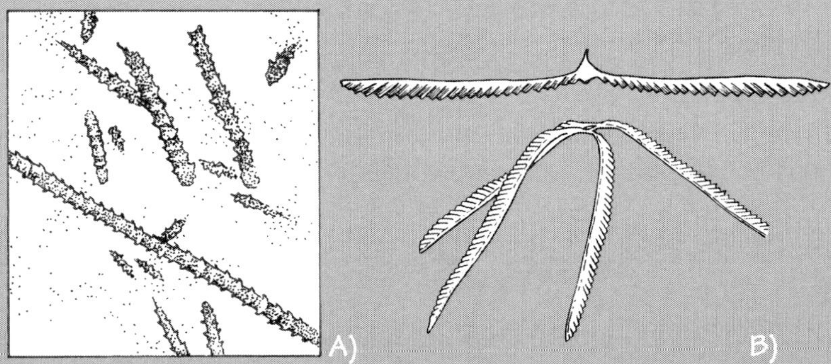

A) Specimen with many fragments of colonies of Climacograptus. *Slab is 7.5 centimeters (3 inches) on each side. B) Restoration of what graptolite colonies may have looked like when they were alive, floating in the water.*

chordate • *an animal that has a notochord, a hollow dorsal nerve cord, pharyngeal gill slits, an endostyle, and a post-anal tail during at least one stage of its development.*

Big Hatchet Mountains of Hidalgo County in southwestern New Mexico. Other Pennsylvanian marine invertebrates found in the Basin and Range include corals, bryozoans, gastropods, bivalves, trilobites, and (as usual) brachiopods. The Pennsylvanian Naco Formation near Payson, Arizona (just south of the southern edge of the Colorado Plateau) also contains the teeth of many types of fossil sharks (*Figure 3.52*). Nonmarine Pennsylvanian fossils in central and southern New Mexico include freshwater fishes, plants, and insects, as well as terrestrial amphibians and reptiles.

As the ocean receded from the Southwest during the Permian, the Basin and Range transitioned to a mainly terrestrial environment. Marine sediments are exposed in a few places—red beds in the Los Pinos Mountains of Socorro County, New Mexico contain abundant fossil lungfish burrows, and Permian outcrops in western Utah contain fusulinids, conodonts, bivalves, and brachiopods (*Figure 3.53*). In New Mexico's Permian rocks, the footprints of terrestrial vertebrates are widespread and abundant. Reptile and amphibian prints in the Abo and Robledo Mountain formations, for example, are the basis

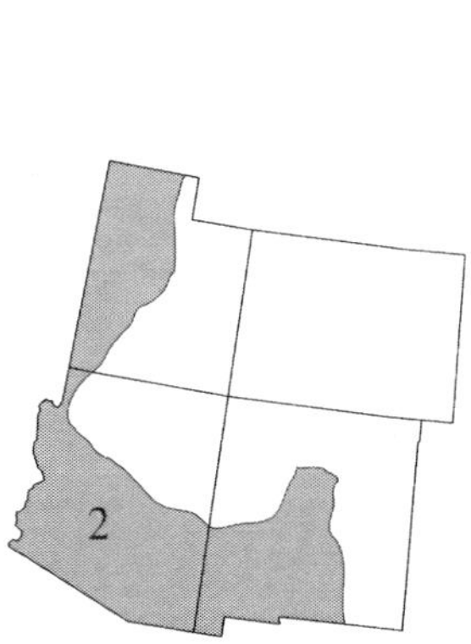

3 Fossils

Region 2

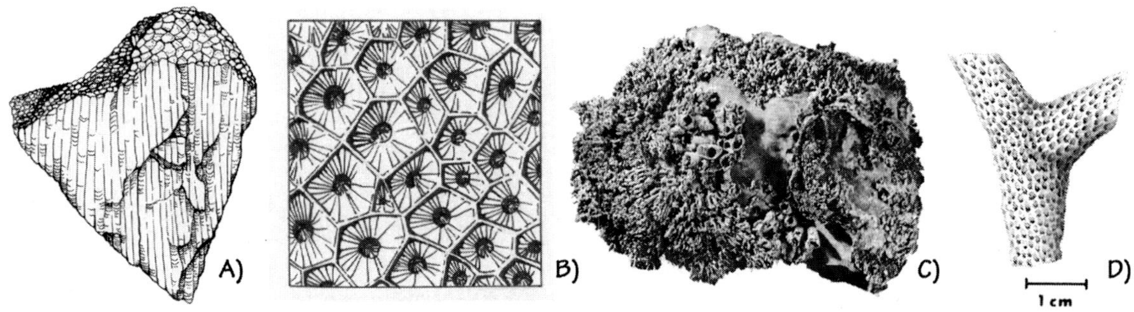

Figure 3.48: Silurian and Devonian corals from the Basin and Range region. A) Colonial tabulate coral, Favosites, *approximately 12 centimeters (5 inches) wide. B) Colonial rugose coral,* Hexagonaria *(polished section), individual corallites approximately 12 millimeters (0.4 inches) in diameter. C) Fossil assemblage consisting of the corals* Diphyllum *and* Syringopora, *and stromatoporoid sponge; approximately 20 centimeters (8 inches) wide. D) Branching tabulate coral,* Cladopora.

sessile • *unable to move, as in an organism that is permanently attached to its substrate.*

rugose coral • *an extinct group of corals that were prevalent from the Ordovician through the Permian.*

tabulate coral • *an extinct form of colonial coral that often formed honeycomb-shaped colonies of hexagonal cells.*

Corals

Corals are *sessile* relatives of jellyfish and sea anemones. They possess stinging tentacles, which they use to feed on small planktonic prey. Each group of coral possesses distinctly shaped "cups" that hold individual animals, or polyps. Colonial corals live in colonies of hundreds or even thousands of individuals that are attached to one another. Solitary coral lives independently, as a single isolated polyp.

Rugose corals were both colonial and solitary. (Solitary forms are often called "horn corals.") *Tabulate corals* were exclusively colonial and produced a variety of shapes, including sheetlike and chainlike forms. These corals receive their name from the table-like horizontal partitions within their chambers. Both rugose and tabulate corals went extinct at the end of the Permian. Modern corals—*scleractinians*—first appeared in the Triassic, and include both solitary and colonial species. Many scleractinian corals have photosynthetic symbiotic algae in their tissues, called zooxanthellae. These algae provide nutrition to the coral polyps, helping them to grow more rapidly.

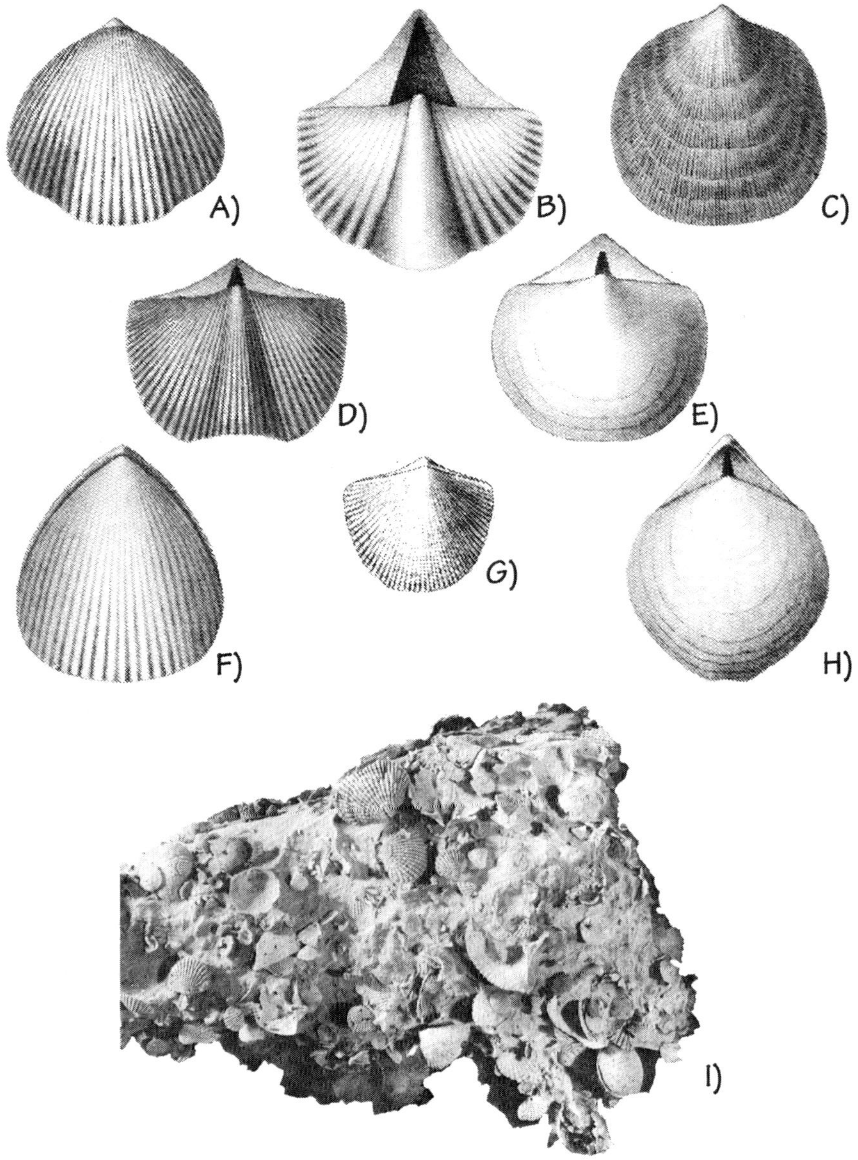

Figure 3.49: Devonian brachiopods from the Basin and Range region. A) Camarotechia, *3 centimeters (1.2 inches) wide. B)* Allanaria, *3 centimeters (1.2 inches) wide. C)* Rhipodomella, *3 centimeters (1.2 inches) wide. D)* Cyrtiopsis, *3 centimeters (1.2 inches) wide. E)* Ambothyris, *1 centimeter (0.4 inches) wide. F)* Paurorhynchia, *3 centimeters (1.2 inches) wide. G)* Atrypa, *1.5 centimeters (0.6 inches) wide. H)* Stringocephalus, *3 centimeters (1.2 inches) wide. I) Cluster of Devonian brachiopods from Arizona (*Schizophoria iowensis*), approximately 20 centimeters (8 inches) wide. The original calcite shells have been replaced with silica.*

for the recent establishment of Prehistoric Trackways National Monument near Las Cruces in Doña Ana County (*Figure 3.54*). This area contains one of the most abundant assemblages of nonmarine Paleozoic trace fossils in the world, with more than 20 different kinds of traces, and it has been called an "ichnofossil **Lagerstätte**."

scleractinian coral *• a colonial or solitary marine invertebrate animal characterized by an encrusting calcareous skeleton enclosing polyps that capture prey with small tentacles equipped with stinging cells (nematocysts).*

Lagerstätte *• fossil deposit containing animals or plants that are preserved unusually well, sometimes even including the soft organic tissues.*

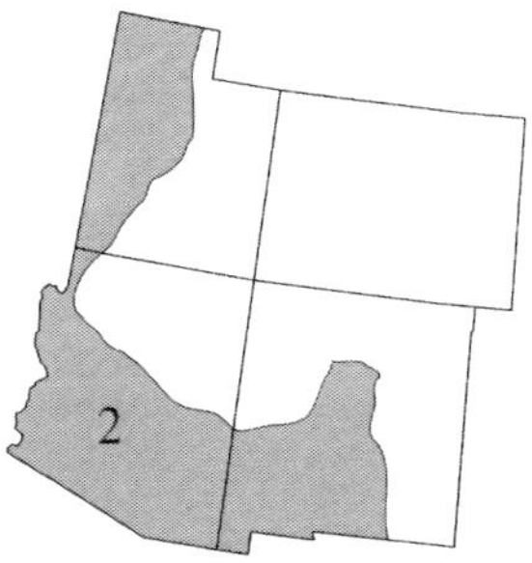

Brachiopods

Brachiopods are filter-feeding animals that have two shells and are superficially similar to bivalves (such as clams). Instead of being mirror images between shells (symmetrical like your hands), brachiopod shells are mirror images across each shell (symmetrical like your face). There are two major types of brachiopod shells, distinguished by how the two valves connect to each other: articulate brachiopods have tooth-and-socket hinges that tightly interlock, whereas inarticulate brachiopod shells lack hinge structures entirely. Internally, brachiopods are substantially different from bivalves, with a lophophore (filter-feeding organ made of thousands of tiny tentacles), and a small and simple gut and other organs. Bivalves, in contrast, have a fleshier body and collect their food with large gills.

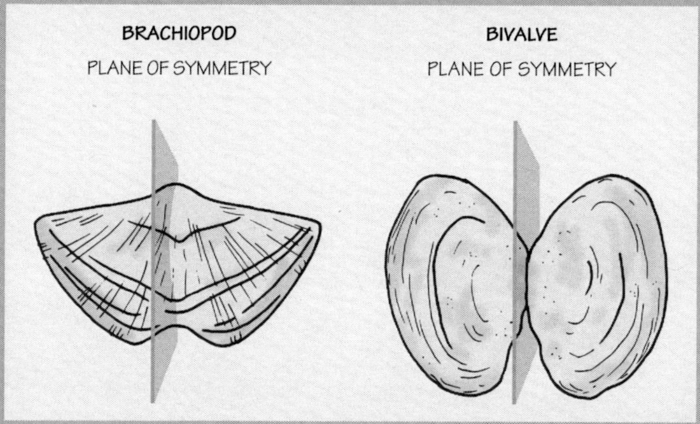

The difference between the shells of a typical brachiopod (left) and a typical bivalve mollusk (right). Most brachiopods have a plane of symmetry across the valves (shells), whereas most bivalves have a plane of symmetry between the valves.

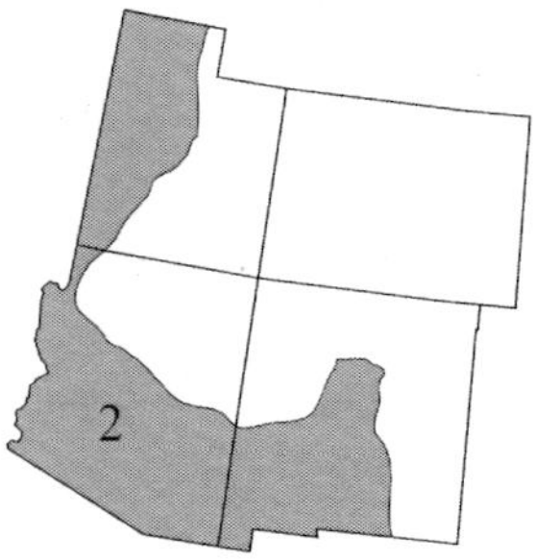

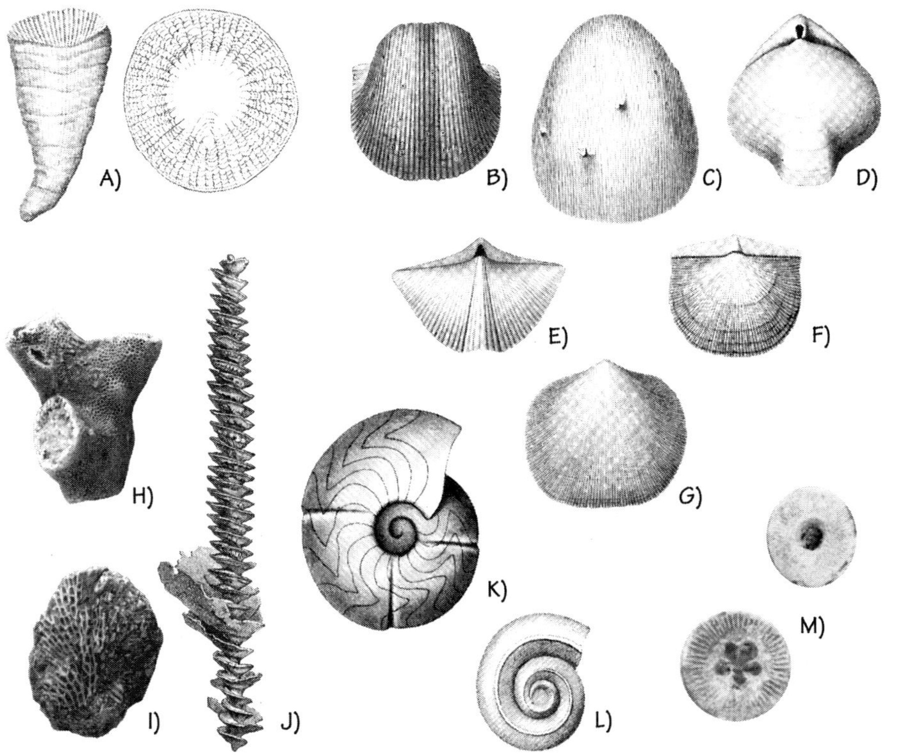

Figure 3.50: Carboniferous marine invertebrates from the Basin and Range. A) Solitary rugose coral, Caninia, *side and top view, approximately 2.75 centimeters (1.08 inches) tall. B) Mississippian brachiopod,* Inflatia inflatus, *approximately 3 centimeters (1.2 inches) wide. C) Mississippian brachiopod,* Ovatia (Linoproductus) ovatus, *approximately 1.6 centimeters (0.6 inches) wide. D) Pennsylvanian brachiopod,* Composita trilobata, *approximately 2 centimeters (0.8 inches) wide. E) Mississippian brachiopod,* Spirifer centronatus, *approximately 3.5 centimeters (1.4 inches) wide. F) Pennsylvanian brachiopod,* Derbya crassa, *approximately 2.5 centimeters (1 inch) wide. G) Pennsylvanian brachiopod,* Schizopora texana, *approximately 2.5 centimeters (1 inch) wide. H) Bryozoan,* Anisotrypa, *approximately 2 centimeters (0.8 inches) tall. I) Bryozoan,* Fenestella, *approximately 2 centimeters (0.8 inches) tall. J) Bryozoan,* Archimedes, *approximately 5 centimeters (2 inches) tall. K) Ammonoid,* Cravenoceras hesperium, *approximately 1.75 centimeters (0.7 inches) in diameter. L) Gastropod,* Euomphalus utahensis, *approximately 1.3 centimeters (0.5 inches) diameter. M) Crinoid stem columnals, approximately 1 centimeter (0.4 inches) in diameter.*

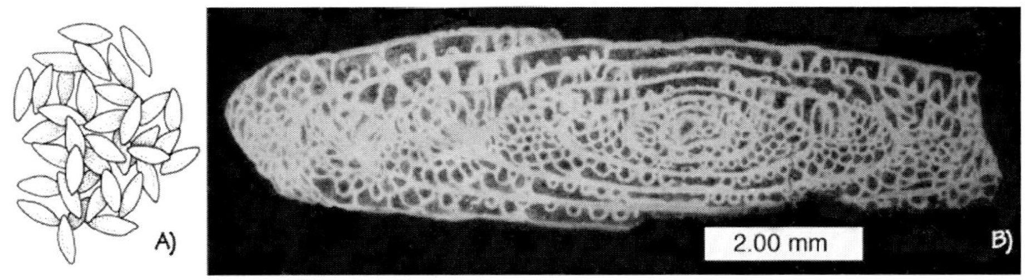

Figure 3.51: Single-celled fusulinid foraminifera from the Pennsylvanian. A) A cluster of the shells, roughly the size and shape of large rice grains. B) Photograph of a cross-section through a single fusulinid, as seen through a microscope.

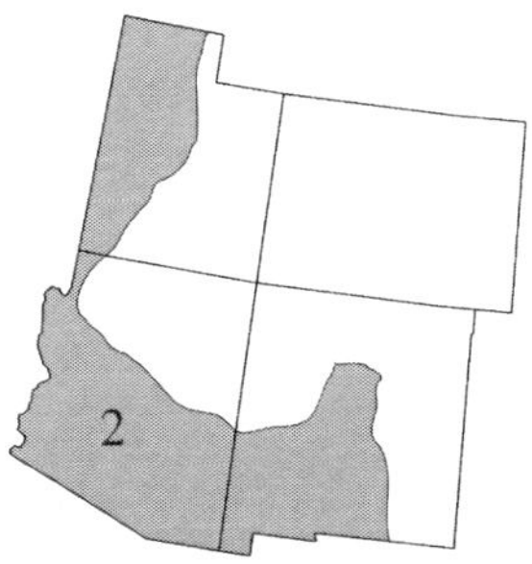

3 Fossils

Region 2

rudists • an extinct group of box- or tube-shaped bivalves that arose during the Jurassic.

tetrapod • the first four-limbed animals (early land vertebrates) and all of their descendants.

Figure 3.52: Pennsylvanian shark, Petalodus. *A) Tooth. B) Restoration, approximately 1 meter (3 feet) long.*

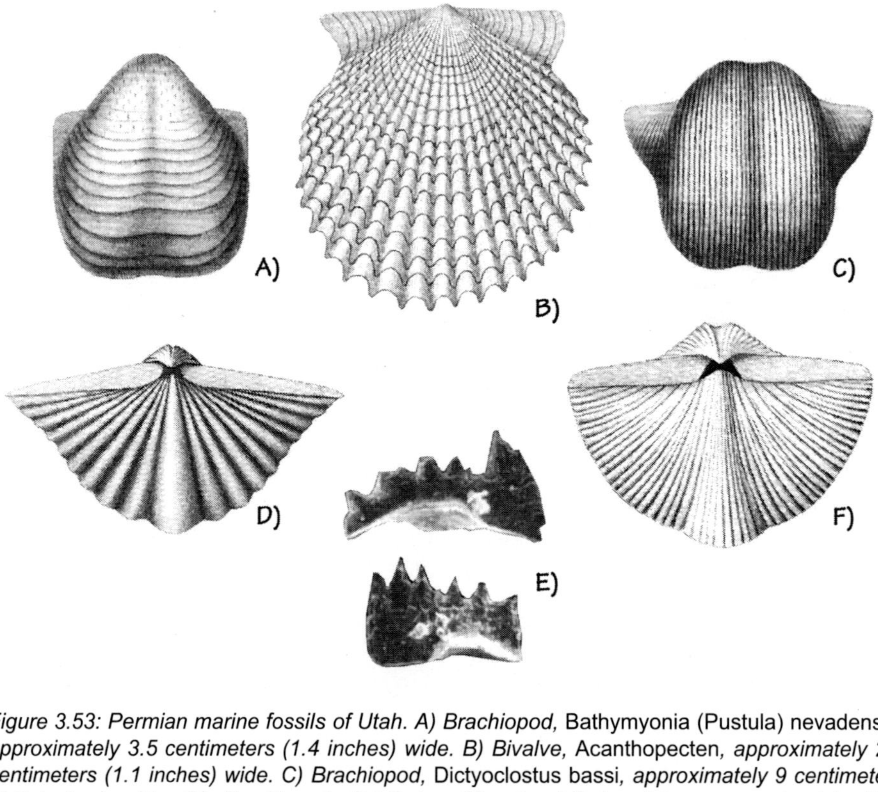

Figure 3.53: Permian marine fossils of Utah. A) Brachiopod, Bathymyonia (Pustula) nevadensis, *approximately 3.5 centimeters (1.4 inches) wide. B) Bivalve,* Acanthopecten, *approximately 2.6 centimeters (1.1 inches) wide. C) Brachiopod,* Dictyoclostus bassi, *approximately 9 centimeters (3.5 inches) wide. D) Brachiopod,* Spiriferina (Punctospirifer) pulcher, *approximately 5.25 centimeters (2 inches) wide. E) Conodont elements, approximately 0.4 millimeters long. F) Brachiopod,* Neospirifer triplicatus, *approximately 4 centimeters (1.6 inches) wide.*

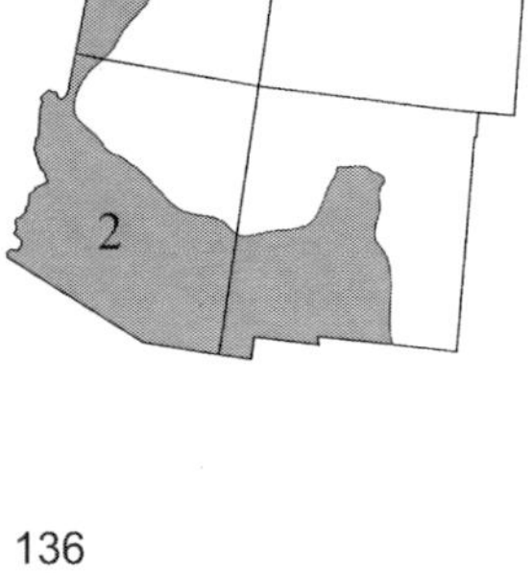

Region 2

Neogene • *the geologic time period extending from 23 to 2.6 million years ago.*

mastodon • *an extinct terrestrial mammal belonging to the Order Proboscidea, characterized by an elephant-like shape and size, and massive molar teeth with conical projections.*

mammoth • *an extinct terrestrial mammal belonging to the Order Proboscidea, from the same line that gave rise to African and Asian elephants.*

Figure 3.54: Possible footprint of Dimetrodon *(see also Figure 3.12), from the Abo Formation at Prehistoric Trackways National Monument.*

Mesozoic marine rocks are uncommon in the Basin and Range. While marine environments spread over the area multiple times during the Mesozoic, many of the corresponding rocks have been destroyed by erosion or buried by other sediments. Outcrops containing marine fossils include the Triassic Moenkopi, Dinwoody, and Thaynes formations; the Jurassic Twin Creek, Carmel, and Curtis formations; and the Cretaceous Aspen-Mowry Shale and Dakota beds. Ammonoids, snails, fish scales, and bivalves are found in these rocks (*Figures 3.55 and 3.56*); **rudistid** bivalves are uncommon, but have been found at a number of Cretaceous localities in New Mexico. In Washington County, southwesternmost Utah, late Triassic-aged rocks exposed at the St. George Dinosaur Discovery Site preserve an extraordinary diversity and abundance of trackways, as well as the skeletal remains of dinosaurs and other **tetrapods**, fish, and terrestrial plants (*see Figures 3.17–3.19 and 3.21*).

Above the Cretaceous-Paleogene (K-Pg) boundary, the rocks of southeastern and west-central Arizona and central and southern New Mexico are rich in **Neogene** land mammals (*Figures 3.57–3.59*). These deposits include proboscideans such as *Gomphotherium, Cuvieronius,* and *Stegomastodon,* as well as the American **mastodon** *(Mammut americanum*) and at least two species of **mammoth** (*Mammuthus columbi* and *M. meridionalis*). Also present are camels (*Stenomylus, Protolabis, Michenia, Megacamelus, Megatylopus, Hemiauchenia, Procamelus)*, rhinos (*Diceratherium*, *Teleoceras*), carnivores (*Borophagus*), horses (*Dinohippus, Onohippidium*, *Nannippus*, *Equus giganteus, E. simplicidens,* and *E. occidentalis*), ground sloths (*Megalonyx leptostomus, Glossotherium chapadmalense*), and the glyptodont *Glyptotherium texanum*, as well as many kinds of rodents (including the earliest known species of porcupine) and rabbits.

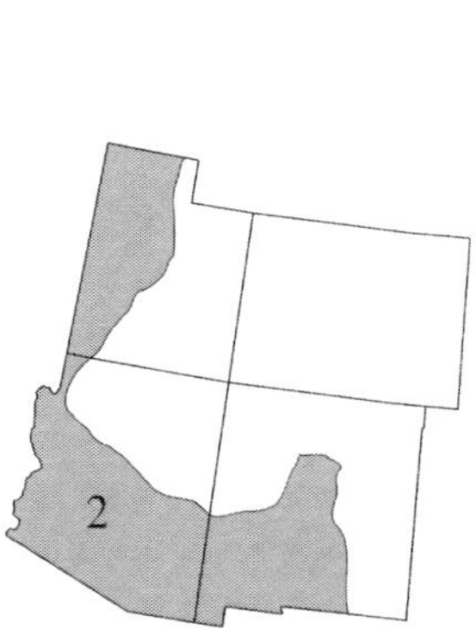

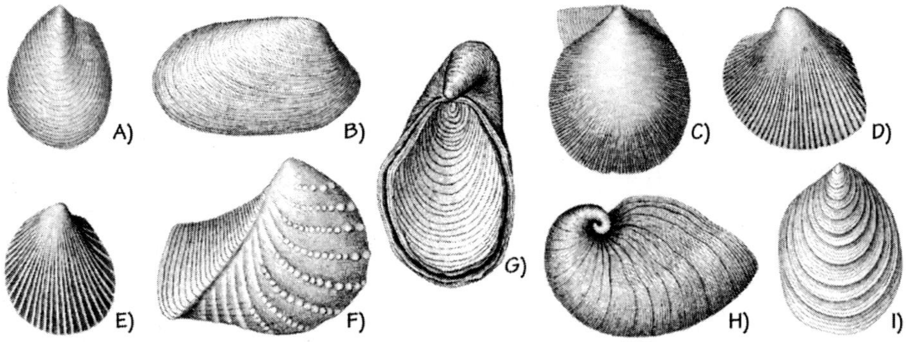

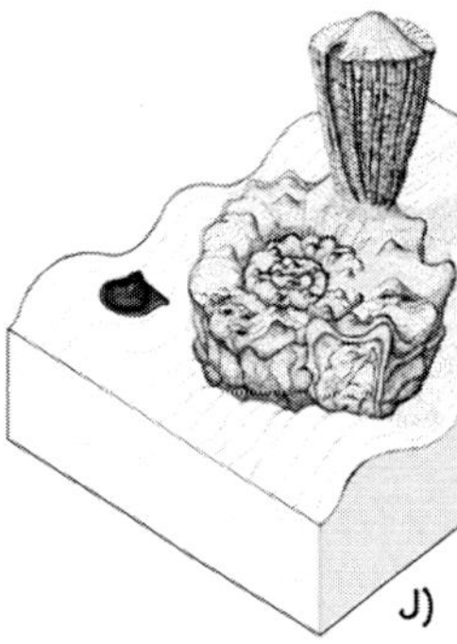

Figure 3.55: Mesozoic bivalves of the Basin and Range region. A) Claria, *Triassic, 2.25 centimeters (0.9 inches) long. B)* Unio, *Triassic, 3.5 centimeters (1.4 inches) long. C)* Camptonectes platessiformis, *Jurassic, 3 centimeters (1.3 inches) long. D)* Cardium curtum, *Cretaceous, 1.5 centimeters (0.6 inches) across. E)* Meleagrinella (Eumicrotis) curta, *Jurassic, 2.5 centimeters (1 inch) across. F)* Myophorella montanensis, *Jurassic, 5 centimeters (2 inches) across. G)* Gryphaea nebrascensis, *Jurassic, 3 centimeters (1.3 inches) across. H)* Exogyra columbella, *Cretaceous, 3 centimeters (1.3 inches) across. I)* Inoceramus labiatus, *Cretaceous, 3.5 centimeters (1.4 inches) across. J) Reconstruction of a rudistid bivalve growing on a dead ammonoid shell on the sea floor.*

Bivalves

Clams and their relatives, such as mussels, scallops, and oysters, are mollusks possessing a pair of typically symmetrical shells. Most are filter feeders, collecting food with their gills. Paleozoic bivalves typically lived on the surface of the sediment ("epifaunally"), but in the Mesozoic they evolved the ability to burrow more deeply into the sediment, becoming "infaunal." This innovation led to the rapid evolution of a large number of groups present in today's oceans.

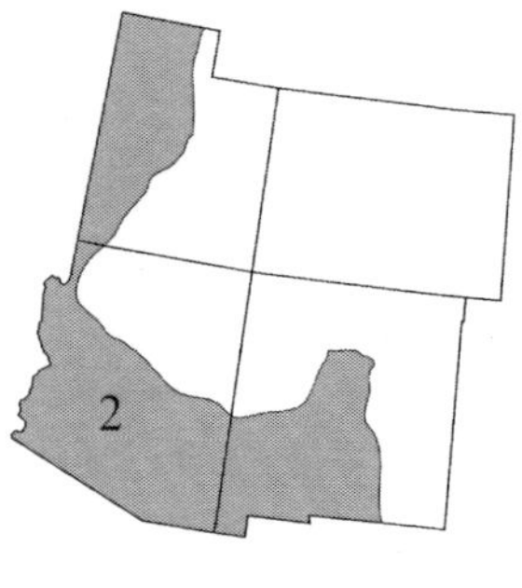

Figure 3.56: Mesozoic ammonoids of Utah. A) Columbites, *Triassic, 4.75 centimeters (1.9 inches) in diameter. B)* Anasibrites, *Triassic, 2.5 centimeters (1 inch) in diameter. C)* Flemingites, *Triassic, 3 centimeters (1.2 inches) in diameter. D)* Meekoceras, *Triassic, 3 centimeters (1.2 inches) in diameter. E)* Scaphites warreni, *Cretaceous, 3 centimeters (1.2 inches) in diameter. F)* Cadoceras, *Jurassic, 5 centimeters (2 inches) in diameter. G)* Scaphites ventricosus, *Cretaceous, 4.75 centimeters (1.9 inches) in diameter. H)* Clioscaphites vermiformis, *Cretaceous, 4.75 centimeters (1.9 inches) in diameter. I)* Collignicera woolgari, *Cretaceous, 3 centimeters (1.3 inches) in diameter.*

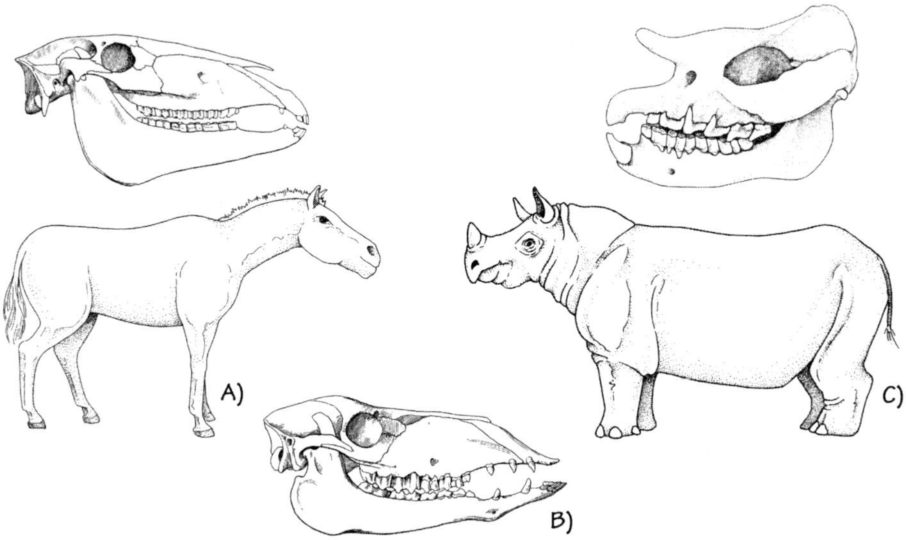

Figure 3.57: Mammals of the Basin and Range region. A) Horse, Equus simplicidens, *skull and reconstruction, height 110–145 centimeters (43–57 inches) at the shoulder. B) Camel,* Procamelus, *skull, approximately 28 centimeters (11 inches) long. C) Rhinoceros,* Teleoceras, *skull and restoration, body approximately 4 meters (13 feet) long.*

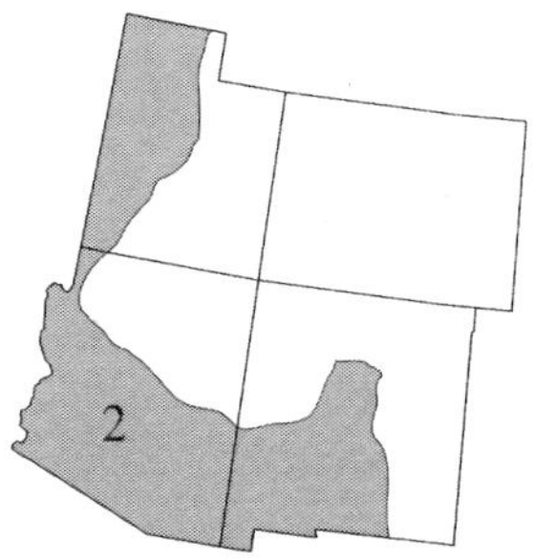

Region 2

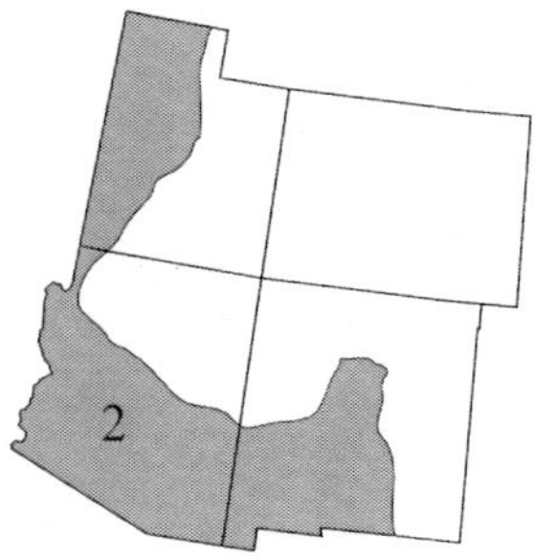

Ammonoids

Ammonoids are a major group of cephalopods that lived from the Devonian to the end of the Cretaceous. Both nautiloids (the group that today contains the chambered nautilus) and ammonoids have chambered shells subdivided by walls, or septa (plural of septum). These shells are frequently, but not always, coiled. The term "ammonoid" refers to the larger group of these extinct cephalopods, distinguished by complex, folded septa. Within ammonoids, "ammonites" is a smaller sub-group, distinguished by the extremely complex form of their septa. Ammonites were restricted to the Jurassic and Cretaceous periods. The form of the septa in nautiloids and ammonoids is not visible in a complete shell; it is most often seen in the trace of the intersection between the septum and the external shell. This trace is called a suture. Sutures are usually visible in fossils when sediment has filled the chambers of a shell, and the external shell has been broken or eroded away.

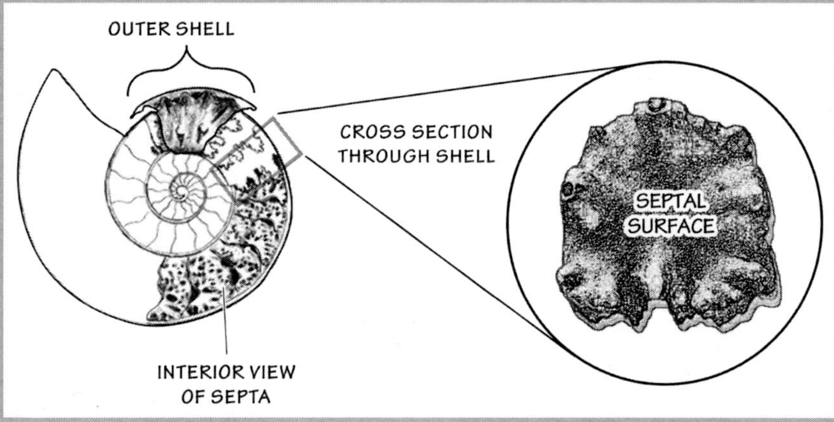

Ammonite shell break-away cross-section; surface plane of a septum and sediment-filled chamber.

Figure 3.58 (AT RIGHT): Glyptodont. A) Skeleton, with and without the external armor. B) Detail of the bony scutes that formed the solid outer armor. Glyptodonts reached lengths of up to 3 meters (10 feet).

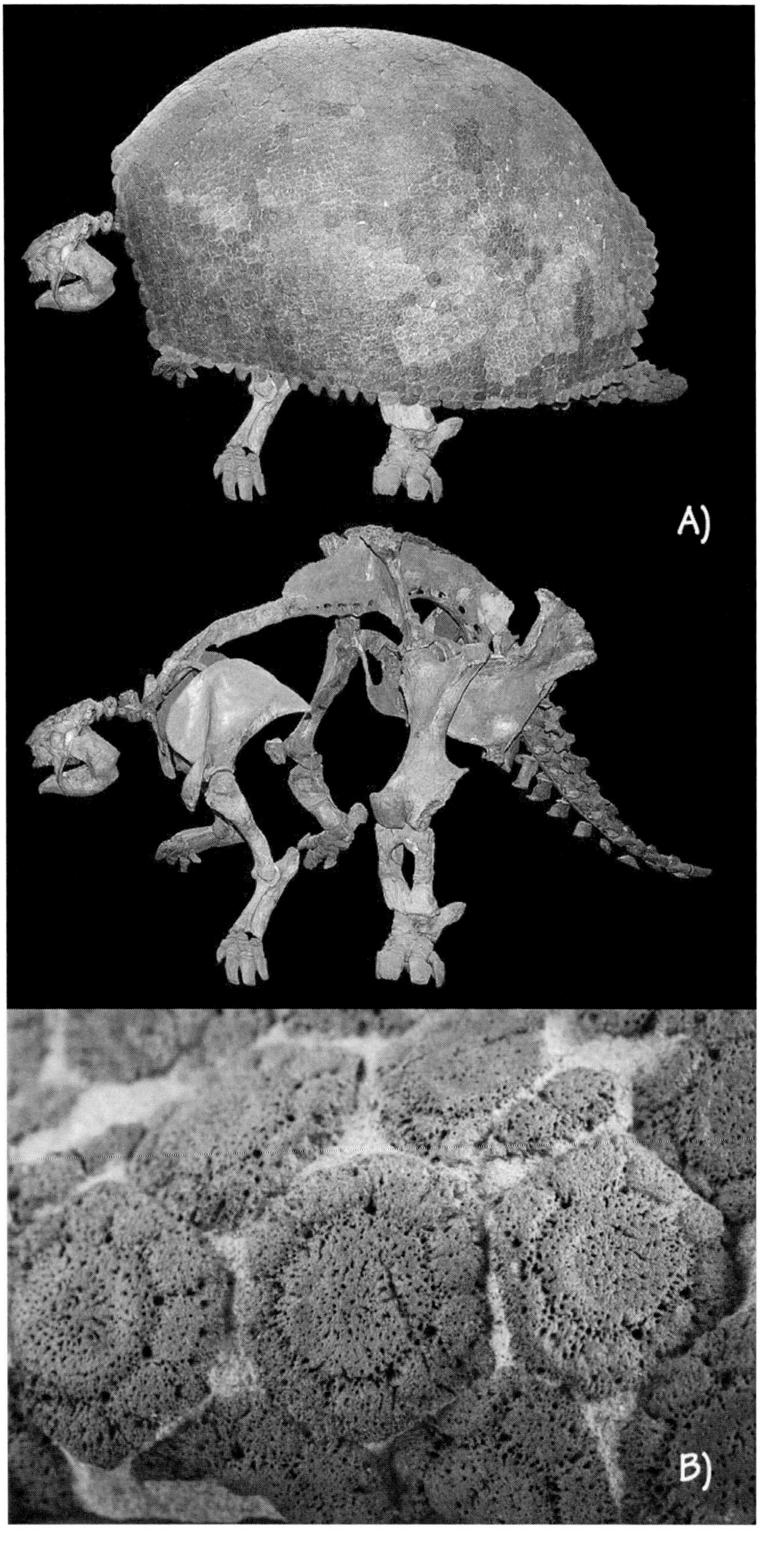
A)
B)

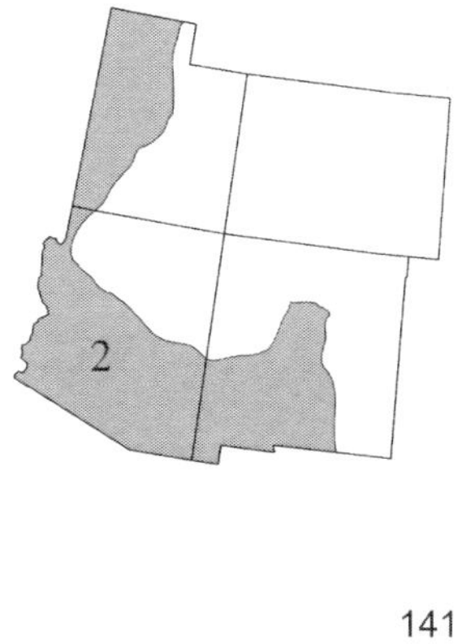
2

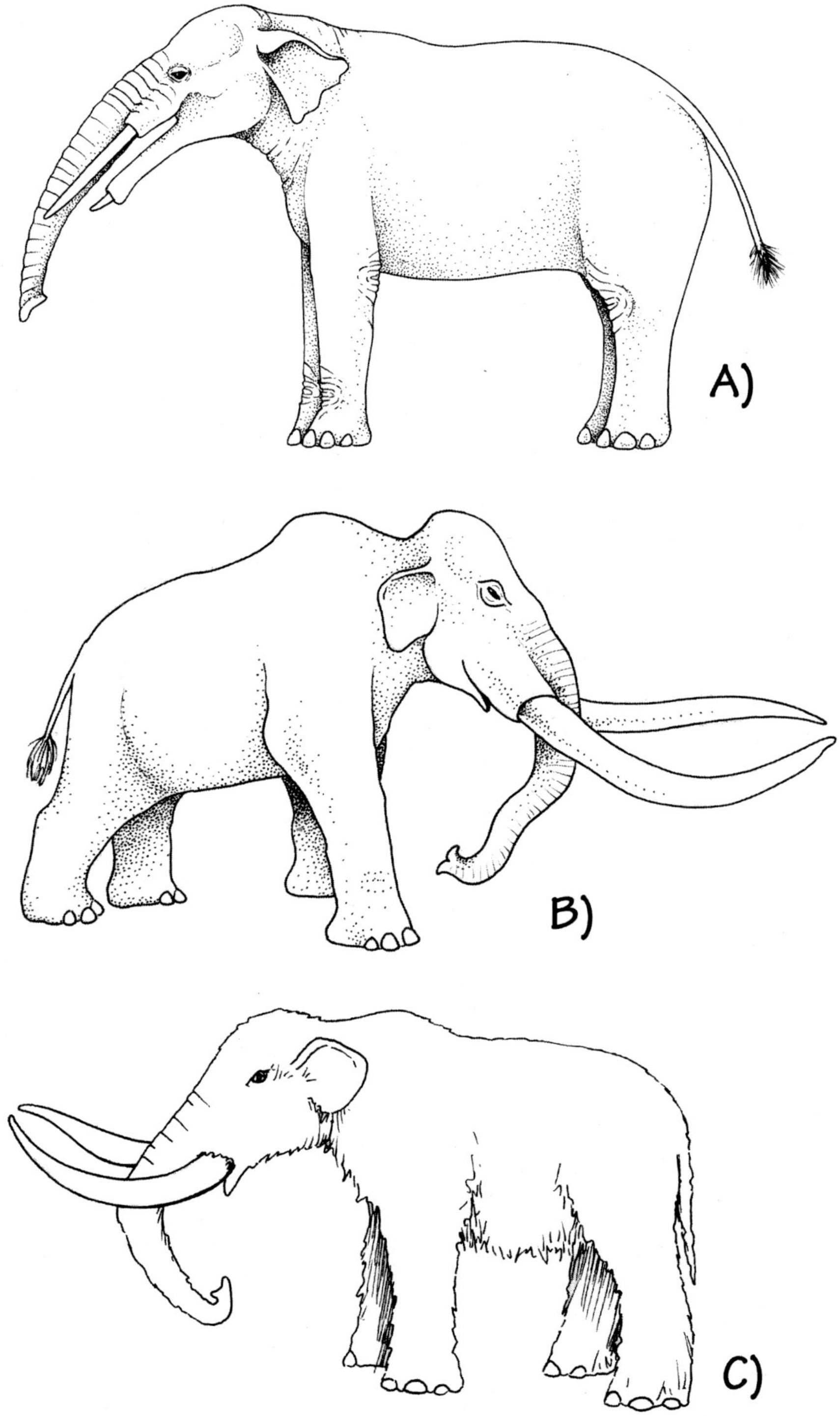

Figure 3.59: Restorations of A) Gomphotherium, *approximately 2.3 meters (7.5 feet) high at the shoulder. B) Columbian mammoth,* Mammuthus columbi, *approximately 4 meters (13 feet) high at the shoulder. C) American mastodon,* Mammut americanum, *approximately 2.3 meters (7.5 feet) high at the shoulder.*

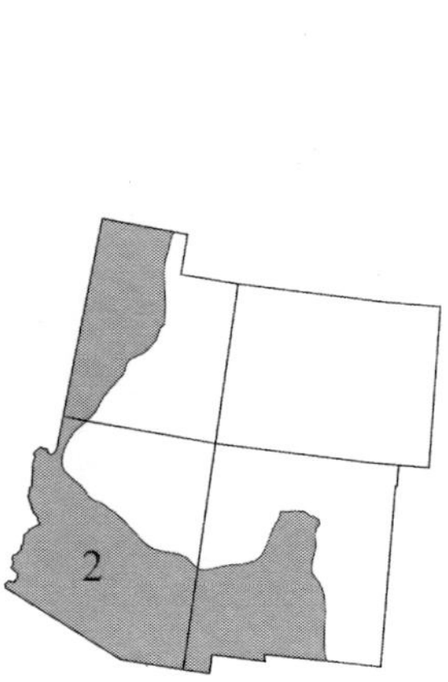

Mastodons and Mammoths

These two kinds of ancient elephants (or, more technically, proboscideans) are frequently confused. Both were common during the Pleistocene, but they had different ecological preferences and are usually found separately. Mammoths are close cousins of modern African and Asian elephants; mastodons are more distant relatives, from a separate line of proboscideans that branched off from the modern elephant line in the Miocene. Mastodons have a shorter, stockier build and longer body; mammoths are taller and thinner, with a rather high "domed" skull. In skeletal details, the quickest way to tell the difference is by the teeth: mastodons have teeth with conical ridges, a bit like the bottom of an egg carton; mammoths, in contrast, have teeth with numerous parallel rows of ridges. The teeth are indicative of the two species' ecological differences. Mastodons preferred to bite off twigs of brush and trees, while mammoths preferred tough siliceous grasses. Thus, mastodon teeth are more suitable for cutting, while mammoth teeth are more suitable for grinding. Both mammoths and mastodons became extinct around 10,000 years ago.

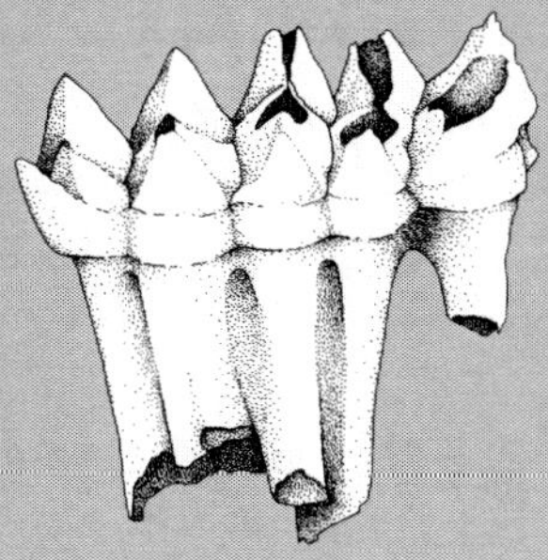

A mastodon tooth, suitable for chewing twigs and tree leaves. Approximately 20 centimeters (8–9 inches) long.

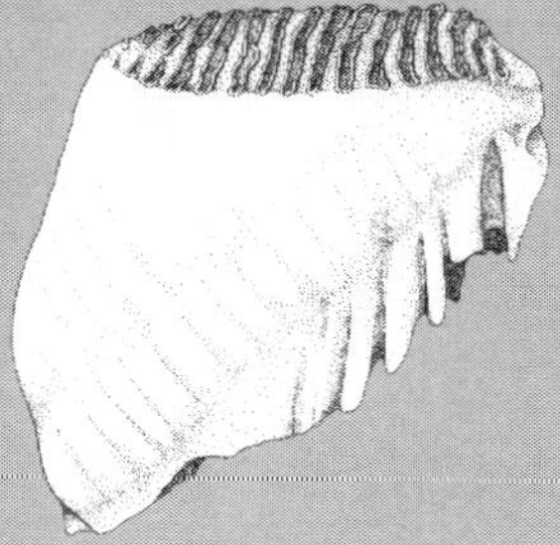

A mammoth tooth, suitable for grinding grass and softer vegetation. Approximately 25 centimeters (1 foot) long.

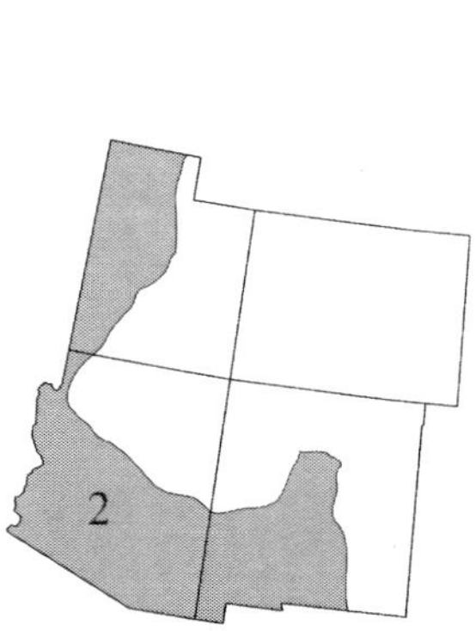

3 Fossils

Region 3

Fossils of the Rocky Mountains
Region 3

Precambrian • *a geologic time interval that spans from the formation of Earth (4.6 billion years ago) to the beginning of the Cambrian (541 million years ago).*

eukaryotes • *organisms with complex cells containing a nucleus and organelles.*

pteraspidomorph • *a member of a group of jawless fish with extensive head armor that lived mostly in coastal marine, and possibly freshwater, bottom environments.*

Late **Precambrian** rocks (approximately 1.8 billion years old) are exposed in northeastern Utah, where they contain what may be the oldest known fossils in the region: acritarchs. These unique fossils are probably the resting stages of single-celled **eukaryotic** plankton (*see Figure 3.4B*). Precambrian rocks of the Rocky Mountains also contain stromatolites, layered domes of sediment formed by mats of bacteria living in a shallow sea (*see Figure 3.3*).

Shallow marine waters continued to cover most of this area through the early part of the Paleozoic (Cambrian-Silurian), supporting a great diversity of life including trilobites, graptolites, brachiopods, and cephalopods (*see Figures 3.43, 3.45, and 3.46*). The Ordovician Harding Sandstone of central Colorado is famous for containing some of the oldest known fossil bone in the world, which belonged to fishes called **pteraspidomorphs** (*Figure 3.60*). These animals lacked jaws, but much of their bodies were covered with a bony shell made of numerous small plates. The Harding Sandstone's depositional environment was once thought to be fresh or brackish water, but now most geologists believe it was part of a shallow marine setting. The late Paleozoic saw these seas retreat, and some Pennsylvanian layers in central Colorado, including the area around Vail and Miturn in Eagle County, accumulated in terrestrial environments such as rivers, lakes, swamps, and floodplains (*Figure 3.61*). Some of these sediments include significant coal deposits (*see Figures 3.9 and 3.35*). The Permian Lyons Sandstone, deposited as a series of desert sand dunes, contains abundant fossil insects and the footprints of various reptiles and amphibians (*see Figure 3.11*).

See Chapter 6: Energy for more information about Colorado's coal deposits and mines.

The Jurassic Morrison Formation is exposed across much of central Colorado, including around Denver, at a site known as Dinosaur Ridge. This National Natural Landmark is located in Morrison, Colorado, just west of Denver. The site includes tilted exposures of the Morrison Formation (on the west side of the Ridge), from which many dinosaur skeletons were collected in the late nineteenth century (*see Figures 3.24 and 3.25*). The rocks on the east side of the Ridge are part of the Cretaceous Dakota Formation and contain hundreds of dinosaur footprints (*see Figure 3.27*). The Dakota Sandstone, which is also exposed at Red Rock Canyon in Colorado Springs, overlies both the Purgatoire and Morrison formations and stretches into the Colorado Plateau. The Dakota is well known for its terrestrial and marine fossils. The formation contains abundant terrestrial

The early Cretaceous Purgatoire Formation is the source of a recently described hadrosaur dinosaur: the bones of *Theiophytalia kerri* had been misidentified as *Camptosaurus* when they were found in 1878.

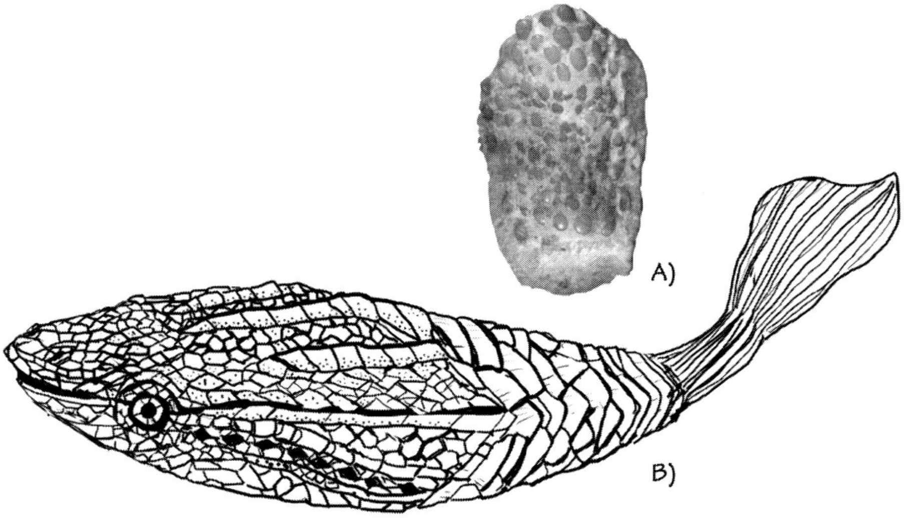

Figure 3.60: The jawless fossil fish Astraspis *from the Ordovician Harding Sandstone of Colorado. A) One of the bony scutes that covered the fish's articulated exoskeleton, approximately 1 centimeter (0.4 inches) long. B) Restoration, about 30 centimeters (1 foot) long.*

Figure 3.61: The Pennsylvanian period in the ancestral Rockies. Artist's reconstruction of the landscape of central Colorado during the Pennsylvanian period, with large stands of giant club mosses (lycopods).

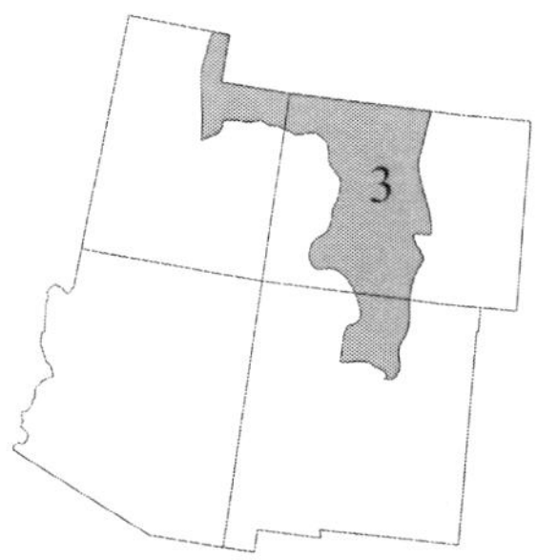

3 Fossils

Region 3

plants, occasional dinosaur footprints (attributed to *Iguanodon* and *Ankylosaurus*) and skeletal remains, and abundant marine invertebrates, especially mollusks and fish (*see Figure 3.39*). The layers above the Dakota are also rich in marine fossils. The Fort Hayes Limestone and Codell Sandstone, exposed in Red Rock Canyon and many other areas, contain abundant ammonites, bivalves, and shark teeth.

The Cretaceous Raton Formation of northern New Mexico preserves an abundant fossil record from the latest Cretaceous period. It is also the source of what is widely accepted as the only known footprint of *Tyrannosaurus rex* (*see Figure 3.32B*). The Raton Basin in northern New Mexico is also the only area of New Mexico where the K-Pg boundary is preserved—this boundary is also visible on South Table Mountain south of Denver.

Above the K-Pg boundary, **Paleocene** mammals (mostly represented by teeth) are found near Golden, Colorado. Nearby at Castle Rock, on the margin of the Denver Basin, is a very high-diversity assemblage of fossil leaves, which appears to represent the oldest known tropical rainforest (*Figure 3.62*), in existence just 1.4 million years after the end-Cretaceous extinction.

The Green River Formation is a layer (600–2000 meters [1970–6560 feet] thick) of brown to cream-colored **shale**, with occasional layers of **chert** and **limestone**, which outcrops across a large area of southwest Wyoming, northwest Colorado, and northwest Utah. The Green River comprises the largest known accumulation of **lacustrine** sedimentary rock in the world. Its sediments accumulated in a **system** of lakes that covered this area during the **Eocene**, between 58 and 40 million years ago (*Figure 3.63*). The Green River is famous for the great number of well-preserved fossils found in its lake and river sediments, especially aquatic organisms such as fish, gastropods, and algae, but also many terrestrial plants and animals (*Figure 3.64*), including insects, birds, and mammals. The formation also contains a large amount of oil shale.

See Chapter 6: Energy to learn more about the organic-rich Green River Formation and the oil shale it contains.

The Uinta Formation, exposed in the Uinta Mountains of northeastern Utah (about 42–45 million years old), is famous for its middle Eocene mammals, including marsupials, insectivores, primates, rabbits, rodents, hoofed mammals, condylarths (*see Figure 3.40*), and primitive carnivores known as **creodonts** (*Figure 3.65*).

One of the most amazing fossil occurrences in the Southwest is found at Florissant Fossil Beds National Monument in Teller County, central Colorado. There, 35-million-year-old lake sediments from the late Eocene epoch contain a huge diversity of extraordinarily preserved plant and animal fossils (*Figure 3.66*). Florissant's fossils were the result of massive **volcanic** eruptions nearby, which deposited a variety of volcanic layers including **lahars**, **pyroclastic** ash, and **pumice**. The layers dammed streams to form a lake that was then filled with

Paleocene • *a geologic time interval spanning from about 66 to 56 million years ago.*

shale • *a dark, fine-grained, laminated sedimentary rock formed by the compression of successive layers of silt- and clay-rich sediment.*

chert • *a sedimentary rock composed of microcrystaline quartz.*

limestone • *a sedimentary rock composed of calcium carbonate ($CaCO_3$).*

lacustrine • *of or associated with lakes.*

system • *a set of connected things or parts forming a complex whole.*

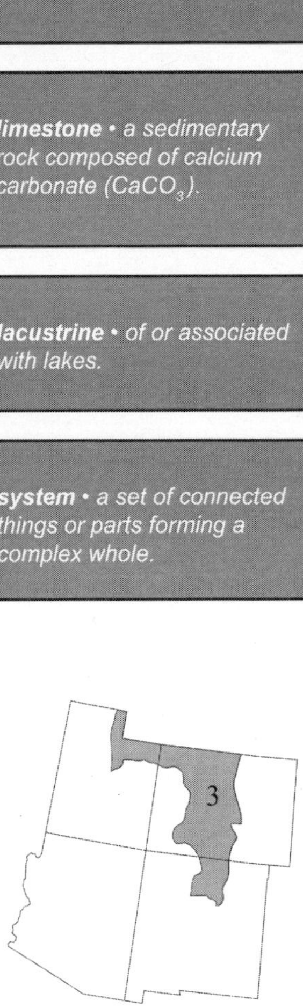

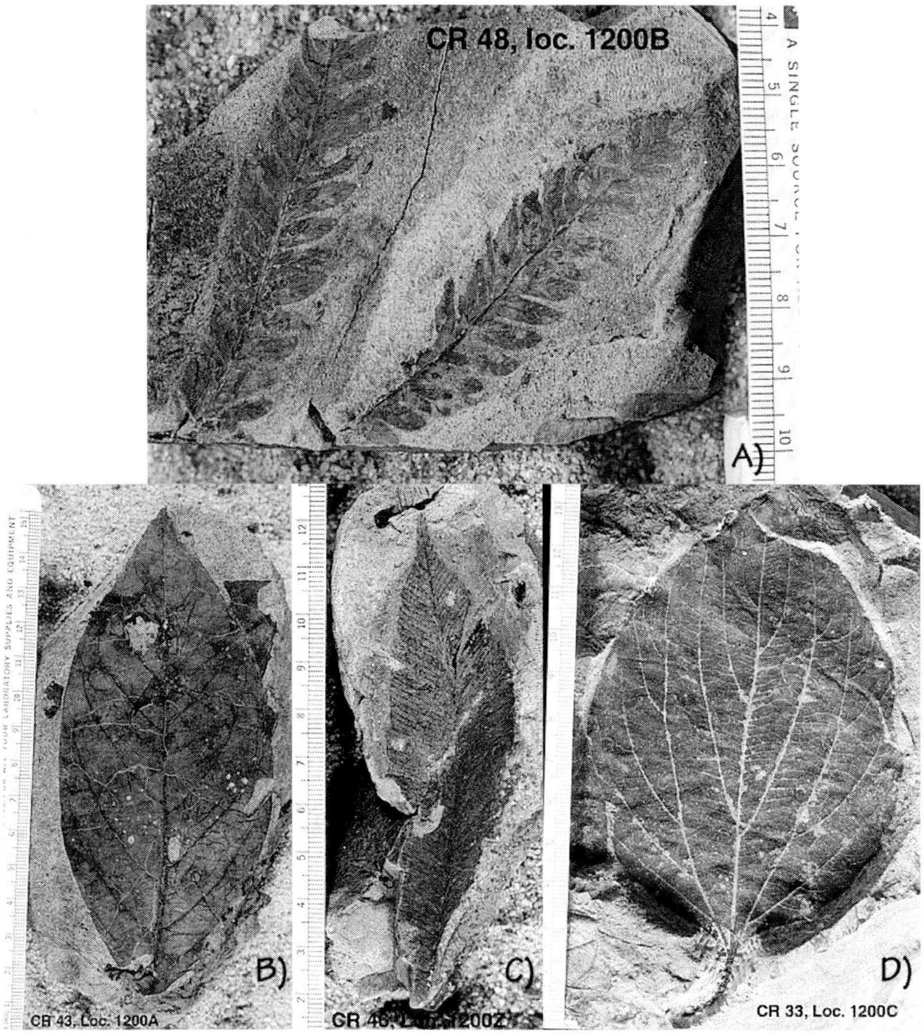

Figure 3.62: Fossils from an early Paleocene rainforest, at Castle Rock, alongside Interstate 25 near Denver, Colorado. Some of the fossil leaves are so well preserved that original leaf material is present. A) Undescribed fern leaves. This split slab of rock shows the abundance of fossil leaves at the site. B) An undescribed angiosperm species. This specimen shows insect damage. C) Allantodiopsis erosa. *The elongated tip of this leaf is called a "drip tip," which helps the leaf shed water in very wet conditions. Drip tips are common in modern tropical rainforests. D)* "Zizyphus" fibrillosus. *This leaf may be related to the modern buckthorn.*

additional volcanic deposits, as well as layers of **diatoms**, forming sediments known as diatomite. The extraordinary preservation of fossils in this lake may have been caused by an interaction of the **volcanic ash** with algal mats in the lake, inhibiting decomposition. Most conspicuous among the Florissant fossils are numerous large petrified tree stumps similar to modern sequoias, which are among the largest-diameter fossilized trees in the world. Fossils of fruits, seeds, cones, and flowers are also abundant. Among the animals preserved at Florissant are more than 1500 species of spiders and insects, including grasshoppers, butterflies, moths, wasps, bees, ants, mosquitos, crickets, flies, and many other types of beautifully preserved arthropods. Vertebrate remains

Eocene • *a geologic time period extending from 56 to 33 million years ago.*

creodont • *a member of an informal (not evolutionary) group of carnivorous mammals.*

volcanism • *the eruption of molten rock onto the surface of the crust.*

lahar • *a pyroclastic debris flow or mudflow that typically flows down river valleys after a volcanic eruption.*

pyroclastic rocks • *rocks that form during explosive volcanic eruptions, and are composed of a variety of different volcanic ejecta.*

pumice • *a pyroclastic rock that forms as frothing and sputtering magmatic foam cools and solidifies.*

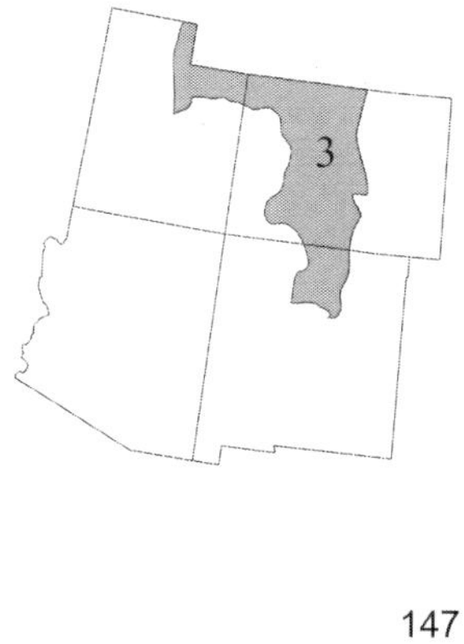

climate *• a description of the average temperature, range of temperature, humidity, precipitation, and other atmospheric/hydrospheric conditions a region experiences over a period of many years (usually more than 30).*

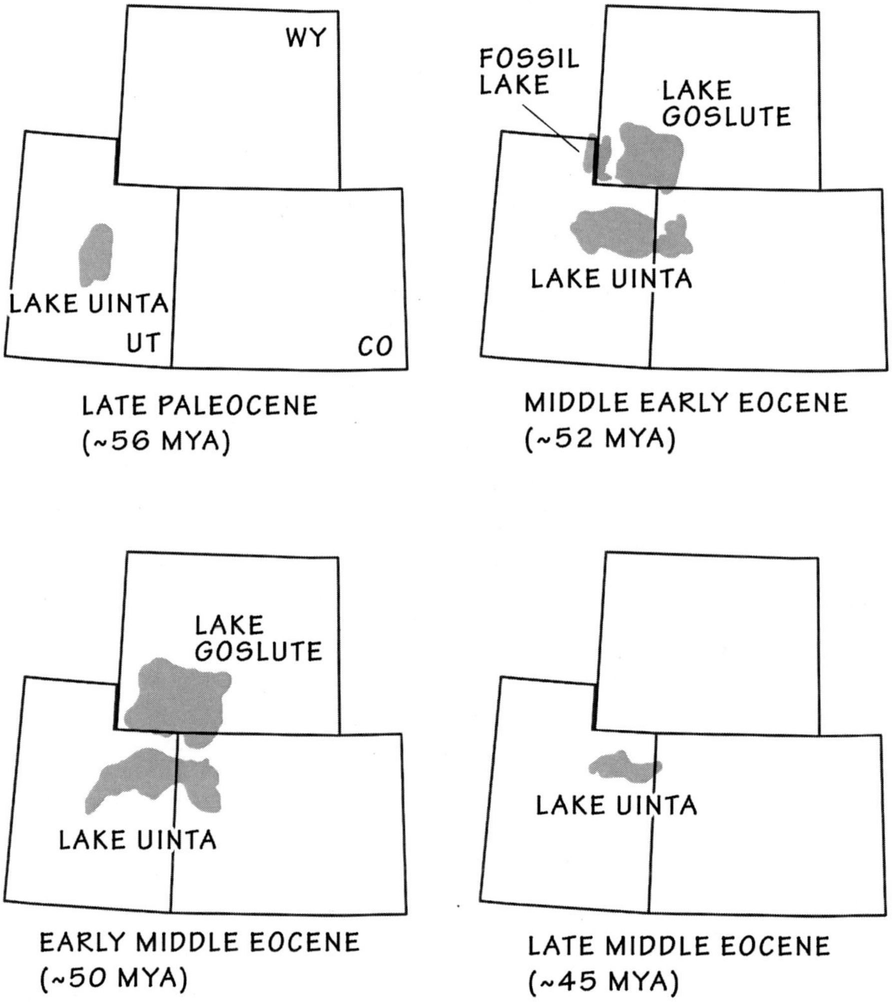

Figure 3.63: The size and location of various lakes in which the Green River sediments were deposited at different times during the Eocene epoch.

include fish, birds, and a few mammals. Based on these fossil remains—the plants, in particular—scientists estimate that the **climate** of the Eocene was significantly warmer than it is today. The modern mean annual temperature at Florissant is around 4°C (39°F), while the estimated mean annual temperature during the Eocene was 13°C (55°F).

See Chapter 8: Climate to learn how changes in climate have influenced life in the Southwest throughout geologic time.

In 2010, discoveries near Snowmass Village, Colorado revealed one of the most significant sites of large Pleistocene mammals ever found in North America. There, more than 4500 bones were recovered, representing more

Figure 3.64: Fossil insects and plants from the Green River Formation of Utah and Colorado. A) Unidentified beetle, about 1 centimeter (0.4 inches) long. B) Beetle (weevil), 13 millimeters (0.5 inches) long. C) Scorpionfly, 2.5 centimeters (1 inch) long. D) Robber fly, 11 millimeters (0.4 inches) long. E) Maple leaf, Acer *sp., 6.25 centimeters (2.5 inches) wide. F) Planetree leaf,* Platanus wyomingensis, *14.5 centimeters (5.7 inches) wide. G) Poplar leaf,* Populus willmattae, *7.3 centimeters (2.9 inches) long. H) Legume leaf,* Leguminosites lesquereuxiana, *3 centimeters (1.2 inches) wide.*

ice age • a period of global cooling of the Earth's surface and atmosphere, resulting in the presence or expansion of ice sheets and alpine glaciers.

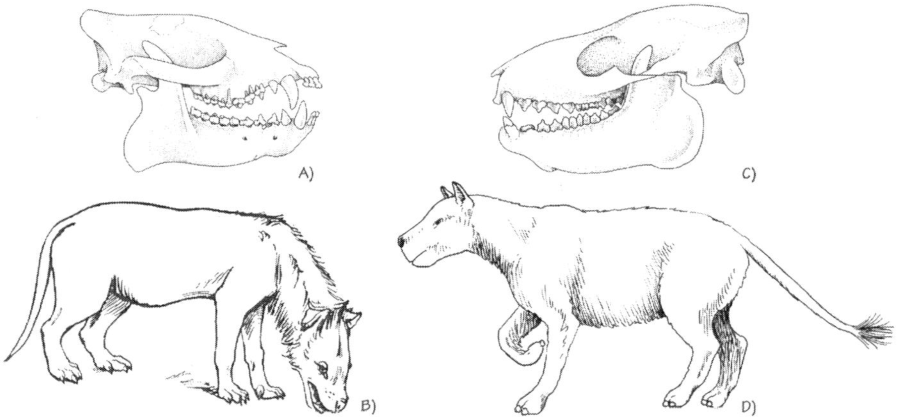

Figure 3.65: Uinta Formation mammals. A) Skull of the phenacodont Tetraclaenodon, *approximately 23 centimeters (9 inches) long. B) Restoration of the creodont* Hyaenodon, *approximately 2 meters (6 feet) long. C) Skull of the artiodactyl* Protoreodon, *approximately 17 centimeters (6.7 inches) long. D) Restoration of* Protoreodon, *approximately 70 centimeters (28 inches) long.*

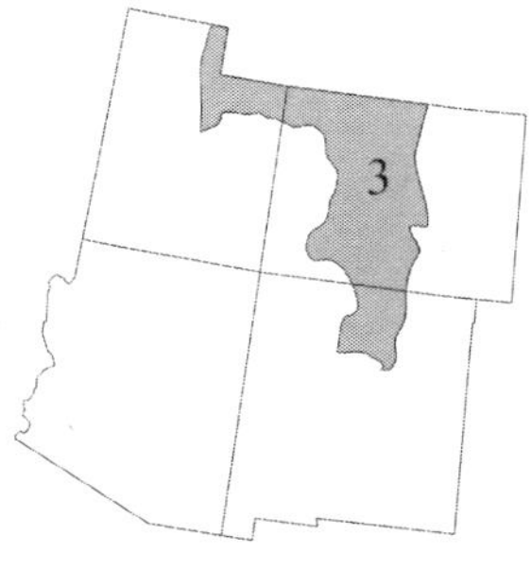

than 40 types of **ice age** animals. Larger mammals included mastodons and mammoths, bison, deer, horses, camels, and ground sloths. Smaller mammals included otters, beavers, chipmunks, rabbits, muskrats, and mice. Birds, snakes, and lizards were also discovered. This site provides an amazingly well-preserved snapshot into Colorado's ice age environment.

3 Fossils

Regions 3-4

silica • a chemical compound also known as silicon dioxide (SiO_2).

calcium carbonate • a chemical compound with the formula $CaCO_3$, commonly found in rocks in the mineral forms calcite and aragonite, as well as in the shells and skeletons of marine organisms.

Figure 3.66: Fossil plants and insects from the Florissant Fossil Beds of Colorado. A) Fossil redwood stump, approximately 2 meters (6 feet) in diameter. B) Fossil angiosperm leaf, slab approximately 5 centimeters (2 inches) across. C) Fossil wasp, Palaeovespa florissantia, *approximately 2 centimeters (0.8 inches) long. D) Crane fly, approximately 2.5 centimeters (1 inch) long. E) Wasp and cypress leaves, approximately 5 centimeters (2 inches) long.*

Fossils of the Great Plains
Region 4

The Great Plains of eastern Colorado, northern New Mexico, Oklahoma, Kansas, and northern Texas are dominated by Mesozoic- and Cenozoic-aged rocks. In the southeastern corner of New Mexico, however, is an extension of the Permian Basin. During the Permian period, this area contained a marine environment. Today, it is home to the largest, best preserved, most accessible, and most intensively studied Paleozoic reef in the world—the Capitan Reef (*Figure 3.67*)—which stood about 600 meters (2000 feet) above the adjacent deep sea basin. Many of the region's reef fossils are **silicified**; that is, they were preserved through replacement of their original **calcium carbonate** ($CaCO_3$) shells by silica (SiO_2). The late Permian rocks of westernmost Texas and southeasternmost New Mexico contain almost 1000 species of brachiopods (*Figure 3.68*), sponges, gastropods, foraminifera, algae, and other marine organisms.

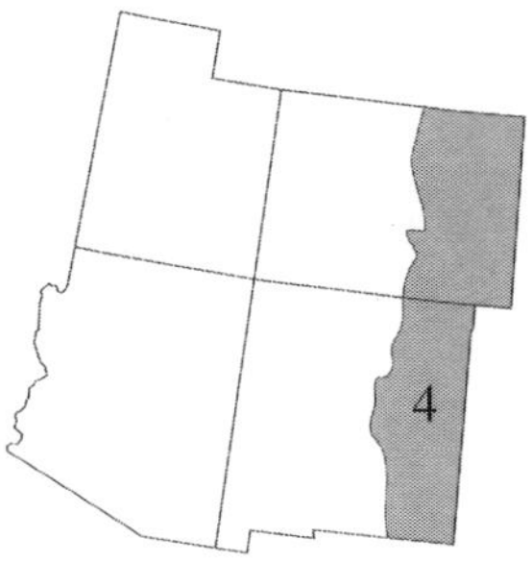

The Cretaceous rocks of Colorado produce an abundance and diversity of marine fossils from the end of the Mesozoic. Fossils of ammonoids, bivalves, fish, and marine reptiles document the life of the shallow sea (*Figure 3.69; see also Figures 3.38 and 3.39*). Along the sea's margins to the west, dinosaur footprints and flowering plants are preserved in what were once forested and

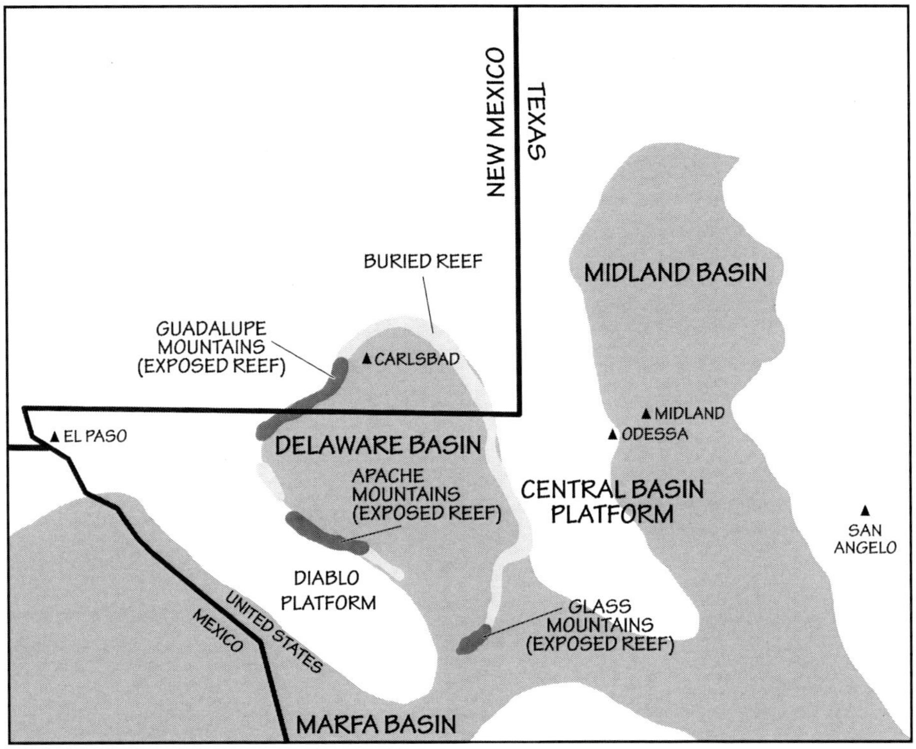

Figure 3.67: Permian reefs of southeastern New Mexico and western Texas. This map illustrates the paleogeographic relationship of the reefs to the topographic lows and highs of the shallow sea that covered the area during the Permian. The reef occupied the rim of the Delaware Basin, an extension of the Permian Basin.

chalk *• a soft, fine-grained, easily pulverized, white-to-grayish variety of limestone, composed of the shells of minute planktonic single-celled algae.*

gravel *• unconsolidated, semi-rounded rock fragments larger than 2 millimeters (0.08 inches) and smaller than 75 millimeters (3 inches) across.*

silt *• fine granular sediment most commonly composed of weathered grains of quartz and feldspar, of grain diameter 1/250 to 1/16 millimeters.*

Pliocene *• a geologic time interval extending from roughly 5 to 2.5 million years ago.*

swampy coastal areas. To the east, the fossils of open-water animals, such as mosasaurs, plesiosaurs, turtles, sharks, and other fishes—including the massive *Xiphactinus*—are preserved. In addition to strictly aquatic animals, the Cretaceous deposits of Colorado also preserve the remains of pterosaurs and toothed birds. These fossils are generally preserved in deposits of **chalk** (which is itself formed mostly from the shells of single-celled plankton) that accumulated at the bottom of the sea 80 to 90 million years ago.

Following the retreat of the Western Interior Seaway and the rise of the Rocky Mountains to the west, the Paleogene-era Great Plains saw deposition by streams carrying **gravel**, sand, and **silt** eroding off the Rockies to the west. Wide, shallow valleys were filled with sediment, creating a broad, gently dipping plain. Tropical forests (including the world's oldest known tropical rainforest at Castle Rock, Colorado; *see Figure 3.62*) growing along the margins of the new Rockies were home to mammals, crocodiles, and turtles. Fossils of fishes, plants, and insects are preserved within the lake deposits. Later in the Cenozoic, during the Miocene to **Pliocene** epochs, the climate cooled and dried, and this led to major changes in the region's fauna. For example, the Ogallala Formation of New Mexico contains abundant fossils of primitive horses and land tortoises (*Figure 3.70*), as well as mammal footprints.

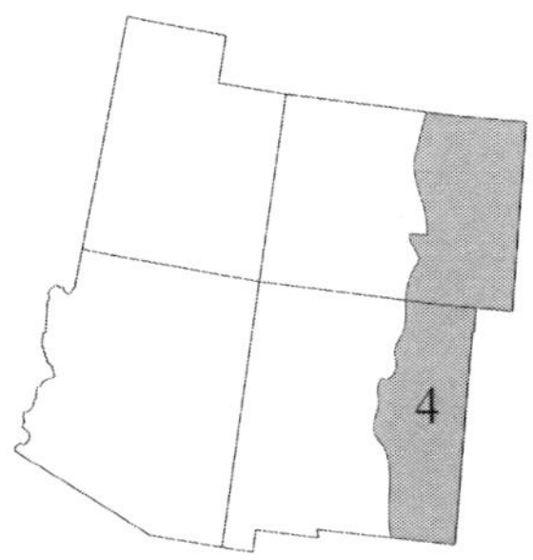

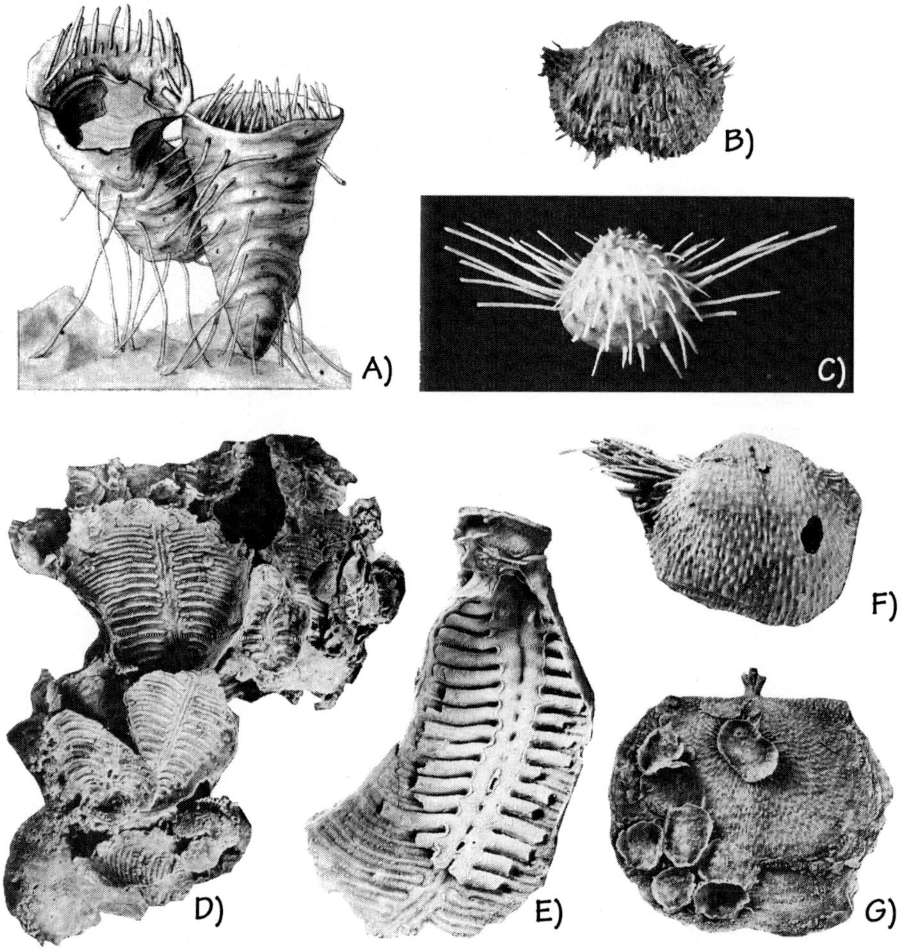

Figure 3.68: Capitan reef fossil brachiopods. A) Prorichtofenia permiana, *approximately 6 centimeters (2.5 inches) tall. B)* Ametoria residua, *3.7 centimeters (1.5 inches) wide. C)* Grandaurispina *sp., 8.5 centimeters (3.3 inches) wide. D)* Collematra gregaria, *cluster of large pedicle valves, 15 centimeters (6 inches) long. E)* Collematra elongata, *pedicle valve, 6.6 centimeters (2.6 inches) long. F)* Penicularis subcostata, *5.7 centimeters (2.25 inches) wide. G)* Bathymyonia *sp., 4.6 centimeters (1.8 inches) wide.*

Tortoises are a group of turtles that live on land, and have short, strong legs used for support and digging burrows. In contrast, most turtles live in the water and have webbed feet to help them swim efficiently, but will venture onto land occasionally to lay eggs.

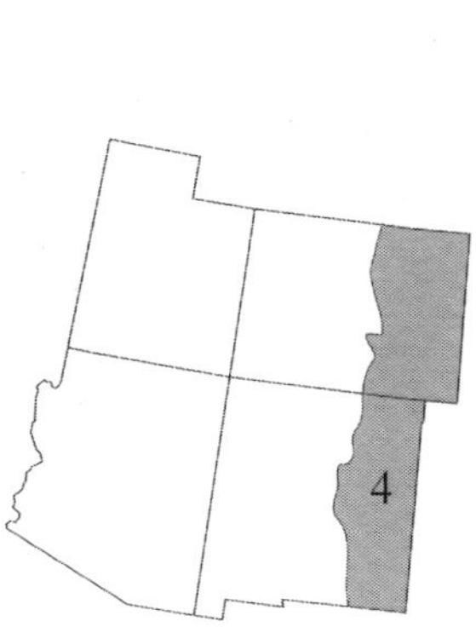

Figure 3.69: *Skeleton of the giant Cretaceous fish* Xiphactinus, *about 5 meters (16 feet) long.*

Figure 3.70: *Tortoise,* Geochelone *sp., about 1 meter (3 feet) long.*

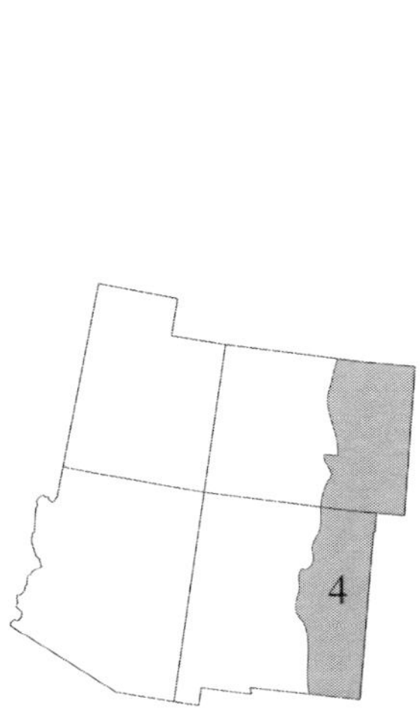

Fossils

State Fossils

Arizona
Araucarioxylon arizonicum (conifer wood; Triassic) (*Figure 3.20*).

Colorado
Stegosaurus stenops (plated dinosaur; Jurassic) (*Figure 3.25C*).

New Mexico
Coelophysis bauri (carnivorous dinosaur; Triassic) (*Figure 3.18*).

Utah
Allosaurus fragilis (carnivorous dinosaur; Jurassic) (*Figure 3.25B*).

Resources

General Books on the Fossil Record and Evolution

Allmon, W. D., 2009, *Evolution & Creationism: a Very Short Guide, 2nd edition*, Paleontological Research Institution, Ithaca, NY, 128 pp.
Benton, M. J., 2008, *The History of Life: a Very Short Introduction*, Oxford University Press, Oxford, UK, 170 pp.
Fenton, C. L., & M. A. Fenton, 1958, *The Fossil Book: A Record of Prehistoric Life*, Doubleday, Garden City, NY, 482 pp. (A well-illustrated classic.)
Fortey, R. A., 1998, *Life: A Natural History of the First Four Billion Years of Life on Earth*, Alfred A. Knopf, New York, 346 pp.
Knoll, A. H., 2003, *Life On a Young Planet: The First Three Billion Years of Evolution on Earth*, Princeton University Press, Princeton, NJ, 277 pp.
Lockley, M.G., 1991, *Tracking Dinosaurs: a New Look at an Ancient World*, Cambridge University Press, Cambridge, 238 pp.
Nudds, J.R., & P.A. Selden, 2008, *Fossil Ecosystems of North America: a Guide to the Sites and their Extraordinary Biotas*, University of Chicago Press, Chicago, 288 pp.
Prothero, D.R., 2006, *After the Dinosaurs: the Age of Mammals*, Indiana University Press, Bloomington, 362 pp.
Sampson, S. D., 2011, *Dinosaur Odyssey: Fossil Threads in the Web of Life*, University of California Press, Berkeley, CA, 352 pp.
Shubin, N., 2009, *Your Inner Fish: a Journey into the 3.5-Billion-Year History of the Human Body*, Vintage Books, NY, 256 pp.
Switek, B., 2010, *Written In Stone: Evolution, the Fossil Record, and Our Place In Nature*, Bellevue Literary Press, New York, 320 pp.
Thomson, K. S., 2005, *Fossils: a Very Short Introduction*, Oxford University Press, Oxford, UK, 147 pp.

Guides to Collecting and Identifying Fossils

Arduini, P., G. Teruzzi, & S. S. Horenstein, 1986, *Simon & Schuster's Guide to Fossils*, Simon and Schuster, New York, 317 pp.
Garcia, F. A., & D. S. Miller, 1998, *Discovering Fossils: How To Find and Identify Remains of the Prehistoric Past*, Stackpole Books, Mechanicsburg, PA, 212 pp.
Lichter, G., 1993, *Fossil Collector's Handbook: Finding, Identifying, Preparing, Displaying*, Sterling Publishing Company, New York, 160 pp.
Macdonald, J. R., 1983, *The Fossil Collector's Handbook: a Paleontology Field Guide*, Prentice-Hall, Englewood Cliffs, NJ, 193 pp.
Murray, M., 1967, *Hunting for Fossils: a Guide to Finding and Collecting Fossils in All Fifty States*, Macmillan Company, Toronto, Canada, 348 pp.
Nudds, J. R., & P. A. Selden, 2008, *Fossil Ecosystems of North America: a Guide to the Sites and their Extraordinary Biotas*, University of Chicago Press, Chicago, 288 pp.
Parker, S., 1990, *The Practical Paleontologist: a Step-By-Step Guide To Finding, Studying, and Interpreting Fossils*, Simon and Schuster, New York, 159 pp.
Parker, S., 2007, *Fossil Hunting: an Expert Guide to Finding and Identifying Fossils and Creating a Collection*, Southwater, London, 96 pp.
Ransom, J. E., 1964, *Fossils In America: their Nature, Origin, Identification and Classification and a Range Guide To Collecting Sites*, Harper and Row, New York, 402 pp.
Thompson, I., 1982, *The Audubon Society Field Guide To North American Fossils*, Knopf, New York, 846 pp.
Walker, C., D. Ward, & C. Keates, 2009, *Fossils*, Dorling Kindersley (Smithsonian Handbooks), New York, 320 pp.

Books, Articles, and Websites on Fossils of Specific Areas of the Southwest

Multistate areas

Foster, J., 2007, *Jurassic West: The Dinosaurs of the Morrison Formation and Their World*, Indiana University Press, Bloomington, 416 pp.
Lockley, M. G., & A. P. Hunt, 1995, *Dinosaur Tracks and Other Fossil Footprints of the Western United States*, Columbia University Press, New York, 338 pp.
Ratkevich, R. P., 1976, *Dinosaurs of the Southwest*, University of New Mexico Press, Albuquerque, 115 pp.
Sadler, C., 2007, *Life in Stone: Fossils of the Colorado Plateau*, Grand Canyon Association, Grand Canyon, AZ, 72 pp.
Tidwell, W. D. 1998, *Common Fossil Plants of Western North America*, Smithsonian Institution Press, Washington, DC, 299 pp.

Arizona

Fossils, Grand Canyon National Park, National Park Service, http://www.nps.gov/grca/learn/nature/fossils.htm.
Fossils, Petrified Forest National Park, National Park Service, http://www.nps.gov/pefo/naturescience/fossils.htm.
Heckert, A. B. & S. G. Lucas, eds., 2005, Vertebrate paleontology in Arizona, *New Mexico Museum of Natural History and Science Bulletin*, 29: 210 pp. [Technical]
Ratkevich, R. P., 1979, *Field Guide to Arizona Fossils*, Dinograph Southwest, Alamogordo, NM, 114 pp.
Sedimentary Rock Formations of the Grand Canyon, In the Playground of Giants, by Ken Bevis (Department of Geology, Hanover College), http://intheplaygroundofgiants.com/geology-of-the-grand-canyon-region/sedimentary-rock-formations-of-the-grand-canyon-region/.
Thayer, D., 2009, *An Introduction to Grand Canyon Fossils*, Grand Canyon Association, Grand Canyon, AZ, 64 pl.
Timmons, M. J., & C. A. Manduca, eds., 2012, Grand Canyon geology: Two billion years of Earth's history, *Geological Society of America Special Paper* 489, 156 pp. [Technical]

Colorado

Ancient Denvers, Denver Museum of Nature and Science, 2002, http://www.dmns.org/main/minisites/ancientDenvers/landscapes.html.
Dinosaur National Monument, National Park Service, http://www.nps.gov/dino/index.htm.
Grande, L., 2013, *The Lost World of Fossil Lake: Snapshots from Deep Time*, University of Chicago Press, Chicago, 432 pp.
Jenkins, J. T., & J. L. Jenkins, 1993, Colorado's dinosaurs, *Colorado Geological Survey Special Publication* 35, 74 pp.
Johnson, K. R., & R. G. Raynolds, 2006, *Ancient Denvers: Scenes from the Past 300 Million Years of the Colorado Front Range*, Denver Museum of Science and Nature, Denver, CO, 32 pp.
Leopold, E. B., & H. W. Meyer, 2012, *Saved in Time: the Fight to Establish Florissant Fossil Beds National Monument, Colorado*, University of New Mexico Press, Albuquerque, 139 pp.
Lockley, M. G., 2001, *A Field Guide to Dinosaur Ridge, 3rd edition*, Friends of Dinosaur Ridge/University of Colorado at Denver Dinosaur Tracker Research Group, Denver, CO, 32 pp.
Lockley, M., 2003, *Fossil Footprints of the Dinosaur Ridge and Fossil Trace Areas, 2nd edition*, Friends of Dinosaur Ridge/University of Colorado at Denver Dinosaur Tracker Research Group, Denver, CO, 66 pp.
Meyer, H. W., 2003, *Fossils of Florissant*, Smithsonian Institution Press, Washington, DC, 272 pp.
Milito, S., 2010, A survey of fossils and geology of Red Rock Canyon Open Space, Colorado Springs, Colorado, *The Mountain Geologist*, 47(1): 1–14.
Wright, J. L., 2004, Walking with dinosaurs (and other extinct animals) along Colorado's Front Range: a field trip to Paleozoic and Mesozoic terrestrial localities, *GSA Field Guides*, 5: 219–234. [Technical]

New Mexico

Colbert, E. H., 1995, *The Little Dinosaurs of Ghost Ranch*, Columbia University Press, New York, 250 pp.
Ghost Ranch Geologic Tour, New Mexico Bureau of Geology and Mineral Resources, https://geoinfo.nmt.edu/tour/landmarks/ghost_ranch/home.html.
Kues, B., 2008, *The Paleontology of New Mexico*, University of New Mexico Press, Albuquerque, 446 pp.
Lucas, S. G., & A. B. Heckert, eds., 2000, Dinosaurs of New Mexico, *New Mexico Museum of Natural History and Science Bulletin* 17, 230 pp. [technical]
Lucas, S. G., G. S. Morgan, & K. E. Ziegler, eds., 2005, New Mexico's ice ages, *New Mexico Museum of Natural History and Science Bulletin* 28, 280 pp. [technical]

Utah

Barnes, F. A. 1997, *Canyon Country Dinosaur Tracks and Trackers*, Canyon Country Publications, Moab, UT, 176 pp.
Cleveland-Lloyd Dinosaur Quarry, Utah, Geologic Site Seeing, Bureau of Land Management, http://www.blm.gov/wo/st/en/res/Education_in_BLM/Learning_Landscapes/For_Travelers/go/geology/cleveland-lloyd.print.html.
DeCourten, F., 2014, *Dinosaurs of Utah, 2nd edition*, University of Utah Press, Salt Lake City, 336 pp.
Dinosaurs and Fossils, Utah Geological Survey, http://geology.utah.gov/popular/general-geology/dinosaurs-fossils/.
Gillette, D. D., ed., 1999, Vertebrate paleontology in Utah, *Utah Geological Survey Miscellaneous Publication* 99-1, 554 pp. [Technical]
Harris, J. D., & A. R. C. Milner, 2015, *Tracks in Deep Time: the St. George Dinosaur Discovery Site at Johnson Farm*, University of Utah Press, Salt Lake City, 99 pp.
Mesozoic Fossils, Glen Canyon National Recreation Area, National Park Service, https://www.nps.gov/glca/learn/nature/copy-of-mesozoicfossils.htm.
Robison, R. A., L. E. Babcock, & V. G. Gunther, 2015, Exceptional Cambrian fossils from Utah: a window into the age of trilobites, *Utah Geological Survey Miscellaneous Publication* 15-1, 97 pp.
Squatting Dinosaur Tracks at Johnson's Farm in St. George (Basal Jurassic Dinosaur Footprints, Fishes, and Plants, Whitmore Point Member of the Moenave Formation), by Nancy Carruthers, Utah Geological Survey, http://www.sunstar-solutions.com/AOP/fieldtrips/Juratracks/BasalJurassic.htm.
Titus, A. L., & M. A. Loewen, 2013, *At the Top of the Grand Staircase: the Late Cretaceous of Southern Utah*, Indiana University Press, Bloomington, 634 pp. [technical]
Wilson, J. R., 1995, A collector's guide to rock, mineral and fossil localities in Utah, *Utah Geological Survey Miscellaneous Publication* 95-4, 148 pp.

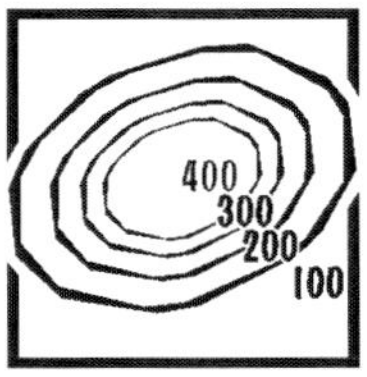

Chapter 4: Topography of the Southwestern US

Does your region have rolling hills? Mountainous areas? Flat land where you never have to bike up a hill? The answers to these questions can help others understand the basic topography of your region. The term **topography** is used to describe the shape of the land surface as measured by how elevation—height above sea level—varies over large and small areas. Over **geologic time**, topography changes as a result of weathering and erosion, as well as the type and structure of the underlying bedrock. It is also a story of **plate tectonics**, volcanoes, folding, **faulting**, **uplift**, and mountain building.

The Southwest's topographic zones are under the influence of the destructive surface processes of weathering and erosion. **Weathering** includes both the mechanical and chemical processes that break down a rock. There are two types of weathering: physical and chemical. Physical weathering describes the physical or mechanical breakdown of a rock, during which the rock is broken into smaller pieces but no chemical changes take place. Water, ice, and **wind** all contribute to physical weathering, sculpting the landscape into characteristic forms determined by the **climate**. In most areas, water is the primary agent of **erosion**. Streams are constantly eroding their way down through bedrock to sea level, creating valleys; Arizona's Grand Canyon provides a dramatic example of this process. Given sufficient time, streams can cut deeply and develop wide flat **floodplains** on valley floors. The pounding action of ocean waves on a coastline contributes to the erosion of coastal rocks and sediments, while the emptying of a river can lead to the formation of a **delta**. Rock material is carried by rivers and streams to the oceans or to an inland lake or basin where it is eventually deposited. In the case of the Basin and Range and parts of the desert in southern New Mexico and Arizona, streams end in basins within the region. The rest of the Southwest is drained by rivers that reach either the Gulf of Mexico (the Platte, Arkansas, Canadian, Pecos, and Rio Grande rivers) or the Gulf of California (the Colorado River and its major tributaries, the Green and San Juan rivers).

Pressure release can also cause rocks to crack. Growing plant roots can exert many pounds per square inch of pressure on rocks—think of **tree** roots uplifting and cracking a sidewalk. Additionally, since rocks buried kilometers (miles) beneath the surface are under considerable pressure, if those rocks become exposed at the Earth's surface (where the rock is under less pressure), the rock may expand and crack in a process called **exfoliation** (*Figure 4.1*). At higher elevations, ice can also change the landscape due to frequent episodes of freezing and thawing, causing both temperature and pressure differentials within a rock. As water trapped in **fractures** within the rock freezes and thaws, the fractures continue to widen (*Figure 4.2*). This alone can induce significant

topography • *the landscape of an area, including the presence or absence of hills and the slopes between high and low areas.*

geologic time scale • *a standard timeline used to describe the age of rocks and fossils, and the events that formed them.*

plate tectonics • *the process by which the plates of the Earth's crust move and interact with one another at their boundaries.*

fault • *a fracture in the Earth's crust in which the rock on one side of the fracture moves measurably in relation to the rock on the other side.*

uplift • *upward movement of the crust due to compression, subduction, or mountain building.*

weathering • *the breakdown of rocks by physical or chemical means.*

CHAPTER AUTHORS

Bryan L. Isacks
Richard A. Kissel
Warren D. Allmon

Review

Quaternary • a geologic time period that extends from 2.6 million years ago to the present.

mineral • a naturally occurring solid with a specific chemical composition and crystalline structure.

igneous rocks • rocks derived from the cooling of magma underground or molten lava on the Earth's surface.

metamorphic rocks • rocks formed by the recrystallization and realignment of minerals in pre-existing sedimentary, igneous, and metamorphic rocks when exposed to high enough temperature and/or pressure.

limestone • a sedimentary rock composed of calcium carbonate ($CaCO_3$).

marble • a metamorphic rock composed of recrystallized carbonate minerals, most commonly calcite or dolomite.

carbonate rocks • rocks formed by accumulation of calcium carbonate, often made of the skeletons of aquatic organisms.

breakdown of large rock bodies. During the **Quaternary**, ice has been an important agent of erosion throughout the highest parts of the Southwest.

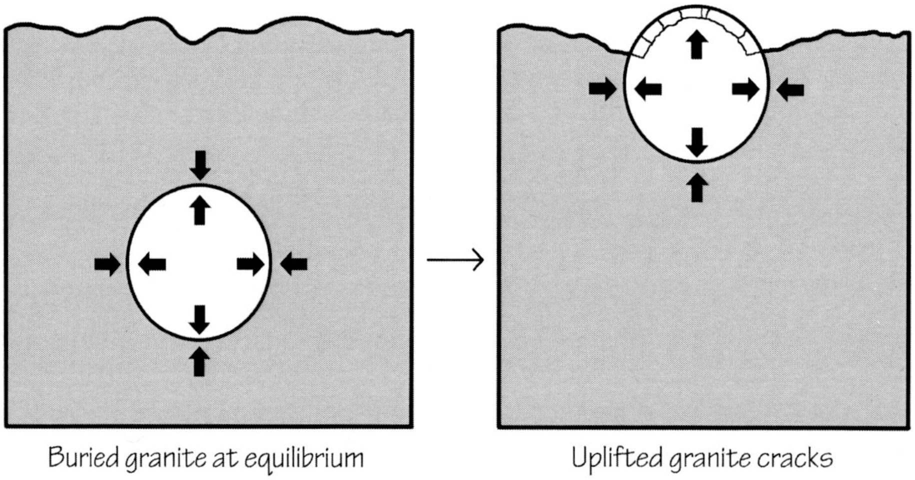

Figure 4.1: Exfoliation as a result of uplift and pressure release.

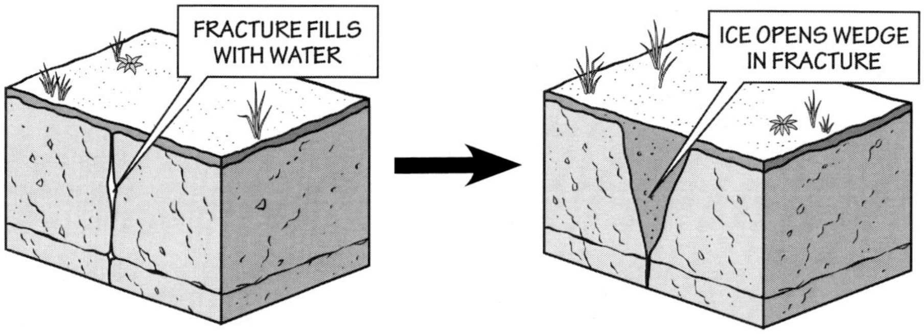

Figure 4.2: Physical weathering from a freeze-thaw cycle.

Working in conjunction with physical (mechanical) weathering, chemical weathering also helps to break down rocks through changes in the chemical composition of their constituent **minerals**. Some minerals contained in **igneous** and **metamorphic** rocks that are formed at high temperatures and pressures (far below the surface of the Earth) become unstable when they are exposed at the surface or placed in contact with water, where the temperature and pressure are considerably lower. Unstable minerals transition into more stable minerals, resulting in the breakup of rock. Weak acids, such as carbonic acid found in rainwater, promote the disintegration of certain types of rocks. **Limestone** and **marble** may be chemically broken down as carbonic acid reacts with the **carbonate** mineral composition of these rocks, forming cavities and caverns in the rock. Other **sedimentary rocks** held together by carbonate **cement** are also particularly susceptible to chemical weathering, which expedites the process of **soil** formation.

Topography 4

Volcanic activity has shaped the land throughout the Southwest. Although there are no active volcanoes there today, evidence of past activity—such as volcanic cones and craters, **lava** flows, **dikes**, and plugs—can be seen in a variety of locations, including volcanic fields throughout the Basin and Range and along the southern edge of the Colorado Plateau, **tuff** beds in and along the Rio Grande Rift, and the Carrizozo Malpais lava field in New Mexico.

The specific rock type at the surface has an important influence on the topography of a region. Certain rocks are able to resist weathering and erosion more easily than are others; resistant rocks that overlie weaker layers act as caps and form ridges. The great continental seas that advanced across the face of the continent during the **Paleozoic** and **Mesozoic** collected and preserved sediments that became sedimentary rocks, such as the deposits exposed along the walls of the Grand Canyon. Sedimentary rocks weather and erode differently than do crystalline (and generally **harder**) igneous and metamorphic rocks, such as those found at the base of the Grand Canyon and in the Rocky Mountains. **Silica**-rich igneous rocks have a crystalline nature and mineral composition that resists weathering far better than do the cemented grains of a sedimentary rock. The metamorphic equivalents of sedimentary and igneous rocks are often even more resistant than the original unmetamorphosed rocks, due to **recrystallization**. There are exceptions, however, such as **schist**, which is much weaker than its pre-metamorphic limestone or **sandstone** state.

See Chapter 2: Rocks for more information about the igneous, metamorphic, and sedimentary rocks of the Southwest.

The underlying structure of the rock layers also plays an important role in the topography at the surface. In the Southwest, with its semi-arid climates and lack of dense vegetative cover, the rock structures that influence the area's topography are clearly revealed. Sedimentary rocks are originally deposited in flat-lying layers that rest on top of one another. The movement of tectonic **plates** creates stress and tension within the **crust**, especially at plate boundaries. **Intrusions** beneath the surface may also cause deformation of the crust. All these different sources of geological stress can deform flat sediment layers through folding, faulting, or overturning. These terms are collectively used to describe rock structure, and they can also be used to determine which forces have affected rocks in the past. The folding of horizontal rock beds followed by erosion and uplift brings layers of rock to the surface. Tilted rocks expose underlying layers; resistant layers stick out and remain as ridges, while surrounding layers of less resistant rock erode away. Faulting likewise exposes surface layers to erosion, as blocks of crust move and tilt along the fault plane. For example, the Basin and Range formed as a result of normal faulting (*Figure 4.3A*), which occurs due to extensional stresses that create uplifted ranges and downdropped basins. The Rocky Mountains provide another regional example of folding and faulting: this range formed as a result of uplift associated with **subduction** along the western edge of the North American plate. The shallow angle of the subducting plate generated thrust (reverse) faults (*Figure 4.3B*) and the onset of the **Laramide Orogeny**.

Review

sedimentary rock • *rock formed through the accumulation and consolidation of grains of broken rock, crystals, skeletal fragments, and organic matter.*

cementation • *the precipitation of minerals that binds together particles of rock, bones, etc., to form a solid mass of sedimentary rock.*

soil • *the collection of natural materials that collect on Earth's surface, above the bedrock.*

volcanism • *the eruption of molten rock onto the surface of the crust.*

lava • *molten rock located on the Earth's surface.*

dike • *a sheet of intrusive igneous or sedimentary rock that fills a crack cutting across a pre-existing rock body.*

tuff • *a pyroclastic rock made of consolidated volcanic ash.*

Paleozoic • *a geologic time interval that extends from 541 to 252 million years ago.*

4 Topography

Review

physiography • *a subfield of geography that studies the Earth's physical processes and patterns.*

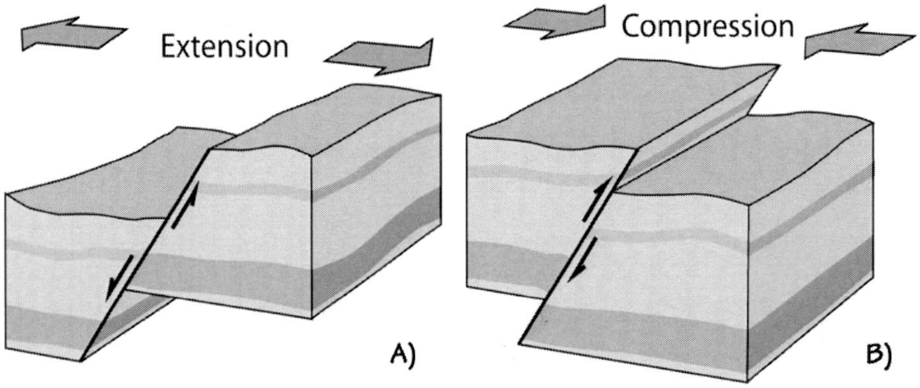

Figure 4.3: A) Normal faulting and B) thrust (reverse) faulting.

Just as we are able to make sense of the type of rocks in an area by knowing the geologic history of the Southwestern US, we are able to make sense of its topography (*Figure 4.4*) based on rocks and structures resulting from past geologic events. Topography is a central element of the broader concepts of geomorphology or **physiography**, which also include consideration of the shape (not just the height) of land forms, as well as the bedrock, soil, water, vegetation, and climate of an area, and how they interacted in the past to form the landscape we see today. A physiographic province is an area in which these features are similar, one in which these features are significantly different from those found in adjacent regions, and/or one that is separated from adjacent regions by major geological features. The "regions" of the Southwest that we use in this book are examples of major physiographic provinces. The topography unique to each region thus provides a set of clues to its extensive geologic history.

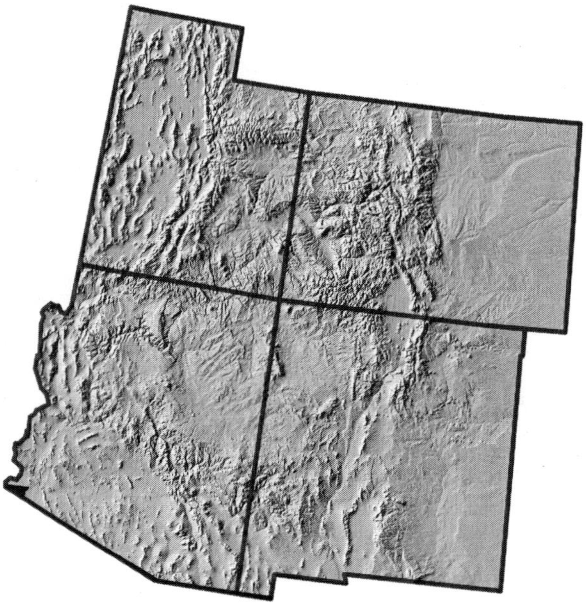

Figure 4.4: Digital shaded relief map of the Southwestern states.

Topography 4

The rocks of the Southwest reach back into the **Precambrian**, but the area's remarkable topography is primarily the result of dramatic Mesozoic and **Cenozoic** plate tectonics. Subduction of the Farallon plate during the late **Cretaceous** and Cenozoic, and the associated **Sevier** and Laramide **orogenies**, shaped the mountains of the Basin and Range as well as uplifiting the Rocky Mountains, Colorado Plateau, and the edge of the Great Plains. Late Cenozoic crustal extension led to faulting and volcanic activity that is most dramatically exhibited in the Basin and Range of Utah and Arizona and the Rio Grande Rift of New Mexico and Colorado. Southwestern topography particularly emphasizes broad **epeirogenic** uplift, one of the more contentious and least understood features of the Earth's topography. In contrast to orogenic (mountain building) uplift, epeirogenic uplift raises the land surface without **compressing** and thus thickening the crust. During the Cenozoic, most of the Southwest was lifted by this process. Today, the elevation of nearly the entire Southwest lies above 1200 meters (4000 feet), except for the southwestern desert in Arizona (*Figure 4.5*). The north-south "backbone" of highest elevation, shown by areas above 1800 meters (6000 feet), stretches from New Mexico through Colorado and into Wyoming. The Rocky Mountains of Colorado define a particularly high area, where elevations exceed 2700 meters (9000 feet) over a large part of the state.

See Chapter 1: Geologic History for more information about the tectonics and mountain building that shaped the Southwest.

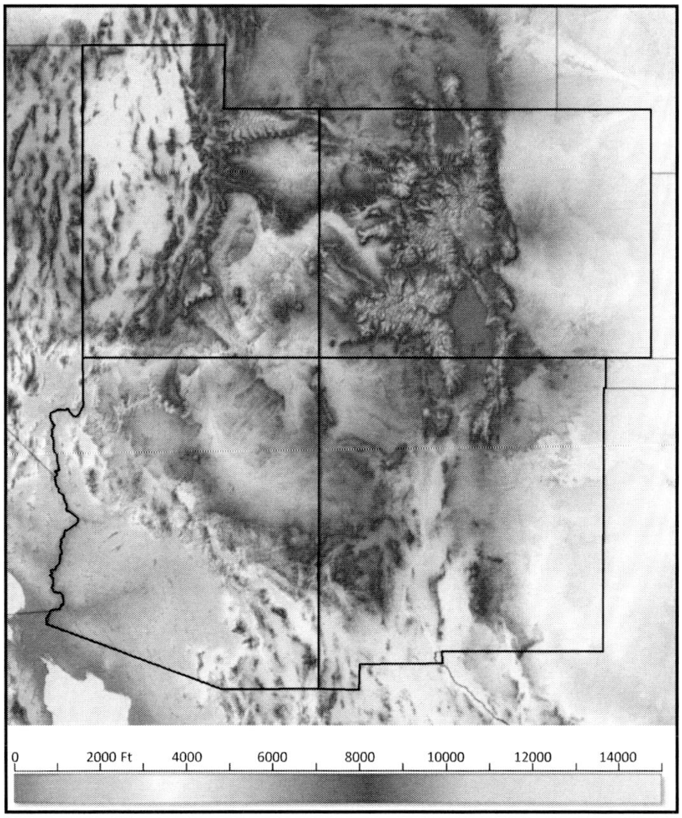

Figure 4.5: Elevation map of the Southwest (See TFG website for full-color version).

Review

Precambrian *• a geologic time interval that spans from the formation of Earth (4.6 billion years ago) to the beginning of the Cambrian (541 million years ago).*

Cenozoic *• the geologic time period spanning from 66 million years ago to the present.*

Cretaceous *• a geologic time period spanning from 144 to 66 million years ago.*

Sevier Orogeny *• a mountain-building event resulting from subduction along the western edge of North America, occurring mainly during the Cretaceous.*

orogeny *• a mountain-building event generally caused by colliding plates and compression of the edge of the continents.*

epeirogenic *• large-scale crustal uplift caused by hot or upwelling mantle underlying the surface.*

compression *• flattening or squeezing as a result of forces acting on an object from all or most directions.*

4 Topography

Region 1

anticline • a layer of rock folded (bent) along an axis, concave side down (i.e., in an upside down "u" or "v" shape).

Topography of the Colorado Plateau
Region 1

The Colorado Plateau covers an area of approximately 335,000 square kilometers (130,000 square miles) close to the center of the Southwest. It is the one region located in all four states, whose borders intersect at the "four corners" near the center of the plateau. It is bordered by the Colorado Rockies in the northeast, the Uinta Mountains in the northwest, and the Basin and Range (including the Rio Grande Rift) along the west, southwest, and southeast. Although the Colorado Plateau is largely semi-arid, the Colorado River and its tributaries access considerable runoff from snowmelt and rain in the Rocky Mountains to the north and east. These rivers and streams have carved numerous canyons throughout the region, culminating in the grandest of canyons on Earth. The Colorado River defines the lowest elevation on the plateau, near 330 meters (1100 feet) where it exits into Lake Mead, but most of the plateau is higher, with an average elevation of approximately 2000 meters (6500 feet) and elevations mainly in the range of 1200 to 2400 meters (4000 to 8000 feet). Higher elevations up to 3300 meters (11,000 feet) are located in the High Plateaus of Utah and the volcanic fields of the Navajo section (*Figure 4.6*).

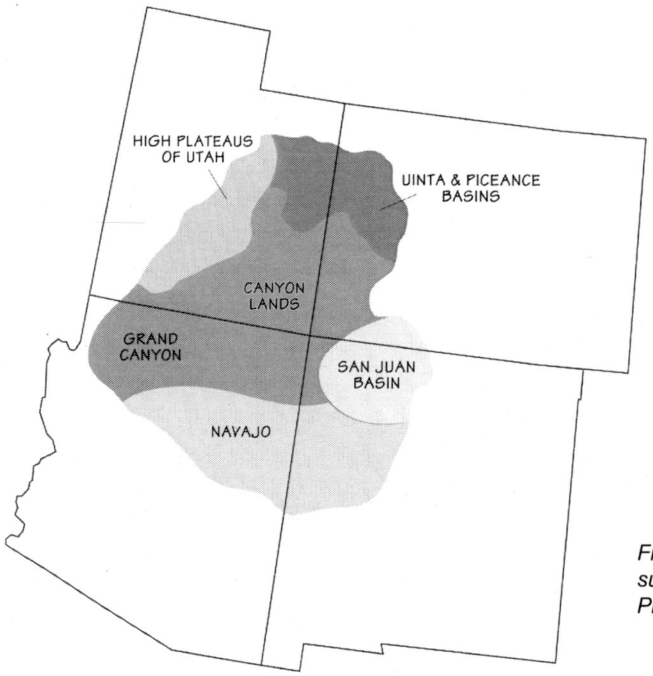

Figure 4.6: Physiographic subdivisions of the Colorado Plateau.

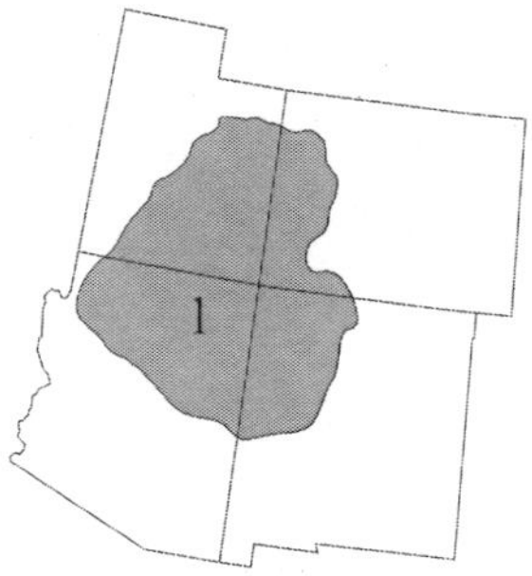

The outstanding and unusual topographic scenery of the Colorado Plateau is a main theme for the 8 National Parks and 25 National Monuments located there. Initially, this topography was shaped when the Laramide Orogeny deformed the thick sequence of Paleozoic and Mesozoic strata that had previously been lying flat near sea level beneath the Western Interior Seaway. The broad arches (**anticlines**), monoclines, domes, and basins formed by the Laramide

Orogeny are beautifully preserved in the plateau as textbook examples of geologic structures. During the Cenozoic, the margins of the plateau were also affected by Basin and Range faulting and **magmatism**. Second, erosion of the plateau, especially during the late Cenozoic, has mostly occurred in a semi-arid environment. This has allowed a dramatic display of both the unique forms of semi-arid erosion (driven by flash floods, frost, and intermittent rain) and the structural features generated by past geologic deformation.

See Chapter 1: Geologic History to learn more about the Laramide and Sevier orogenies.

magma • *molten rock located below the surface of the Earth.*

butte • *an isolated hill with steep, often vertical sides and a small, relatively flat top.*

caprock • *a harder, more resistant rock type that overlies a softer, less resistant rock.*

joint • *a surface or plane of fracture within a rock.*

relief • *the change in elevation over a distance.*

Canyon Lands

The Canyon Lands (or Canyonlands), encompassing the central portion of the Colorado Plateau, are a scenic wonderland where erosion has sculpted some of the most beautiful and geologically fascinating landscapes on Earth. The semi-arid climate prevailing over the plateau during much of the Cenozoic has interacted with the area's mildly deformed layer cake of sedimentary strata to produce fantastic erosional forms. The strata's variation in thickness and erodibility led to the formation of plateaus, mesas, **buttes**, and chimneys, all fringed by cliffs or a staircase of cliffs and slopes topped with more resistant **caprocks**. The slopes between cliffs are formed from eroded rock debris. As plateaus continue to erode, they leave behind progressively smaller remnants, including arches, natural bridges, windows, tents, spires, needles, hoodoos, balanced rocks, fins, and pinnacles that appear in stunning displays (*Figures 4.7-4.9*). The differing patterns of vertical **joints** in the rocks are a major factor in the production of this variety of bizarre features. In thick massive sandstones such as the Navajo Formation exposed in Zion National Park, erosion of vertical joints has produced deep and narrow canyons with up to 600 meters (2000 feet) of **relief** on nearly vertical canyon walls. The same type of erosion has created thousands of spectacular hoodoos in Bryce Canyon.

Figure 4.7: Weathered rock spires in the Needles district of Canyonlands National Park, Utah.

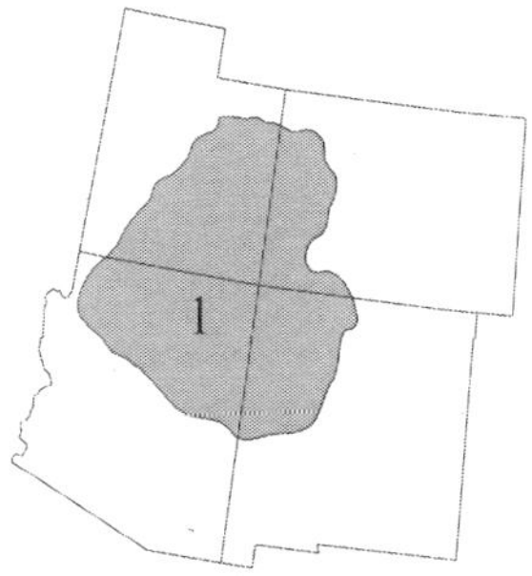

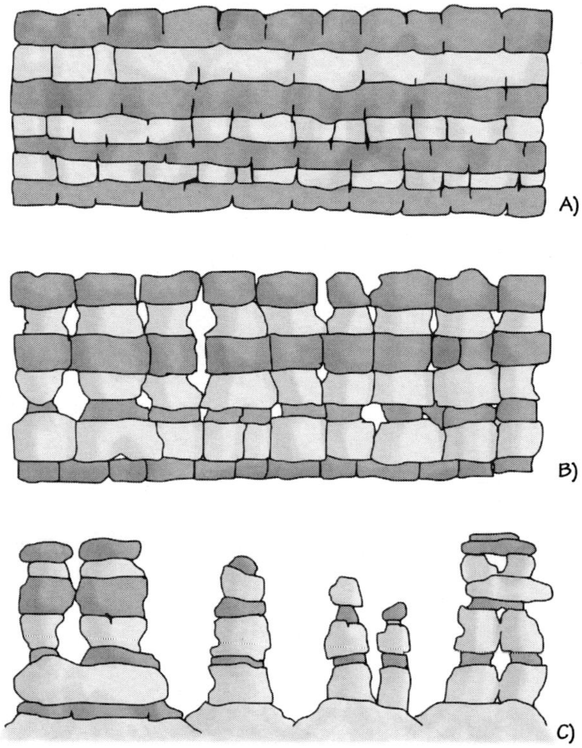

Figure 4.8: Erosive processes in the Colorado Plateau. A) Solid sedimentary strata contain fractures, joints, and other points of weakness. B) Weathering from rain and frost eats away at the sedimentary rock along its weak points. C) Progressively smaller remnants are left behind as erosion continues over time. More resistant rock layers cap the spires.

Figure 4.9: Delicate Arch, Arches National Park, Utah. The La Sal Mountains are visible in the background.

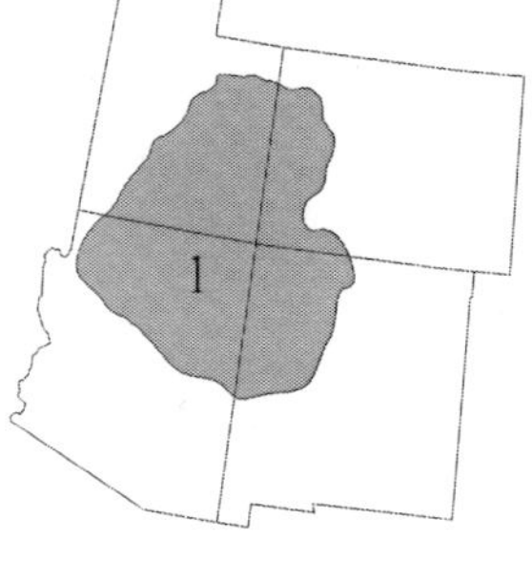

In spite of the plateau's generally semi-arid climate, erosion of such features is aided by localized, intense thunderstorms that produce very energetic (and dangerous) flash floods, transporting considerable masses of rock debris. The gorges, slots, and chasms of the Canyon Lands are a particularly prominent example of the erosive actions of flash floods. The area boasts the highest concentration of slot canyons in the world—these features are formed by the wear of water rushing through rock, and are extremely tall and narrow. Antelope Canyon, a slot canyon located approximately 10 kilometers (6 miles) southwest of the town of Page, Arizona, near the Glen Canyon Dam, is over 30 meters (100 feet) deep but measures only a few meters (yards) in width (*Figure 4.10*) The Upper and Lower Antelope canyons are carved in a 300-meter-long (1000-foot-long) section of Navajo Sandstone that partially blocks Antelope Creek, a stream that flows north to join the Colorado River. During flash floods, water fills the upstream basin, forcing mud- and **sand**-saturated water through a crack in the obstruction. The combination of turbulent flow interacting with the sand dune structures of the Navajo Sandstone has produced intricately carved and incredibly beautiful slot canyons.

sand *• rock material in the form of loose, rounded, or angular grains, and formed as a result of the weathering and decomposition of rocks.*

See Chapter 2: Rocks for more information about the Navajo Sandstone and other Jurassic sand dune deposits.

Figure 4.10: Lower Antelope Canyon (The Corkscrew), Arizona, a slot canyon characterized by the flowing spiral shape of its walls. The canyon was formed by erosion of the Navajo Sandstone due to flash flooding.

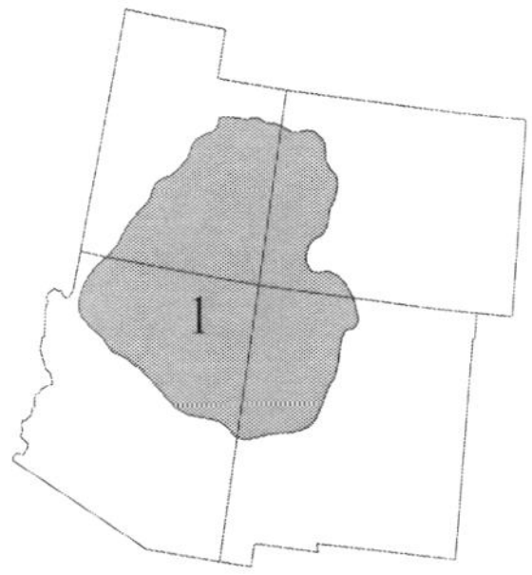

4 Topography

Region 1

Paleocene • a geologic time interval spanning from about 66 to 56 million years ago.

Jurassic • the geologic time period lasting from 201 to 145 million years ago.

Permian • the geologic time period lasting from 299 to 252 million years ago.

laccolith • an intrusive igneous rock body that forms from magma that has, through pressure, been forced between two sedimentary layers.

In places, the Colorado Plateau has been deformed by sharp, fault-cored folds called monoclines. These geologic landforms are step-like folds in rock strata, where an otherwise horizontal rock sequence dips steeply. The Waterpocket Fold, a monocline that defines Capitol Reef National Park, extends for nearly 160 kilometers (100 miles) across the Canyon Lands of central Utah. It formed between 70 and 50 million years ago, when the Laramide Orogeny activated a buried fault. The overlying rock layers on the fold's west side were lifted more than 2100 meters (7000 feet) higher than those on the east side (*Figure 4.11*). Today, the erosion of these tilted rock layers, which dip sharply downwards to the east, has resulted in colorful cliffs with fantastical domes, spires, and monoliths. Capitol Reef National Park is named for its white domes of Navajo Sandstone, which resemble the dome of the Capitol Building (*Figure 4.12*).

In addition to monoclines, compression during the Laramide Orogeny also formed convex uplifted domes, or anticlines (*Figure 4.13*). The San Rafael Swell of central Utah is a 120-kilometer-long (75-mile-long) sandstone anticline that was pushed up during the **Paleocene**. Today, its sedimentary rocks have eroded into valleys, canyons, and mesas, exposing the **Jurassic** Navajo and Wingate sandstones and the **Permian** Coconino and Kaibab formations. I-70 cuts across this area, exposing the rock strata and providing beautiful vistas.

Cenozoic magmatic activity has also left major imprints on the Colorado Plateau's topography. During the Laramide Orogeny, magma intruded into the region's sedimentary strata and spread out, creating **laccoliths**—domed intrusions of magma injected into the sedimentary rock layers like jelly in a jelly donut. The laccoliths pushed up overlying sediment, which later eroded away to expose the igneous rock below. The Henry, La Sal, Abajo, and Carrizo mountains, as well as the Sleeping Ute and Lone Cone peaks, are all examples of laccolithic mountain ranges in the Canyon Lands.

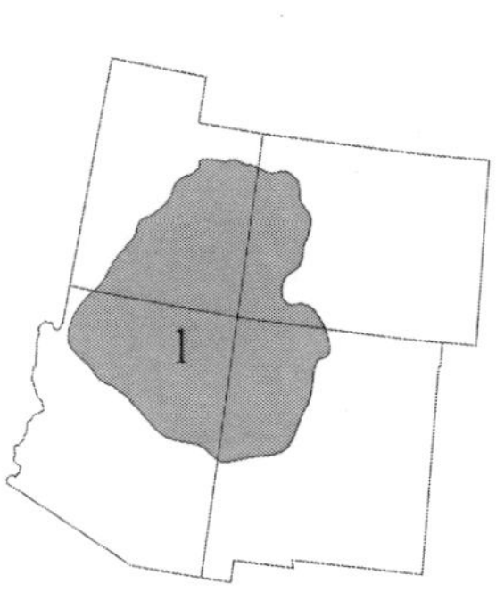

Figure 4.11: A view south along the Waterpocket Fold, Utah, which slopes steeply down across the white Navajo Sandstone. The formation is named for its "waterpockets," small depressions that form when rain erodes the sandstone.

Topography 4

Region 1

Figure 4.12: Capitol Dome and other white domes of Navajo Sandstone are common geologic features in Capitol Reef National Park, Utah.

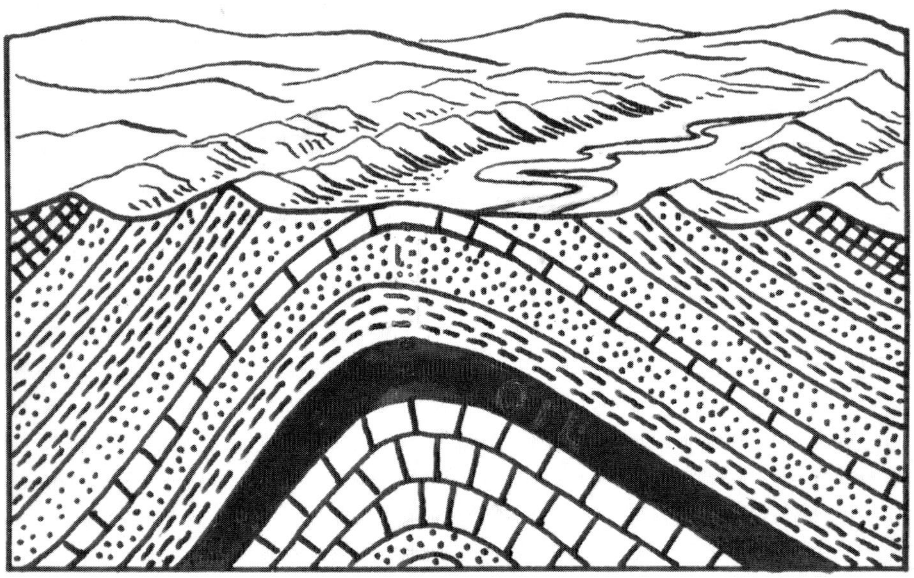

Figure 4.13: An anticline, an upward fold in layered rocks.

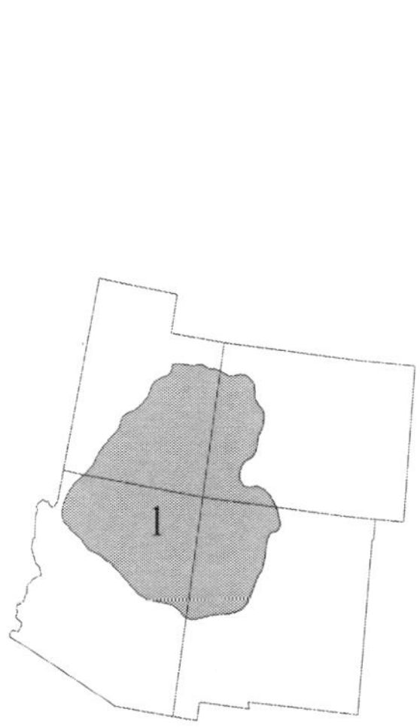

4 Topography

Region 1

basement rocks • *the foundation that underlies the surface geology of an area, generally composed of igneous or metamorphic crystalline rock.*

fluvial • *see outwash plain: large sandy flats created by sediment-laden water deposited when a glacier melts.*

Grand Canyon

The most extensive erosion anywhere on the Colorado Plateau has taken place at the region's southwest corner, in the area of the Grand Canyon. Permian-aged strata are found at much of the surface, revealing the extensive erosion that has taken place during the Cenozoic. An entire section of Mesozoic strata, estimated to be perhaps 450 meters (1500 feet) thick, has been eroded from the Plateau's southwest corner.

It is at the Grand Canyon itself that the erosional power of water is so dramatically demonstrated (*Figure 4.14*). The Colorado River has incised through all of the area's Mesozoic and Paleozoic strata, reaching the Precambrian metamorphic and igneous **basement rock**. This process has formed a canyon with dramatic topographic relief—the river's elevation is over a kilometer (nearly a mile) lower than the south rim of the canyon, and two kilometers (over a mile) lower than the north rim. An enormous array of tributary canyons is also visible from the canyon's rims. These tributaries exhibit remnant mesas and chimneys, and a step-like erosional pattern of cliffs, flats, and steep piles of rock rubble, resulting from variations in the rock layers' resistance to erosion. In contrast, in areas where the river has incised into the very hard Precambrian basement rock, the cross-section becomes V-shaped. V-shaped streambeds form in areas where rivers cut downwards into rock that is relatively uniform in resistance.

The rise of the Colorado Plateau is closely connected to the formation of the Grand Canyon. How and when this gigantic example of **fluvial** erosion occurred has remained hotly debated since geologist John Wesley Powell's trip by boat through the canyon in 1867. Many generations of geologists have questioned whether the canyon is very old, dating back to or before the Laramide Orogeny, or is geologically relatively young. Much of the literature available in past years and in tourist guides speaks of "hundreds of millions of years" as the time it took to carve the canyon, but this is incorrect. In fact, major portions of the Grand Canyon were downcut in response to the region's uplift, starting eight million years ago, in possible combination with a lower base elevation associated with Basin and Range crustal stretching.

The Colorado River does, however, have a complex ancestry that extends further back in time. Rivers have been running through the Grand Canyon area for the tens of millions of years it has been above sea level, and some initially carved their own canyons through normal erosional processes. One such canyon on the western side, Hurricane Canyon, may have been formed by a river that flowed northwest 70 to 50 million years ago; another on the eastern side may have formed approximately 25 to 15 million years ago. Segments of the Grand Canyon probably did develop to perhaps half their current depth during an early- and mid-Cenozoic time frame. The westernmost portion upstream from Lake Mead, and the northernmost segment, Marble Canyon, formed in just the past six million years. As these young river segments integrated with older ones, they formed the modern Colorado River that widened and deepened the Grand Canyon we know today.

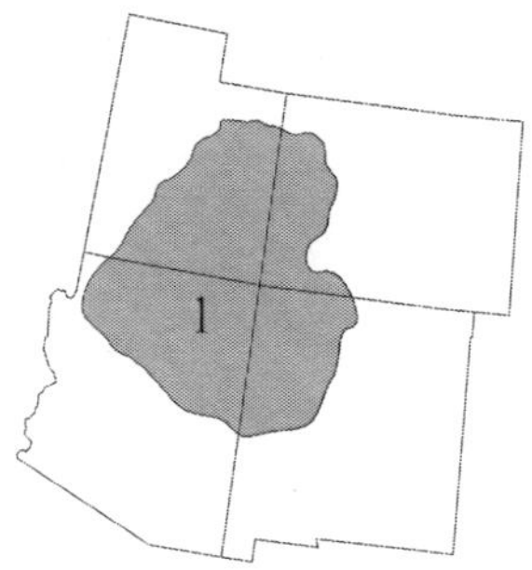

The entire series of Mesozoic and Cenozoic strata is preserved farther north and east of the Grand Canyon, and is progressively revealed along the Grand Staircase. Here, successively younger strata each form higher steps. South-

Region 1

color (mineral) • *a physical property determined by the presence and intensity of certain elements within the mineral.*

Figure 4.14: A view of the Colorado River flowing through the Grand Canyon, from Pima Point, Arizona.

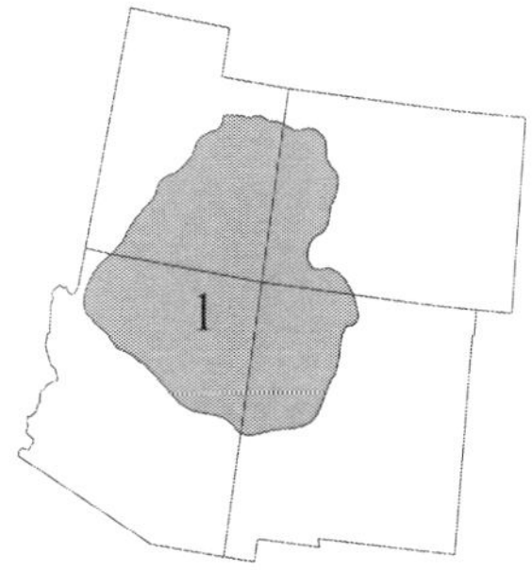

ward facing risers—the more sloping or vertical parts between the steps, also known as scarps—are formed by south-facing cliffs (cuestas) of the northward dipping strata. The cliffs have names describing their **colors**, including Chocolate, Vermillion, White, Gray, and Pink cliffs (*Figure 4.15*).

See Chapter 2: Rocks for more on the stratigraphy of the Grand Staircase and Grand Canyon.

4 Topography

Region 1

The Colorado River

The Colorado River and its tributaries drain most of the Colorado Plateau, as well as the Uinta Mountains, the southwestern Wind River Mountains, the eastern Wyoming Range, and the western Rockies. The river exits the plateau into the lower elevation desert of Arizona and southern Nevada and then travels southwards, as the boundary between Arizona and California, to the Gulf of California. The river descends steeply through the Grand Canyon from Lees Ferry, at an elevation of 940 meters (3100 feet), to about 330 meters (1100 feet), where it flows into Lake Mead.

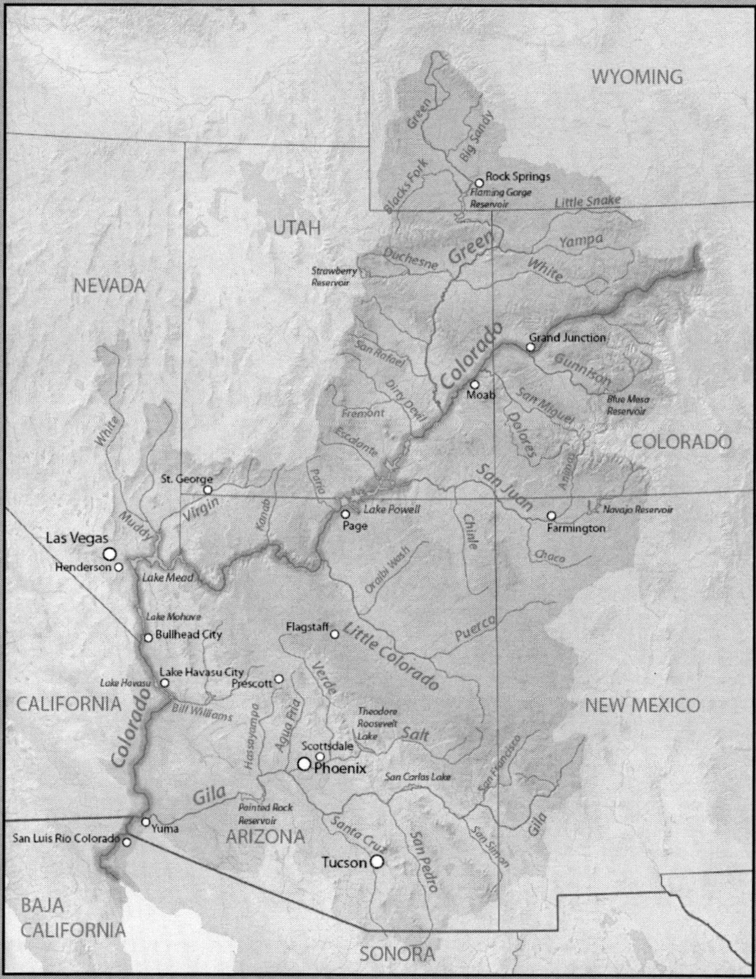

The Colorado River and its tributaries.

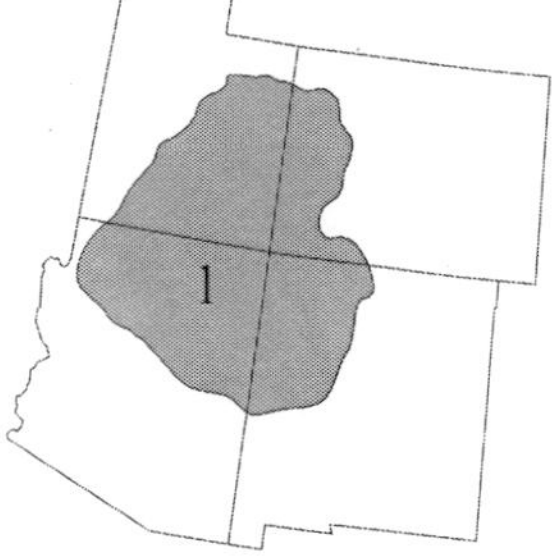

Figure 4.15: Colorful mesas of the Vermillion Cliffs near Para, Utah, in the Grand Staircase-Escalante National Recreation Area.

The Uinkaret Volcanic Field, which includes numerous Quaternary **basaltic cinder** cones and flows, is located near and on the northern rim of the western Grand Canyon. The age of these flows has provided key data for estimating the Grand Canyon's incision rates during the past 100 to 600 thousand years. The most recent eruption in Uinkaret occurred approximately 1000 years ago; some of the lava flows spilled into the gorge and temporarily dammed the Colorado River.

High Plateaus of Utah

North of the Grand Canyon lie the High Plateaus of Utah, a part of the Colorado Plateau where extensive volcanism has covered much of the landscape. The plateaus, separated by north-south trending normal faults, form a transition zone between the Colorado Plateau and the more fully developed extensional structures of the Basin and Range. The Marysvale Volcanic Field, which arose in the mid- and late Cenozoic, forms the caprocks of the High Plateaus in this area.

See Chapter 2: Rocks to learn more about volcanic fields in the Southwest.

On the easternmost Aquarius Plateau, the highest forested plateau in North America, Boulder Mountain forms a high elevation mesa with a volcanic cap protecting the more erodible Mesozoic sediments beneath. The Boulder Mountain mesa, at an elevation of approximately 3350 meters (11,000 feet), had a small **ice cap** and **glaciers** during the last **ice age**. A combination of Quaternary glaciation and the susceptibility of the underlying sedimentary rocks to erosion has produced remarkable **landslides** around much of the mesa.

basalt • *an extrusive igneous rock, and the most common rock type on the surface of the Earth.*

cinder • *a type of pyroclastic particle in the form of gas-rich lava droplets that cool as they fall.*

ice cap • *an ice field that lies over the tops of mountains.*

glacier • *a body of dense ice on land that does not melt away annually and has sufficient mass to move under its own weight.*

ice age • *a period of global cooling of the Earth's surface and atmosphere, resulting in the presence or expansion of ice sheets and alpine glaciers.*

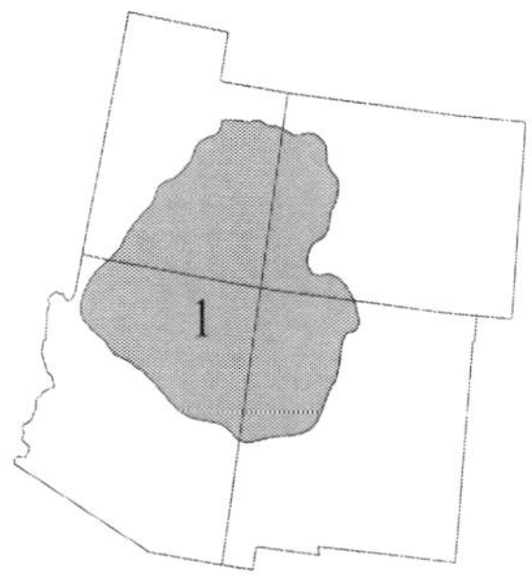

4 Topography

Region 1

oil • See petroleum: a naturally occurring, flammable liquid found in geologic formations beneath the Earth's surface and consisting primarily of hydrocarbons

shale • a dark, fine-grained, laminated sedimentary rock formed by the compression of successive layers of silt- and clay-rich sediment.

coal • a combustible, compact black or dark-brown carbonaceous rock formed by the compaction of layers of partially decomposed vegetation.

downwarp • a segment of the Earth's crust that is broadly bent downward.

badlands • a type of eroded topography that forms in semi-arid areas experiencing occasional periods of heavy rainfall.

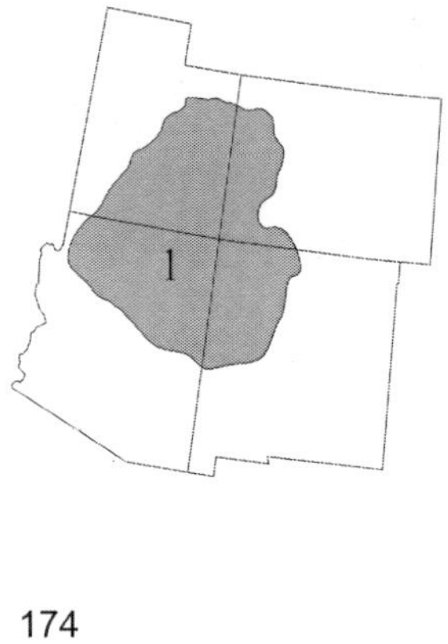

Uinta and Piceance Basins

The Uinta and Piceance basins encompass a diverse landscape of forests, canyons, and high mountains, which overlie deep basins formed in association with the Laramide basement uplift of the Rocky Mountains. The Uinta and Piceance basins are two parts of an ancient east-west trending basin that developed across the Utah-Colorado state boundary during the late Cretaceous. They are known for their accumulation of **oil shale** from the deep parts of the basin, and **coal** beds from the basin edges. Nearly the entire sequence of Mesozoic and Cenozoic strata, through the late Cenozoic, is preserved in this area.

See Chapter 6: Energy to learn more about the extraction of oil in the Uinta and Piceance Basins.

San Juan Basin

The San Juan Basin is an area of rugged plains and valleys, covering approximately 12,000 square kilometers (4600 square miles) of northwestern New Mexico and southeast Colorado. The basin encompasses a large, **downwarped** area of Mesozoic and Cenozoic rocks, formed in association with the Laramide Orogeny, and it is an important reservoir for oil and gas. Countless archaeological sites of the Ancestral Puebloan people can also be found throughout the San Juan's sandstone canyons. The landscape is punctuated by buttes, mesas, and **badlands**, including Chacra Mesa, Fajada Butte, and the Bisti Badlands (*Figure 4.16*).

Figure 4.16: Sandstone hoodoos in the Ah-Shi-Sle-Pah Wilderness Study Area, San Juan County, New Mexico. A hoodoo forms when weathering erodes a softer material out from underneath a mass of harder capstone, leading to "mushroom" formations.

Navajo

The Navajo section, along the southern edge of the Colorado Plateau in Arizona and New Mexico, is dominated by Cenozoic volcanism. The San Francisco, Hopi Buttes, Springerville-Red Hill, Zuni-Bandera, and Mount Taylor volcanic fields are all located in this part of the region. Volcanism in this area of the Colorado Plateau is related to tectonic activity that stretched the neighboring Basin and Range region.

The San Francisco Volcanic Field, a young volcanic field dating from six million years ago to the **Holocene**, is located between the Grand Canyon and Flagstaff, Arizona. The field is dotted with 600 large and small volcanic cinder cones; most are basaltic cinder cones, but the largest edifice is an eroded **stratovolcano** that forms the San Francisco Peaks (*Figure 4.17*). An **aquifer** that supplies much of Flagstaff's water is located within the volcano's ancient **caldera**. The San Francisco Peaks contain the six highest peaks in Arizona, including Humphreys Peak (the highest point in Arizona at 3851 meters [12,633 feet]), and the mountain range is a popular site for hiking and skiing. The most recent eruption within the San Francisco Volcanic Field occurred between the years 1064 and 1180, forming Sunset Crater—the focus of the Sunset Crater Volcano National Monument—and nearby smaller cones and flows.

Figure 4.17: The San Francisco Peaks of north central Arizona are a remnant of San Francisco Mountain, a former stratovolcano that collapsed around 200,000 years ago.

There are striking remains of volcanic **maars**, craters left by volcanic explosions, in the Hopi Buttes Volcanic Field. The maars are approximately 80 kilometers (50 miles) east of the San Francisco Volcanic Field, in the Navajo Reservation near the town of Dilkon, Arizona. The volcanism there is older than the San Francisco field and has been dated to seven million years ago. Erosion at Hopi Buttes exposes approximately 300 deep vertical volcanic pipes called **diatremes** and their associated maars. These exposures of diatremes and maars have become a mecca for geologists studying monogenetic volcanism—locations where many small volcanoes erupt only a single time each.

Holocene • *the most recent portion of the Quaternary, beginning about 11,700 years ago and continuing to the present.*

stratovolcano • *a conical volcano made up of many lava flows as well as layers of ash and breccia from explosive eruptions.*

aquifer • *a water-bearing formation of gravel, permeable rock, or sand that is capable of providing water, in usable quantities, to springs or wells.*

caldera • *a collapsed, cauldron-like volcanic crater formed by the collapse of land following a volcanic eruption.*

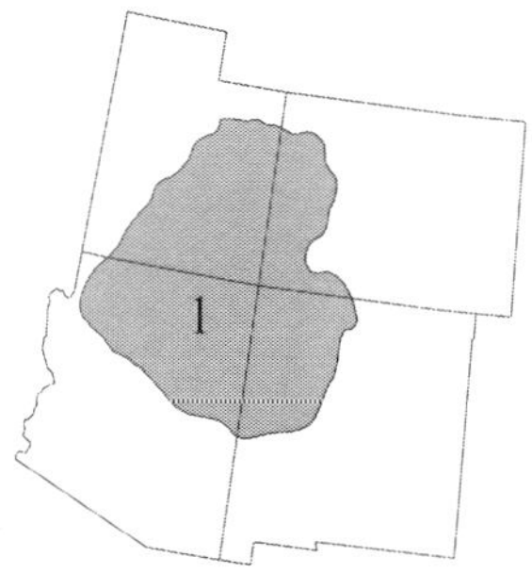

4 Topography

Region 1

Pliocene • *a geologic time interval extending from roughly 5 to 2.5 million years ago.*

Pleistocene • *a subset of the Quaternary, lasting from 2.5 million to about 11,700 years ago.*

shield volcano • *a volcano with a low profile and gradual slope, so named for its likeness to the profile of an ancient warrior's shield.*

meteorite • *a stony or metallic mass of matter that has fallen to the Earth's surface from outer space.*

nickel • *a ductile, silvery-white metallic element (Ni) generally found in combination with iron.*

iron • *a metallic chemical element (Fe).*

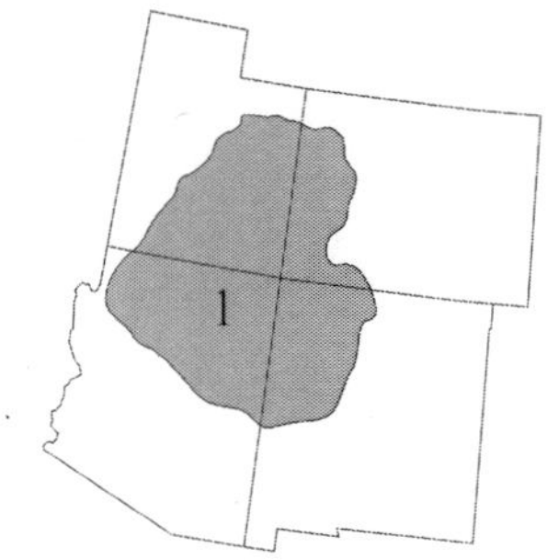

Scattered cones and basaltic flows of late Cenozoic age leave their topographic imprint in a zone trending northeast along the southern border of the Colorado Plateau including the Springerville–Red Hill, Zuni-Bandera, and Mount Taylor volcanic fields. These areas, similar to the San Francisco Volcanic Field, are characterized by basaltic flows and cinder cones plus a large eroded stratovolcano, Mount Taylor. Cabezon Peak, a prominent feature in the northwestern New Mexico landscape, is a volcanic plug—the solidified remains of magma trapped in the neck of a volcano—that is also part of the Mount Taylor Volcanic Field (*Figure 4.18*).

Figure 4.18: Cabezon Peak, a basalt volcanic plug that rises to 2373 meters (7785 feet) in elevation. It is part of the Mount Taylor Volcanic Field in New Mexico.

The Springerville–Red Hill Volcanic Field is another large area of numerous **Pliocene-Pleistocene** cinder cones and flows. It is located close to the southern border of the Colorado Plateau, approximately 160 kilometers (100 miles) south-southeast of the Hopi Buttes. Just to the south of this field is Mount Baldy, a deeply eroded **shield volcano** of late Cenozoic age.

Approximately 55 kilometers (35 miles) east of Flagstaff is a small topographic feature that at first glance might be mistaken for a volcanic feature, such as a maar. This crater, approximately 1200 meters (3900 feet) in diameter and 45 meters (148 feet) deep, is in fact the result of a **meteorite** impact. It is, appropriately, called Meteor Crater (*Figure 4.19*). A **nickel-iron** meteorite generated the crater approximately 50,000 years ago; it is estimated to have been approximately 45 meters (150 feet) across and was mostly vaporized upon impact. The largest recovered fragment is called the Holsinger Meteorite and is exhibited at the crater's visitor center. It weighs 639 kilograms (1409 pounds) and is approximately 1.2 meters (4 feet) long.

Figure 4.19: Meteor Crater (also called Barringer Crater), Coconino County, Arizona.

Topography of the Basin and Range
Region 2

In the Southwest, the Basin and Range stretches along the western sides of Utah and Arizona as well as the southern parts of Arizona and New Mexico (*Figure 4.20*). The entire Basin and Range region covers a massive expanse of land, stretching from Idaho and southeastern Oregon through all of Nevada, southeastern California, and extending into western Texas and Mexico.

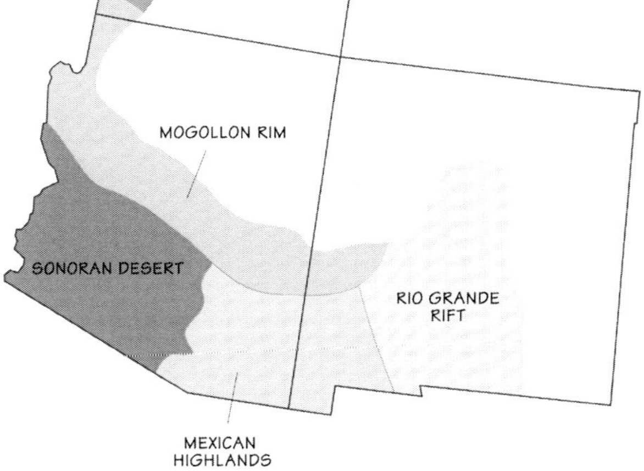

Figure 4.20: Physiographic subdivisions of the Basin and Range.

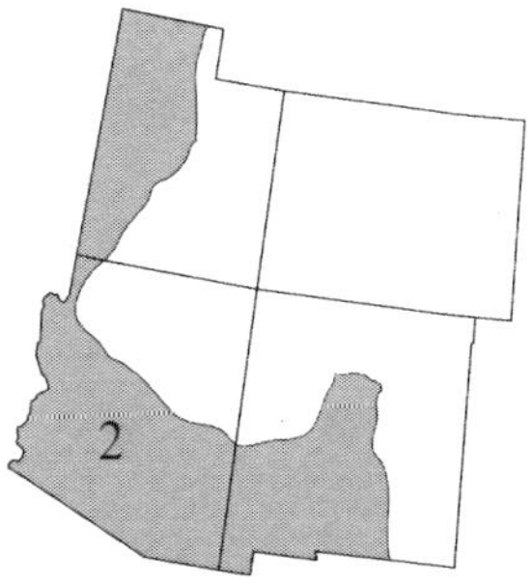

Region 2

Paleogene *• the geologic time period extending from 66 to 23 million years ago.*

system *• a set of connected things or parts forming a complex whole.*

lithosphere *• the outermost layer of the Earth, comprising a rigid crust and upper mantle broken up into many plates.*

mantle *• the layer of the Earth between the crust and core.*

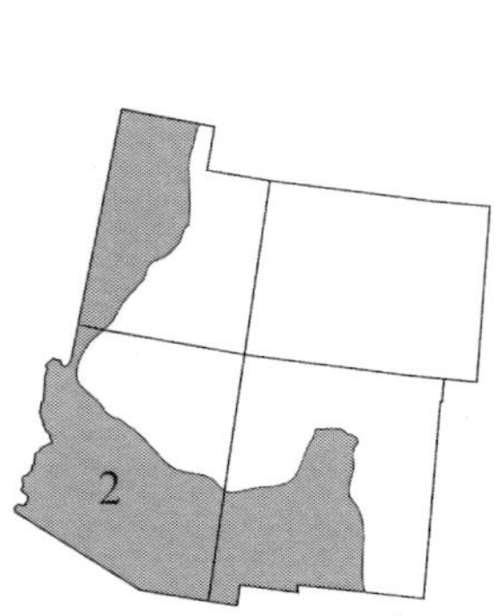

The Basin and Range is characterized by rapid changes in elevation alternating from flat and dry basins to narrow, faulted mountains. This pattern of many parallel, north-south mountain ranges found throughout the region inspired geologist Clarence Dutton to famously observe that the topography of the Basin and range appeared "like an army of caterpillars crawling northward." The formation of this topography is directly related to tectonic forces that led to crustal extension (pulling of the crust in opposite directions). After the Laramide Orogeny ended in the **Paleogene**, tectonic processes stretched and broke the crust as California's San Andreas fault **system** began to develop, and the upward movement of magma weakened the **lithosphere** from underneath. Approximately 20 million years ago, the crust along the Basin and Range stretched, thinned, and faulted into some 400 mountain blocks. The pressure of the **mantle** below uplifted some blocks, creating elongated peaks and leaving the lower blocks below to form down-dropped valleys. The boundaries between the mountains and valleys are very sharp, both because of the straight faults between them and because many of those faults are still active.

These peaks and valleys are also called horst and graben landscapes (*Figure 4.21*). Such landscapes frequently appear in areas where crustal extension occurs, and the Basin and Range is often cited as a classic example thereof. In the Basin and Range, the crust has been stretched by up to 100% of its original width. The distance between the Sierra Mountains of California and the Colorado Plateau increased by over 320 kilometers (200 miles) as a result of this crustal extension. Due to the stretching, the average crustal thickness of the Basin and Range is 30–35 kilometers (19–22 miles), compared with a worldwide average of approximately 40 kilometers (25 miles).

There is a marked change in elevation between the northern and southern portions of the Basin and Range. In Utah, the Great Basin is in the range of 1200 to 2100 meters (4000 to 7000 feet) in elevation, while the Sonoran Desert in Arizona is mostly 150 to 600 meters (500 to 2000 feet) above sea level.

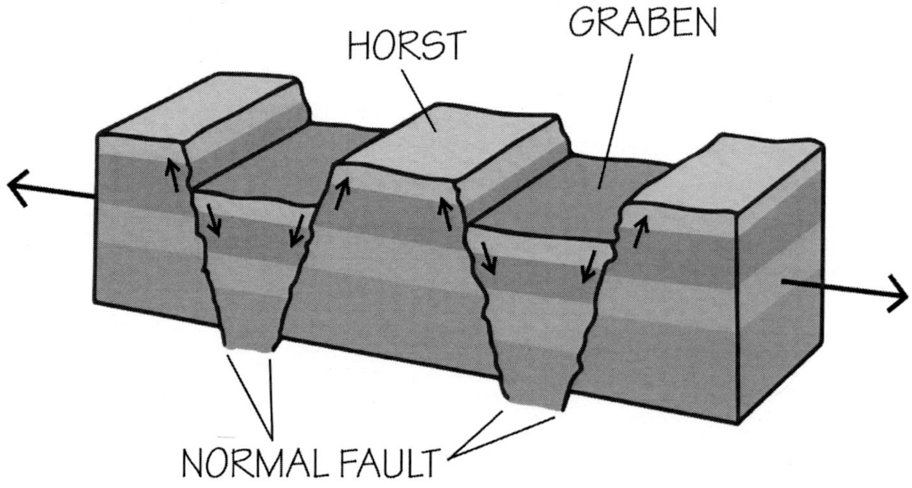

Figure 4.21: A horst and graben landscape occurs when the crust stretches, creating blocks of lithosphere that are uplifted at angled fault lines.

Great Basin

Utah's western half is part of the Great Basin, an area that spans most of Nevada and reaches into California, Oregon, and Idaho (*Figure 4.22*). The Great Basin is bounded on all sides by topographical features—the Wasatch Range in Utah, the Sierra Nevada and Cascades in California, and the Snake River Basin to the north—that prevent it from having any outlet to the ocean. As such, the Basin drains internally; that is, all water in the region evaporates, flows into lakes, or sinks underground. Because there are no flow outlets and water typically leaves lakes via evaporation (allowing minerals to concentrate), many of the area's lakes are saline.

In Utah, the Great Basin largely drains into the Great Salt Lake, which receives considerable runoff from the Wasatch Range. Today, the Great Salt Lake is the largest saltwater lake in the Western Hemisphere and the largest remnant of

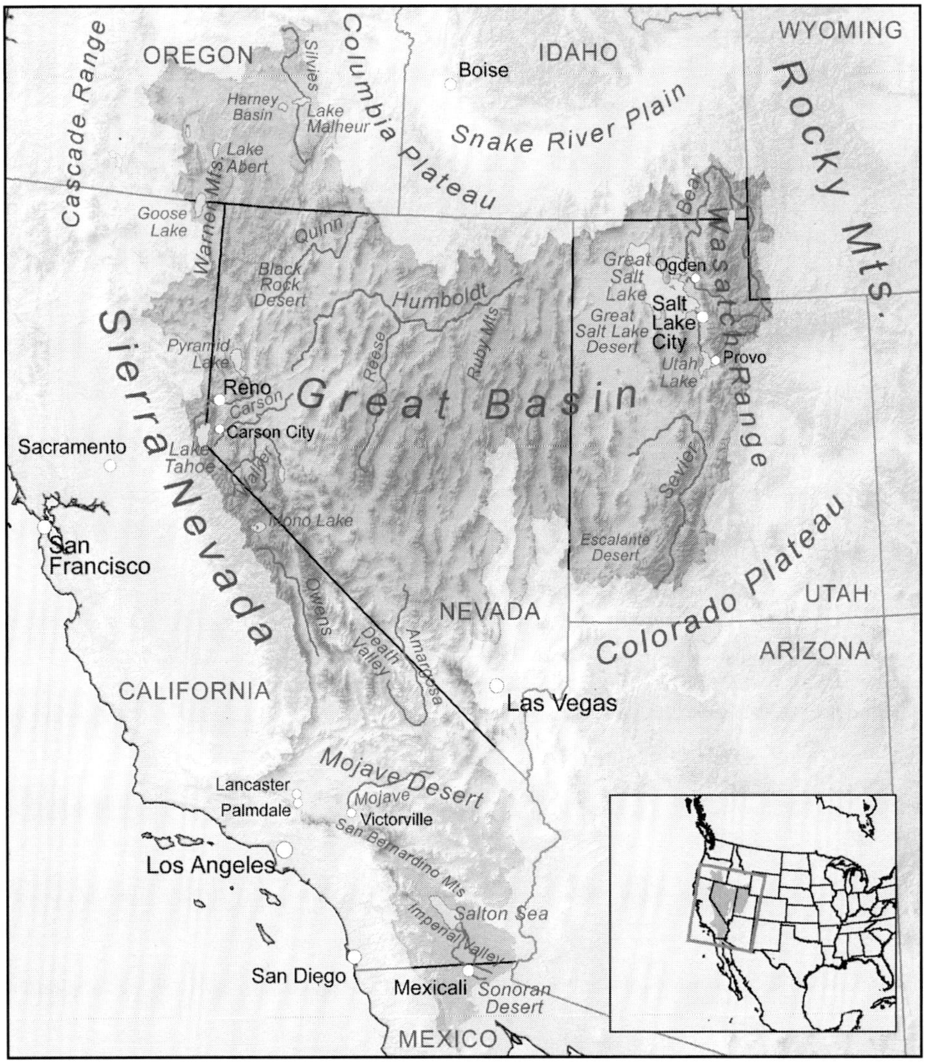

Figure 4.22: The extent of the Great Basin.

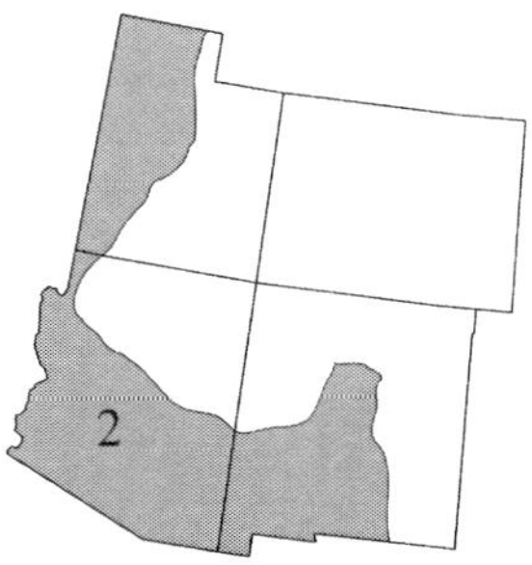

pluvial lake • a landlocked basin that fills with rainwater or meltwater during times of glaciation.

salt • a mineral composed primarily of sodium chloride (NaCl).

isostasy • an equilibrium between the weight of the crust and the buoyancy of the mantle.

Carboniferous • a geologic time period that extends from 359 to 299 million years ago.

Lake Bonneville, a large, deep **pluvial lake** that covered much of western Utah during the Pleistocene. As the lake evaporated, **salts** were concentrated in the remaining water, then left behind with the muds as the final water evaporated from low areas. The present day Great Salt Lake has a saline content of approximately 13% (depending on lake level) and contains approximately 4.1 billion metric tons (4.5 billion tons) of salt. Deltas and shoreline deposits from Lake Bonneville remain visible on the slopes of the Wasatch Mountains, and their elevations provide measurements of the area's **isostatic** rebound after the lake receded and the weight of its water was removed from the land. Other, smaller remnants of Lake Bonneville include Rush Lake, Sevier Lake, and Utah Lake, as well as the Bonneville Salt Flats and the Escalante Desert Valley.

See Chapter 8: Climate to learn more about the formation and extent of glacial Lake Bonnevile.

Mogollon Rim

The Mogollon Rim is a major escarpment that forms the southern edge of the Colorado Plateau, extending approximately 320 kilometers (200 miles) through Arizona and into New Mexico. Erosion and faulting have carved dramatic canyons into the rim, and **Carboniferous** and Permian sandstone and limestone form spectacular high cliffs. The Mogollon Rim rises approximately 1200 meters (4000 feet) above the land to the south (*Figure 4.23*).

Escarpments, or "scarps," form when faulting or erosion acts to create a cliff or steep slope that separates two level or gently sloping topographical surfaces.

Figure 4.23: The Mogollon Rim in Coconino National Forest, Arizona.

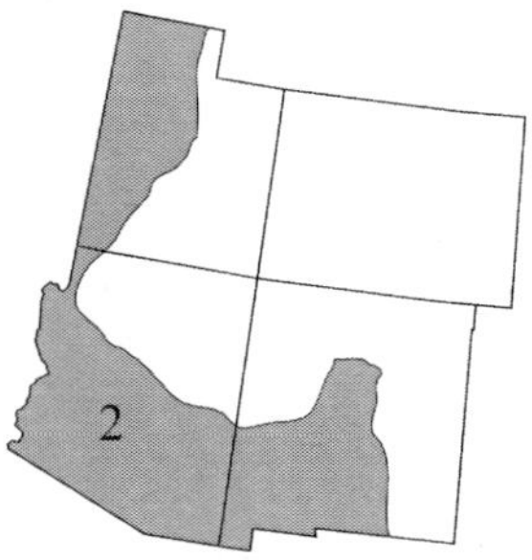

Topography 4

Region 2

Isostasy

The elevation of Earth's surface topography, on scales of a few kilometers (miles), may seem to result just from the height of the rigid rock that the crust is made of. However, over hundreds or thousands of kilometers, surface elevation is mainly influenced by how the weight of the thin rigid lithosphere (which includes the crust and uppermost mantle) floats buoyantly on the more fluid-like part of the mantle (the *asthenosphere*). Since this relationship is complex and good data from the deep subsurface are difficult to acquire, it is still not well understood.

The weight of the lithosphere in a unit area is controlled by its thickness and the *density* of its rocks. All else being equal, the heavier the lithosphere, the more it will sink into the mantle. For example, water and sediment may accumulate in a basin, increasing weight and pressing downward, or water may be drained and rocks eroded away. In higher latitudes, Pleistocene continental glaciers pushed the crust downward, and when they melted, the land surface slowly rose again in a process called isostatic rebound. With tectonic movement, the lithosphere may thicken by horizontal compression or thin by extension (as in the Basin and Range). Since the boundary between the lithosphere and asthenosphere is controlled by where rigid rocks become fluid (about 1300°C [2370°F]), the lithosphere's thickness and density are also controlled by how quickly temperature increases with depth. These and yet other factors can make a detailed understanding of Earth's topography remarkably challenging.

asthenosphere • *a thin semifluid layer of the Earth, below the outer rigid lithosphere, forming much of the upper mantle.*

density • *a physical property of minerals, describing the mineral's mass per volume.*

alluvial • *a layer of river-deposited sediment.*

Sonoran Desert

Arizona's southwestern corner is occupied by the Sonoran Desert, the hottest desert in North America. The area is characterized by steep, blocky mountain ranges and intervening flat desert basins, as well as intermittent streams and **alluvial** fans formed when heavy rainstorms transport eroded material. During extension of the Basin and Range in southern Arizona, the overlying sedimentary rocks were stretched so much that underlying metamorphic rocks were exposed. These rocks are called metamorphic core complexes (MCCs), and they consist of metamorphic rock bodies that formed deep in the crust before moving to the surface along low-angle faults. MCCs are usually the

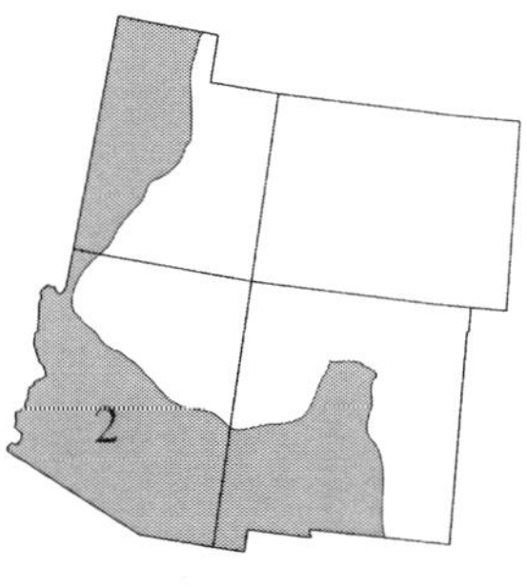

andesite • *a fine-grained extrusive volcanic rock, with a silica content intermediate between that of basalt and dacite.*

rhyolitic • *a felsic volcanic rock high in abundance of quartz and feldspar.*

rift • *a break or crack in the crust that can be caused by tensional stress as a landmass breaks apart into separate plates.*

lava tube • *a natural tube formed by lava flowing beneath the hardened surface of a lava flow.*

gypsum • *a soft sulfate mineral that is widely mined for its use as fertilizer and as a constituent of plaster.*

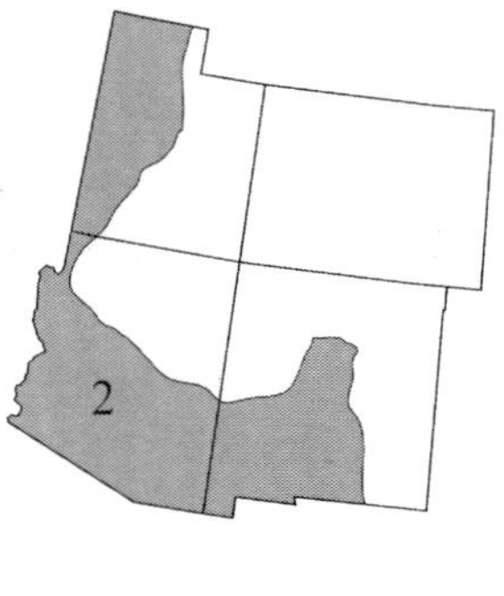

highest topography in the area, but are also domed and frequently do not match the form of other nearby mountains, which have modest relief (300 to 600 meters [1000 to 2000 feet]) and are nearly buried in their eroded detritus. This extremely resistant metamorphic rock, along with patches of volcanics, are responsible for many small mountain ranges that do not follow the mostly north-south pattern of narrow ridges so characteristic of the rest of the Basin and Range. Large intervening basins, filled with eroded sediment from the mountains, are drained by the Gila River into the Colorado River. Saguaro National Park is located close to an especially prominent example, forming the Rincon and Santa Catalina mountains near Tucson (*Figure 4.24*). A band of MCCs stretches from southeastern Arizona to the junction of its borders with California and Nevada (*Figure 4.25*); the pattern continues north through eastern Nevada, then reemerges in Utah's northwestern corner.

Mexican Highlands

The Mexican Highlands, extending from eastern Arizona into western New Mexico, contains Basin and Range structures with higher relief and higher basin elevations, similar to those of the Great Basin. Desert grasslands surround the highlands' isolated mountain ranges, and the area also includes a number of volcanic structures.

The Mogollon-Datil Volcanic Field in southeastern Arizona and southwestern New Mexico includes mid-Cenozoic volcanic **andesites**, tuff, **rhyolites**, and basalts that extend into the complexly faulted area where the Basin and Range melds into the Rio Grande Rift zone. The deeply eroded remains of a number of silicic calderas have been identified, including the Mogollon Mountains whose peaks reach nearly 3350 meters (11,000 feet). This volcanic field includes part of the Continental Divide between the Gila and the Rio Grande rivers.

Rio Grande Rift

The Basin and Range of New Mexico is dominated by the Rio Grande Rift, a zone of extension in the continental crust running from New Mexico north into Colorado. A small amount of true crustal spreading occurred in this narrow area, resulting in an elongate depression bounded by faults, which also controls the course of the Rio Grande River. Magma erupting through the **rift** resulted in adjacent volcanic fields, as well as volcanic fill, that cover the rift valley floor. Faulting of the Rio Grande Rift also reaches into and affects the topography of the Colorado-New Mexico Rocky Mountains.

The Rio Grande Rift began to form approximately 30 million years ago, peaked from 16 to 10 million years ago, and remains tectonically active today. One recent eruption, the 5000-year-old Carrizozo Malpais, resulted in a 75-kilometer-long (47-mile-long) flow of black basalt that filled the Tularosa Basin and contains **lava tubes**, collapse pits, wrinkles, and fissures (*Figure 4.26*). The Carrizozo Malpais is considered to be one of the youngest lava flows in the continental US.

Directly south of the Carrizozo Malpais lies a famous non-volcanic feature, the White Sands National Monument. This is the largest **gypsum** dune field in the world, covering an area of 710 square kilometers (275 square miles) in southern

Figure 4.24: The heavily eroded Santa Catalina Mountains near Tucson, Arizona are surrounded by forests of saguaro cacti. The highest point in these mountains has a relief of 1572 meters (5157 feet) over the surrounding landscape, and the mountains are tall enough to receive snowfall.

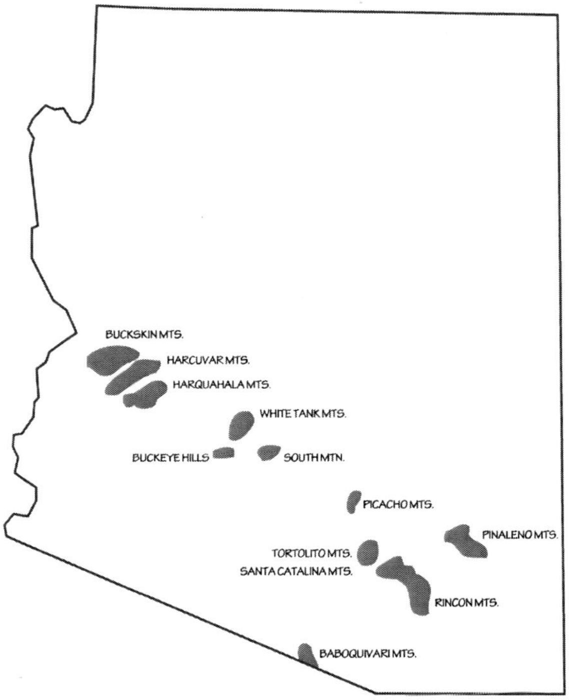

Figure 4.25: Distribution of metamorphic core complexes in Arizona.

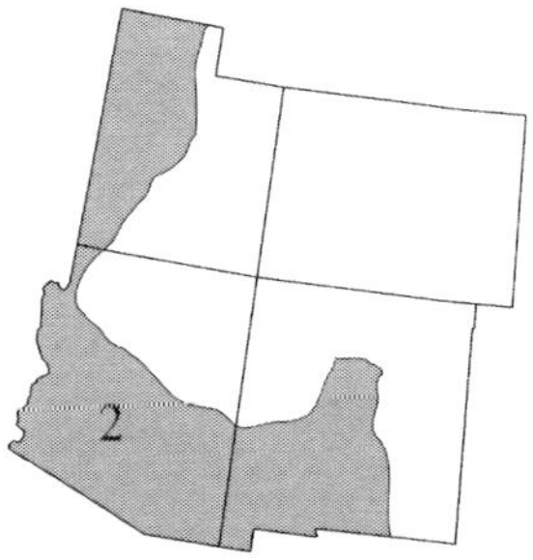

New Mexico (*Figure 4.27*). The dunes are composed of windblown gypsum grains, eroded from Permian Basin **evaporite** deposits in the surrounding San Andres Mountains. Gypsum is water-soluble, and is found in sand form only rarely; White Sands has endured thanks to the area's aridity and the fact that the Tulsarosa Basin has no outlet to the sea.

evaporite • *a sedimentary rock created by the precipitation of minerals directly from seawater, including gypsum, calcite, dolomite, and halite.*

Figure 4.26: A characteristic ropy lava flow covers the ground at the Valley of Fires Recreation Area in New Mexico, directly adjacent to the Carrizozo Malpais.

Figure 4.27: Windblown gypsum dunes of the White Sands National Monument, New Mexico. The dune field began forming 25,000 years ago.

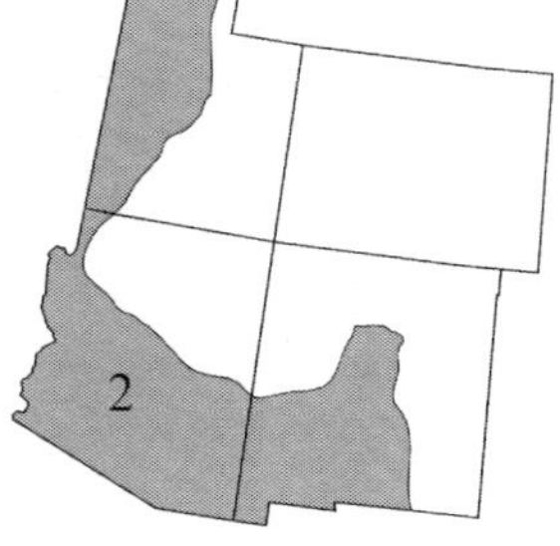

Topography of the Rocky Mountains
Region 3

The Rocky Mountains of the Southwest consist of multiple uplifted ranges resulting from the Sevier and Laramide orogenies, as well as from Cenozoic volcanism and extension. The mountains extend more than 4800 kilometers (3000 miles) from northern British Columbia southward through Alberta, Idaho, Montana, Wyoming, and—in the Southwest—into central and western Colorado, northeastern Utah, and north-central New Mexico (*Figure 4.28*). They include the high and rugged mountain ranges of the Southern Rocky Mountains, located between the Colorado Plateau and the Great Plains; the Uinta Mountains, bordering the northwestern part of the Colorado Plateau; and the Wasatch Mountains, located in Utah along the eastern border of the Basin and Range. Because of their high elevations, all the ranges capture rain and snow, and have been subjected to both fluvial and glacial erosion during the Quaternary. Evidence for this varies from the **cirques** and U-shaped valleys of the Front Range to the massively glaciated San Juan Mountains and their numerous huge rock glaciers (glaciers buried in **talus**).

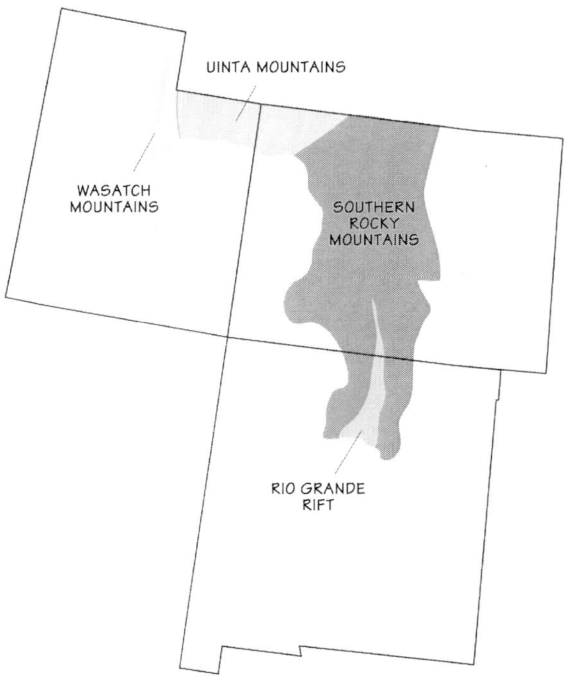

Figure 4.28: Physiographic subdivisions of the Rocky Mountains.

The Rocky Mountains, from the Wyoming border to northern New Mexico, boast 53 peaks over 4200 meters (14,000 feet) high—the "Fourteeners" of mountain climbing fame. The highest part of the Rocky Mountain uplift is centered in Colorado; the state has the highest concentration of mountains over 4200 meters (14,000 feet) high in the continental US, as well as the highest base elevation of any area of the continent. Forty of these peaks are located in Colorado's Sawatch, Sangre de Cristo, and San Juan ranges, and were uplifted by faulting or magmatic activity during the mid- to late Cenozoic.

Region 3

cirque *• a large bowl-shaped depression carved by glacial erosion and located in mountainous regions.*

talus *• debris fields found on the sides of steep slopes, common in periglacial environments.*

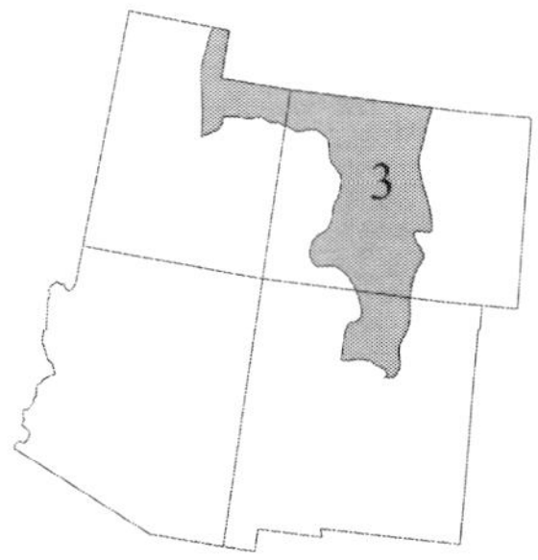

Region 3

seismic waves • the shock waves or vibrations radiating in all directions from the center of an earthquake or other tectonic event.

ice field • an extensive area of interconnected glaciers spanning less than 50,000 square kilometers (19,305 square miles).

Wasatch Mountains

The Wasatch Mountains, located at the western edge of the Rockies in between the Basin and Range and the Colorado Plateau, stretch approximately 260 kilometers (160 miles) south from the Utah-Idaho border. This mountain range owes its current high relief mainly to faulting that encroached into the Colorado Plateau during the Cenozoic. Normal faulting is still active along the Wasatch Fault today, posing **seismic** risks to Salt Lake City.

See Chapter 9: Earth Hazards to learn more about earthquakes along the Wasatch Fault.

There are no exposures of Precambrian basement rock in the Wasatch Range, except for a small area near the westward extension of the Uinta uplift—the mountains are largely made up of Paleozoic strata. The mountain range is punctuated by valleys, into which it drops steeply on its western side and more gently to the east. Much of the Wasatch Range exhibits glacially sculpted landforms (*Figure 4.29*).

Figure 4.29: Mount Timpanogos in Utah's Wasatch Range displays the sculpting power of moving ice. Alpine glaciers shaped the mountain's knife-edge ridges and its wide U-shaped amphitheaters.

Uinta Mountains

The Uinta Mountains are the only the only part of the Rockies where the mountains run east-west instead of generally north-south. This range is a classic example of Laramide uplift, when the Precambrian basement rock and its cover of Paleozoic and Mesozoic sedimentary strata were pushed up to form an anticline bounded on either side by low-angle thrust faults. The faults dipped beneath the Precambrian basement and thrust it above sedimentary cover, deforming the strata into basins adjacent to the mountain range. During and subsequent to these crustal deformations, erosion removed the entire series of Paleozoic and Mesozoic strata above the uplift, exposing the more resistant Precambrian rocks beneath. The exposed Precambrian block eroded slowly, forming relatively smooth relief during the Cenozoic, and was nearly buried by the accumulation of eroded sediments. During glaciation in the Quaternary, the highest parts of the Uinta Mountains were covered by an **ice field**. High peaks poked through the ice as isolated points called "nunataks," while glaciers moving down on either side carved cirques and U-shaped valleys, generating the alpine ruggedness we are familiar with today.

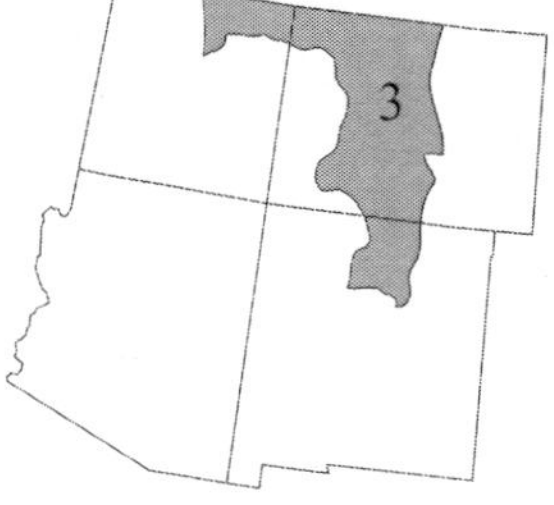

During the mid- to late Cenozoic, the eastern part of the Uinta uplift played a key role in shaping the flow of the Colorado River. In the mid-Cenozoic, the ancestral Green River flowed south from the Wind River Mountains in northern Wyoming and then east towards the Great Plains. After the Laramide Orogeny ended, faulting in the low-elevation and sediment-covered eastern segment of the Uintas led to a complex series of stream captures that eventually connected the southward flow of the Green River to the then-northern headwaters of the Colorado River. With this added contribution of water from the Wind River and Wyoming ranges and the northern slopes of the Uintas, the connection cut through the sedimentary cover to carve the dramatic Lodore Canyon into the Precambrian core of the Uinta uplift (*Figure 4.30*).

Figure 4.30: The Green River flows through the Gates of Lodore near Dinosaur National Monument, Colorado.

Southern Rocky Mountains

The Rocky Mountains of Colorado and New Mexico have a complex topographical history. Their high relief is a product of four components: uplift of the late Paleozoic Ancestral Rockies, compressional uplift of Precambrian basement rock during the Laramide Orogeny, igneous activity during the mid- to late Cenozoic, and extensional tectonics associated with the Rio Grande Rift. The effects of volcanism and the Rio Grande Rift were most important at the region's southern end, where the rift dominates the landscape.

Pennsylvanian • *a subperiod of the Carboniferous, spanning from 323 to 299 million years ago.*

The Ancestral Rockies date to the **Pennsylvanian**, during which time thrust faulting uplifted the Precambrian basement rock in Colorado and northern New Mexico. These ancient mountains were covered beneath the Western Interior Seaway during the Mesozoic, but they generated weaknesses in the crust that led to greater faulting during the Laramide Orogeny. During the late Cretaceous

outwash plain • *large sandy flats created by sediment-laden water deposited when a glacier melts.*

terrace • *a flat or gently sloped embankment or ridge occurring on a hillside, and often along the margin of (or slightly above) a body of water, representing a previous water level.*

pluton • *a body of intrusive igneous rock that formed under the Earth's surface through the slow crystallization of magma.*

and early Cenozoic, the Rockies were rejuvenated by compressional uplift, and erosion began anew, cutting stream valleys and leaving the more resistant rock as topographic highs. Alpine glaciers later carved cirques into the high peaks, leaving the rugged sharp peaks we see today. This is best exhibited in the northern Colorado Rockies, including the Front Range north of Colorado Springs (*Figure 4.31*), the Gore Range near Vail, and the Park Range, which extends north into Wyoming. In between these ranges lie the North Park and Middle Park basins, which are broad, lower areas of relatively low relief floored by Precambrian basement rocks. Meltwater from glaciers deposited **outwash plains** and **terraces** downstream—Rocky Mountain National Park in northern Colorado contains spectacular examples of these features.

Figure 4.31: The view from Mount Evans, one of the highest points in Colorado's Front Range at 4350 meters (14,271 feet).

Farther south, the Rocky Mountains become increasingly affected by Cenozoic faulting and volcanism associated with the Rio Grande Rift. On the western side of the rift, the Elk, Sawatch, and Mosquito ranges are cut by a rift valley (graben) bordered by normal faults between the Sawatch and Mosquito ranges. The valley forms the headwaters of the West Arkansas River. The Nacimiento Mountains are another example of Laramide-aged basement uplift; their Precambrian rocks thrust towards the west. The northern part of the range includes the San Pedro Parks Wilderness Area, which has the highest elevation of the mountain range, reaching up to approximately 3230 meters (10,600 feet). The area's high relief is due to the resistance of its Precambrian **plutonic** rocks. The San Juan Mountains, also located on the rift's western side, are primarily mid-Cenozoic volcanoes that are now deeply eroded into rugged mountains. The ruggedness of the terrain, especially in contrast to mountains formed from Precambrian metamorphics, highlights the difference in erodability between the two types of rocks.

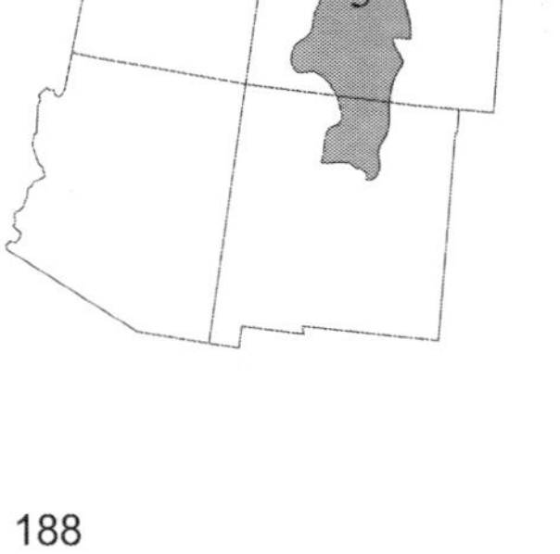

East of the rift across the South Park Basin, uplifted Precambrian basement rocks of the Rampart Range and Pikes Peak Massif face the Great Plains. Exposures of uplifted Precambrian rocks continue in the Wet Mountain Range and the long, narrow Sangre de Cristo Range, which ends south of Santa Fe, New Mexico. The Sangre de Cristo Mountains were uplifted in the mid- to late Cenozoic through a process similar to Basin and Range extension. South of the Sangre de Cristo Range, the mountains are slightly lower in elevation. The Sierra Blanca Mountains are the eroded remains of a large mid-Cenozoic volcano with peak elevation reaching nearly 3700 meters (12,000 feet). The Sandia Mountains, reaching peak elevations just under 3000 meters (10,000 feet), comprise uplifted late Paleozoic and Mesozoic strata.

Rio Grande Rift

The Rio Grande Rift is a zone of extension that reaches from New Mexico's Basin and Range up through the Rocky Mountains to central Colorado, near Leadville. Recent studies indicate traces of the rift may extend farther north, almost to the Wyoming border. Cenozoic volcanic activity within the rift had a major impact on the topography of the surrounding Rocky Mountains, including the flow of the Rio Grande River. When the rift formed, it captured the river, which initially began as a stream trickling from the mountains near Leadville.

The Jemez Lineament in northern New Mexico is thought to be an elongate, ancient crustal weakness through which magmas erupted as part of the Basin and Range extension. Where the Jemez Lineament crosses the Rio Grande Rift, more intense volcanism occurred, forming the Jemez Volcanic Field. This area includes the Valles Caldera, a **supervolcano** that last erupted approximately 1.5 and 1.2 million years ago. Its crater measures approximately 19 by 24 kilometers (12 by 15 miles) across; the younger caldera, 22 kilometers (14 miles) wide, partially buried the older. The supervolcano eruptions produced thick tuffs (i.e., the Bandelier Formation), which cover the eastern side of the caldera and form the Pajarito Plateau. The tuffs slope gently eastward, and are eroded into narrow, vertically walled canyons with relief of up to 240 meters (800 feet). Hoodoos, such as those found at the Kasha-Katuwe Tent Rocks National Monument in Sandoval County, New Mexico, are also common formations found in Jemez **pyroclastic** flows. The northeast trend of late Cenozoic volcanism continues farther across the Rio Grande Rift and into the Great Plains. This linear movement has been attributed to a **hot spot** track, but the dates of the eruptions do not travel in a line from oldest to youngest, as do those associated with other well-known hot spots such as Yellowstone or Hawai'i.

See Chapter 2: Rocks to learn more about the Jemez Lineament, supervolcanoes, and volcanic fields in the Southwest.

The Rio Grande Rift is also occupied by a series of high-altitude basins: the upper Arkansas graben and San Luis Basin in Colorado, and the Espanola, Albuquerque, and Socorro basins in New Mexico. Each basin is filled with sediment eroded from the nearby mountains and interbedded with **volcanic ash** and lava. The San Luis Valley covers approximately 21,000 square kilometers (8000 square miles) of Colorado and a small portion of New Mexico.

supervolcano • *an explosive volcano capable of producing more than 1000 cubic kilometers (240 cubic miles) of ejecta.*

pyroclastic rocks • *rocks that form during explosive volcanic eruptions, and are composed from a variety of different volcanic ejecta.*

hot spot • *a volcanic region thought to be fed by underlying mantle that is anomalously hot compared with the mantle elsewhere.*

volcanic ash • *fine, unconsolidated pyroclastic grains under 2 millimeters (0.08 inches) in diameter.*

One of the valley's most prominent features is the Great Sand Dunes National Park and Preserve, which contains the tallest sand dunes in North America (*Figure 4.32*). The sand in these dunes formed from Cenozoic deposits left by the Rio Grande and its tributaries, which flow through the valley. The dunes rise approximately 230 meters (750 feet) from the valley floor, near the western base of the Sangre de Cristo Range.

Figure 4.32: Great Sand Dunes National Park and Preserve, at the foot of the Sangre de Cristo Mountains, Colorado.

Topography of the Great Plains
Region 4

The Great Plains is a lowland area underlain by flat-lying sedimentary rocks, and it extends in a north-south band across the entire United States from Montana and North Dakota to Texas and New Mexico. In the Southwest, the Great Plains lie along the eastern margins of Colorado and New Mexico (*Figure 4.33*). The region has a basement of flat-lying Precambrian metamorphic and igneous rocks, overlain by Paleozoic and Mesozoic sedimentary rocks. The Mesozoic sediments consist largely of materials eroded from the Rocky Mountains and deposited in the Western Interior Seaway, which covered this area during the Cretaceous. The Laramide uplift that affected the Colorado Plateau and Rocky Mountains also raised the western Great Plains, but did not significantly buckle or disturb the sedimentary layers. For this reason, the Great Plains region is not entirely flat, but changes in elevation from 1830 meters (6000 feet) on its western edge to 460 meters (1500 feet) on its eastern edge. During much of the Cenozoic, the Great Plains' gentle eastward slope has received sediments deposited by streams issuing from the western highlands. The top of that sequence of continental strata is the Ogallala Formation, deposited between 15 and 5 million years ago in the **Miocene**. This formation houses the famous Ogallala (or High Plains) Aquifer, which supplies water for most of the Great Plains.

See Chapter 9: Earth Hazards to learn about drought's impact on the Ogalalla Aquifer.

Miocene • a geological time unit extending from 23 to 5 million years ago.

subsidence • the sinking of an area of the land surface.

Phanerozoic • a generalized term used to describe the entirety of geological history after the Precambrian, from 541 million years ago to the present.

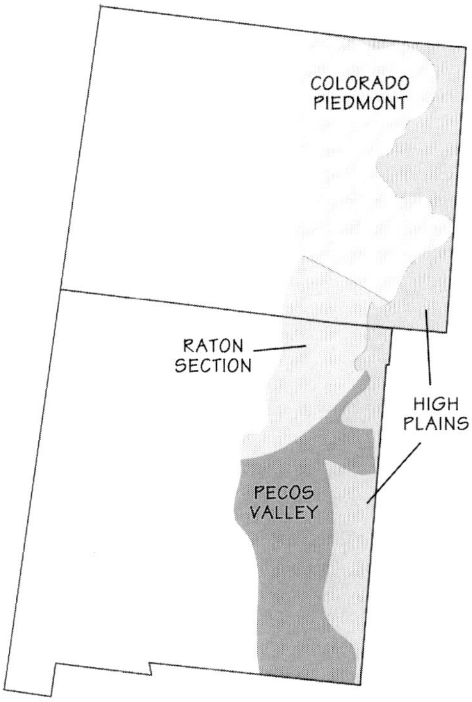

Figure 4.33: Physiographic subdivisions of the Great Plains.

There are deep sedimentary basins in the Great Plains, which have **subsided** throughout most of the **Phanerozoic** while at the same time collecting thick layers of sediment. The Denver Basin covers northeastern and eastern Colorado, into Wyoming, Nebraska, and Kansas. The Raton Basin spans southeastern Colorado and northeastern New Mexico, and the Permian Basin is located in southeastern New Mexico and into west Texas. Multiple strata in these sedimentary basins are important for oil and gas, and certain sandstones contain uranium deposits.

See Chapter 6: Energy for more information about the Southwest's sedimentary basins and major sources of fossil fuels.

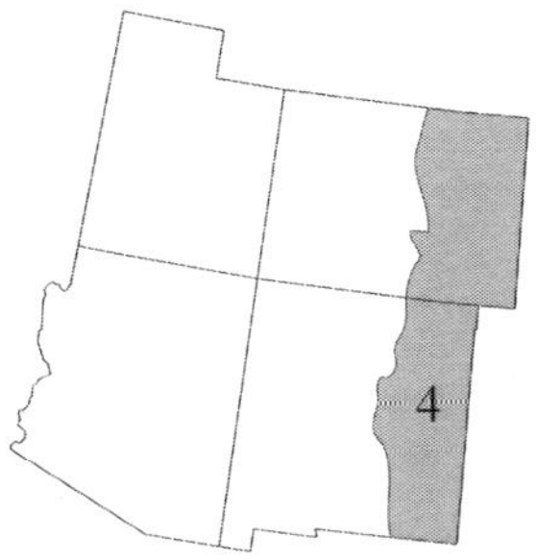

Colorado Piedmont

The Colorado Piedmont, which runs along the foothills of the Front Range in the Rocky Mountains, is a broad hilly valley that serves as the locus of Colorado's urban and agricultural activity. The area hosts the South Platte and Arkansas rivers and their tributaries, which have cut deeply into the Paleogene

dendritic drainage • a drainage pattern where many smaller streams join and contribute to ever larger streams.

clay • the common name for a number of very fine-grained, earthy materials that become plastic (flow or change shape) when wet.

sediments deposited after the uplift of the Rocky Mountains. A less eroded segment between the river valleys is located east of Pikes Peak. Here, the middle Cenozoic Castle Rock Conglomerate is preserved in many locations, forming the resistant, protective cap seen on mesas and buttes. Along the eastern edge of the Rockies, older limestones and sandstones upturned during the Laramide Orogeny form a series of hogback ridges and flatirons (steeply sloping triangular landforms) that run parallel to the great mountain chain (*Figure 4.34*).

Figure 4.34: The Flatirons rise from the hilly grasslands of the Colorado Piedmont near Boulder, Colorado.

Raton

South of the Colorado Piedmont, the Raton section comprises a less eroded, higher segment between the Arkansas and Canadian Rivers. The area borders the southern end of the Rockies and exhibits interesting erosional and igneous topography.

The Park Plateau is formed from the surface outcrop of the Raton Formation, deposited from material eroded off the Sangre de Cristo Range during Laramide mountain building. The area covered by the formation straddles the divide between the Arkansas and Canadian Rivers and the border between Colorado and New Mexico. The tributaries of the two rivers have eroded the landscape in an intricate **dendritic** pattern, forming a finely dissected hilly area with a relief of approximately 100 meters (330 feet) between hills and valleys. The formation is also of great interest because it includes an iridium-rich **clay**

See Chapter 3: Fossils to learn more about the Cretaceous-Paleogene extinction.

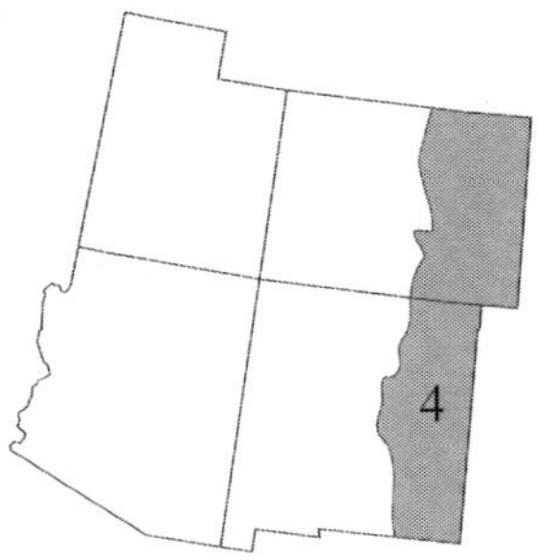

unit that marks the great meteor impact at the Cretaceous-Cenozoic boundary. The Spanish Peaks, a 27- to 14-million-year-old intrusion, are located in south-central Colorado just east of the Sangre de Cristo Range at the northern edge of the Raton Formation. These eroded igneous plutons lie at the border between the Great Plains and Rocky Mountains, forming two large, high peaks with elevations of 4153 meters (13,626 feet) and 3866 meters (12,683 feet). A set of radial dikes extends out from the main body of rock, along with various shapes and sizes of intruding igneous rocks. The dikes range from 1 meter (3 feet) to 30 meters (100 feet) wide, and are kilometers (miles) long.

On the eastern edge of the Park Plateau are topographic expressions of late Cenozoic volcanic flows and eruptions. The Raton-Clayton Volcanic Field is located within a southwest-northeast trending zone, termed the Jemez Lineament, that stretches from the Basin and Range across the Rio Grande Rift and into the Great Plains. The age of volcanism in the area ranges from Capulin Mountain, a cinder cone that erupted nearly 60,000 years ago (*Figure 4.35*), to eruptions and lava flows that occurred between eight and two million years ago. Quaternary basaltic flows can be found farther east, extending almost to the borders of Oklahoma and Texas. In some cases where the flows filled older streambeds, subsequent erosion preserved the flow while removing the surrounding strata to produce what is termed a "**topographic inversion.**"

Figure 4.35: Capulin Mountain at Capulin Volcano National Monument, New Mexico. The volcanic field surrounding the monument contains more than 100 recognizable volcanoes.

The Raton section also includes landforms related to the submergence of the Great Plains by the Western Interior Seaway during the Cretaceous. Methane seeps are vents on the seafloor through which hydrocarbons emanate, producing cements that **lithify** the surrounding rocks and are thus elevated above the surrounding seafloor. Seeps of this type formed just west of Boone, Colorado during the late Cretaceous, approximately 79 to 69 million years ago.

Region 4

topographic inversion • *a landscape with features that have reversed their elevation relative to other features, most often occurring when low areas become filled with lava or sediment that hardens into material that is more resistant to erosion than the material that surrounds it.*

lithification • *the process of creating sedimentary rock through the compaction or sementation of soft sediment.*

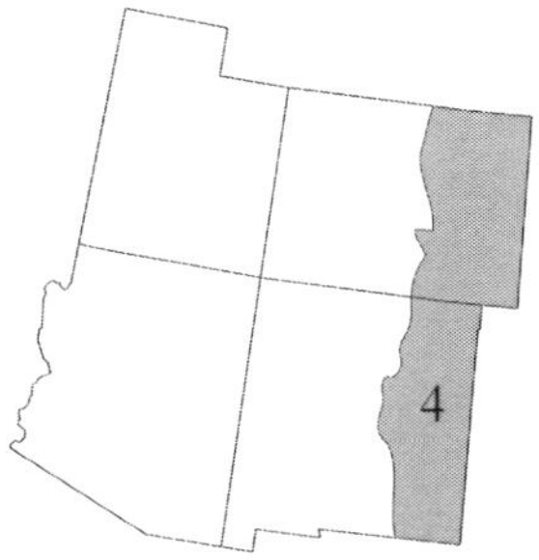

4 Topography

Region 4

weather • *the measure of short-term conditions of the atmosphere such as temperature, wind speed, and humidity.*

fossil • *preserved evidence of ancient life.*

craton • *the old, underlying portion of a continent that is geologically stable relative to surrounding areas.*

Topographic Inversions

Typically, synclines (U-shaped folds) form valleys and anticlines (A-shaped folds) form ridges. However, the reverse can also be true. In a phenomenon called topographic inversion, topographic lows (valleys) may form from the structural high (top of an anticline)—the term "structure" refers to the form of the rock layers. At the top of the anticline, a layer may erode away because of cracks caused by bending of the rock at the top of the fold. Fracturing at the top of the fold allows increased water penetration, and topographic highs are subjected to more severe weather. Once exposed, the less resistant layers below the eroded top quickly *weather* away to form a valley. The limbs of the resistant layer, however, are generally still intact. This leaves two ridges of resistant rock on either side of a valley floored by softer, less resistant layers.

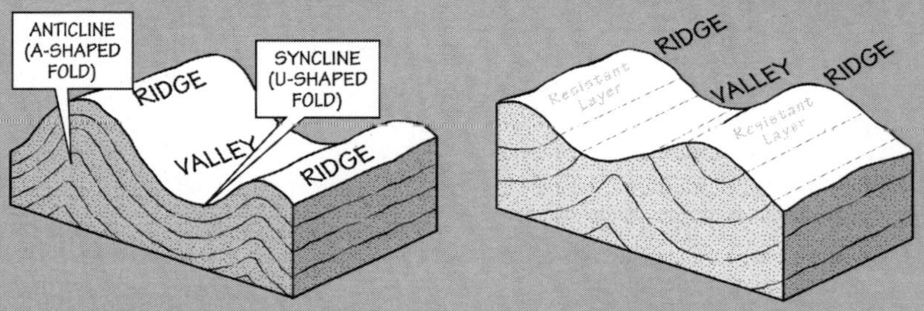

Normal erosion of a fold. *Topographic inversion.*

They are referred to as the Tepee Buttes, and they form a linear belt extending in a northeast-southwest direction. The buttes are approximately 10 meters (30 feet) high and are composed of well-cemented **fossiliferous** limestone surrounded by Cretaceous gray shales of the Pierre Formation.

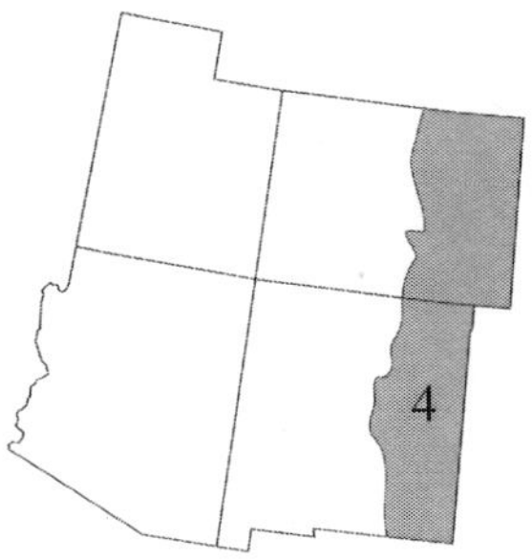

High Plains

The High Plains is an area of flat relief that reflects 500 million years of **cratonic** stability in the continent's interior. Much of this area was submerged by the Cretaceous Western Interior Seaway, leading to the deposition of sediment eroded from the mountains to the west. As such, the Ogallala Formation—a series of porous Miocene strata—defines the extent of the High Plains. The western limits of the Ogallala Formation are located 80 to 240 kilometers (50 to 150 miles) east of the Rocky Mountains because approximately five million years ago, rivers flowing eastwards from the mountains began to erode rather than deposit material in the Great Plains. This drastic change removed the Ogallala and much of the region's older Cenozoic strata in swaths around major east-flowing streams—the North Platte, South Platte, Arkansas, and Canadian rivers. From the rim of the High Plains, the Ogallala caprock dips eastward to form the Llano Estacado or Staked Plains of Texas and New Mexico— a large area of dry, treeless, nearly flat, high prairie.

Pecos Valley

The Pecos Valley is a broad valley formed by erosion driven by the Pecos River, a tributary of the Rio Grande, which drains the area. The valley follows the river from the Sangre de Cristo Mountains of New Mexico southward through the Edwards Plateau to join the Rio Grande River in Texas. The Pecos Valley is relatively flat, and is distinguished by the Ogallala Formation on its eastern side, where it forms a rim rock at the top of the Mescalero escarpment.

At the northern end of the Pecos Valley lies another field of late Cenozoic basalt flows, the Ocate Volcanic Field, which is also a part of the Jemez Lineament. The mesa-capping flows in this field have been dated and reveal a pattern of ages similar to those of the Raton-Clayton Volcanic Field. Flows at higher elevations are older than those found at lower elevations. This pattern can be explained by progressive erosion that lowered the elevation of the sedimentary strata upon which younger flows were deposited.

There is an extensive amount of **karst topography** throughout the Pecos Valley, leading to the formation of some of New Mexico's most spectacular caves (including Carlsbad Caverns, Fort Stanton Cave, Torgac Cave, and Lechuguilla Cave) (*Figure 4.36*). Though most karst is formed when weak carbonic acid in rain or groundwater dissolves the sedimentary bedrock, New Mexico's caverns have been shaped through a unique process called **sulfuric** acid dissolution. In this process, deep sources of hydrogen sulfide—usually derived from gas and oil reservoirs—interact with ground or rainwater to form sulfuric acid, a strong acid that is able to aggressively dissolve large cave passages out of the limestone.

Karst Topography

Karst topography refers to a region where the landscape's features are largely the result of chemical weathering by water, resulting in caves, sinkholes, disappearing and reappearing streams, cliffs, and steep-sided hills called towers. These structures typically form when water picks up carbon dioxide from the *atmosphere* and ground to form carbonic acid. Even this fairly weak and dilute acid dissolves carbonate rocks (such as limestone) relatively easily, resulting in dramatic features while other rock is comparatively unaffected. Karst is found in every state except Hawai'i, and as an aquifer it is the source of a significant amount of our drinking water.

karst topography • *a kind of landscape defined by bedrock that has been weathered by dissolution in water, forming features like sinkholes, caves, and cliffs.*

sulfur • *a bright yellow chemical element (S) that is essential to life.*

atmosphere • *a layer of gases surrounding a planet.*

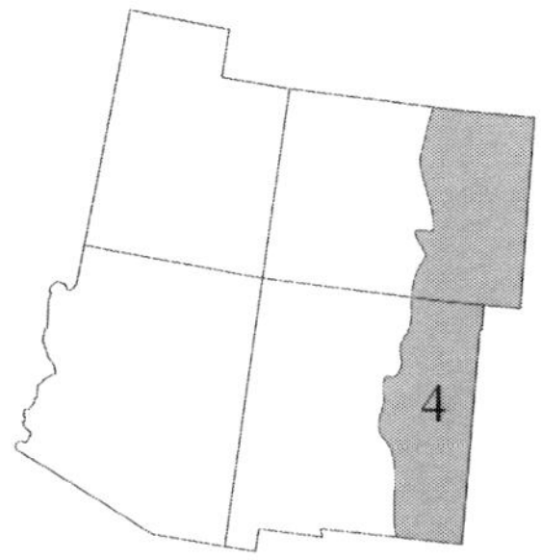

Karst Topography (continued)

While common, karst is not always easily identifiable since it is often not expressed at the surface or its topography has been affected by other factors. Karst topography is a relatively mature type of landscape, taking many tens of thousands of years to develop, and it can indicate that a region has been free of other forms of erosion or deposition for an extended period. Karst topography in the Southwest is present wherever water has eroded the limestone bedrock, especially in the southern Great Plains and around the Grand Canyon. In addition, some karst in the Southwest is generated by the action of sulfuric acid.

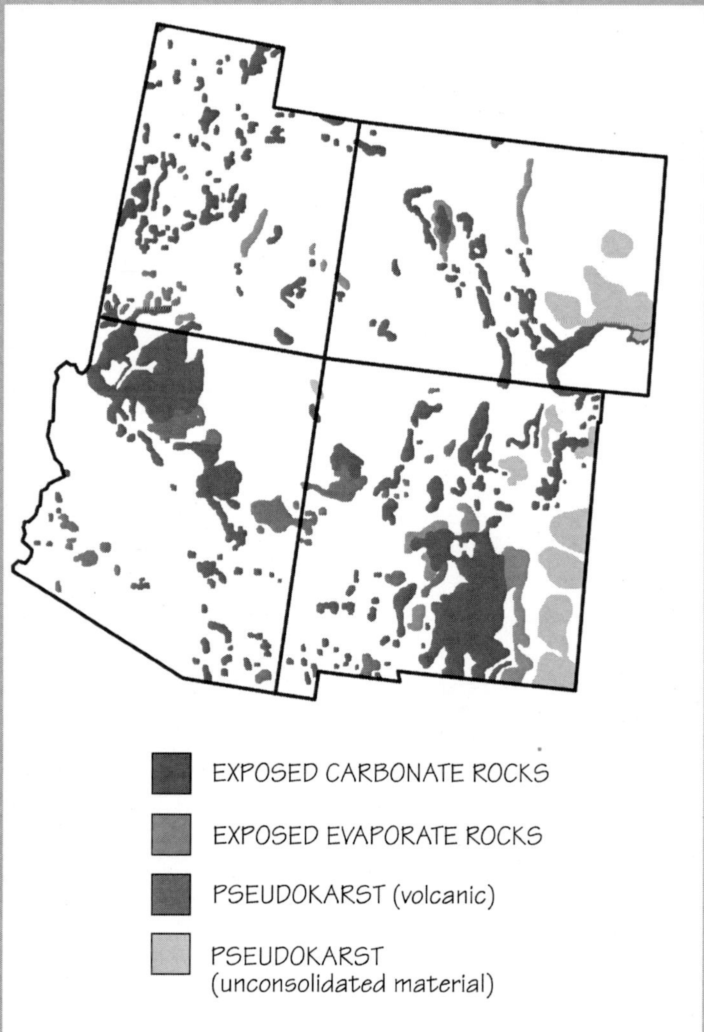

(See TFG website for full-color version.)

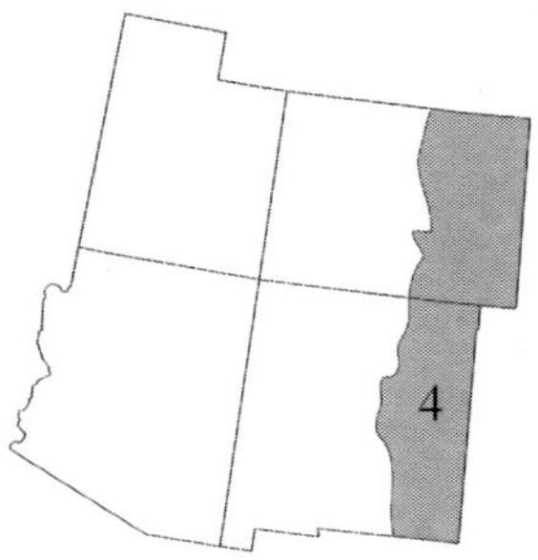

Region 4

Figure 4.36: Stalagmites, stalactites, and speleothems in Lechuguilla Cave, Eddy County, New Mexico. It is the deepest cave in the continental US.

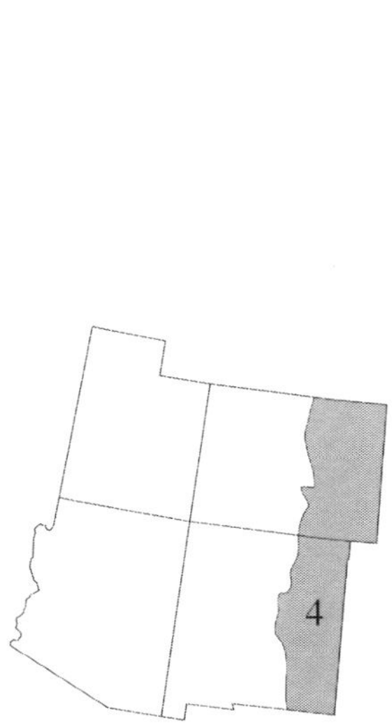

4 Topography

Elevations

Highest and Lowest Elevations by State

Arizona

Arizona's highest point is Humphreys Peak, a 3852-meter-high (12,637-foot-high) mountain in the Kachina Peaks Wilderness 18 kilometers (11 miles) north of Flagstaff. It is the highest of the San Francisco Peaks, remnants of an extinct stratovolcano. The state's lowest point is along the Colorado River near San Luis, which lies at only 22 meters (72 feet) above sea level.

Colorado

Mount Elbert, rising 4401 meters (14,440 feet) above sea level, is Colorado's highest point as well as the highest summit in the North American Rockies. The mountain, located 19 kilometers (12 miles) southwest of Leadville, is also the second-highest summit in the continental United States. The lowest point in Colorado is found on the Arikaree River where it flows into Kansas, at an elevation of 1011 meters (3317 feet). This spot is also the highest low point of any US state.

New Mexico

The Sangre de Cristo Mountains are home to Wheeler Peak, New Mexico's highest mountain. At 4011 meters (13,161 feet), the mountain is a popular hiking and climbing destination. It was named in honor of George Montague Wheeler, an American pioneering explorer and naturalist. New Mexico's lowest point is the Red Bluff Reservoir on the Pecos River, with an elevation of 866 meters (2842 feet).

Utah

At 4125 meters (13,534 feet), Kings Peak is Utah's highest point, located in the Uinta Mountains of north-central Duchesne County. The peak is regarded as the hardest state high point to climb without special rock climbing skills or a guide; the easiest trail to the summit requires a 47-kilometer (29-mile) round-trip hike. Beaver Dam Wash, at the Utah-Arizona state line in Washington County, is the lowest point in the state at 664 meters (2178 feet) above sea level.

Topography

4

Elevations

4 Topography

Resources

Books on Topography

de Blij, H. J., P. O. Muller, J. E. Burt, & J. A. Mason, 2013, *Physical Geography of the Global Environment*, Oxford University Press, New York, 626 pp.

Trimble, D.E., 1980, *The Geologic Story of the Great Plains*, Geological Survey Bulletin 1493, US Government Printing Office, Washington, D.C., http://library.ndsu.edu/exhibits/text/greatplains/text.html.

Wyckoff, J., 1999, *Reading the Earth: Landforms in the Making*, Adastra West, Mahwah, NJ, 352 pp.

Topographic Maps

Color Landform Atlas of the US, http://fermi.jhuapl.edu/states/states.html. (Low resolution shaded relief maps of each state.)

Topoquest, https://www.topoquest.com/.

Websites on Topography

OpenLandform Catalog, Education Resources, OpenTopography. [High resolution topographic images that may be useful in teaching.] http://www.opentopography.org/index.php/resources/lidarlandforms.

Physiographic Provinces, in: Earth Science Concepts: Geology by Region, National Park Service, http://www.nature.nps.gov/geology/education/concepts/concepts_regional_geology.cfm.

Teaching Geomorphology in the 21st Century, On the Cutting Edge, Strong Undergraduate Geoscience Teaching, SERC. [A set of resources for college level, some of which may be adaptable to secondary education.] http://serc.carleton.edu/NAGTWorkshops/geomorph/index.html.

Teaching with Google Earth, On the Cutting Edge, Starting Point: Teaching Entry Level Geoscience, SERC, http://serc.carleton.edu/introgeo/google_earth/index.html.

United States Geography, by S.S. Birdsall & J. Florin, http://countrystudies.us/united-states/geography.htm.

US Geography: The Land Geography, Landscape, and Landforms of the US, 50 State Guide, eReference Desk. [Includes state-by-state information on geography.] http://www.ereferencedesk.com/resources/state-geography/.

Topography Resources for the Southwest

Davis, W., 2013, *River Notes: a Natural and Human History of the Colorado*, Island Press, Washington, D.C., 176 pp.

Owen, D., 2015, Where the river runs dry: the Colorado and America's water crisis, *The New Yorker* (May 25, 2015), http://www.newyorker.com/magazine/2015/05/25/the-disappearing-river.

Physiographic Provinces, Virtual Geologic Tour of New Mexico, New Mexico Bureau of Geology and Mineralogical Resources, http://geoinfo.nmt.edu/tour/provinces/home.html.

Sand Dunes of the Southwest, by Peter Olsen, http://sand.xboltz.net/.

For additional resources relevant to topography, see the section "General Geology Resources by State" at the end of this volume.

Chapter 5: Mineral Resources of the Southwestern US

What is a mineral?

A **mineral** is a naturally occurring inorganic solid with a specific chemical composition and a well-developed crystalline structure. Minerals provide the foundation of our everyday world. Not only do they make up the rocks we see around us in the Southwest, they are also used in nearly every aspect of our lives. The minerals found in the rocks of the Southwest are used in industry, construction, machinery, technology, food, makeup, jewelry, and even the paper on which these words are printed.

Minerals provide the building blocks for rocks. For example, **granite**, an **igneous rock**, is typically made up of crystals of the minerals feldspar, quartz, **mica**, and **amphibole**. In contrast, **sandstone** may be made of **cemented** grains of feldspar, quartz, and mica. The minerals and the bonds between the crystals define a rock's color and resistance to **weathering**.

Several thousand minerals have been discovered and classified according to their chemical composition. Most of them are **silicates** (representing approximately a thousand different minerals, of which quartz and feldspar are two of the most common and familiar), which are made of silicon and oxygen combined with other elements (with the exception of quartz, SiO_2). **Carbonate rocks** are made of carbon and oxygen combined with a metallic element; **calcium carbonate** ($CaCO_3$) is the most common example, and most of it today originates as skeletal material precipitated by organisms. Other mineral categories include native elements (such as gold), **oxides** and **sulfur**-bearing minerals, and **salts**.

Metallic minerals are vital to the machinery and technology of modern civilization. However, many metals occur in the **crust** in amounts that can only be measured in parts per million (ppm) or parts per billion (ppb). A mineral is called an **ore** when one or more of its elements can be profitably removed, and it is almost always necessary to process ore minerals in order to isolate the useful element. For example, **chalcopyrite** ($CuFeS_2$), which contains copper, **iron**, and sulfur, is referred to as a copper ore when the copper can be profitably extracted from the iron and sulfur. Ores are not uniformly distributed in the crust of the Earth, but instead occur in localized areas where they are concentrated in amounts sufficient to be economically extracted by mining.

Non-metallic minerals do not have the flash of a metal, though they may have the brilliance of a **diamond** or the silky appearance of **gypsum** ($CaSO \cdot 2H_2O$). Generally much lighter in color than metals, non-metallic minerals can transmit light, at least along their edges or through small fragments.

granite • *a common and widely occurring type of igneous rock.*

igneous rocks • *rocks derived from the cooling of magma underground or molten lava on the Earth's surface.*

mica • *a large group of sheetlike silicate minerals.*

amphibole • *a group of dark colored silicate minerals, or either igneous or metamorphic origin.*

sandstone • *sedimentary rock formed by cementing together grains of sand.*

cementation • *the precipitation of minerals that binds together particles of rock, bones, etc., to form a solid mass of sedimentary rock.*

CHAPTER AUTHOR

Thomas R. Fisher

5 Mineral Resources

Review

copper • a ductile, malleable, reddish-brown metallic element (Cu).

gold • a soft, yellow, corrosion-resistant element (Au), which is the most malleable and ductile metal on Earth.

feldspar • an extremely common group of rock-forming minerals found in igneous, metamorphic and sedimentary rocks.

quartz • the second most abundant mineral in the Earth's continental crust (after the feldspars), made up of silicon and oxygen (SiO_2).

hardness • a physical property of minerals, specifying how hard the mineral is, and its resistance to scratching.

erosion • the transport of weathered materials.

calcite • a carbonate mineral, consisting of calcium carbonate ($CaCO_3$).

Elements: The Building Blocks of Minerals

Elements are the building blocks of minerals. The mineral quartz, for example, is made of the elements silicon and oxygen, and, in turn, is also a major component of many rocks. Most minerals present in nature are not composed of a single element, though there are exceptions such as gold. Elements such as *copper* (Cu), lead (Pb), zinc (Zn), and even silver (Ag), *gold* (Au), and diamond (C) are not rare, but they are usually widely dispersed throughout rocks and occur at very low average concentrations. Eight elements make up (by weight) 99% of the Earth's crust, with oxygen being the most abundant (46.4%). The remaining elements in the Earth's crust occur in very small amounts, some in concentrations of only a fraction of one percent. Since silicon (Si) and oxygen (O) are the most abundant elements in the crust by mass, it makes sense for silicates (e.g., *feldspar*, *quartz*, and garnet) to be some of the most common minerals in the Earth's crust and to therefore be found throughout the Southwest.

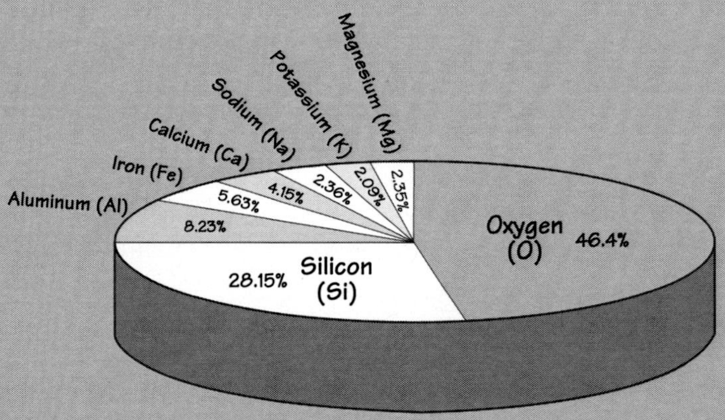

Mineral percentage by mass in the Earth's crust.

Mineral Identification

Although defined by their chemical composition and crystal structure, minerals are identified based on their physical properties. A variety of properties must usually be determined when identifying a mineral, with each such property eliminating possible alternatives.

Hardness is a very useful property for identification, as a given mineral can only exhibit a narrow range of hardnesses, and since it is easily testable, this property can be used to quickly and simply minimize the number of possibilities. Hardness is important because it helps us understand why some rocks are more or less resistant to weathering and **erosion**. Quartz, with a rating of 7 on the **Mohs scale**, is a relatively hard mineral, but **calcite** ($CaCO_3$), rating 3 on the Mohs scale, is significantly softer. Therefore, it should be no surprise that quartz sandstone is much more resistant to erosion and weathering than is **limestone**, which is primarily made of the mineral calcite. Quartz is a very common mineral in the Earth's crust, and it is quite resistant due to its hardness and relative insolubility. Thus, quartz grains are the dominant mineral type in nearly all types of **sand**.

Mohs Scale of Hardness

In 1824, the Austrian mineralogist Friedrich Mohs selected ten minerals to which all other minerals could be compared to determine their relative hardness. The scale became known as the Mohs scale of hardness, and it remains very useful as a means for identifying minerals or for quickly determining their hardness. Everyday items can be used to determine hardness if the minerals in the scale are not available. These include a streak plate or piece of unglazed porcelain (hardness 7), a piece of glass (hardness 5), a penny (hardness 3), and a fingernail (hardness 2).

1	Talc
2	Gypsum
3	Calcite
4	Fluorite
5	Apatite
6	Feldspar
7	Quartz
8	Topaz
9	Corundum
10	Diamond

Color is helpful in identifying some minerals such as sulfur, but it is uninformative or even misleading in others such as garnet. **Luster** describes how light is reflected from a mineral's surface and it can range from adamantine, seen in diamonds, to dull or earthy (effectively no luster), such as in **kaolinite**. **Crystal form**, if visible, can also be diagnostic. For example, **fluorite** and calcite may appear superficially similar, but fluorite forms cubic crystals while calcite forms trigonal-rhombohedral crystals.

Relatedly, crystals may have planes of weakness that cause them to break in characteristic ways, called **cleavage**. Or they may not, but instead display **fracture** when broken. For example, mica and **graphite** have very strong cleavage, allowing them to easily be broken into thin sheets, while quartz and

Review

limestone *• a sedimentary rock composed of calcium carbonate ($CaCO_3$).*

sand *• granular sediment most commonly composed of weathered grains of quartz and feldspar, of grain diameter 1/16 to 2 millimeters.*

color (mineral) *• a physical property determined by the presence and intensity of certain elements within the mineral.*

luster *• a physical property of minerals, describing the appearance of the mineral's surface in reflected light, and how brilliant or dull it is.*

kaolinite *• a silicate clay mineral, also known as china clay.*

crystal form *• a physical property of minerals, describing the shape of the mineral's crystal structure.*

fluorite *• the mineral form of calcium fluoride (CaF_2).*

5 Mineral Resources

Review

density • *a physical property of minerals, describing the mineral's mass per volume.*

streak • *a physical property of minerals, obtained by dragging the mineral across a porcelain plate and effectively powdering it.*

magnetic • *affected by or capable of producing a magnetic field.*

luminescence • *the emission of light.*

radioactivity • *the emission of radiation by an unstable atom.*

thorium • *a radioactive rare earth element.*

effervesce • *to foam or fizz while releasing gas.*

double refraction • *the result of light passing through a material that splits it into two polarized sets of rays, doubling images viewed through that material.*

glass (the latter not being a mineral) have no cleavage, instead displaying a distinctive curved fracture form known as conchoidal. The **density** of a mineral may also aid in identifying it (e.g., metals tend to be very dense). Finding the exact density is straightforward, but it does require measuring the volume of the sample. Placing an unknown mineral in water (or other liquid) to find its volume by displacement can be a risky undertaking since several minerals react violently with water, and many more break down with exposure. A mineral's **streak** is obtained by dragging it across a porcelain plate, effectively powdering it. The color of the powder eliminates conflating variables of external weathering, crystal habit, impurities, etc. Some minerals are **magnetic** (affected by magnetic fields), while a few are natural magnets (capable of producing a magnetic field).

Most minerals can be identified through the process of elimination after examining a few of these properties and consulting a mineral identification guide. Mineral testing kits often include several common objects used to test hardness: a porcelain streak plate, a magnet, and a magnifying glass. Some minerals have rare properties, which may be more difficult to test. For example, there are minerals that exhibit **luminescence** of all types, giving off light due to a particular stimulus. Some minerals are **radioactive**, usually due to the inclusion of significant amounts of uranium, **thorium**, or potassium in their structure. Carbonate minerals will **effervesce** when exposed to hydrochloric acid. **Double refraction** describes the result of light passing through a material that splits it into two polarized sets of rays, doubling images viewed through that material. For example, a single line on a sheet of paper will appear as two parallel lines when viewed through a clear calcite crystal.

There are many more interesting and distinguishing properties that minerals may possess, and there are many more elaborate and precise means for identifying them. The branch of geology that studies the chemical and physical properties and formation of minerals is called *mineralogy.*

What Are Minerals Used For?

Mineral resources fall into many different categories, including industrial minerals, construction materials, **gemstones**, and metallic and non-metallic ores. Some minerals and rocks are abundant and are used in the construction industry or in the manufacturing of many of the products we commonly find in stores. Construction materials include **dimension stone** (e.g., sandstone, limestone, and granite), which is used for the exterior or interior of structures.

Minerals used in manufacturing include kaolinite for ceramics, gypsum for wallboard, fluorite for the fluoride in toothpaste, and halite for common table and rock salt. We also seek out specific rock types and sediment to use in the construction of buildings, highways, and bridges. Decorative statues are commonly constructed of marble, jade, or soapstone. Granite, travertine, and other decorative stones are increasingly used to beautify our home interiors and

What distinguishes a regular mineral from a gem?

Minerals are assigned to the category of gemstones based primarily on our interpretation of what has value. Typically, the beauty, durability, and rarity of a mineral qualify it as a gemstone. Beauty refers to the luster, color, transparency, and brilliance of the mineral, though to some degree it is dependent on the skillfulness of the cut. Not all gems are prized for these reasons; for example, scarcity may be artificially inflated, or a mineral may be valued for its unusual color.

Gemstones can be further categorized as precious or semiprecious stones. Precious stones, including diamond, topaz, and sapphire, are rare and translucent to light. They are more durable because they are hard, making them scratch resistant. On the Mohs scale of hardness, the majority of precious gemstones have values greater than 7. Semi-precious stones are generally softer, with hardness scale values between 5 and 7. The minerals peridot, jade, garnet, amethyst, citrine, rose quartz, tourmaline, and turquoise are examples of semi-precious stones that can be cut and used in jewelry.

Gems may have common names that differ from their geological ones, and these names may be dependent on mineral color. For example, the mineral beryl is also referred to as emerald, aquamarine, or morganite depending on its color. *Corundum* can also be called sapphire or ruby, and peridot is another name for *olivine*.

to make art, in addition to being used in public buildings. Some minerals are considered to be precious or semi-precious and are used in jewelry, including diamond and some crystalline forms of quartz.

Metallic minerals have many applications and are used to manufacture many of the items we see and use every day. For example, iron comes from **hematite** and **magnetite**, and from it we make steel. **Lead**, from the mineral **galena**, is used in the manufacture of batteries and in the solder found in electronic devices. **Titanium**, from the mineral **ilmenite**, is used in airplanes, spacecraft, and even white nail polish. **Aluminum** comes from **bauxite** and is known for being both lightweight and strong—many of the parts that make up today's

Review

gem • *a mineral that has aesthetic value and is often cut and polished for use as an ornament.*

dimension stone • *the commercial term applied to quarried blocks of rock cut to specific dimensions.*

corundum • *an aluminum oxide mineral (Al_2O_3) that is, after diamond, the hardest known natural substance.*

olivine • *an iron-magnesium silicate mineral ($(Mg,Fe)_2SiO_4$) that is a common constituent of magnesium-rich, silica-poor igneous rocks.*

hematite • *a mineral form of iron oxide (Fe_2O_3), with vivid red pigments that make it valuable as a commercial pigment.*

magnetite • *a mineral form of iron oxide (Fe_3O_4) with naturally occurring magnetic properties.*

lead • *a metallic chemical element (Pb).*

Review

magma • molten rock located below the surface of the Earth.

intrusive rock • a plutonic igneous rock formed when magma from within the Earth's crust escapes into overlying strata.

rift • a break or crack in the crust that can be caused by tensional stress as a landmass breaks apart into separate plates.

volcanism • the eruption of molten rock onto the surface of the crust.

felsic • igneous rocks with high silica content and low iron and magnesium content.

gabbro • a coarse-grained, mafic and intrusive igneous rock.

mafic • igneous rocks that contain a group of dark-colored minerals, with relatively high concentrations of magnesium and iron.

automobiles are made of this metal. Copper comes from a variety of copper-bearing minerals, including chalcopyrite, and is used to make electrical wire, tubing, and pipe.

Mineral Formation

Economically recoverable mineral deposits are formed by geologic processes that can selectively concentrate desirable elements in a relatively small area. These processes may be physical or chemical, and they fall into four categories:

Magmatic processes separate minor elements of **magma** from the major elements and concentrate them in a small volume of rock. This may involve either the early crystallization of ore minerals from the magma while most other components remain molten or late crystallization after most other components have crystallized. Magmatic processes responsible for the formation of mineral deposits are usually associated with igneous **intrusions** (formed during mountain building events, **rifting**, and **volcanic** activity), which can range in composition from granite (**felsic**) to **gabbro** (**mafic**). **Metamorphism** may also cause **recrystallization** of minerals and concentration of rare elements. Under conditions of extreme high-temperature metamorphism, minerals with the lowest melting temperatures in the crust may melt to form small quantities of **pegmatite** magmas.

Hydrothermal processes involve **hydrothermal solutions** that dissolve minor elements dispersed through large volumes of rock, transport them to a new location, and precipitate them in a small area at a much higher concentration. Hydrothermal solutions are commonly salty, acidic, and range in temperature from over 600°C (~1100°F) to less than 60°C (140°F). Some of these fluids may travel very long distances through **permeable sedimentary rock**. Eventually, the hydrothermal fluids precipitate their highly dissolved load of elements, creating concentrated deposits.

Sedimentary processes gather elements dispersed through large volumes of water and precipitate them in a sedimentary environment, such as in sedimentary layers on the ocean floor or on lakebeds. Sedimentary mineral deposits form by direct precipitation from the water.

Weathering and erosion break down large volumes of rock by physical and chemical means and gather previously dispersed elements or minerals into highly concentrated deposits. **Residual weathering deposits** are mineral depos-its formed through the concentration of a weathering-resistant mineral, as a result of surrounding minerals being eroded and dissolved. In contrast, mineral deposits formed by the concentration of minerals in moving waters are called **placer deposits**.

Mineral Resources 5

Region 1

A mineral is not necessarily restricted to one method of concentration or environment of formation. For example, economically important deposits of gypsum may form as a precipitate from evaporating water. In contrast, gypsum formation may also be associated with volcanic regions where limestone and sulfur gases from the volcano have interacted with one another, or even from other areas as a product of the chemical weathering of pyrite.

Minerals in the Southwest

Each region of the Southwest (the Colorado Plateau, Basin and Range, Rocky Mountains, and Great Plains) contains significant economic metallic and non-metallic mineral deposits. The distribution and occurrence of these deposits are not always restricted to one region—the geology, geologic history, and associated mineral resources of the Southwestern US are intimately intertwined across regional and state boundaries. This cross cutting of the regions by mineral deposits reflects not only the type of deposit but also how and when the minerals were **emplaced**, and how geology controlled their emplacement. In many parts of the Southwest, the dry environment also allows for relatively easy access to both bedrock surfaces and accumulations of weathered rock for the exploration of minerals that lie within or that have been weathered out of the region's rocks.

Both the Southwestern and Northwest Central states are major contributors to mineral production in the United States. In some cases, these states produce the majority of a particular mineral used by the US and may even contain the largest deposits in the world of certain mineral types. For example, most of the country's uranium deposits are located in the Southwest or the Northwest Central. Ninety percent of the copper produced in the US comes from two states in the Southwest: Arizona and Utah. Significant quantities of gold, silver, and molybdenum are also produced here, along with industrial minerals such as potash and soda ash. Throughout the Southwest, the deposition of sediment has also left behind an abundance of deposits useful as construction materials. River systems deposited sand and gravel, while ancient seas that spread across the area left behind thick deposits of halite and gypsum. The advance of **inland seas** and the subsequent deposition of marine detritus also made possible the widespread existence of **energy** resources (**fossil fuels**) throughout the area, most notably **oil**, **natural gas**, and **coal**. Some of the natural gas produced in the Southwest also contains **helium** in sufficient concentrations to be profitably extracted—it originates from the decay of radioactive elements in the source rocks of accumulated natural gas.

See Chapter 6: Energy to learn more about the extraction of fossil fuel resources in the Southwest.

metamorphic rocks *• rocks formed by the recrystallization and realignment of minerals in pre-existing sedimentary, igneous, and metamorphic rocks when exposed to high enough temperature and/or pressure.*

recrystallization *• the change in structure of mineral crystals that make up rocks, or the formation of new mineral crystals within the rock.*

pegmatite *• a very coarse-grained igneous rock that formed below the surface.*

hydrothermal solution *• hot, mineral-rich water moving through rocks.*

permeability *• a capacity for fluids and gas to move through fractures within a rock, or the spaces between its grains.*

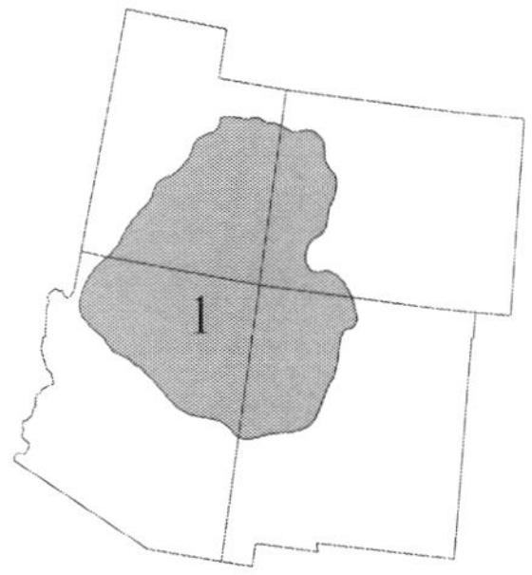

5 Mineral Resources

Region 1

Mineral Resources of the Colorado Plateau
Region 1

An area about the size of the state of Montana, the Colorado Plateau is an enigmatic block of continental crust some 45 kilometers (28 miles) thick that has remained relatively stable for over 500 million years. It began rising around 100 million years ago, rose further in the **Eocene** (56-33 million years ago), and then once more during an **epeirogenic** event approximately 8 to 6 million years ago. The Grand Canyon began to incise into the Plateau beginning in the Eocene, eventually cutting down far enough to expose 500 million years of horizontal sedimentary strata and underlying tilted metamorphic and igneous **Precambrian basement rocks**. There have been no major magmatic events in the Colorado Plateau, so the metallic resources found there are relatively minor, although a few younger igneous and volcanic intrusive rocks were important to metal emplacement. Most of the Plateau's mineral ore deposits have sedimentary origins and are associated with the region's depositional basins (*Figure 5.1*).

See Chapter 4: Topography for more information about the formation of the Grand Canyon.

Metallic Resources

Although the US has **perennially** imported over 90% of its uranium ore from foreign sources, the Paradox Basin in the Colorado Plateau is seen as an important source for US uranium supplies (*Figure 5.2*), which are mined for use in **nuclear** energy. These uranium deposits occur largely as "roll-front" deposits, in which groundwater leaches uranium from the source rock (usually igneous or metamorphic basement rock or **volcanic ash** deposits), and carries it through a **porous** and permeable rock, typically sandstone or **conglomerate**. Uranium oxide minerals are precipitated when the uranium-bearing groundwater is reduced by contact with organic materials within the rock. The uranium minerals carnotite and coffinite account for the majority of the ores in these deposits. **Vanadium** minerals such as corvurite and doloresite are also found with these ores. Most commonly, the region's uranium- and vanadium-bearing minerals are hosted in **fluvial** and **lacustrine** sandstones and limestones of the **Triassic** Chinle, **Jurassic** Morrison and Todilto, and **Cretaceous** Dakota formations. **Lithium**, a major component of high-tech batteries, is recovered from the Paradox Basin.

See Chapter 6: Energy for more information about nuclear power in the Southwest.

Of the US states, New Mexico currently ranks second in uranium resources after Wyoming. Most of the state's deposits are located in the Grants Mineral Belt to the northwest, although no mining has taken place there since 2002. Other major uranium deposits are found near Moab, Utah in the Paradox Basin; on the San Rafael Swell in central Utah; and near Uravan, Colorado, where the

Eocene • *a geologic time period extending from 56 to 33 million years ago.*

epeirogenic • *large-scale crustal uplift caused by hot or upwelling mantle underlying the surface.*

Precambrian • *a geologic time interval that spans from the formation of Earth (4.6 billion years ago) to the beginning of the Cambrian (541 million years ago).*

basement rocks • *the foundation that underlies the surface geology of an area, generally composed of igneous or metamorphic crystalline rock.*

perennial • *continuous; year-round or occurring on a yearly basis.*

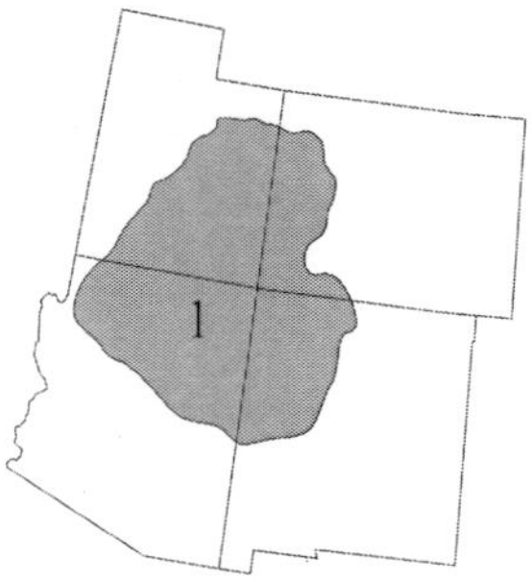

Mineral Resources 5

Region 1

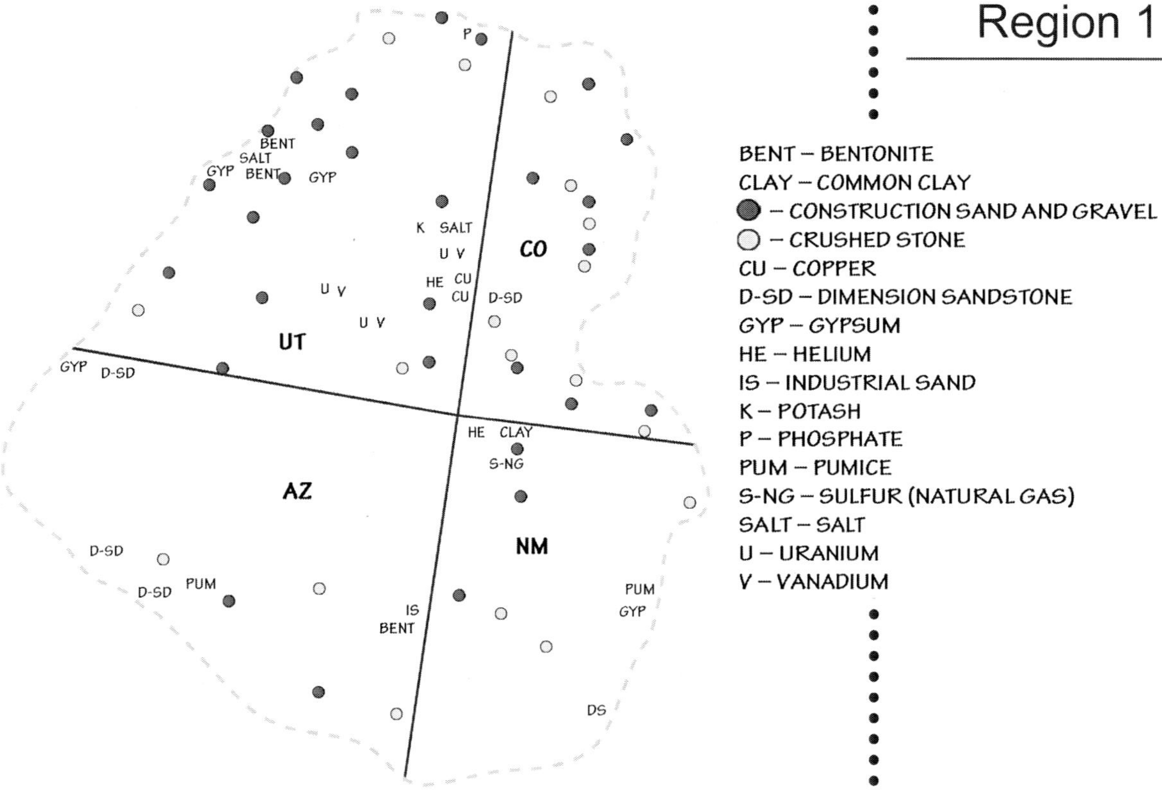

Figure 5.1: Principal mineral resources of the Colorado Plateau region.

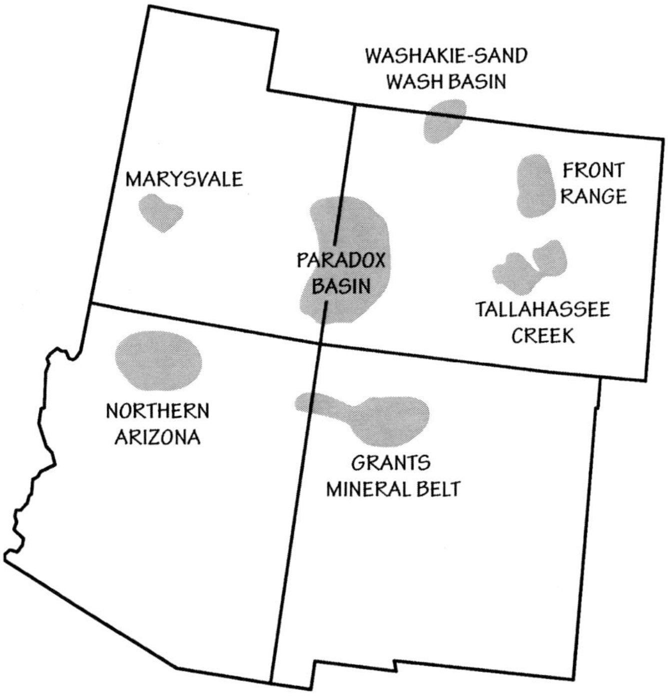

Figure 5.2: Distribution of uranium deposits in the Southwestern US.

nuclear • *pertaining to a reaction, as in fission, fusion, or radioactive decay, that alters the energy, composition, or structure of an atomic nucleus.*

volcanic ash • *fine, unconsolidated pyroclastic grains under 2 millimeters (0.08 inches) in diameter.*

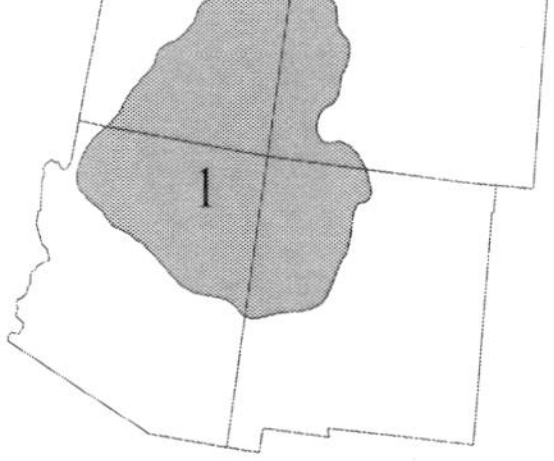

power • the rate at which energy is transferred, usually measured in watts or, less frequently, horsepower.

stratiform deposit • a mineral deposit that forms when a hydrothermal solution moves through a sedimentary deposit such as an aquifer, and deposits minerals parallel to the sedimentary layers.

laccolith • an intrusive igneous rock body that forms from magma that has, through pressure, been forced between two sedimentary layers.

Oligocene • a geologic time interval spanning from about 34 to 23 million years ago.

Miocene • a geological time unit extending from 23 to 5 million years ago.

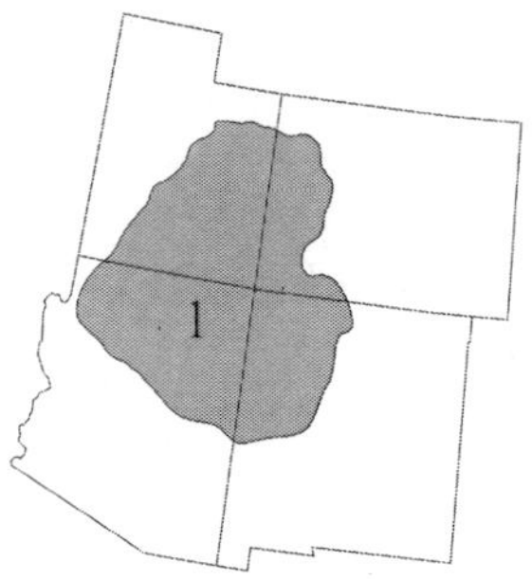

last uranium mine was closed in 2009 due to a drop in uranium prices. New mines are being planned for the Colorado Plateau should demand increase for nuclear **power**.

Similar to uranium deposits, sediment-hosted **stratiform** copper deposits are found in the Triassic, Jurassic, and Cretaceous rocks of the Plateau, mainly in Utah's Paradox Basin. The Salt and Lisbon valleys near Moab, Utah host several copper deposits in porous and permeable fluvial sandstones. These deposits formed from warm saline and copper-bearing fluids that deposited copper minerals, such as chalcocite, malachite, and azurite.

The only gold deposits reported on the Colorado Plateau are of placer origin. These are found largely in the Abajo, La Sal, and Henry mountains of Utah and along the nearby Green, Colorado, and San Juan rivers and their tributaries. The Abajo, La Sal, and Henry mountains are all **laccolithic** intrusions of relatively late **Oligocene** to early **Miocene** in age (29 to 22 million years ago). The region's placer gold was likely concentrated by the erosion of these intrusive rocks.

Non-Metallic Resources

In Colorado and Utah, the Colorado Plateau hosts the Eocene Green River Formation, which represents an almost continuous six-million-year record of very thinly bedded lacustrine (lake) sediments, including salt deposits from times of high salinity. The formation is restricted to the Green River basins of Wyoming, the Uinta and Washakie basins of Utah, and the Piceance Basin of Colorado (*Figure 5.3*). These basins contain the largest known deposit of the mineral nahcolite (sodium bicarbonate) in the world. Sodium bicarbonate, commonly known as baking soda, is used in cooking, fire extinguishers, "green" bio-pesticides, hygiene products, cleaning agents, and numerous medical applications. The Green River Formation also contains the world's largest deposit of trona, a non-marine **evaporite** mineral that is mined as a primary source of sodium carbonate. Trona is a common food additive and water softener, and it also has applications in the manufacturing of paper, textiles, glass, and detergents.

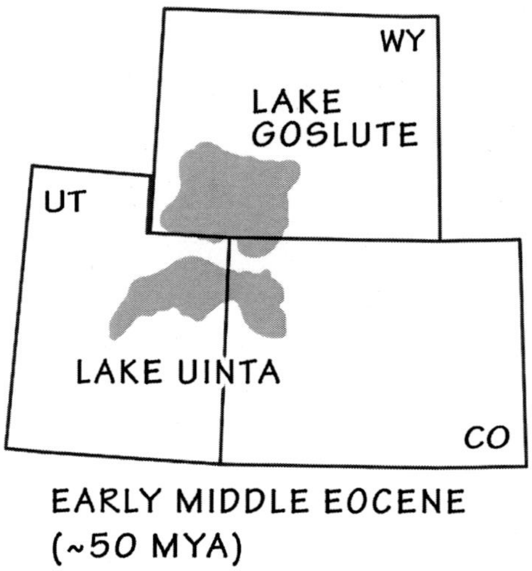

Figure 5.3: Distribution of the Eocene Green River Formation. Blue areas mark the location of middle Eocene lakebeds that bear very large oil shale, nahcolite, and trona deposits.

Mineral Resources 5

Region 1

Thick repetitive sequences of salts formed by the evaporation of shallow seas are found in the **Pennsylvanian** Paradox Formation. Great amounts of these evaporites are recovered from deep beneath the surface of the Paradox Basin in Utah and Colorado, where the formation yields potash and other salts such as magnesium chloride, sylvite, carnalite, and halite (rock salt). These evaporite minerals are mined in two different ways. When deposited in thick beds, salt can be excavated by mechanically carving and blasting it out. This method, called "room and pillar" mining, usually requires that pillars of salt be left at regular intervals to prevent the mine from collapsing (*Figure 5.4*). Another method, called **solution mining**, involves drilling a well into a layer of salt. In some cases, the salt exists as part of a brine that can then be pumped to the surface, where the water is then removed, leaving the salt behind. In others, fresh water is pumped down to dissolve the salt, and the solution is brought back to the surface where the salt is removed (*Figure 5.5*). Most evaporites in the Colorado Plateau are extracted by solution mining. Gypsum has also been mined in large quantities from the Paradox Formation and several Jurassic-age formations.

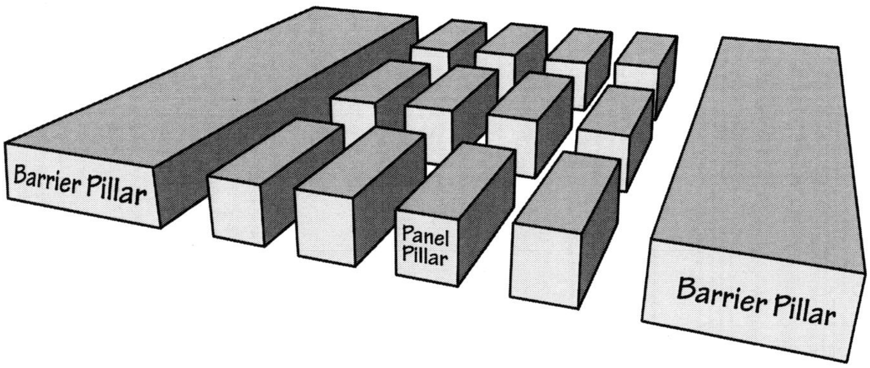

Figure 5.4: In room and pillar mining, the mine is divided up into smaller areas called "panels." Groups of panels are separated from one another by extra-large (barrier) pillars that are designed to prevent total mine collapse in the event of the failure of one or more regular-sized (panel) pillars.

The Paradox Basin salt beds represent a massive resource of potash, a name used for a variety of salts containing potassium, with mined potash being primarily potassium chloride. The USGS has estimated that the basin contains up to 1.8 billion metric tons (2 billion tons) of the mineral, the majority of which is used as fertilizer. Today, an increasing amount of potash is being used in a variety of other ways: for water softening, for snow melting, in a variety of industrial processes, as a medicine, and to produce potassium carbonate. Intrepid Potash, Inc., a Denver-based corporation, is the largest producer of potash in the United States and operates three mines in the Southwest, with the primary site located west of Moab, Utah (*Figure 5.6*).

Bentonite is currently mined in the Southwest, and is often found interbedded with volcanic ash. Bentonites are formed from weathered volcanic ash and glass, and are used in a variety of water-absorbing applications such as drilling mud and cat litter. In the Colorado Plateau, bentonites are found in **Pliocene** beds in Chito, Arizona as well as in the Triassic Chinle and Morrison formations.

evaporite *• a sedimentary rock created by the precipitation of minerals directly from seawater, including gypsum, calcite, dolomite, and halite.*

Pennsylvanian *• a subperiod of the Carboniferous, spanning from 323 to 299 million years ago.*

solution mining *• the extraction of soluble minerals from subsurface strata by the injection of fluids.*

bentonite *• a clay, formed from decomposed volcanic ash, with a high content of the mineral montmorillonite.*

Pliocene *• a geologic time interval extending from roughly 5 to 2.5 million years ago.*

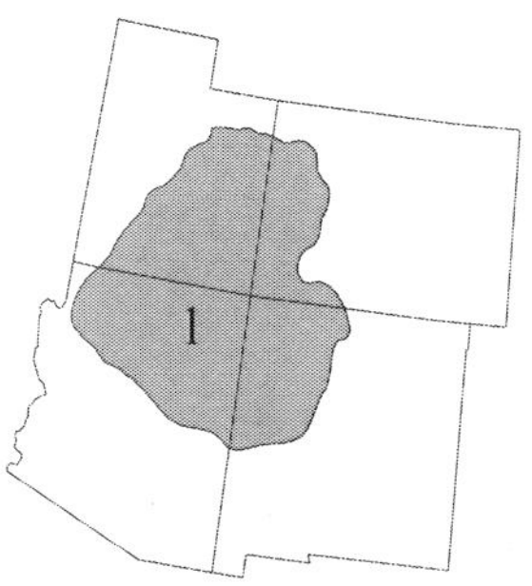

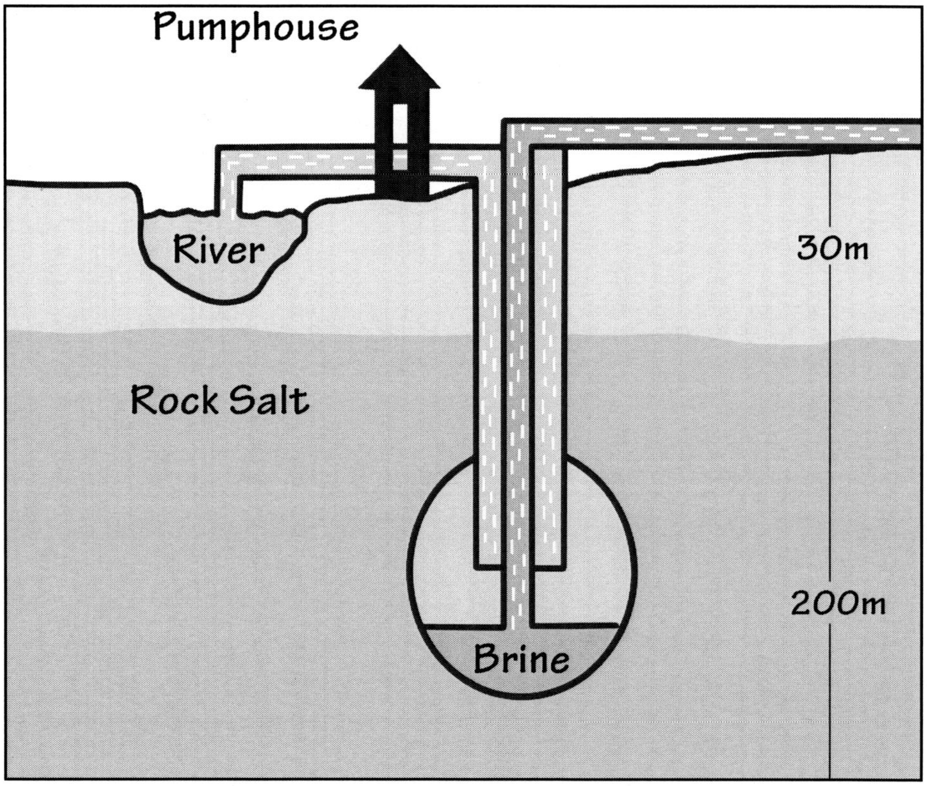

Figure 5.5: An example of solution mining that involves the pumping of fresh water through a borehole drilled into a subterranean salt deposit.

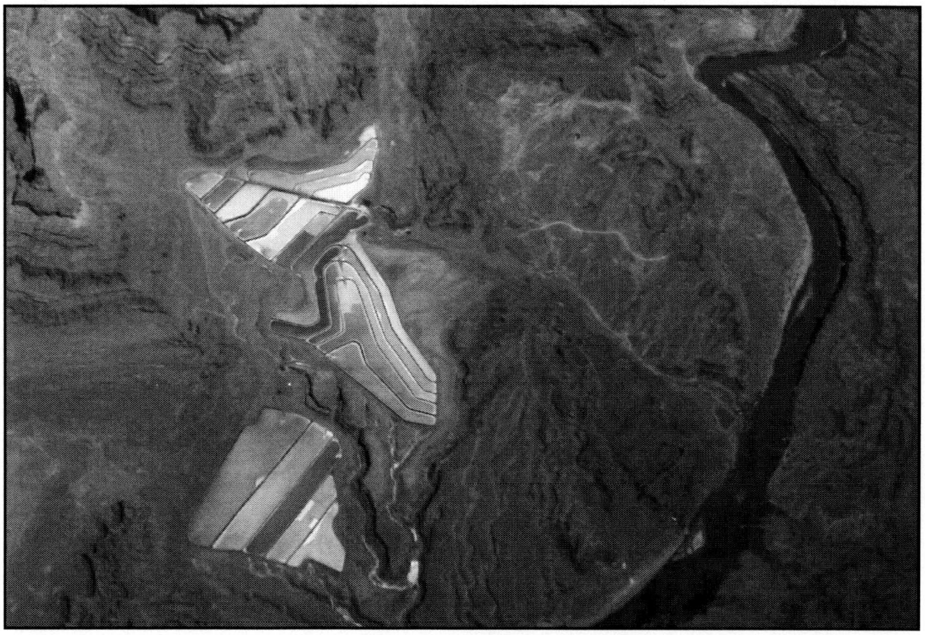

5.6: Potash evaporation ponds at the Intrepid Potash mine near Moab, Utah. The Colorado River is located to the right of the ponds, which are dyed blue to enhance evaporation.

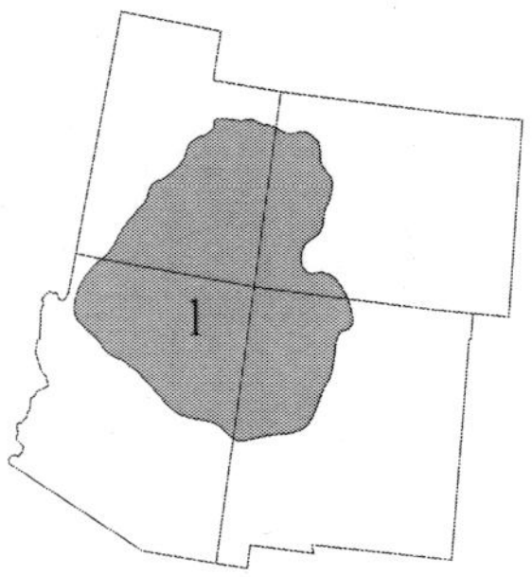

Large quantities of sand, gravel, and crushed stone are locally quarried for use as construction **aggregate**, which is used to strengthen concrete, make blacktop, produce building materials, and as road and dam foundations. Many of these are gathered from old stream **terrace** deposits, deposited after ancient streams eroded the Rocky Mountains to the north and carried gravels made up of quartz and **quartzite**, limestone, sandstone, and harder igneous rocks. Some types of sand are quartz rich, which makes them useful for other industrial purposes. This "industrial sand" is used in sandblasting, filtering, and the manufacturing of glass. Dimension stone, used for buildings, monuments, curbing and facing, is also quarried around the Plateau's rim. Ash Fork, in Yavapai County, Arizona, is the center of production for Arizona Flagstone, a warm-colored dimension sandstone dominated by quartz and silica (*Figure 5.7*).

Ancient sedimentation patterns favored the placement of widespread fossil fuel resources in the Colorado Plateau. Processing plants in Utah and New Mexico recover helium gas, an important byproduct of natural gas extraction.

Figure 5.7: Slabs of Arizona Flagstone sit upright in a supply yard near Ash Fork, Arizona.

Mineral Resources of the Basin and Range

Region 2

The Basin and Range region covers significant portions of Utah, Arizona, and New Mexico, as well as Nevada and parts of adjacent states. Around 30 million years ago in the early Oligocene, the North American **plate** began to override hot upwelling **mantle**, resulting in extensional forces that pulled the continental crust apart to form the region's distinctive "horst and graben" or "basin and range" structure of alternating, roughly north-south oriented valleys and mountain ranges. This extension influenced the shape of geological structures

aggregate • *crushed stone or naturally occurring unlithified sand and gravel.*

terrace • *a flat or gently sloped embankment or ridge occurring on a hillside, and often along the margin of (or slightly above) a body of water, representing a previous water level.*

quartzite • *a hard metamorphic rock that was originally sandstone.*

plates • *large, rigid pieces of the Earth's crust and upper mantle, which move and interact with one another at their boundaries.*

mantle • *the layer of the Earth between the crust and core.*

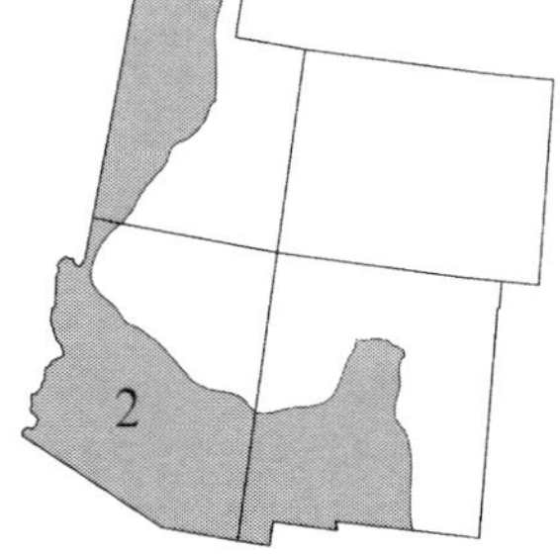

Sevier Orogeny • *a mountain-building event resulting from subduction along the western edge of North America, occurring mainly during the Cretaceous.*

fault • *a fracture in the Earth's crust in which the rock on one side of the fracture moves measurably in relation to the rock on the other side.*

zinc • *a metallic chemical element (Zn, atomic number 30).*

from earlier episodes of deformation, such as those of the **Sevier Orogeny**, which folded and **faulted** the region from the early Cretaceous (120 million years ago) through the Oligocene (52 million years ago). The events that formed the Basin and Range had a significant hand in emplacing the region's considerable mineral resources (*Figure 5.8*), which include major copper deposits as well as large quantities of gold, silver, lead, **zinc**, and molybdenum.

See Chapter 4: Topography to learn more about the formation of the Basin and Range.

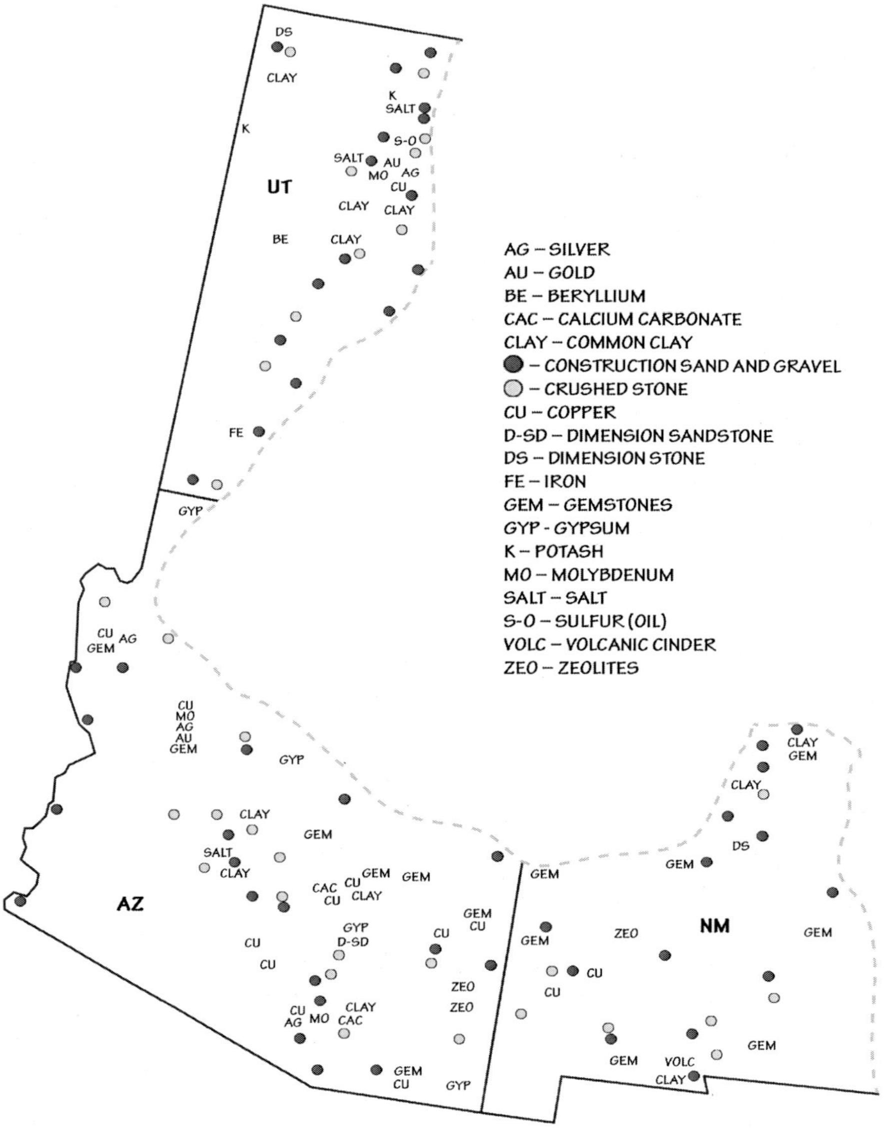

Figure 5.8: Principal mineral resources of the Basin and Range region.

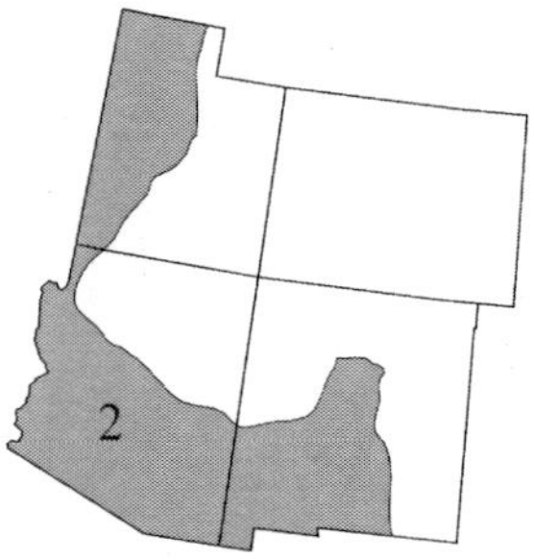

Mineral Resources 5

Region 2

Some of the region's mineral deposits predate **Cenozoic** extension of the landscape. In western Utah, for example, deposits are often concentrated in semi-parallel northeast-trending belts, reflecting deep crustal structures or zones of weakness. The three belts in Utah are, north to south, the Uinta, the Tintic, and the Pioche belts. These structures cross **physiographic** region boundaries (e.g., from the Basin and Range into the Colorado Plateau), indicating they predate the development of these regions, and may date back to the Precambrian. Similar northeast-trending zones extend across other parts of the western US. These zones served as natural pathways for magmas and associated hydrothermal fluids to rise through the crust, from areas where **heat** from the hot, upwelling mantle melted the lower crust. This magmatic activity peaked from about 35 to 17 million years ago, and has continued at a decreasing rate nearly to the present day.

Cenozoic • the geologic time period spanning from 66 million years ago to the present.

physiography • a subfield of geography that studies the Earth's physical processes and patterns.

heat • a form of energy transferred from one body to another as a result of a difference in temperature or a change in phase.

Paleogene • the geologic time period extending from 66 to 23 million years ago.

porphyry • an igneous rock consisting of large grained crystals, or phenocrysts, cemented in a fine-grained matrix.

Metallic Resources

The Basin and Range is a hotbed of metallic ore deposits, especially along orogenic belts—the sites of mountain formation. The Sevier Orogenic Belt, a major geologic feature of the Basin and Range (and of the much larger Cordilleran Fold and Thrust Belt), marks the transition from the Colorado Plateau to the Basin and Range. Formed in the late Cretaceous and **Paleogene**, its structures provided fluid paths and hosts for mineral deposits all along this transition line. Economic deposits of gold, silver, copper, tungsten, and oil and gas are found within the margins of the Sevier Belt.

Thanks to the prolific amounts of copper ore found in Utah and Arizona, both states have taken copper as their state mineral. The Uinta, Tintic, and Pioche belts in Utah produce copper, gold, silver, lead, molybdenum, tungsten, uranium, and beryllium (*Figure 5.9*). The Tintic Mining District, although largely defunct today, was an important source of silver, gold, copper, and bismuth during the late 19th and early 20th centuries. The Pioche Belt also contains the world's largest deposit of alunite, a sulfate mineral that is a source of both potassium and aluminum. The Bingham Canyon deposit, located within the Uinta Belt, contains significant quantities of copper as well as considerable amount of gold and other minerals. Kennecott Copper's Bingham Canyon Mine, near Salt Lake City, Utah, is one of the largest open pit mines in the world and produces 25% of the US domestic copper supply (*Figure 5.10*). So large that it is visible from space, the mine is about a kilometer (over half a mile) deep and produces 408,000 metric tons (450,000 tons) of material per day. Since extraction began in 1863, the mine has produced over 15.4 billion kilograms (17 million tons) of copper, 715,000 kilograms (23 million troy ounces) of gold, 5.9 million kilograms (190 million troy ounces) of silver, and 317 million kilograms (850 million pounds) of molybdenum. This quantity exceeds all of the metals ever produced from the famous Comstock Lode, the Klondike, and the California Gold Rush combined.

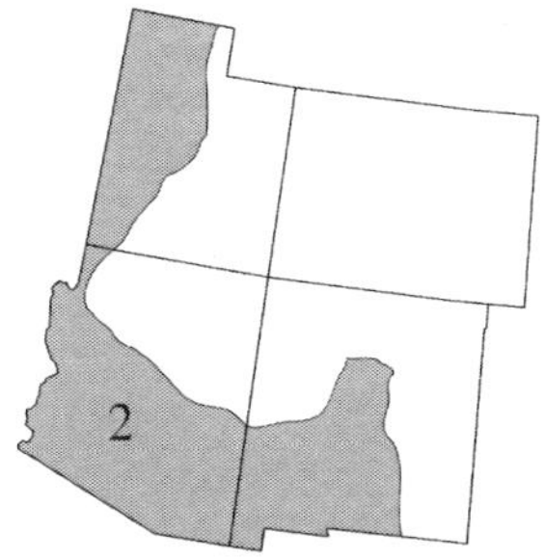

Southeastern Arizona is dominated by **porphyry** copper and associated lead, zinc, gold, and silver deposits in granitic rocks. These deposits were emplaced between 75 and 55 million years ago, predating the development of the Basin and Range's modern structure. Metal ores along the state's western side are dominated by gold deposits in 25- to 15-million-year-old volcanic rocks. Although gold was the first mineral to be mined in Arizona, the state quickly

Figure 5.9: Gold and quartz from the Saw Tooth Mountains, near Salt Lake City, Utah.

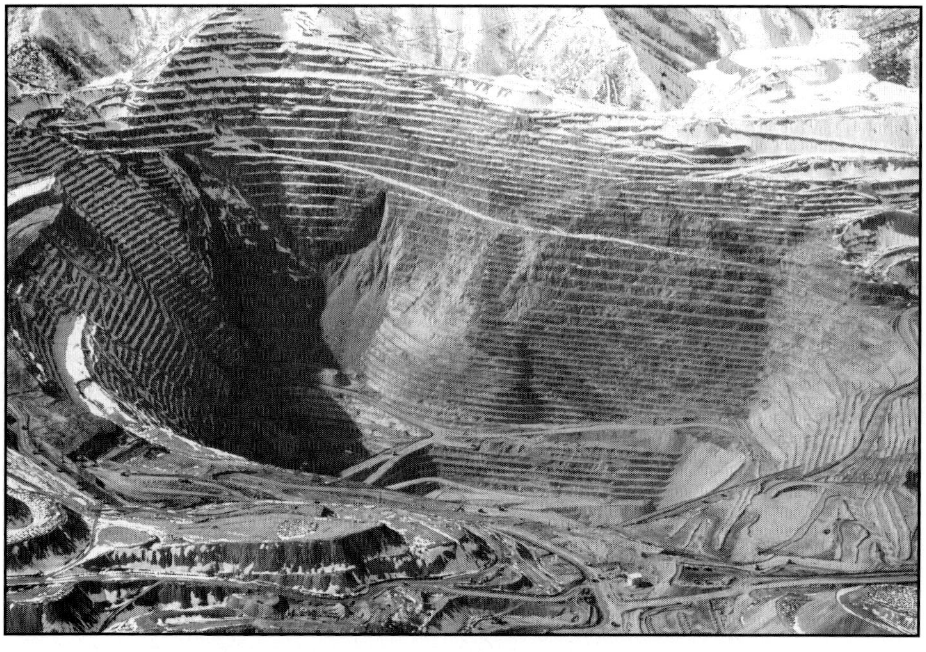

Figure 5.10: The Bingham Canyon Mine in Salt Lake County, Utah, is one of the largest man-made excavations in the world and the second largest copper producer in the US.

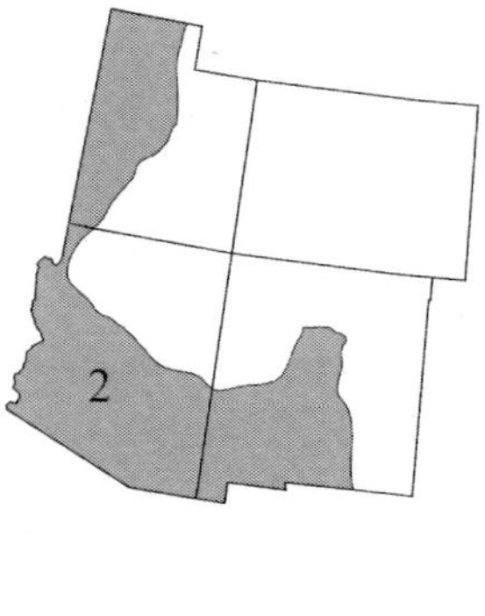

became known for its copper, and by 1910 it was the top copper-producing state in the nation—a title it still holds today. Enormous mines in Arizona's southeastern corner account for over 60% of US copper production (*Figures 5.11 and 5.12*); the Morenci Mine in Greenlee County has reserves estimated at 2.9 billion metric tons (3.2 billion tons) of ore. There is so much copper in Arizona that its nickname is "the Copper State."

The Rio Grande Rift, which extends from Mexico into Colorado and the southern Rocky Mountains, was formed by extensional forces that pulled the continental crust apart in an east-west direction to form the Rio Grande Valley. Like the

Mining

Mining is a profit-focused undertaking. The profitability of mining minerals or rocks depends on a number of factors, including the concentrations of recoverable elements or material contained in the deposit, the anticipated amount of the deposit that can be mined, its accessibility using current mining methods and technologies, its marketability, and lastly the cost of returning the site to its original state once the extraction phase of mining has ended (reclamation). All these factors determine the choice of mining method. Types of mining include underground (tunnel or shaft), surface (open pit or quarry), hydraulic operations (placer), solution using hot water, and seawater evaporation ponds. Once a mineral resource has been removed from the ground, the next step is to process it in order to recover its useful elements or to transform it so that it can be used in manufacturing or other industrial processes.

Modern mining is accomplished in three phases: exploration, extraction, and reclamation. Exploration is performed to determine the extent of the mineral resource and usually involves extensive use of drilling and geophysical techniques to determine the shape, size, and quality of the resource. Extraction involves removing the mineral resource from the ground. Reclamation is undertaken when mining ceases and is designed to restore the land to a condition where it can be used for other purposes. This last phase usually involves removing sources of contamination, which can be considerable depending on the scope of the mining activity.

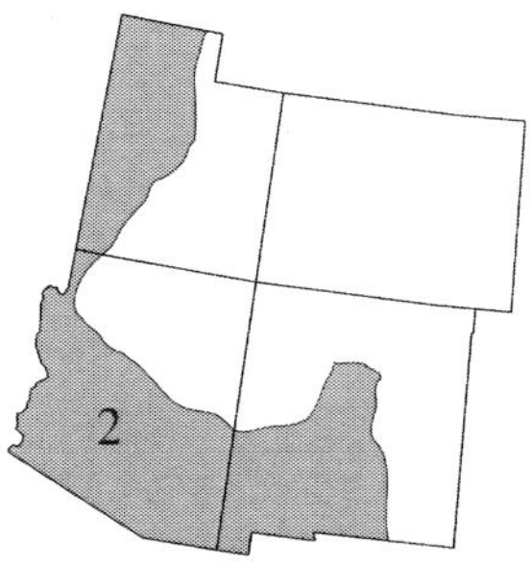

5 Mineral Resources

Region 2

manganese • *a metallic chemical element (Mn).*

rare earth elements • *a set of 17 heavy, lustrous elements with similar properties, some of which have technological applications.*

wind • *the movement of air from areas of high pressure to areas of low pressure.*

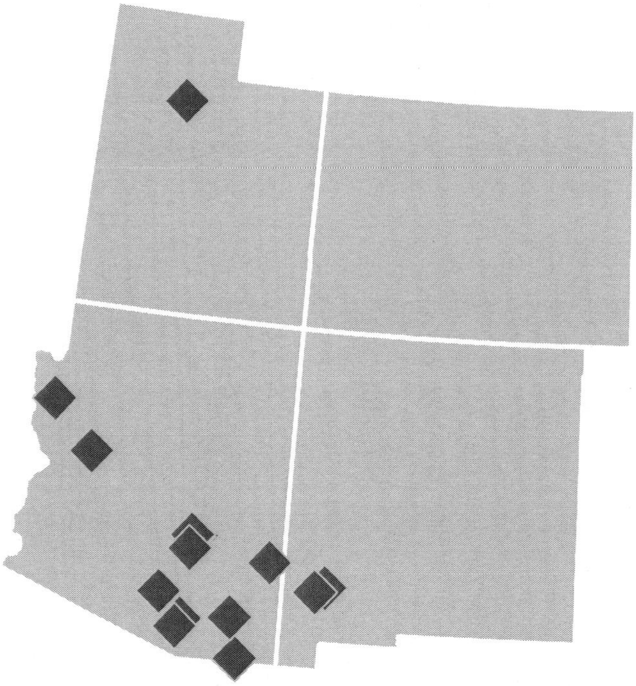

Figure 5.11: Distribution of copper mines in the Southwestern US.

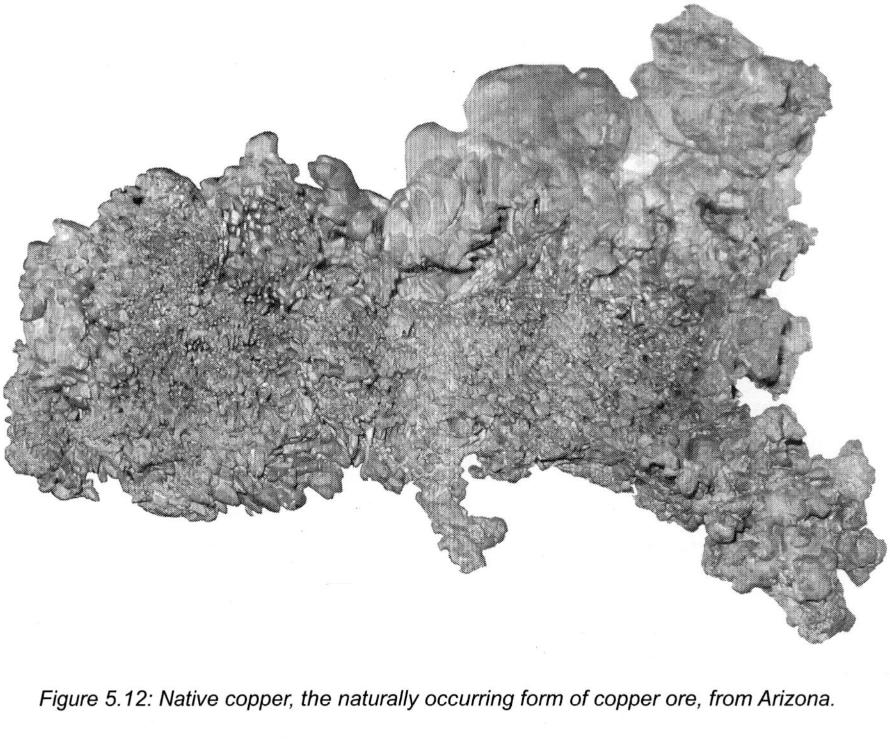

Figure 5.12: Native copper, the naturally occurring form of copper ore, from Arizona.

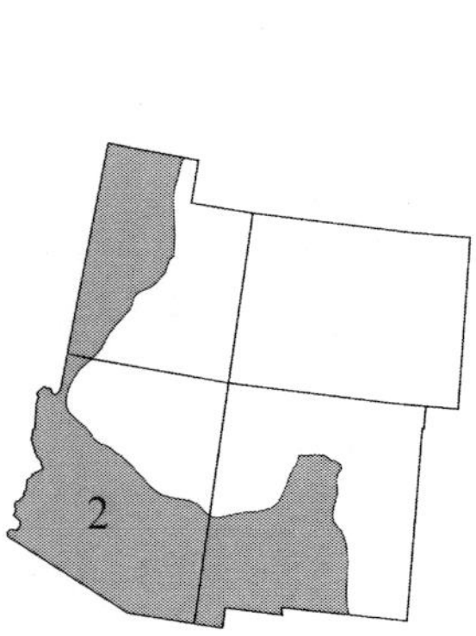

rest of the Basin and Range, ore in the Rio Grande Rift was deposited from hydrothermal solutions associated with magmatic bodies that rose after heat from the hot upwelling mantle melted them near the lower crust. The Rift is known for its production of copper, gold, silver, lead, zinc, **manganese**, and a host of other minerals, as well as molybdenum from the Questa Mine near Red River. New Mexico is currently the nation's number three copper-producing state, with two large open pit mines in Grant County: the Tyrone mine and the Santa Rita mine. Together, these mines have produced as much as 113 million kilograms (249 million pounds) of copper, 370,000 grams (13,000 ounces) of gold, and 5.9 million grams (209,000 ounces) of silver per year. The Santa Rita mine is the oldest copper mine in the western US, and was used by Spaniards as early as 1800.

The Rio Grande Rift is also home to a great number of **rare earth elements** vital to developing technologies, including thorium, lanthanum, yttrium, and the cerium-rich mineral bastnasite. These valuable metals are useful in a range of technological industries, with applications ranging from manufacturing processes to use in electronics such as HDTVs, computers, hybrid and electric vehicles, solar and **wind** power generators, compact fluorescent lamps, and LEDs.

extrusion *• an igneous rock formed by the cooling of lava after magma escapes onto the surface of the Earth through volcanic craters and cracks in the Earth's crust.*

opal *• a silicate gemstone lacking a rigid crystalline structure.*

beryl *• a white, blue, yellow, green, or pink mineral, found in coarse granites and igneous rocks.*

Non-Metallic Resources

The Southwestern US, particularly the Basin and Range, is especially well-known for its gemstones and precious stones. This region, with its combination of faulted sedimentary rocks and intrusive and **extrusive** igneous rocks, provides a rich set of geochemical environments for a diversity of minerals to grow along fracture surfaces, in cavities, and within the igneous bodies themselves. Thus, while the individual minerals form under specific conditions (e.g., chemistry, heat, pressure, and space for growth), the region has a sufficiently broad mix of conditions to allow many different precious gems to form. Gems and precious stones found in the Basin and Range include turquoise, peridot, amethyst, garnet, jade, **opal**, **beryl**, topaz, and many others.

See Chapter 2: Rocks to learn about the types and origins of different igneous rocks.

The Basin and Range is especially famous for turquoise; it is the state gemstone of both Arizona and New Mexico (*Figure 5.13*). This copper-bearing precious mineral is found in areas with substantial copper deposits, and is sometimes removed during copper mining. In the Basin and Range, turquoise is found where copper sulfide deposits weather around certain intrusive igneous rocks. It was one of the first gems to be mined, generally for use in jewelry or sculpture, with extraction dating back over one thousand years. Arizona and parts of New Mexico are among the largest turquoise producing areas of the US, and Arizona still produces the most valuable turquoise in the country, though many mines there have now been depleted. The precious stones azurite and malachite, spectacular blue and green by-products of copper ore weathering, are also common in this area and highly prized by collectors (*Figure 5.14*).

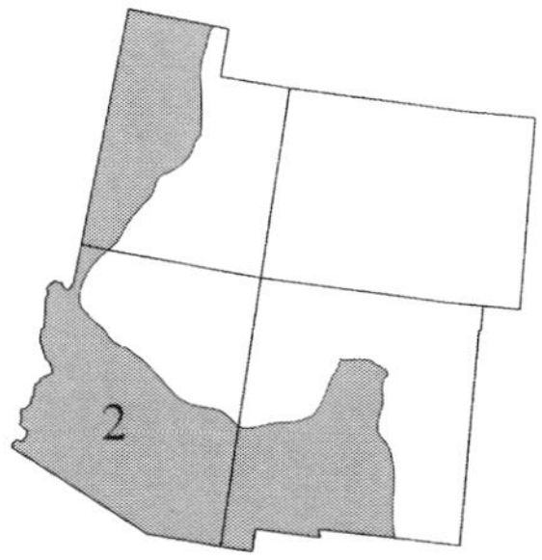

5 Mineral Resources

Region 2

What are hydrothermal solutions?

Hot water enriched in salts such as sodium chloride (NaCl), potassium chloride (KCl), and calcium chloride ($CaCl_2$) is called a hydrothermal solution, or simply "brine." The brine is as salty or even saltier than seawater, and may contain minute bits of dissolved minerals such as gold, lead, copper, and zinc. The presence of salt in the water stops the metallic minerals from precipitating out of the brine because the chlorides in the salt preferentially bond with the metals. Additionally, because the brine is hot, the minerals are more easily dissolved, just as hot tea dissolves sugar more easily than cold tea does.

Hot water brines can have varying origins. Most bodies of magma contain mineral-enriched, superheated water, which is released into the surrounding rock as the magma cools. Rainwater can become a hydrothermal solution as it filters through rocks and picks up soluble materials along its path. Seawater, which is already enriched in salt, often becomes a hydrothermal solution in the vicinity of volcanic activity on the ocean floor where tectonic plates are pulling apart.

Hydrothermal solutions move away from their source of heating through cracks, faults, and solution channels into the adjacent cooler rocks. As the water moves quickly through fractures and openings in the rock (where it experiences changes in pressure or composition and dilution with groundwater), it can cool rapidly. This rapid cooling over short distances allows concentrations of minerals to be deposited. When a hydrothermal solution cools sufficiently, the dissolved salts form a precipitate, leaving behind minerals in a vein or strata-bound deposit.

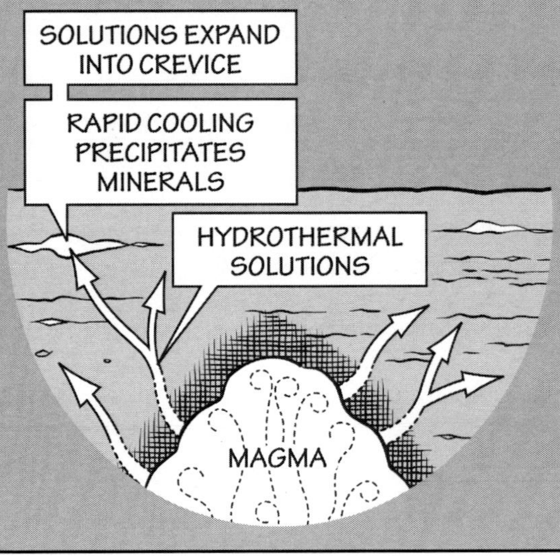

Figure 5.13: A pile of polished natural turquoise from New Mexico.

Figure 5.14: A polished specimen of azurite and malachite from the Bisbee Mine in Cochise County, Arizona. Specimen is 9 centimeters (3.5 inches) wide.

rhyolitic • *a felsic volcanic rock high in abundance of quartz and feldspar.*

matrix • *a fine-grained mass of material around and embedding larger grains or crystals.*

Topaz and beryl form in the fractures and cavities of silica-rich igneous rocks such as granite and **rhyolite**. Topaz is hard (Mohs scale 8) and resistant to erosion, often weathering out of its **matrix** to be found as pebbles in streams. In Utah, where it is the state gem, topaz can be collected from the rhyolites of the Thomas Range, specifically at Topaz Mountain. The mineral is fairly rare, as it contains

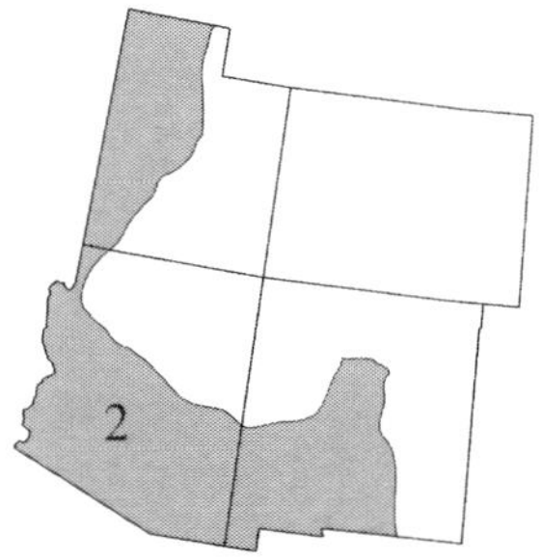

hectare • a metric unit of area defined as 10,000 square meters.

Pleistocene • a subset of the Quaternary, lasting from 2.5 million to about 11,700 years ago.

pluvial lake • a landlocked basin that fills with rainwater or meltwater during times of glaciation.

cinder • a type of pyroclastic particle in the form of gas-rich lava droplets that cool as they fall.

pumice • a pyroclastic rock that forms as frothing and sputtering magmatic foam cools and solidifies.

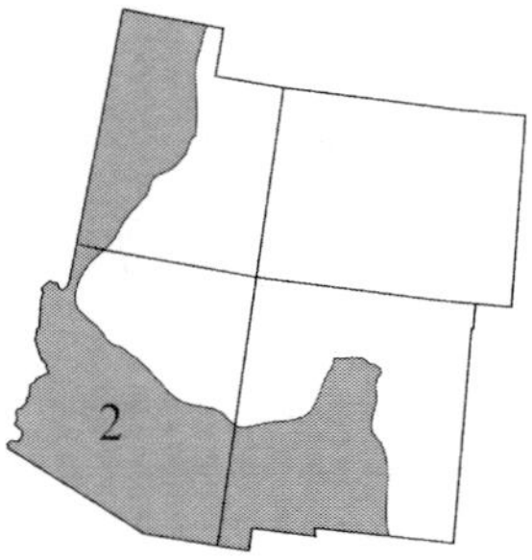

fluorine, which does not occur frequently in quantities sufficient for mineral formation. Other gemstones can be found at Topaz Mountain, including red beryl. This rare mineral contains the element beryllium, and is colored red due to trace amounts of manganese.

The Basin and Range in Utah is known for the expansive Bonneville Salt Flats, a 12,000-**hectare** (30,000-acre) salt pan along the western margin of the Great Salt Lake Basin (*Figure 5.15*). The salt crust is as much as 1.5 meters (5 feet) thick toward the center of the salt flats, and its total volume has been estimated at 133 million metric tons (147 million tons) of salt. This massive, salty expanse is the evaporative remnant of **Pleistocene** Lake Bonneville—a huge **pluvial lake** that once covered much of Utah. The salt is mined at Sevier Lake and the Great Salt Lake, where it is recovered for industrial and commercial use via evaporation in man-made ponds. The salt beds and brines of the Great Salt Lake produce salt, potassium sulfate, and magnesium chloride, and are reported to contain lithium chloride. A potash development project is also underway at Sevier Lake.

See Chapter 8: Climate for more about Lake Bonneville and the Southwest's Pleistocene environment.

Figure 5.15: The Bonneville Salt Flats near Interstate 80, Utah.

Volcanic deposits associated with Basin and Range tectonics provide industrial minerals such as volcanic **cinder** and **pumice**, as well as **clays** from weathered volcanic ash. The combination of volcanic glass and ash with saline and alkaline water in **rift basins** gave rise to natural **zeolites**, which are mined in New Mexico and Arizona. These porous alumino-silicate minerals have cation-exchange properties that can transform hard water into soft water.

Mineral Resources 5

Mineral Resources of the Rocky Mountains
Region 3

The Rocky Mountains are a discontinuous chain of mountain ranges that extend from northern New Mexico northward into Alberta and British Columbia, Canada. In the Southwestern US, the Southern Rocky Mountains are a formally recognized physiographic division that contains the southernmost extension of these ranges, extending from New Mexico through Colorado and into the southern half of Wyoming. The Southern Rocky Mountains began to rise during the **Laramide Orogeny** (which peaked about 68 to 65 million years ago). The region is rich in minerals and mining lore, and continues to produce a variety of mineral resources (*Figure 5.16*). This was one of the earliest areas exploited for minerals during westward expansion across the US. Minerals, especially gold, were known to early Native Americans in this region, and later mined in small amounts by Spanish explorers—perhaps as early as the mid-16th century, and certainly in the 17th and 18th centuries.

See Chapter 4: Topography for more about the Southwest's physiographic regions and divisions.

Metallic Resources

The Rocky Mountains saw its greatest development with the advent of the Great Pikes Peak Gold Rush of 1859, when gold and silver were discovered in the mountains just west of Denver, Colorado. With the Gold Rush came the discovery and development of rich mineral resources all along the Colorado Mineral Belt, which trends northeastward from the southwest corner of the state in the La Plata Mountains to the Front Range near Denver before disappearing at a depth of around 4200 meters (13,000 feet) beneath the sedimentary cover of the Denver Basin (*Figure 5.17*). The Colorado Mineral Belt generally follows a pre-existing and deep-seated crustal structure or a zone of crustal weakness that is Precambrian in age (about 1.4 billion years old). It lies within an ancient **terrane** formed by the crustal **accretion** of central Colorado 1.8 to 1.7 billion years ago. During magmatic events in the Cretaceous and Paleogene, minerals (e.g., gold, silver, and uranium) were carried from deeper Precambrian rocks and deposited as veins and other ore bodies. Abundant quantities of gold, silver, molybdenum, lead, zinc, and other minerals have been found in the Colorado Mineral Belt. Most of Colorado's major mining districts have been located along this belt, with a major exception being Cripple Creek (*Figure 5.18*). Here, the Newmont Mining Corporation currently produces gold from an open pit mine in the mountains west of Colorado Springs. This mine is the largest producer of gold in Colorado; it was incorporated in 1892, and was maintained as an underground operation until 1995. The surrounding Cripple Creek Mining District is now a National Historic Landmark.

See Chapter 1: Geologic History to learn how Cenozoic volcanism shaped the Southwest.

Region 3

clay • *the common name for a number of very fine-grained, earthy materials that become plastic (flow or change shape) when wet.*

rift basin • *a topographic depression caused by subsidence within a rift.*

zeolites • *porous alumino-silicate minerals, often formed some time after sedimentary layers have been deposited, or where volcanic rocks and ash react with alkaline groundwater.*

Laramide Orogeny • *a period of mountain building that began in the Late Cretaceous, and is responsible for the formation of the Rocky Mountains.*

terrane • *a piece of crustal material that has broken off from its parent continent and become attached to another plate.*

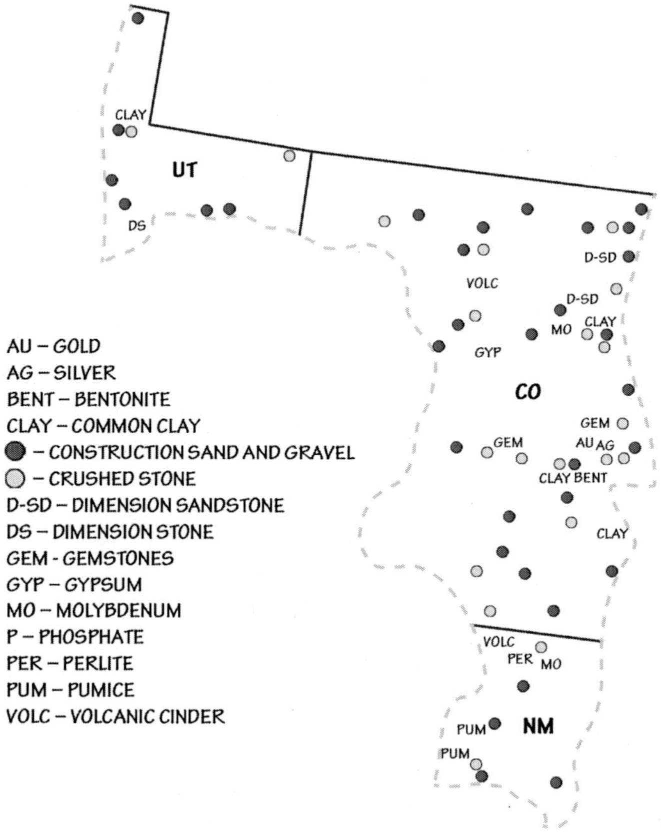

Figure 5.16: Principal mineral resources of the Rocky Mountains region.

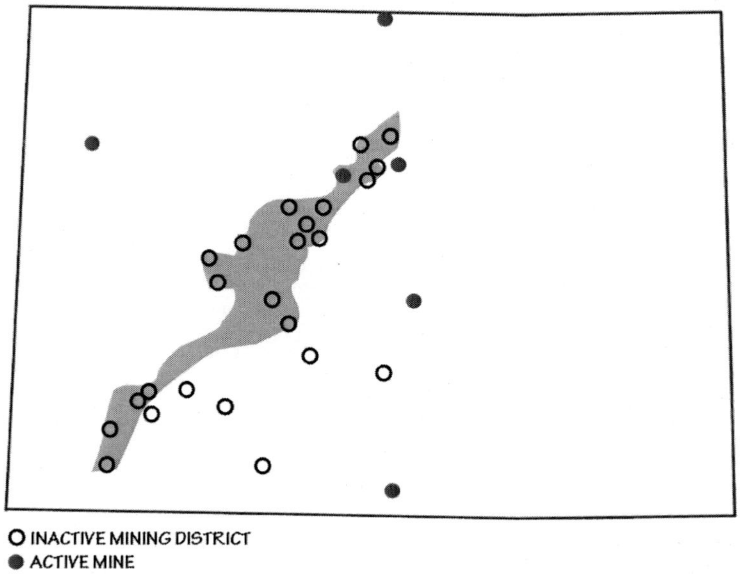

Figure 5.17: Map of Colorado showing position of the Colorado Mineral Belt and nearby sites of mining activity. (See TFG website for a full-color version.)

Mining Districts

Mines in the American West are grouped into "mining districts," defined by their mineral resources as well as by natural boundaries such as rivers. Districts were originally defined informally by miners, but in the late 19th century the US federal government developed regulations for staking claims, property ownership, and mining itself. Some districts have been defined by political boundaries.

Figure 5.18: Visitors can tour the underground tunnels of the Mollie Kathleen Gold Mine, a historic vertical shaft mine in Colorado's Cripple Creek Mining District that descends 300 meters (1000 feet) into the mountain. Many abandoned mining structures and pieces of old equipment are preserved on site, both above and below ground.

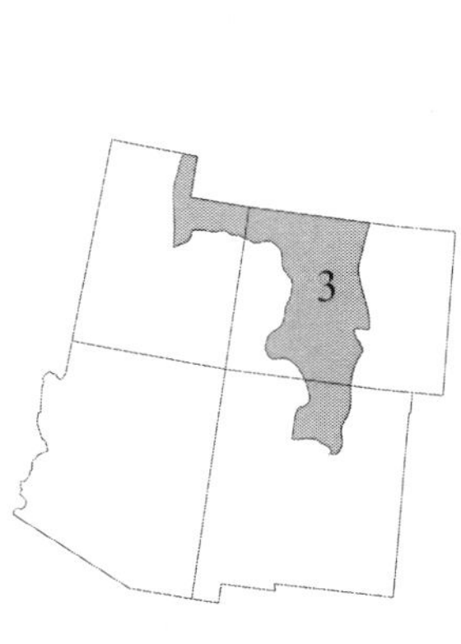

The first discoveries of gold and silver in the Colorado Mineral Belt opened the mining districts of Idaho Springs, Central City, and Georgetown, Colorado. These were followed closely by the opening of the Gold Hill area west of the city of Boulder, Colorado, known for its gold and tungsten deposits. The belt includes the once famous mines of the Leadville area, known for rich silver, gold, and lead ores, and for the eccentric personalities associated with these mines. The story of the legendary Matchless Mine, owned by Horace Tabor and his wife "Baby Doe," inspired the opera *The Ballad of Baby Doe*. The famed Little Johnny Gold Mine owned by J. J. Brown and his wife, the "Unsinkable" Molly Brown (known for her bravery during the sinking of the *RMS Titanic*), inspired the musical play and movie of that name. The gold mines of Aspen, Telluride, Ouray, Creede, and Silverton are also located in the Colorado Mineral

Belt. Well over 25 million troy ounces of gold (approximately 708 metric tons [780 tons]) have been extracted from the Colorado Mineral Belt to date (*Figure 5.19*), along with millions of grams (ounces) of silver, and millions of kilograms (pounds) of lead, zinc, and molybdenum.

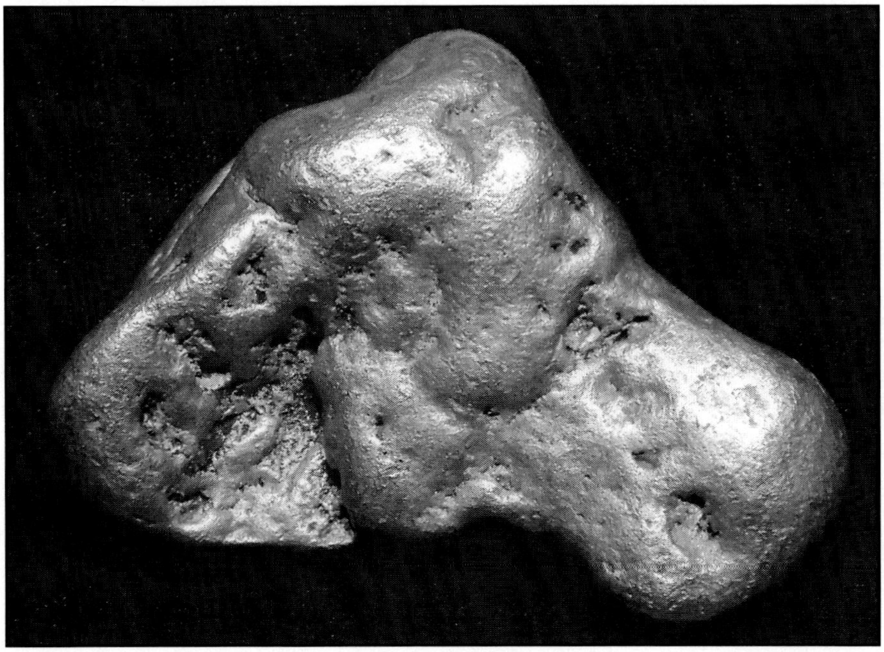

Figure 5.19: A nugget of placer gold from Pennsylvania Mountain in the Alma Mining District, Park County, Colorado. This is Colorado's largest known gold nugget, weighing in at 373 grams (12 troy ounces).

How is gold mined?

Gold can be extracted using a wide variety of methods. Placer mining searches stream bed deposits for minerals moved from their original source by water. Placer deposits can be mined in several different ways: panning, which uses a small, hand-held pan to manually sort the gold from sand and rock fragments; sluicing, in which water is sent through a manmade stepped channel that traps particles of gold; or dredging, where a large machine uses mechanical conveyors or suction to pull loads of material from the river bottom and then dump smaller fragments into a sluice box. Gold that is trapped in layers of rock may be excavated through underground mining, where tunnels or shafts are used to locate the ore, or by open pit mining, which is used when deposits are relatively close to the surface.

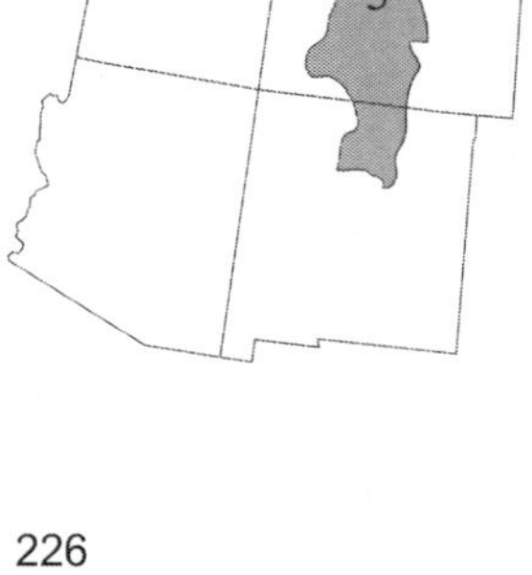

The Climax Mine north of Leadville, Colorado, and the Henderson Mine west of Empire, Colorado, in the northern part of the Rio Grande Rift, contain some of the largest molybdenum ore bodies on Earth. This ore was emplaced between 33 and 24 million years ago, during the Oligocene. The mines extract resources from molybdenum-quartz veins and molybdenum sulfide ore related to certain granites and rhyolite porphyry bodies. The Climax Mine supplied 75% of the world's molybdenum for many years during its period of highest output, and today its ore reserves are estimated at about 227 million kilograms (500 million pounds).

The Colorado Mineral Belt is not the only area of the Rocky Mountains to have hosted large ore mines. Park City, Utah was a site of a major silver rush in the 1860s thanks to the finding of silver, gold, and lead; the town flourished for nearly one hundred years before silver prices dropped in the late 1950s. Park City Mountain and the neighboring Deer Valley are major ski resorts today, but beneath those slopes lie more than 1600 kilometers (1000 miles) of old silver mine workings and tunnels.

Non-Metallic Resources

In the 1960s and 1970s, diamond-bearing **kimberlite diatremes** were found in the State Line District of the northern Front Range, Colorado and across the state line in Wyoming. These formations are igneous pipes of **ultramafic rock** that erupted directly from the mantle at high speeds, bringing diamonds—which form at great depth—up toward the surface. Two of the kimberlite pipes were mined at Kelsey Lake Diamond Mine, the only modern diamond mine in the US during the time of its operation, from 1976 to 2000. Although a few gem-quality diamonds have been recovered (*Figure 5.20*), including one weighing 28.3 carats (about half the size of the massive 46-carat Hope Diamond now in the Smithsonian's gem collection), the Colorado-Wyoming diatremes have thus far not been commercially successful. Other gemstones, including rhodochrosite (Colorado's state mineral), aquamarine (Colorado's state gem), and amazonite, are also collected in Colorado. The Sweet Home Mine near Alma, Colorado was originally founded as a silver mine, but is now a world-famous locality for the collection of high-quality rhodochrosite crystals, and has produced the world's largest known specimen (14 x 16.5 centimeters [5.5 x 6.5 inches]).

kimberlite • *a kind of rock formed of iron- and magnesium-rich magma, mixed with gases from deep in the Earth's mantle.*

diatreme • *a rock-filled crack in the Earth's crust through which magma traveled upward before contacting groundwater and exploding due to expanding water vapor and volcanic gases.*

ultramafic rocks • *igneous rocks with very low silica content (< 45%), which are composed of usually greater than 90% mafic minerals.*

Figure 5.20: Diamonds from the State Line Kimberlite Field on the Colorado-Wyoming border.

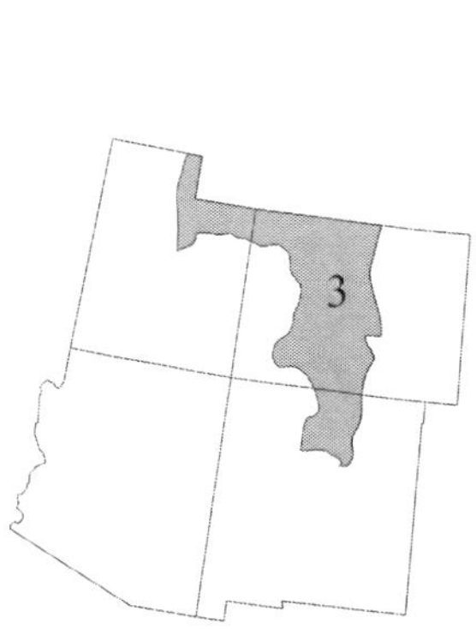

glacier • *a body of dense ice on land that does not melt away annually and has sufficient mass to move under its own weight.*

alluvial • *a layer of river-deposited sediment.*

asphalt • *a black, sticky, semi-solid and viscous form of petroleum.*

Mississippian • *a subperiod of the Carboniferous, spanning from 359 to 323 million years ago.*

contact metamorphism • *the process by which a metamorphic rock is formed through direct contact with magma.*

Industrial stone and construction materials are mined throughout the Rocky Mountains. Crushed granite, quartzite, and rhyolite, along with rounded gravel from rivers, are quarried for landscaping, paving base material, and drainage applications. Sand and gravel aggregate is collected from **glacial** and **alluvial** deposits throughout the Rocky Mountains, and is used as ice control for highways, for filtration purposes, and to aid in the manufacture of concrete, **asphalt**, and other construction projects. The region is also home to the white Yule Marble, used to construct the Lincoln Memorial in Washington DC and the Tomb of the Unknown Soldier at Arlington National Cemetery. The primary quarry for this stone is located at Marble, Colorado, where **Mississippian** limestones underwent **contact metamorphism** from the heat of an adjacent Cenozoic igneous intrusion.

See Chapter 2: Rocks for more information about the Yule Marble.

Volcanic materials contribute to an important mining industry in north-central New Mexico. Perlite, a rapidly cooled volcanic glass with high silica content, is used in horticulture, water filters, **lime**, and cement, as well as in construction-related materials such as ceiling tiles and insulation boards. New Mexico produces the most perlite of any US state, and it is quarried in the No Agua Peaks of Taos County as well as a few other locations with more moderate yields. In addition, pumice and volcanic cinders are quarried in the Rocky Mountains of New Mexico. These deposits are notable for having a lower density and higher porosity than most other rocks (*Figure 5.21*); this makes them commercially valuable for use as a lightweight yet strong construction material. Pumice can also be used as lightweight aggregate, insulators, absorbents, and abrasives. Pumice and cinders are quarried in Rio Arriba County, New Mexico; farther south in Sandova County, pumice is also quarried commercially.

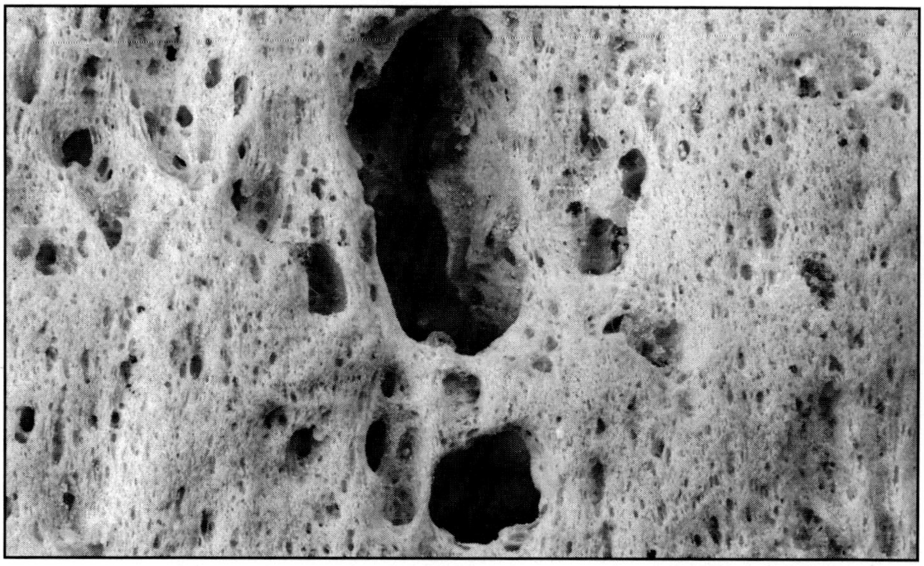

Figure 5.21: A close-up of a pumice stone, revealing its highly porous and vesicular texture. These gas bubbles make the rock so lightweight that it is able to float. Field of view is 2.7 centimeters (1 inch) across.

Mineral Resources of the Great Plains
Region 4

The eastern margin of the Southwest encompasses the Great Plains, a rolling, grassy expanse that slopes eastward to the mid-continent. The Plains are interrupted only by glacial and windblown deposits, river and stream valleys, and other recent erosional features. Beneath a surface cover of **Quaternary** and **Neogene** sediments lies a series of sedimentary and structural basins. Some of these basins were formed during the Laramide Orogeny, and others during tectonic events of the **Paleozoic** and **Mesozoic**. These basin deposits are important sources of minerals such as potash, but major metallic mineral deposits are unknown in this region (*Figure 5.22*).

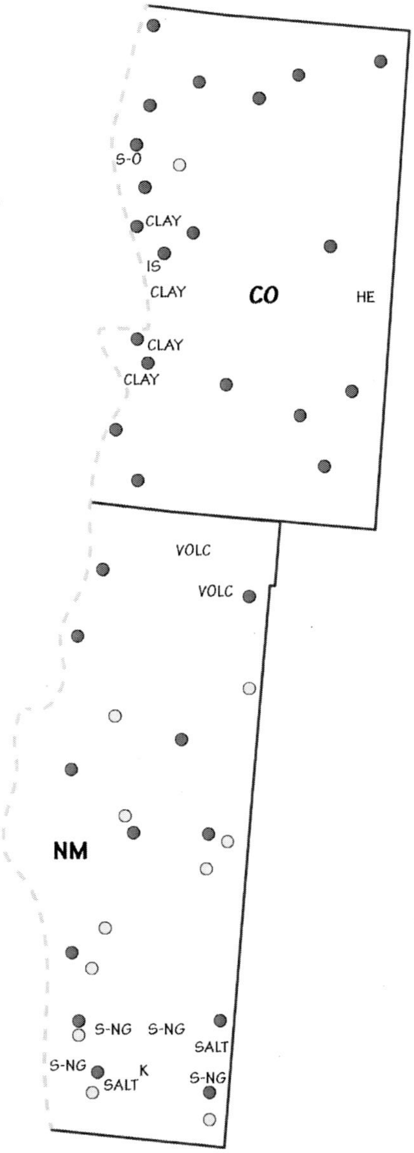

Figure 5.22: Principal mineral resources of the Great Plains region.

Region 4

lime • an inorganic white or grayish-white compound made by roasting limestone (calcium carbonate, $CaCO_3$) until all the carbon dioxide (CO_2) is driven off.

Quaternary • a geologic time period that extends from 2.6 million years ago to the present.

Neogene • the geologic time period extending from 23 to 2.6 million years ago.

Paleozoic • a geologic time interval that extends from 541 to 252 million years ago.

Mesozoic • a geologic time period that spans from 252 to 66 million years ago.

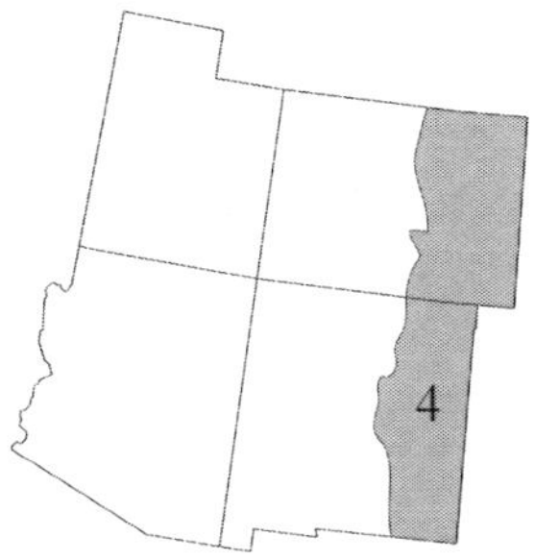

5 Mineral Resources

Region 4

> ***Permian*** *• the geologic time period lasting from 299 to 252 million years ago.*

The Great Plains region contains substantial energy resources, including coal, uranium, abundant oil and gas, and coalbed methane. The region's widespread fossil fuel resources have led to the recovery of several associated elements that are often found alongside gas and oil. Sulfur is extracted from natural gas fields in the Permian Basin (New Mexico's southeastern corner), and from oil fields in the Denver Basin (Adams County, Colorado). Sizeable helium reserves, also associated with the Denver Basin, have been tapped in Cheyenne County, Colorado.

Mountain streams transported and deposited large volumes of eroded sediment onto the plains, resulting in a thick blanket of sand, gravel, **silt**, and clay on top of eroded Mesozoic and **Permian** strata throughout the region. The sands and gravels here are rich in quartz and feldspar from the weathering and erosion of igneous and metamorphic rocks in the Rocky Mountains to the west. Sand, gravel, limestone, dimension stone, and other construction materials are mined throughout the Great Plains. Refractory clay is found in great abundance along the border of the Rocky Mountains in central Colorado, where it is collected for use in the manufacture of brick and ceramics.

Thanks to the Permian Basin's history as part of an ancient inland ocean, it contains some of the world's largest potash deposits, which formed as the seawater evaporated and its potassium salts crystallized. Potash is mined in the vicinity of Carlsbad, New Mexico, and the world's purest mined potash deposit is found in Lea County.

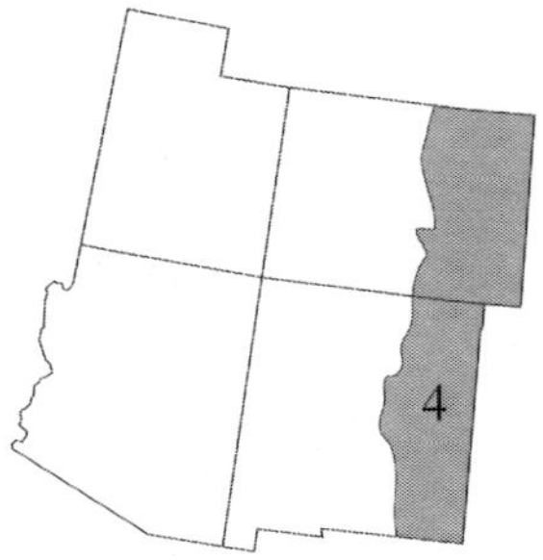

Resources

Books and Articles on Mineral Resources

Frank D., J. Galloway, & K. Assmus, 2005, The life cycle of a mineral deposit – a teacher's guide for hands-on mineral education activities, *US Geological Survey General Information Product* 17, 40 pp.
Johnsen, O., 2002, *Minerals of the World*, Princeton University Press, Princeton, NJ, 439 pp.
Long, K.R., B.S. Van Gosen, N.K. Foley, & D. Cordier, 2010, The principal rare earth elements deposits of the United States: a summary of domestic deposits and a global perspective, *US Geological Survey Scientific Investigations Report* 2010–5220, 96 pp., http://pubs.usgs.gov/sir/2010/5220/pdf/SIR2010-5220.pdf.
Skinner, B. J., 1989, Mineral resources of North America, in A.W. Bally & A.R. Palmer, eds., *The Geology of North America: An Overview, vol. A*, The Geological Society of America: Boulder, CO, pp. 575-584.
Stevens, R., 2010, *Mineral Exploration and Mining Essentials*, Pakawau GeoManagement, Port Coquitlam, BC, 322 pp.
Wark, J., 2012, *The World of Mining*, CRC Press/Balkema, Boca Raton, FL, 250 pp. [Coffee-table book]

For mineral field guides, see the Resources section in Chapter 2: Rocks.

Websites on Mineral Resources

[Copper] Education, Copper Development Association, https://www.copper.org/education/.
Handbook of Mineralogy. [Technical information on 420 minerals available as free individual pdfs.] http://www.handbookofmineralogy.org.
Mineral Data, Hudson Institute of Mineralogy. [Claims to be the world's largest public database of mineral information.] http://www.mindat.org.
Mineral Gallery, Amethyst Galleries, http://www.galleries.com/.
Mineralogy Database, http://webmineral.com.
Mineralpedia—A Mineral Photo Database and Identification Guide, by T. Loomis & V. Loomis, Dakota Matrix Minerals, http://www.dakotamatrix.com.
Mining History Association, http://www.mininghistoryassociation.org/links.htm#museums.
Mining Museums and Tours, National Mining Association, http://www.nma.org/index.php/mining-resources/mining-museums-and-tours.
The New International Mineralogical Association List of Minerals—A Work in Progress, updated July 2016, http://nrmima.nrm.se//IMA_Master_List_%282016-07%29.pdf.
Some Definitions of Common Terms Used in Describing Mineral Deposits, Earth Science Australia, http://earthsci.org/mineral/mindep/depfile/ore_def.htm.
The Gem Hunter, W.D. Hausel, http://gemhunter.webs.com.
US Geological Survey. [A wide range of data on mineral distribution and mining.] http://minerals.usgs.gov/.

Books and Articles on Mineral Resources of Specific Areas of the Southwest

Books and Articles

Anthony, J., 1995, *Mineralogy of Arizona, 3rd edition*, University of Arizona Press, Tucson, AZ, 508 pp.
Arizona Bureau of Mines, 2016, *The Mineral Industries of Arizona: a Brief History of the Development of Arizona's Mineral Resources*, CreateSpace Independent Publishing Platform, North Charleston, SC, 48 pp.

5 Mineral Resources

Resources

Huggard, C. & T. Humble, 2012, *Santa Rita del Cobre: a Copper Mining Community in New Mexico*, Mining the American West, University Press of Colorado, Boulder, CO, 272 pp..

Lundwall, H. & T. Humble, 2012, *Copper Mining in Santa Rita, New Mexico, 1801–1838*, Sunstone Press, Santa Fe, NM, 160 pp.

Northrop, S. & F. LaBruzza, 1996, *Minerals of New Mexico, 3rd edition*, University of New Mexico Press, Albuquerque, 346 pp.

Parry, W., 2004, *All Veins, Lodes, and Ledges Throughout Their Entire Depth: Geology and the Apex Law in Utah Mines*, University of Utah Press, Salt Lake City, 152 pp.

Price, L. G., D. Bland, V. T. McLemore, & J. M. Barker, 2005, *Mining in New Mexico: the Environment, Water, Economics and Sustainable Development*, Decision-Makers Field Conference, New Mexico Bureau of Geology and Mineral Resources, Socorro, 168 pp.

Teacher Resource Packet, by Dave Futey, The Western Museum of Mining and Industry, Colorado Springs, CO, 21 pp., http://www.wmmi.org/_literature_44248/Teacher_Packet.

Voynick, S., 2002, *Colorado Gold: From the Pike's Peak Rush to the Present*, Mountain Press Publishing Company, Missoula, MT, 224 pp.

Whitley, C., 2006, *From the Ground Up: the History of Mining in Utah*, Utah State University Press, Logan, 508 pp.

Wilson, A. B. & P. K. Sims, The Colorado mineral belt revisited—an analysis of new data, *US Geological Survey Open-File Report* 03-046, 4 pp., 2 maps, http://pubs.usgs.gov/of/2003/ofr-03-046/OFR-03-046-508.pdf.

Websites

From Ore to More: a Six-Minute Video on the Story of Copper from Start to Finish, Kennecott Mine, http://www.kennecott.com/content/educator-resources.

Mineral Resources/Mining in Arizona, The Arizona Geological Survey, http://www.azgs.az.gov/minerals_mining.shtml.

Mineral Resources [of Colorado], Colorado Geological Survey, http://coloradogeologicalsurvey.org/mineral-resources/.

Rio Tinto Kennecott Tour, Virtual Tour, Kennecott Mine, http://www.kennecott.com/virtual-tour.

For additional resources on rocks and minerals, see Chapter 2: Rocks.

Chapter 6: Energy in the Southwestern US

All Earth and life processes, including all human activities, depend on flowing **energy**. The energy that drives internal Earth processes (such as tectonics) comes from the internal **heat** generated by **radioactivity**, and generally energy that drives surface and life processes comes from the sun. With the exception of geothermal heat, energy used by humans – to move people and goods, produce electricity, heat our homes and businesses, and manufacture things – comes from the sun.

For most of human history, the way we captured and used energy changed little. With very few exceptions*, materials were moved by human or animal power, and heat was produced largely through the burning of wood. Nearly all the energy to power human society was, in other words, **biomass**. But the transition from brute force and wood burning to the various industrial sources of energy— and the accompanying adoption of energy-intensive lifestyles—has occurred remarkably quickly, in the course of just the last several generations. This has caused changes in virtually every aspect of human life, from economics to war to architecture. Much of the rural US was without access to electricity until the 1930s, and cars have been around for only slightly longer. Our energy **system** (how we get energy and what we use it for) has changed and is changing remarkably quickly, though some aspects of the energy system are also remarkably resistant to change.

The use of **wind** to generate electricity, for example, grew very quickly in the late 2000s and early 2010s. In 2002, wind produced less than 11 million megawatt hours (MWh) of electricity in the US. In 2011, it produced more than 120 million MWh—more than 1000% growth in ten years! That aspect of change stands in contrast to our long-lasting reliance on **fossil fuels**, such as coal, oil, and natural gas. Our reliance on fossil fuels is driven by a number of factors: the low

***Exceptions include the use of sails on boats by a very small percentage of the world's population to move people and goods, and the Chinese use of natural gas to boil brine in the production of salt beginning roughly 2000 years ago.**

Wind and solar power, fossil fuels, nuclear energy, and hydroelectricity are *primary* (natural) *energy sources*: they occur in nature.

Secondary energy sources, also known as *energy carriers*, have been transformed into energy used directly by humans, such as electricity and gasoline.

energy • *the power derived from the use of physical or chemical resources.*

heat • *a form of energy transferred from one body to another as a result of a difference in temperature or a change in phase.*

radioactivity • *the emission of radiation by an unstable atom.*

biomass • *organic material from one or more organisms.*

system • *a set of connected things or parts forming a complex whole.*

wind • *the movement of air from areas of high pressure to areas of low pressure.*

fossil fuels • *fuel for human use that is made from the remains of ancient biomass.*

CHAPTER AUTHORS

Carlyn S. Buckler
Robert M. Ross

Review

British Thermal Unit (BTu or BTU) *• the most commonly used unit for heat energy. One Btu is approximately the amount of heat required to raise one pound of water by one degree Fahrenheit.*

joule *• the energy expended (or work done) to apply a force of one newton over a distance of one meter.*

kinetic energy *• the energy of a body in motion.*

degrade (energy) *• the transformation of energy into a form in which it is less available for doing work, such as heat.*

watt *• a unit of power measuring the rate of energy conversion or transfer designated by the International System of Units as one joule per second.*

chemical reaction *• a process that involves changes in the structure and energy content of atoms, molecules, or ions but not their nuclei.*

upfront cost, very high energy densities, and the reliability and durability of the infrastructure built to use fossil fuels.

Energy production and use not only changes over time, but also with geography, as we will see by looking at energy production and use across different regions of the US.

What do different units of energy mean?

Heat is energy, and heat is at the root of all the ways that we move materials or generate light, so measurements of heat can be thought of as the most basic way to measure energy. The **British thermal unit** (abbreviated Btu or BTU) is the most commonly used unit for heat energy and is approximately the amount of heat required to raise one pound of water by one degree Fahrenheit. A Btu is also roughly 1055 **joules**, or the amount of energy released by burning a single wooden match. A joule is the energy expended (or work done) to apply a force of one newton over a distance of one meter. Since a typical apple weighs approximately one newton, lifting an apple one meter requires approximately a joule of energy. That means that one Btu—the energy contained in a wooden match—is equivalent to the total amount of energy required to lift an apple 1000 meters, or one kilometer.

This comparison of the energy of heat to the energy of motion (**kinetic energy**) might be a little confusing, but energy is transformed from one type to another all the time in our energy system. This is perhaps most obvious with electricity, where electrical energy is transformed into light, heat, or motion at the flip of a switch. Those processes can also be reversed—light, heat, and motion can all be transformed into electricity. The machines that make those transitions in either direction are always imperfect, so energy always **degrades** into heat when it is transformed from one form to another.

The principle of *Conservation of Energy* tells us that energy is neither created nor destroyed, but can be altered from one form to another.

Another measure of energy, the kilowatt-hour (kWh), represents the amount of energy required to light ten 100-**watt** light bulbs for one hour. *Figure 6.1* compares different ways to make and use one kWh.

How do we look at energy in the Earth system?

The concepts used to understand energy in the Earth system are fundamental to all disciplines of science; energy is an interdisciplinary topic. One cannot study physics or understand biomes, photosynthesis, fire, evolution, seismology, **chemical reactions**, or genetics without considering energy. In the US, every successive generation has enjoyed the luxury of more advanced technology (e.g., the ability to travel more frequently, more quickly, and over greater distances). Especially as the global population grows and standards of living increase in some parts of the world, so too does global energy demand continue to grow.

Energy 6

Review

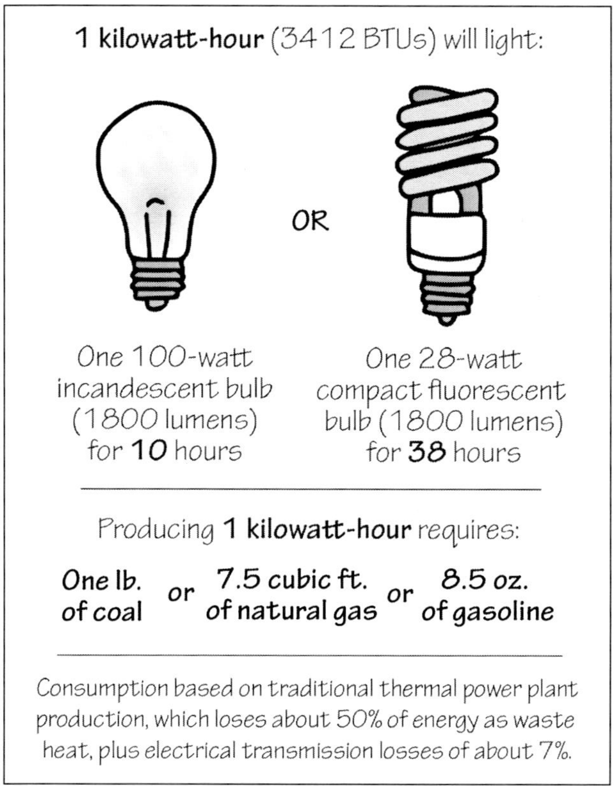

Figure 6.1: Examples of uses and sources of 1 kilowatt-hour.

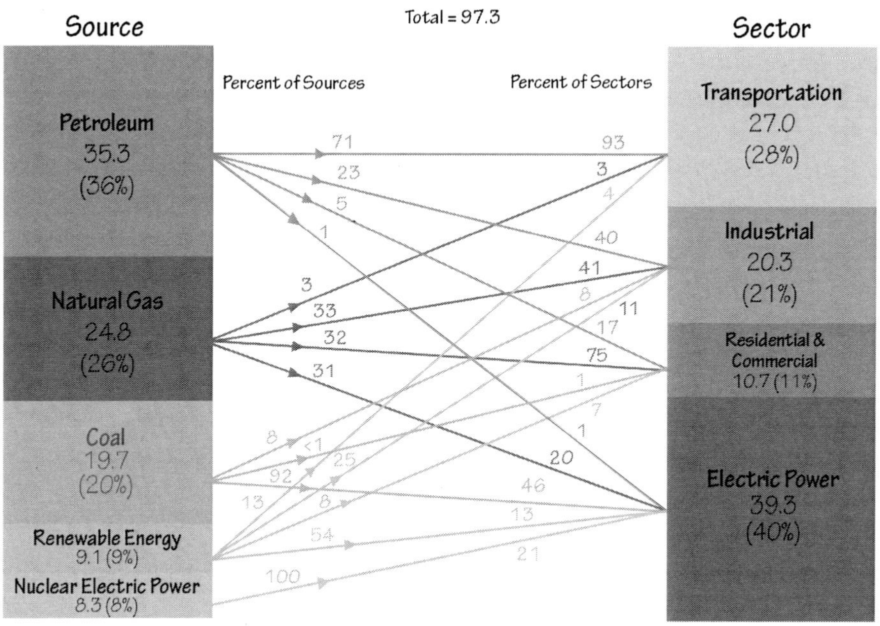

Figure 6.2. US energy production sources and use sectors for 2011. Petroleum provides more energy than any other source, and most of it is used for transportation. More energy is used to generate electricity than for any other use, and electricity is generated by all five energy sources. Nuclear is unique among sources in that all of the energy it generates goes to a single sector: electric generation.

6 Energy

Review

petroleum • *a naturally occurring, flammable liquid found in geologic formations beneath the Earth's surface.*

natural gas • *a hydrocarbon gas mixture composed primarily of methane (CH4), but also small quantites of hydrocarbons such as ethane and propane.*

coal • *a combustible, compact black or dark-brown carbonaceous rock formed by the compaction of layers of partially decomposed vegetation.*

renewable energy • *energy obtained from sources that are virtually inexhaustible (defined in terms of comparison to the lifetime of the Sun) and replenish naturally over small time scales relative to human life spans.*

nuclear • *pertaining to a reaction, as in fission, fusion, or radioactive decay, that alters the energy, composition, or structure of an atomic nucleus.*

power • *the rate at which energy is transferred, usually measured in watts or, less frequently, horsepower.*

Figure 6.2 shows the sources and uses of energy in the US, by sector. The Energy Information Administration (EIA) categorizes energy as coming from one of five sources (**petroleum**, **natural gas**, **coal**, **renewable energy**, and **nuclear** electric **power**) and being used in one of four energy sectors (transportation, industrial, residential & commercial, and electric power). All of the energy that powers our society comes from one of these five sources and is used in one of these four sectors.

The more we come to understand the Earth system, the more we realize that there is a finite amount of consumable energy, and that harvesting certain resources for use in energy consumption may have wide ranging and permanent effects on the planet's life. Understanding energy within the Earth system is the first step to making informed decisions about energy transitions.

Becoming "Energy Literate"

Energy is neither lost nor gained within the universe, but rather is constantly flowing through the Earth system. In order to fully understand energy in our daily lives—and make informed decisions—we need to understand energy in the context of that system. Becoming energy literate gives us the tools to apply this understanding to solving problems and answering questions. The Seven Principles of Energy, as detailed in *Energy Literacy: Energy Principles and Fundamental Concepts for Energy Education* are as follows:

Energy Literacy: Energy Principles and Fundamental Concepts for Energy Education **is a publication of the US Department of Energy. It can be accessed for free online; see Resources for more information.**

1	**Energy is a physical quantity that follows precise natural laws.**
2	**Physical processes on Earth are the result of energy flow through the Earth system.**
3	**Biological processes depend on energy flow through the Earth system.**
4	**Various sources of energy can be used to power human activities, and often this energy must be transferred from source to destination.**
5	**Energy decisions are influenced by economic, political, environmental, and social factors.**
6	**The amount of energy used by human society depends on many factors.**
7	**The quality of life of individuals and societies is affected by energy choices.**

Each principle is defined by a set of fundamental concepts that can help clarify ties to curricula. Keeping these energy principles in mind when we teach others about energy can help us contextualize and make relevant our own energy consumption and its effect on the Earth system.

Energy in the Southwestern Regions

Within its majestic mountain ranges, roaring rivers, and expansive plateaus, the Southwestern US is replete with energy resources. Coal, oil, and gas deposits are extensive (*Figures 6.3 and 6.4*), especially in the the Colorado Plateau and Great Plains regions. Utah, Colorado, and New Mexico all rank within the top dozen producing states of both oil and natural gas, and within the top 15 in coal production. The distribution of these resources is linked to the area's many **sedimentary** basins, which have been host to the formation and trapping of hydrocarbons at multiple intervals since the late **Paleozoic** (*see Figure 6.6*).

Uranium (the raw material used for fission in nuclear power plants) is locally abundant (*Figure 6.5*), and it has been mined extensively, although nearly all of the uranium collected in the Southwest is shipped to other states. There is only one nuclear power plant in the Southwest, located near Wintersburg, Arizona.

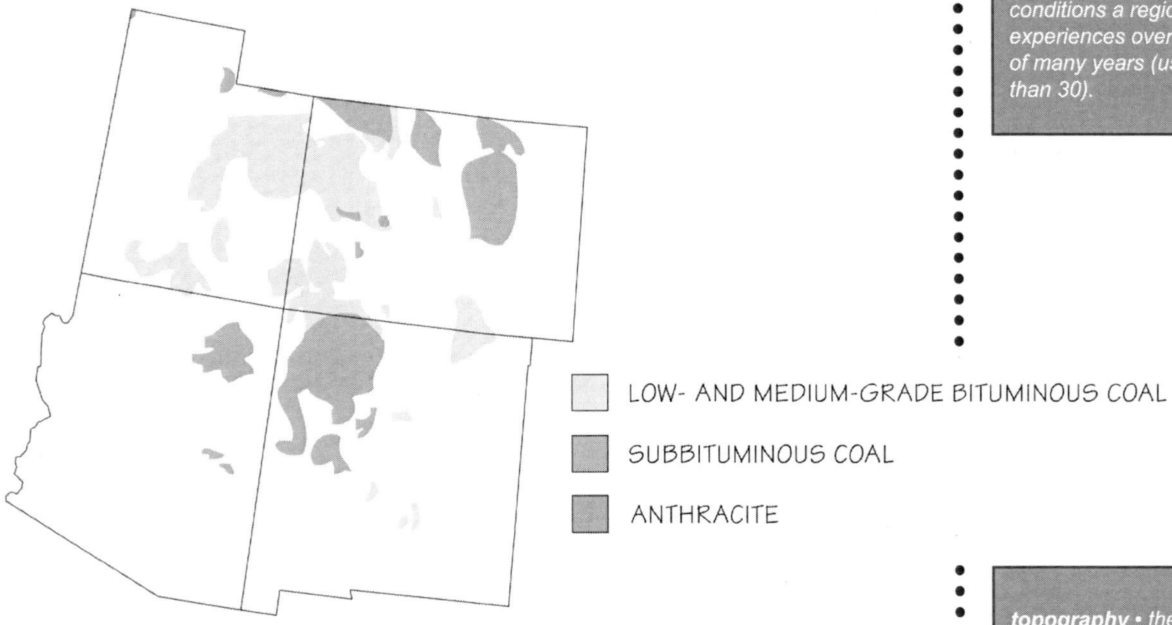

Figure 6.3: Coal-producing areas of the Southwestern US. The Colorado Plateau is a particularly significant coal-producing region. (See TFG website for a full-color version.)

Given its **climate** and **topography**, the Southwest has a reasonably high capacity for generating solar and wind energy. Hydropower along major river systems is also an important source of energy in some areas. Though these sources are increasing rapidly, they remain a relatively small part of energy

Regions

sedimentary rock • *rock formed through the accumulation and consolidation of grains of broken rock, crystals, skeletal fragments, and organic matter.*

Paleozoic • *a geologic time interval that extends from 541 to 252 million years ago.*

climate • *a description of the average temperature, range of temperature, humidity, precipitation, and other atmospheric/hydrospheric conditions a region experiences over a period of many years (usually more than 30).*

topography • *the landscape of an area, including the presence or absence of hills and the slopes between high and low areas.*

Figure 6.4: Areas of oil and gas production in the Southwestern US. The Colorado Plateau and Great Plains are both significant fuel-producing regions.

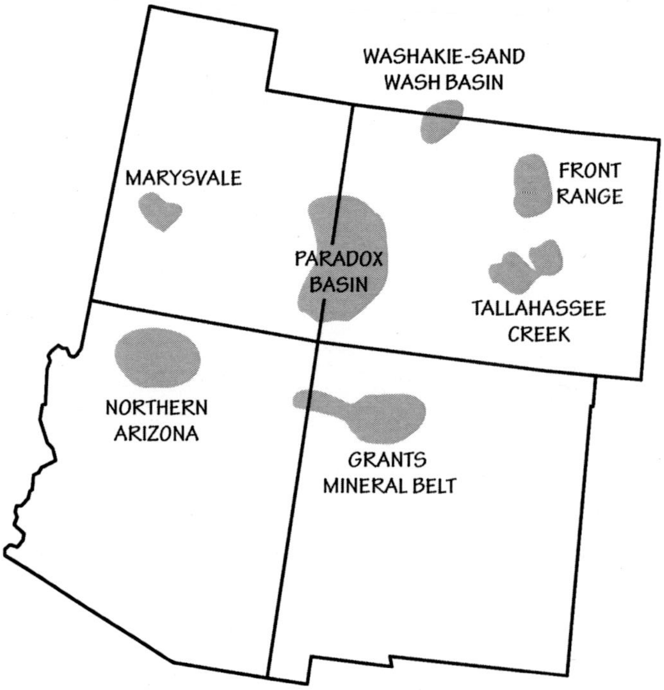

Figure 6.5: Distribution of uranium deposits in the Southwestern US.

Regions

tree • *any woody perennial plant with a central trunk.*

Fossil Fuels

Fossil fuels—oil, natural gas, and coal—are made of the preserved organic remains of ancient organisms. Coal and lignite result from the burial, compaction, and heating of preserved plant matter, whereas petroleum and natural gas originate deep underground through a slow process involving the low-grade heating of sedimentary source rocks that contain an abundance of organic matter. In either case, organic matter is only preserved when the rate of accumulation is higher than the rate of decay. This happens most often when the oxygen supply is sufficiently low that oxygen-loving bacteria cannot thrive, greatly slowing the breakdown of organic matter. In this way, organic matter can be incorporated into the buried sediment. The organics are compacted and heated with the rest of the rock, eventually transforming into fossil fuels.

The history of surface environments, evolution of life, and geologic processes beneath the surface have all influenced where fossil fuel deposits formed and accumulated. The largest oil and gas reserves were at one time nutrient-rich seas with abundant surface phytoplankton and organic-rich bottom sediments; the largest coal beds were swampy environments where fallen forest *trees* and leaves were buried in stagnant muds.

production in the Southwestern US. Some parts of the Southwest also have a high potential for deep geothermal energy due to the proximity of areas of active tectonism.

The majority of the Southwest's land is owned by the federal government, and managed primarily by the Bureau of Land Management (BLM), National Park Service, US Forest Services, and Bureau of Indian Affairs (BIA). Exploration and mining for natural resources in these areas is extensively regulated, and no energy is developed in National Parks, National Monuments, or a variety of other wilderness areas. Because the Southwest is home to some of the most magnificent National Parks and Monuments in the world, many local governments and citizens have gone to great lengths to preserve it by preventing and regulating energy development and mining projects.

6 Energy

Region 1

Cambrian • a geologic time period lasting from 541 to 485 million years ago.

carbonate rocks • rocks formed by accumulation of calcium carbonate, often made of the skeletons of aquatic organisms.

Carboniferous • a geologic time period that extends from 359 to 299 million years ago.

Permian • the geologic time period lasting from 299 to 252 million years ago.

weathering • the breakdown of rocks by physical or chemical means.

silt • fine granular sediment most commonly composed of quartz and feldspar crystals.

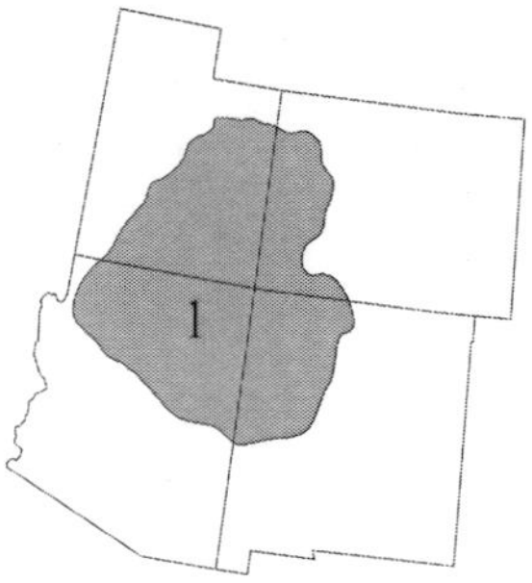

Energy of the Colorado Plateau
Region 1

The Colorado Plateau's sedimentary basins have yielded oil, gas, and coal for the past century, and they contain large reserves of additional oil that could potentially be tapped through hydraulic fracturing and horizontal drilling. The high average wind speed and solar intensity on the Colorado Plateau make wind and solar energy potential future sources of electrical energy, but they currently remain minor compared to existing coal and natural gas-powered plants. A number of power plants are associated with hydroelectric dams along the Colorado River system, and the region, particularly southeastern Utah, has long been a source of uranium for use in nuclear power.

Oil and Gas

The Colorado Plateau contains numerous sedimentary basins, each of which contains many organic-rich layers (*Figure 6.6*). It is possible to make sense of why we find petroleum and natural gas in these areas by understanding the region's geologic history. Mud with relatively high organic matter content tends to accumulate in shallow continental seas and in coastal marine environments. The history of the Southwest's sedimentary basins extends back to the **Cambrian** period, when a broad shallow sea covered much of the area. Thick sequences of **carbonate rocks** accumulated in these basins. During the **Carboniferous** and **Permian**, as sea levels dropped and tectonic changes affected the landscape, parts of these basins became more restricted. As land emerged and **weathered** into **silt** and **sand**, layers of **sandstones** and organic-rich **shales** were laid down and organic material was preserved on the seabed. When seawater in the basins evaporated, **evaporites** were also deposited. With time, pressure, and heat, organic material in the shale was changed into petroleum and gas, and the organic-rich shales became source beds for hydrocarbons. Later deposition of non-marine sandstones in the **Mesozoic** created additional reservoirs for the oil. The **Cretaceous** saw the development of the Western Interior Seaway, which accumulated additional organic-rich shales, along with coastal coals, sandstones, and deeper marine **limestones**. Finally, terrestrial sedimentation in large lake basins during the early **Cenozoic** trapped yet more organic-rich sediments in **lacustrine** shales. The thick set of sediments that built up over millions of years created heat and pressure in deeper layers, compacting and "cooking" much of the organic matter into forms that are now used as **fuels**.

See Chapter 1: Geologic History for more information about the Western Interior Seaway.

Conventionally, finding oil and gas has not been as simple as finding organic-rich rock layers. Oil and gas can flow both within and between rock layers, wherever the number and size of paths between pores, **fractures**, and other spaces (**permeability**) is large enough. Because oil and gas are under pressure and are more buoyant than pore-filling waters, they will move gradually upward to areas of lower pressure and will rise all the way to seeps at the surface unless they are blocked by a **caprock** or seal—that is, one or more layers with

Energy 6

Region 1

Oil and Gas

Oil and gas form from organic matter in the pores of sediments subjected to heat and pressure. The organic matter is primarily composed of photosynthetic plankton that die and sink to the bottom of large water bodies in vast numbers. Shale in particular is often organic rich, because organic matter settles and accumulates in the same places that mud (*clay* and silt particles) settles out of the water. In most environments, organic matter is recycled by bacteria before it can be buried, but the quiet waters where mud accumulates are often relatively stagnant and low in oxygen. In these places, the bacterial decay rate is low relative to the rate of organic matter sinking and to the rate that the organic matter becomes buried in muddy sediments. Under such conditions, organic matter may accumulate enough to make up several percent or more of the deposited sediment.

Because oil and gas are under pressure, they will move gradually upward to areas of lower pressure through tiny connections between pore spaces and natural fractures in the rocks. Reservoir rocks typically have a considerable amount of pore space, and to be viable there must be a way of trapping the oil and gas, such as through a geologic structure or a change in rock type that will prevent the resource from escaping. Often, natural gas and oil are trapped below the surface under impermeable layers that do not have sufficient spaces for liquids and gases to travel through. Folds or "arches" in impermeable layers, or faults in rock layers, are common ways of trapping oil and gas below the surface.

sand • *granular sediment most commonly composed of weathered grains of quartz and feldspar, of grain diameter 1/16 to 2 millimeters.*

sandstone • *sedimentary rock formed by cementing together grains of sand.*

shale • *a dark, fine-grained, laminated sedimentary rock formed by the compression of successive layers of silt- and clay-rich sediment.*

evaporite • *a sedimentary rock created by the precipitation of minerals directly from seawater, including gypsum, calcite, dolomite, and halite.*

lacustrine • *of or associated with lakes.*

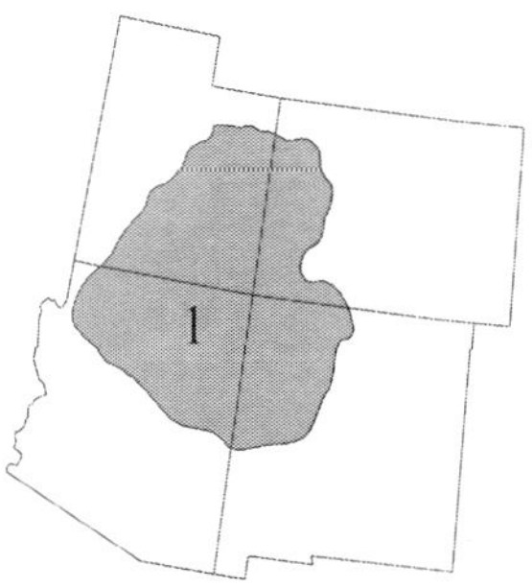

6 Energy

Region 1

anticline • a layer of rock folded (bent) along an axis, concave side down (i.e., in an upside down "u" or "v" shape).

fault • a fracture in the Earth's crust in which the rock on one side of the fracture moves measurably in relation to the rock on the other side.

porosity • the percentage of openings in a body of rock such as pores, joints, channels, and other cavities, in which gases or liquids may be trapped or migrate through.

Devonian • a geologic time period spanning from 419 to 359 million years ago.

Eocene • a geologic time period extending from 56 to 33 million years ago.

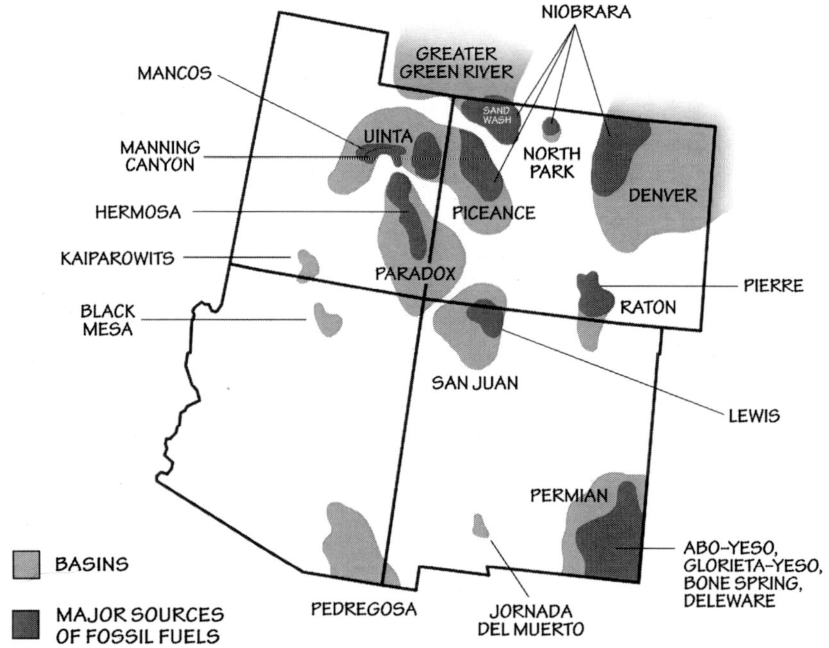

Figure 6.6: Sedimentary basins across the Southwestern US contain significant fossil fuel accumulations. (See TFG website for a full-color version.)

permeability so low that they effectively block the flow of liquids and gases. If the fossil fuel happens to rise beneath a caprock in the shape of a concave surface (such as an **anticline** or certain **faults**), the fossil fuels may accumulate in what geologists call a "petroleum trap." Petroleum traps accumulate oil and gas in **porous** sedimentary layers with thin natural fractures, called reservoirs. Most oil and gas has been extracted using the conventional technique of searching for such reservoirs and then drilling vertically into them, which allows the gas or oil to come to the surface through the well pipe. Reservoirs in the Southwest range from **Devonian** to **Eocene** in age.

The Paradox Basin of southeast Utah and southwest Colorado produces oil and gas, as well as a variety of other **mineral** resources. The basin contains a thick sequence of Paleozoic and Mesozoic sediments. Shallow marine carbonate rocks of Devonian, **Mississippian**, and **Pennsylvanian** age act as primary reservoirs for the basin's oil. The source rock for this oil is largely organic-rich shale that is interbedded with the Pennsylvanian carbonates; the shale accumulated from the eroded sediments of the Ancestral Rockies that were **uplifted** immediately to the east. As oil migrated **stratigraphically** into the overlying sandstone, it ultimately pooled in reservoirs trapped under a variety of impermeable sedimentary deposits such as **gypsum**, anhydrite, **limestone**, and **dolomite**.

See Chapter 5: Mineral Resources to learn about other resources found in the Paradox Basin, including salts and uranium.

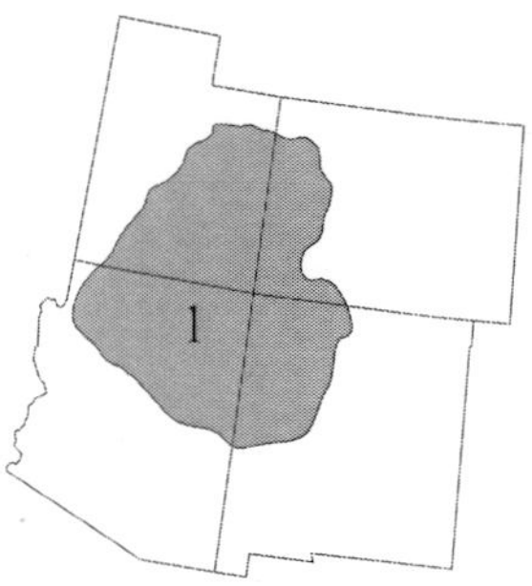

The Paradox Basin contains substantial quantities of evaporites, particularly halite, which greatly influenced the structure of petroleum traps in the basin. These evaporites formed during an interval of repeated basin restrictions during Pennsylvanian and Permian sea level fluctuations. These underlying **salt** structures explain the geographic distribution of many oil and gas reservoirs in the Paradox Basin. Impermeable rocks pushed up by **salt domes** became caprock where oil could be trapped—this is the case in the Hermosa Formation of southeastern Utah. Collapsed domes are also responsible for many of the region's landscape features. For example, the name "Paradox Basin" came from the paradoxical observation that the Doloros River cuts across instead of flowing down the landscape. The river is superimposed upon a buried salt anticline that collapsed, leaving a long dry valley now occupied by the river (*Figure 6.7*).

See Chapter 4: Topography to learn more about anticlines and other structures that influence the Southwestern landscape.

mineral *• a naturally occurring solid with a specific chemical composition and crystalline structure.*

uplift *• upward movement of the crust due to compression, subduction, or mountain building.*

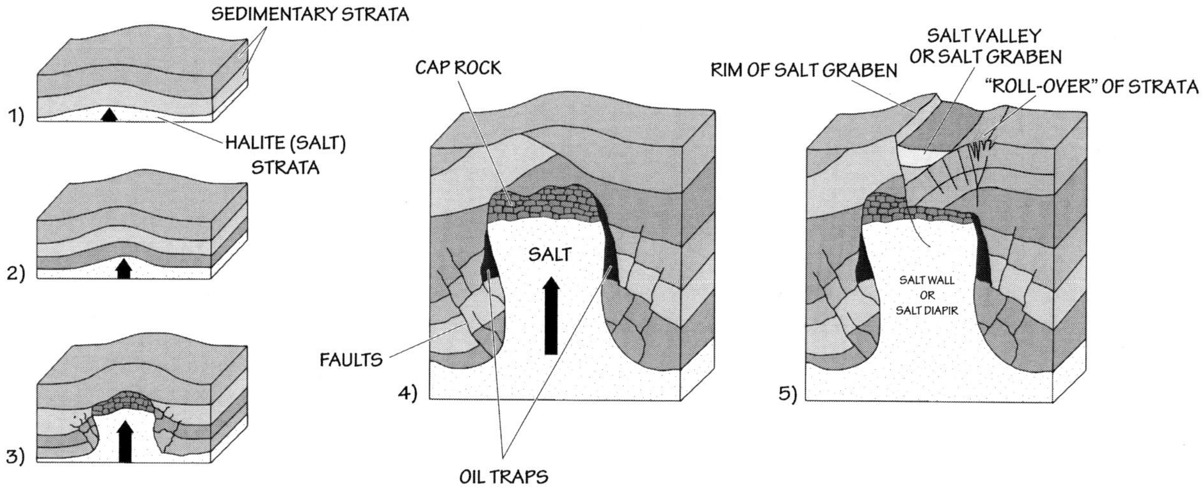

Figure 6.7: The formation and partial collapse of a salt dome can trap fossil fuels and affect the landscape.

Salt domes can also be used to store large quantities of oil and gas. Storage caverns are created by injecting the salt with water to dissolve a cavity within the salt structure—a process called **solution mining** (*Figure 6.8*). Many salt domes along the Gulf Coast are used for this purpose; in the Southwest, however, the first such storage complex is being developed by the Magnum Gas Storage Project. The caverns, hollowed out of a 2100-meter-thick (7000-foot-thick) salt dome in Millard County, Utah, will be capable of storing 1.5 billion cubic meters (54 billion cubic feet) of natural gas and will be connected to an interstate natural gas pipeline system.

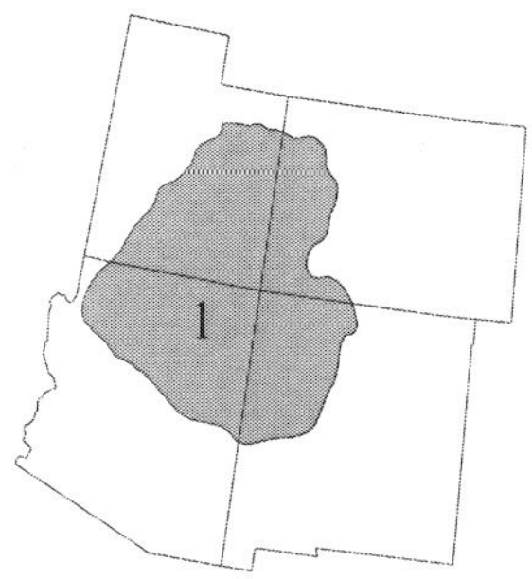

6 Energy

Region 1

fluvial • see outwash plain: large sandy flats created by sediment-laden water deposited when a glacier melts.

Jurassic • the geologic time period lasting from 201 to 145 million years ago.

aeolian • pertaining to, caused by, or carried by the wind.

cross-bedding • layering within a bed in a series of rock strata that does not run parallel to the plane of stratification.

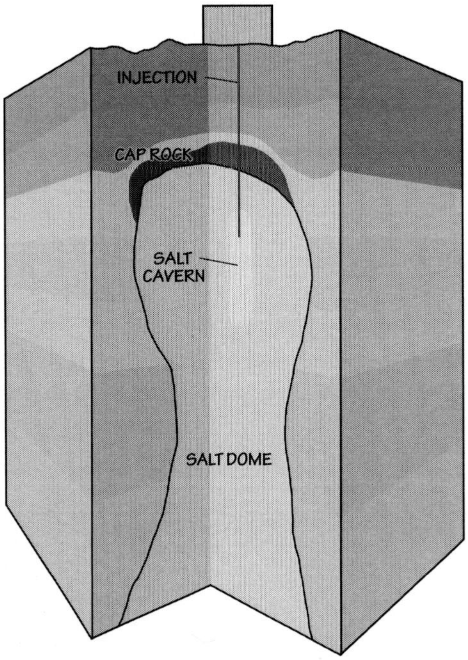

Figure 6.8: Solution mining is used to create a storage cavern inside a salt dome.

The San Juan Basin of northwest New Mexico and southwest Colorado contains important reservoir rocks for oil and gas, primarily late Cretaceous marine sandstones associated with the Western Interior Seaway. The basin's reservoir rocks also include late Cretaceous **fluvial** sandstones and coals associated with the filling of the Seaway, **Jurassic aeolian** sandstones, and limestones and sandstones in the shallow marine Pennsylvanian-age Paradox Formation. Source rocks in the San Juan Basin are equally varied, including organic-rich marine shales from the Paradox Formation and the Mancos and Lewis shales of the Western Interior Seaway.

The Piceance Basin, located in northwest Colorado, and the Uinta Basin, located in northeast Utah, also contain a wide variety of fossil fuel-bearing deposits dating from the late Paleozoic to the early Cenozoic. These two basins share much of their geological history—they are effectively part of the same east-west trending basin that developed across the Utah-Colorado state boundary during the late Cretaceous. The Rangely oil field, near the boundary between the Uinta and Piceance basins, has been producing oil since the 1940s. Oil was discovered at the site in 1901, but the field was not developed until after World War II due to its remoteness. Since development began there, Rangely has produced nearly 800 million barrels of oil, and it was the most prolific oil field in Colorado for many years. The oil at Rangely comes primarily from the late Pennsylvanian and early Permian Weber Sandstone, which includes permeable and porous **cross-bedded** aeolian dune deposits. The source for this oil is probably Carboniferous strata deep in the Uinta Basin, or the shales of the early Permian Phosphoria Formation. Some oil has also been produced from the late Cretaceous Mancos Shale, deposited as part of the Western Interior Seaway. Today, carbon dioxide (CO_2) is injected into the field to increase pressure in the depleted reservoir and enable more oil extraction.

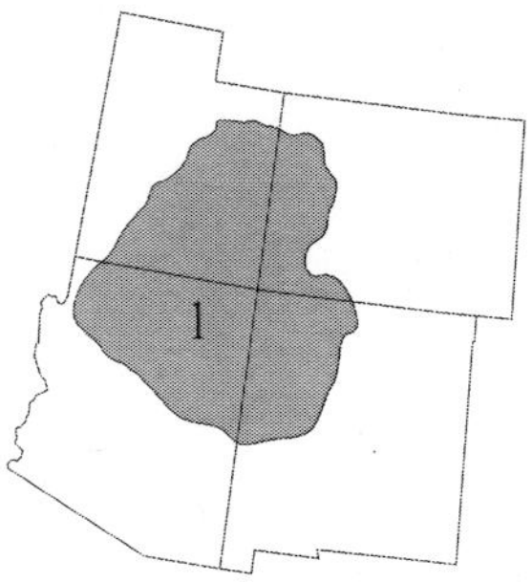

Energy 6

Region 1

Salt Domes

Rock salt (the mineral halite) is solid and impermeable, but when it is under very high pressure it can flow like a thick liquid. When a layer of salt is buried under thousands of meters (feet) of overlying sediment, it will start to deform. Because it is less *dense* than the rocks above it, it flows upward toward areas of lower pressure, forming geological structures named for their shapes (e.g., domes, canopies, tables, and lenses). Salt domes are extremely common geologic features in the Colorado Plateau, and their origin lies in the late Paleozoic, when salt was deposited through the evaporation of shallow seas. Today, these salt-bearing rocks are up to 3000 meters (10,000 feet) thick in some areas, and are overlain by as much as 900 meters (3000 feet) of sedimentary rock.

As salt structures grow, they in turn influence the topography of the surrounding landscape, creating zones of uplift surrounding areas of subsidence, fractures, and faults. When salt flows upward, it deforms the surrounding strata, creating gaps in which oil and gas may pool and be trapped. Oil and gas also accumulate under and along the salt structures; salt domes have led to some of the most prolific oil reservoirs in the US. In addition, due to their inherent impermeability, the salt domes themselves are often solution-mined (by pumping water underground to dissolve the salt) to create caverns that have been used to store petroleum, gas, and even chemical waste.

density • a physical property of minerals, describing the mineral's mass per volume.

kerogen • an immature, waxy, solid organic material that must be artificially heated to convert it into synthetic oil.

The Uinta and Piceance basins contain the largest oil shale deposits in the US. These are part of the Green River Formation, where Eocene lacustrine deposits overlie Cretaceous marine sediments from the Western Interior Seaway. These Colorado and Utah oil shales formed during the greatest extent of Lake Uinta in the Eocene, at a time of relatively high salinity in the lake and high algal productivity. Within the Uinta Basin, some organic-rich Green River shales have been buried deeply enough to generate large quantities of **kerogen** (a form sometimes described as "waxy oil"). Some of this kerogen migrated and was

See Chapter 3: Fossils to learn more about the exceptionally preserved fossils found in the Green River Formation.

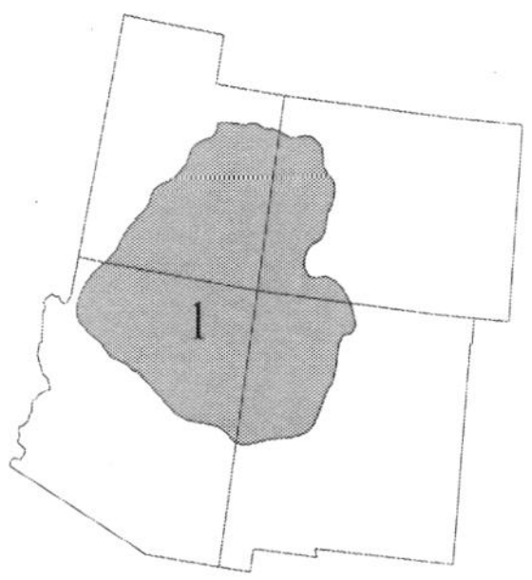

6 Energy

Region 1

derrick • a lifting device in the form of a framework steel tower that is built over a deep drill hole, typically an oil well.

helium • a gaseous chemical element (He), which is the second most abundant and second lightest element in the universe.

peat • an accumulation of partially decayed plant matter.

subsidence • the sinking of an area of the land surface.

alluvial • a layer of river-deposited sediment.

Paleocene • a geologic time interval spanning from about 66 to 56 million years ago.

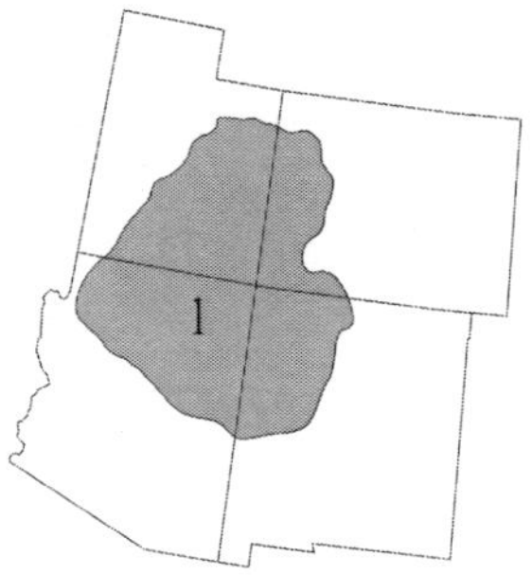

How does oil drilling work?

Once an oil trap or reservoir rock has been detected on land, oil crews will prepare a broad, flat pad for equipment and supplies around the area where the well will be drilled. Once the initial site is prepared, an apparatus called a drilling rig is set up. The rig is a complex piece of machinery designed to drill through rock to a predetermined depth. A typical drilling rig usually contains generators to power the system, motors and hoists to lift the rotary drill, and circulation systems to remove rock from the borehole and lubricate the drill bit with mud. It also contains high-pressure blowout prevention equipment to prevent pressurized oil or gas from rising uncontrollably to the surface after being tapped. The support structure used to hold the drilling apparatus is called a *derrick*. In the early days of oil exploration, drilling rigs were semi-permanent structures and derricks were left onsite after the wells were completed. Today, however, most rigs are mobile and can be moved from well to well. Once the well has been drilled to a depth just above the oil reservoir, steel casing is cemented into the well to structurally reinforce it and prevent leakage of petroleum into shallower aquifers. Once the casing is set and sealed, oil is then allowed to flow into the well, the rig is removed, and production equipment can be put in place to extract the oil. The site is then "reclaimed" (for example, restored to the original habitat), leaving a small area for access to the well and storage tanks.

trapped in porous sandy beds of the Wasatch and Green River formations, while some is still found in its organic-rich Green River source rock. Because the cost of processing oil shale into usable hydrocarbons is so high relative to other forms of producing oil, oil shales have remained mostly untapped.

Gas is produced in the Colorado Plateau as a byproduct of oil extraction, and from reservoirs in porous rock. In the Paradox Basin, gas is extracted from Pennsylvanian and Permian sandstones; **helium** is also a byproduct of gas extraction from the basin's Lisbon Field. The sands of the Cretaceous-aged Mesaverde Group in the Piceance Basin have also yielded substantial volumes of gas. Coals in the Williams Fork Formation of the Mesaverde Group have been a major source of that gas, but the Mancos and Niobrara strata that underlie the northern part of the Piceance Basin may also contribute to gas production there.

Energy 6

Region 1

Oil Shale or Shale Oil?

It is unfortunate that two terms sounding as similar as "shale oil" and "oil shale" are actually quite different kinds of fossil fuel resources. Oil shale is rock that contains an immature, waxy, solid organic material known as kerogen (confusingly, it is not actually oil). Kerogen must be artificially heated to convert it into synthetic oil or a hydrocarbon gas. Thus, the whole rock layer, which may or may not technically be shale, must be mined and/or processed (possibly *in situ*) to produce synthetic oil. In contrast, shale oil is mature oil trapped in the original shale rock in which it formed. In this case, the source rock is also the reservoir rock, because it is so impermeable that the oil never escaped. This type of rock may be fractured (e.g., by hydraulic fracturing) to provide pathways for the oil to escape—ultimately into a drill bore.

intermontane • *between or among mountains.*

bituminous coal • *a relatively soft coal containing a tarlike substance called bitumen, which is usually formed as a result of high pressure on lignite.*

coalification • *the process by which coal is formed from plant materials through compression and heating over long periods of time.*

aquifer • *a water-bearing formation of gravel, permeable rock, or sand that is capable of providing water, in usable quantities, to springs or wells.*

Coal

Billion-ton coal deposits formed in the Colorado Plateau during the late Cretaceous, as the Western Interior Seaway grew and shrank with sea level and local tectonic changes. North American Cretaceous coals are nearly as extensive as those of Carboniferous age, largely due to the widespread extent and long history of the Western Interior Seaway. The thickest coals accumulated as **peat** in coastal swamps while the basins in the area **subsided**, and others accumulated farther inland on **alluvial** plains. Some later **Paleocene** coals were also deposited in **intermontane** basins. These Cretaceous and Paleocene deposits are typically high-volatile (low-grade) **bituminous coals**. Most of the region's coals are deeply buried toward the centers of the major basins, and crop out around the exposed edges of the Piceance Basin in Colorado, the Kaiparowits and Uinta basins in Utah, the Black Mesa Basin in Arizona, and the Raton and San Juan basins in Colorado and New Mexico (*see Figure 6.3*). Coal mining has continued for over 115 years in some of these basins.

Since approximately 1980, large reserves of natural gas have been exploited in tandem with coal seams. This gas, called coalbed methane, is a byproduct of the process of **coalification**, and it accounts for over 5% of US methane production. Coal seams have long been vented to protect underground coal miners from potentially explosive build-ups of methane (CH_4, the primary gas in "natural gas") released from fissures around the coal. While long considered primarily a hazard to be mitigated in subsurface mines, methods have been developed to trap the methane as an additional energy source. In deep subsurface coal seams, water saturates fractures (or cleats) in the seam, making the seam an **aquifer** (which in some places may be clean enough to be part of the local

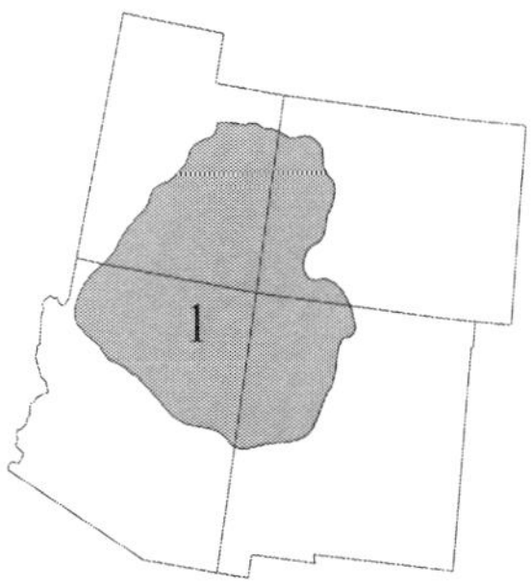

6 Energy

Region 1

compression • *flattening or squeezing as a result of forces acting on an object from all or most directions.*

anthracite • *a dense, shiny coal that has a high carbon content and little volatile matter.*

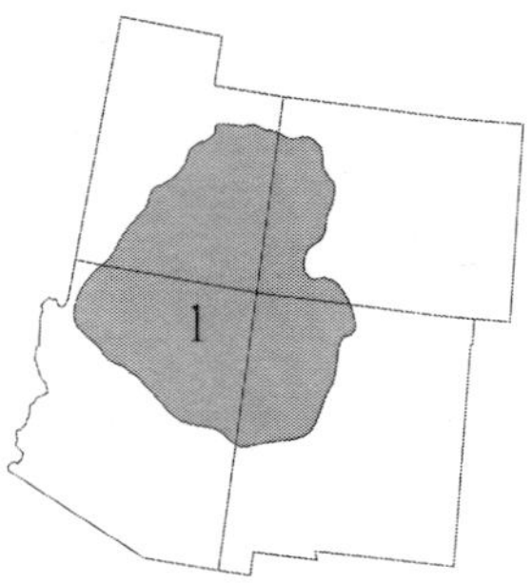

Coal

As leaves and wood become more deeply buried, pressure on them builds from overlying sediments, squeezing and *compressing* them into coal. The coal becomes gradually more enriched in carbon as water and other components are squeezed out: peat becomes lignite, bituminous, and eventually *anthracite* coal, which contains up to 95% carbon. Anthracite has the fewest pollutants of the four types of coal, because it has the highest amount of pure carbon. By the time a peat bed has been turned into a layer of anthracite, the layer is one-tenth its original thickness.

The Carboniferous period takes its name from the carbon in coal. Globally, a remarkable amount of today's coal formed from the plants of the Carboniferous, which included thick forests of trees with woody vascular tissues.

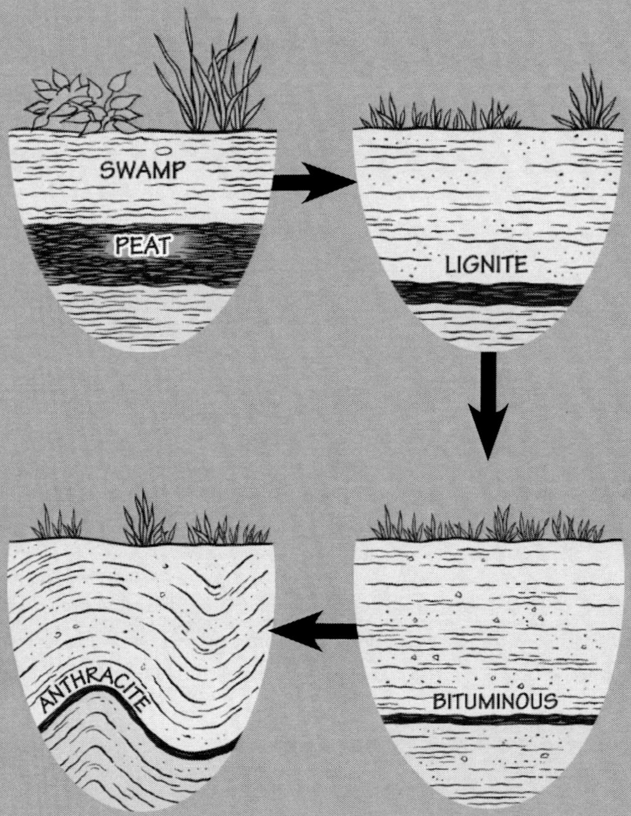

water supply). If there is sufficient water pressure, methane present within the coal fractures may be trapped in the coal. To extract this methane, water can be removed via wells, thereby reducing pressure and allowing methane to escape toward lower pressures along the well bore (*Figure 6.9*). Methane is then separated from the water. After the water is removed it may take some years for the aquifer to be recharged, that is, refilled with water from rain at the surface that filters down to the aquifer. Coal bed methane accounts for over 5% of US methane production. Production rates for coalbed methane climbed steeply beginning in the early 1990s, though in recent years they have decreased both in absolute and relative quantity as shale gas production has increased. Some subsurface coals on the Colorado Plateau are significant sources of coalbed methane. For example, gas from the Fruitland Formation in the San Juan Basin is among the largest non-shale gas sources of natural gas in the US, and is by far the largest coalbed methane source in the country. Another notable coalbed gas-producing area is the Ferron Sandstone Member of the Mancos Shale, in the southwestern part of the Uinta Basin.

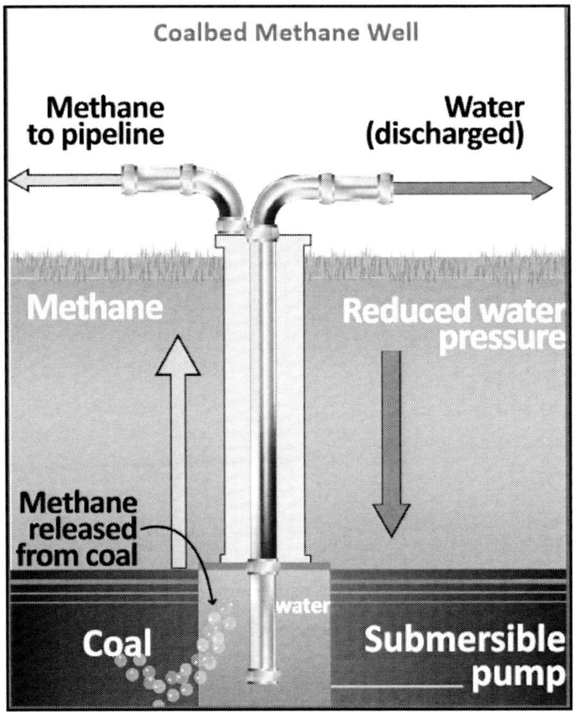

Figure 6.9: Coalbed methane produc-tion involves using water or other fluids to reduce pressure on the coal seam by creating a crack through which the methane can escape into a well.

Alternative Energy

The Paradox Basin contains uranium in the **Triassic** and Jurassic fluvial sandstones of the Chinle and Morrison formations. Other deposits include **breccia** pipes in northwest Arizona, which are sediment-filled cavities in ancient **karst** terrain. Uranium was actively mined in the second half of the 20th century as a source fuel for nuclear energy, though mining has significantly declined since (in part due to more stringent environmental regulations). Uranium mining and processing in the Four Corners area has continued irregularly in the late

Region 1

Triassic *• a geologic time period that spans from 252 to 201 million years ago.*

breccia *• a pyroclastic rock composed of volcanic fragments from an explosive eruption.*

karst topography *• a kind of landscape defined by bedrock that has been weathered by dissolution in water, forming features like sinkholes, caves, and cliffs.*

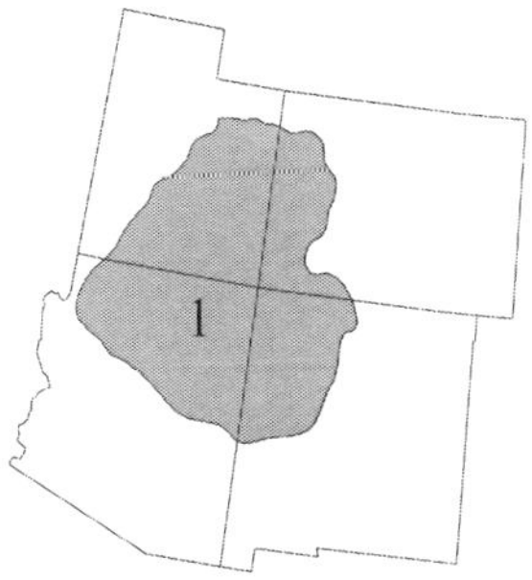

6 Energy

Region 1

20th and early 21st centuries, influenced primarily by uranium prices. Almost all uranium mined in the Southwest is exported; there are no nuclear power plants in the Colorado Plateau.

See Chapter 5: Mineral Resources for more information about uranium mining in the Southwest.

The Colorado River and its tributaries provide the Southwest with the potential for hydroelectric power (*Figure 6.10*), which uses the gravitational force of falling or rushing water to rotate turbines that convert the water's force into energy. There are also several pumped storage facilities within the region, where water is pumped uphill into reservoirs in times of excess production, essentially acting as batteries. Many of the hydroelectric plants scattered through the Colorado Plateau are associated with the topographic drop along the region's rim. The most notable is the Glen Canyon Dam (*Figure 6.11*) near the Arizona-Utah state line, which provides an average of 3.4 billion kWh per year. The massive concrete dam was completed in 1966 and retains Lake Powell, the second largest reservoir in the United States, with a capacity of 30 billion cubic meters (1 trillion cubic feet).

See Chapter 4: Topography to learn more about the Colorado River and its tributaries.

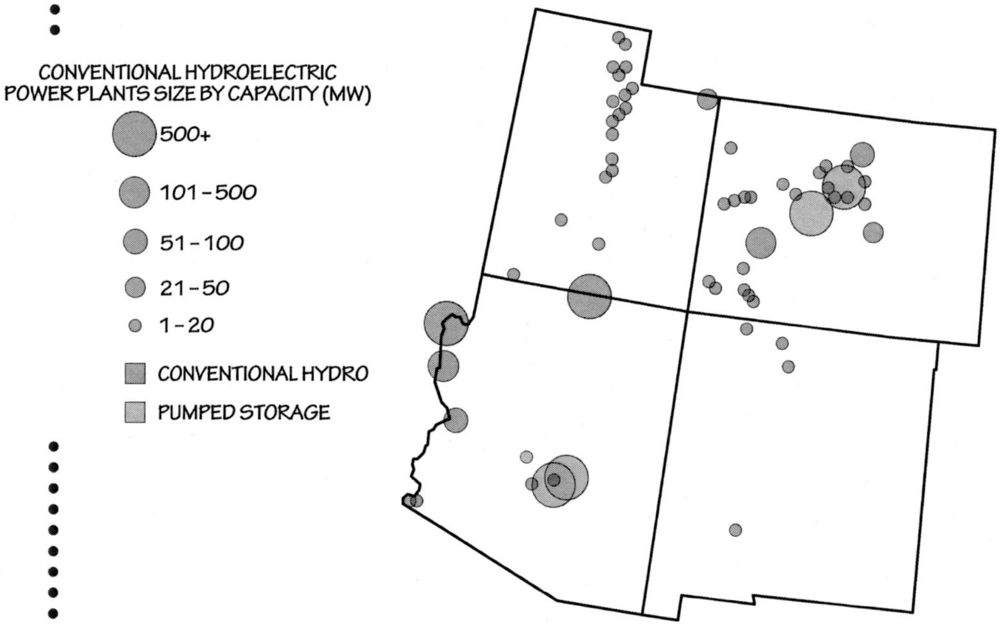

Figure 6.10: Hydroelectric power plants in the Southwestern US. (See TFG website for a full-color version.)

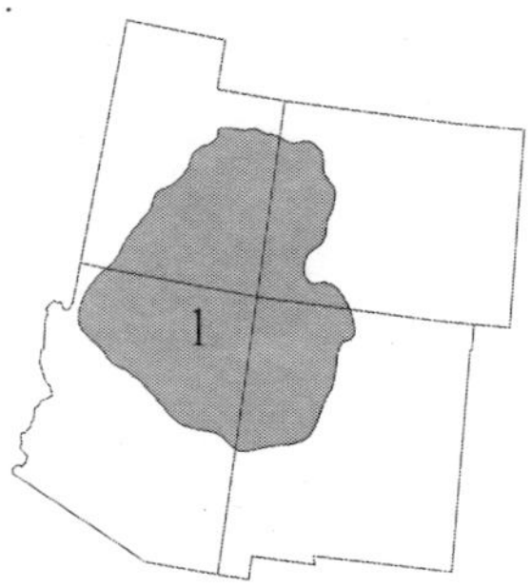

Figure 6.11: The Glen Canyon Dam in Coconino County, Arizona, with Lake Powell in the distance. The concrete arch of the dam is 220 meters (710 feet) high and 48 meters (1560 feet) wide.

The Colorado Plateau has only a fragmented capacity for wind energy (*see Figure 6.22*), but it does host a few wind farms. The Dry Lake Wind Power Project in Navajo County, Arizona was the state's first utility scale wind farm, and generates approximately 127 MW of power. Perrin Ranch Wind Farm in Coconino County, Arizona, also generates approximately 100 MW. A few smaller plants are found in the Colorado Plateau of Utah and New Mexico.

Energy of the Basin and Range
Region 2

rift basin *• a topographic depression caused by subsidence within a rift.*

The Basin and Range produces nuclear energy and has significant potential for both geothermal energy and solar energy, which has been expanding rapidly. The region also develops energy from the Colorado River, which runs along the western boundary of Arizona and Nevada. Due largely to complex active tectonics, much of the Basin and Range region has poor prospects for fossil fuel production. In southern Arizona and New Mexico, the Pedregosa Basin—and the early Cretaceous **rift basins** that have been superimposed upon it—contains a thick sequence of upper Paleozoic rocks, including carbonates and black shales. The rocks in this basin have not been found to have commercially significant petroleum or coal deposits.

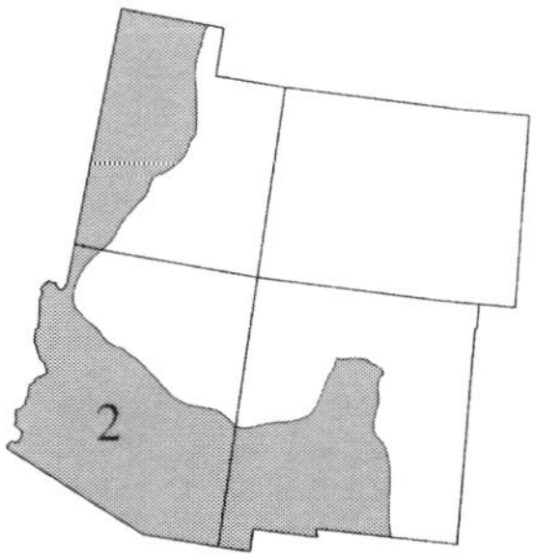

Solar Power
Solar power in the Southwest has grown extremely quickly in recent years, thanks to the solar resources located in the Basin and Range. The Sonoran

mantle • *the layer of the Earth between the crust and core.*

convection • *the rise of buoyant material and the sinking of denser material.*

plate tectonics • *the process by which the plates of the Earth's crust move and interact with one another at their boundaries.*

Desert extends through southern Arizona and the Chihuahuan Desert from southern New Mexico into the southeastern tip of Arizona; both of these deserts make for prime solar energy territory (*Figure 6.12*). Between 2010 and 2015, Arizona increased its installed solar capacity from 110 MW to 2303 MW; today, it produces 13.3% of the country's solar power and is ranked second in the nation for solar electric generation (behind California). By contrast, New Mexico and Utah currently have only 365 and 255 MW of installed solar capacity, respectively. New Mexico expects to install another 1428 MW of solar capacity over the next five years, more than five times the amount installed over the past five years. While production of solar electric power has grown exponentially in Arizona and in the US as a whole, more than half of the country's solar electric production is found in California. The total US output of solar power is also still dwarfed by other sources. For example, in August 2015, five times more electric power was produced in Florida from burning natural gas than the entire country produced from solar power that month, and in Arizona, the explosive growth of solar production brought electric production from non-hydro renewables to just 3.8% of the state's total.

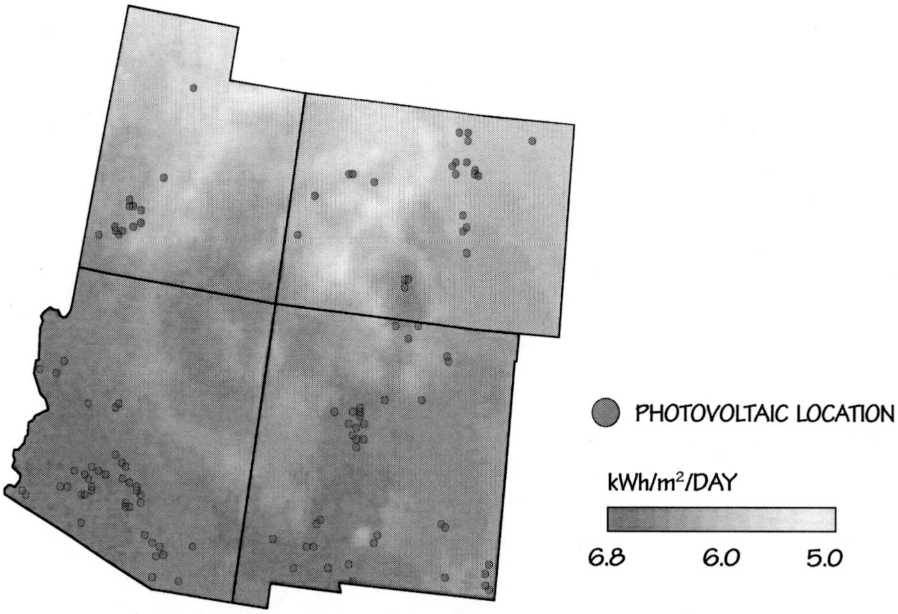

Figure 6.12: Locations of existing photovoltaic power plants overlaid on a map of the annual average solar resource for the Southwestern US. (See TFG website for a full-color version.)

The Basin and Range holds some of the world's largest photovoltaic power plants, though most of these large-scale plants are found in California or Nevada. The Agua Caliente solar project in Yuma County, Arizona currently has a 348 MW capacity, and a much larger project of 1.2 GW has been proposed just across the Arizona-California border in Needles. The Solana Generating Station near Gila Bend, Arizona is a 295 MW plant that uses parabolic mirrors to concentrate a large area of sunlight into a relatively small area (*Figure 6.13*). As

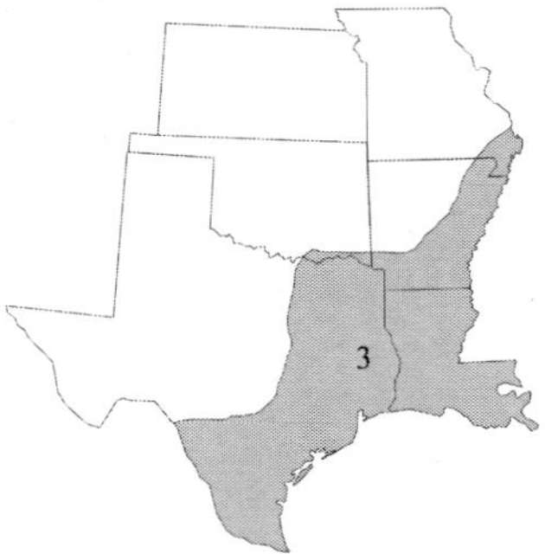

this concentrated light is converted to heat, it drives a turbine that is connected to a power generator. The Solana Generating Station was the first solar plant in the US to use molten salt as a means of storing thermal energy.

Figure 6.13: More than 3200 mirrored parabolic troughs are used to concentrate light and heat at the Solana Plant in Gila Bend, Arizona

Geothermal Energy

Geothermal energy comes from heat within the Earth, which is created on an ongoing basis by radioactivity. This energy powers **mantle convection** and **plate tectonics**. The highest-temperature conditions exist in tectonically active areas, including the Basin and Range, Iceland (a mid-Atlantic ridge), Japan (an area of **subduction**), and Hawaii and Yellowstone (areas with **hot spots**). Warm springs associated with tectonic activity in the Basin and Range have been enjoyed by the region's inhabitants for hundreds of years, beginning with Native Americans. This tectonic thermal energy is associated with a thinning of the **crust**, high heat flow relatively close to the surface, and groundwater that has been heated by cooling **intrusive volcanic** rocks. The heat of the Basin and Range has become the basis for both geothermal power plants and "direct use" operations (that is, use of geothermal energy at the site where it is generated). Typical examples of direct use include geothermally heated greenhouses, swimming pools, and buildings. Western Utah has several geothermal plants and many direct use facilities (*Figures 6.14 and 6.15*), and is the third leading producer of geothermal energy in the US (behind California and Nevada, where geothermal energy is also associated with the Basin and Range). New Mexico is also beginning to develop its geothermal resources.

See Chapter 1: Geologic History for more information about tectonic activity in the Basin and Range.

subduction • *the process by which one plate moves under another, sinking into the mantle.*

hot spot • *a volcanic region thought to be fed by underlying mantle that is anomalously hot compared with the mantle elsewhere.*

crust • *the uppermost, rigid outer layer of the Earth.*

intrusive rock • *a plutonic igneous rock formed when magma from within the Earth's crust escapes into overlying strata.*

volcanism • *the eruption of molten rock onto the surface of the crust.*

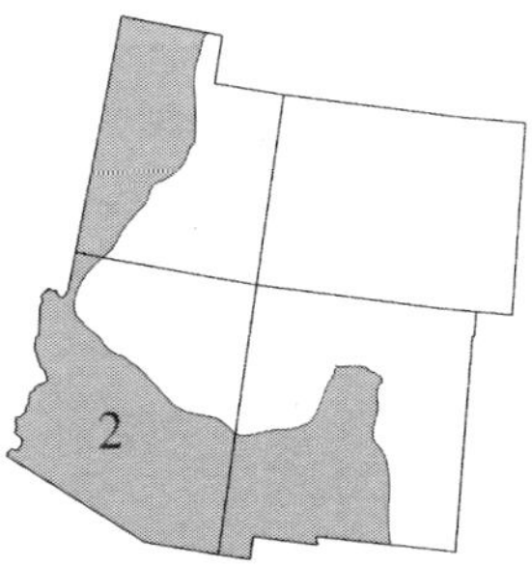

6 Energy

Region 2

Figure 6.14: Researchers from the University of Utah test a geothermal plant's wastewater injection flow in order to optimize steam production and improve the plant's capacity.

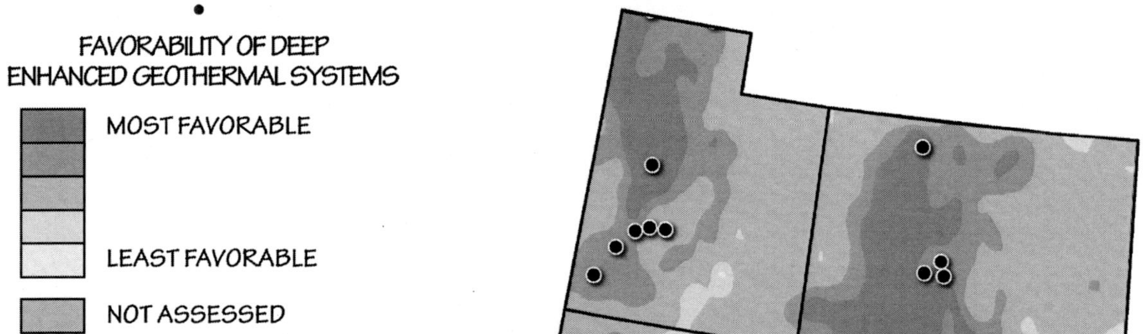

Figure 6.15: Geothermal energy resources in the Southwest.
(See TFG website for a full-color version.)

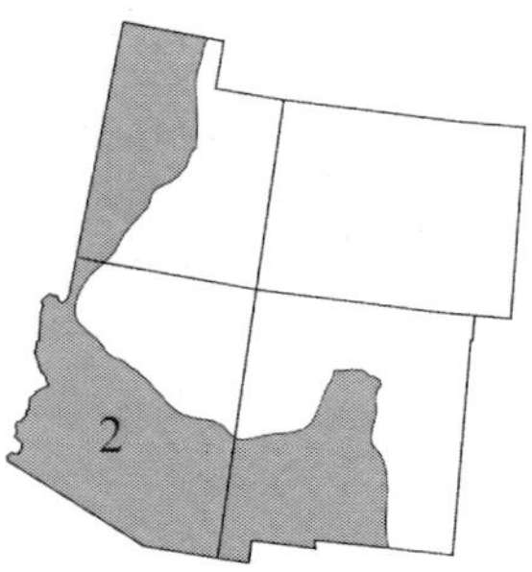

How does geothermal energy work?

Geothermal power stations use steam to power turbines that generate electricity. The steam is created either by tapping a source of heated groundwater or by injecting water deep into the Earth where it is heated to boiling. Pressurized steam is then piped back up to the power plant, where its force turns a turbine and generates power. Water that cycles through the power plant is injected back into the underground reservoir to preserve the resource.

There are three geothermal sources that can be used to create electricity. Geopressurized or dry steam power plants utilize an existing heated groundwater source, generally around 177°C (350°F) in temperature. Petrothermal or flash steam power plants are the most common type of geothermal plant in operation today, and they actively inject water to create steam. Binary cycle power plants are able to use a lower temperature geothermal reservoir by using the warm water to heat a liquid with a lower boiling point, such as butane. The liquid butane becomes steam, which is used to power the turbine.

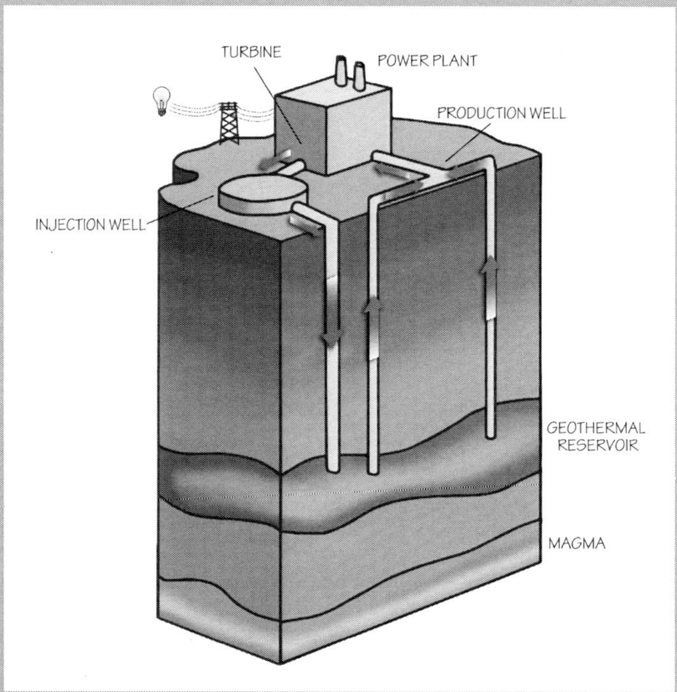

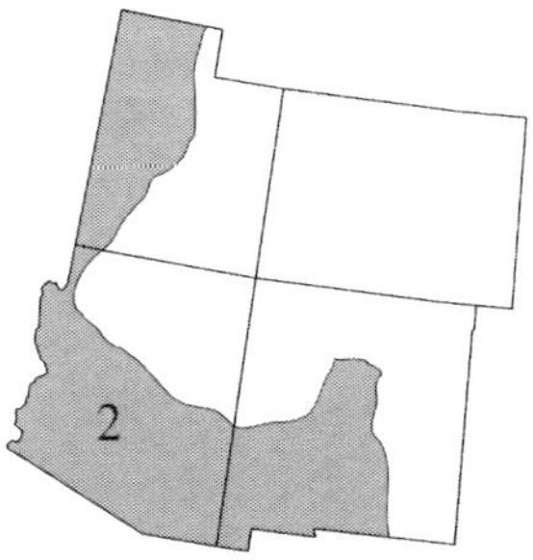

Other Alternative Energy

Arizona ranks 10th in the US for hydropower generation, with 13 hydro and pumped storage facilities producing approximately 663 gigawatt-hours (GWh) of power for the state's energy supply, though this only accounts for approximately 6% of the state's demand. Arizona's largest hydropower plant is the Hoover Dam, located on the Colorado River bordering Arizona and Nevada. When built in the 1930s it was the world's highest dam, largest electric-power generating station, and largest concrete structure. The plant's total power generating capacity is approximately 2 GW, and it is the sixth largest among US hydroelectric power stations. Approximately 20% of the Hoover's generated power goes to Arizona; another 25% goes to Nevada, and the rest is used by California.

The Palo Verde Nuclear Generating Station (*Figure 6.16*) near Tonopah, Arizona is the largest nuclear power facility in the US, producing approximately 3.3 GW of power annually and serving more than four million people. This plant is also unique in that it is the only large nuclear power plant in the world that does not use a nearby body of water for cooling; rather, it evaporates water from the treated sewage of nearby cities. The Palo Verde plant uses 76 billion liters (20 billion gallons) of evaporated treated sewage water per year.

Though space is available for wind farms, wind potential is relatively low in the Basin and Range, especially when compared to the Great Plains (*see Figure 6.22*). The only large-scale wind farm in the Basin and Range region

Figure 6.16: An aerial view of the three reactors at Arizona's Palo Verde Nuclear Generating Station. Each reactor has a generating capacity of 1.27 GW.

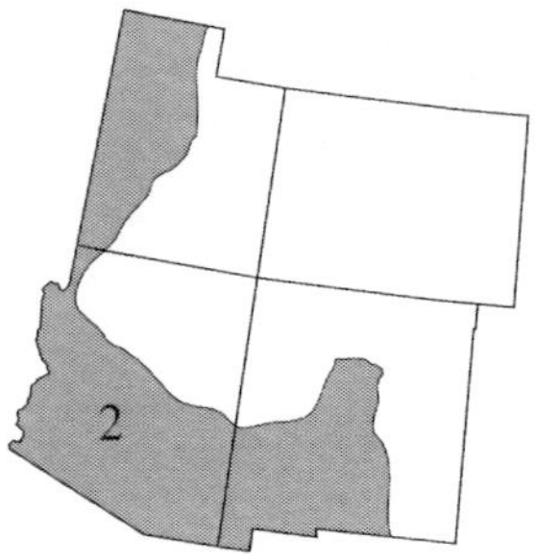

is the Milford Wind Corridor Project in southwestern Utah, with a generating capacity of approximately 300 MW. Other smaller-scale projects are scattered throughout Arizona and New Mexico (*Figure 6.17*).

Figure 6.17: The Macho Springs Wind Farm in Luna County, New Mexico has 28 turbines and produces approximately 50 MW of power.

Wind Energy and Landscape

Economically useful wind energy depends on steady high winds. Variation in wind speed is in large part influenced by the shape and elevation of the land surface. For example, higher elevations tend to have higher wind speeds, and flat areas can allow winds to pick up speed without interruption; thus high plateaus are especially appropriate for large wind farms. Since plateaus with low grass or no vegetation (or water bodies) have less wind friction than do areas of land with higher crops or forests, they facilitate higher winds. For all these reasons, the Great Plains region has high average wind speeds throughout its extent.

The Rockies and the Basin and Range, however, may have *locally* high wind speeds that can support strategically placed wind farms. For example, constricted valleys parallel to wind flow may funnel air into high velocities. Elevated ridges perpendicular to wind flow can also force fast winds across them. Thus, the wind velocities of these areas can vary geographically in quite complicated ways.

6 Energy

Region 3

Laramide Orogeny • *a period of mountain building that began in the Late Cretaceous, and is responsible for the formation of the Rocky Mountains.*

fossil • *preserved evidence of ancient life.*

delta • *a typically wedge-shaped deposit formed as sediment is eroded from mountains and transported by streams across lower elevations.*

marl • *a fine-grained sedimentary rock consisting of clay minerals, calcite and/or aragonite, and silt.*

Energy in the Rocky Mountains
Region 3

The high topography of the Rocky Mountains provides context both for hydroelectric power and wind energy. The same rugged peaks and valleys that contribute to localized high winds also make large-scale wind energy development difficult. The Rocky Mountains region is also known for coal, oil, and gas, in this case from large marine and freshwater sedimentary deposits in the Greater Green River Basin.

Oil and Gas

Petroleum resources are extracted in the Sand Wash Basin (*see Figure 6.6*), a southern lobe of the Greater Green River Basin (the bulk of which is located in Wyoming). The Greater Green River Basin is itself made up of several smaller basins and arches between them, formed during the **Laramide Orogeny** from the end of the Cretaceous period into the Eocene. The basin is known for its Eocene-aged surface rocks that contain both mineral and fossil fuel resources, along with its unusually well-preserved terrestrial **fossils** in the Green River Formation. Fossil fuels, thought to be derived from blue-green algae living in ancient lakes, are found in particularly thick sequences of Eocene oil shale. The Green River Formation hosts the world's largest known oil shale deposits.

The North Park Basin contains Paleozoic and Mesozoic strata, especially deposits laid down by the Western Interior Seaway. Oil and natural gas have long been extracted conventionally at the North and South McCallum oil fields, from the basin's Cretaceous-aged **deltaic** sandstones (*Figure 6.18*). In recent years, unconventional drilling of the late Cretaceous Niobrara Shale has drawn attention to the organic-rich calcareous shale and **marl** in the North Park and Sand Wash basins.

See Region 4: Great Plains later in this chapter to learn more about unconventional drilling methods.

Alternative Energy

Since the Rocky Mountains provide an abundance of water to lower areas in the east and west, hydroelectric power is substantial in this area (*see Figure 6.10*). The Colorado River and its tributaries, including the Gunnison River and the Uncompahgre River, provide the potential for much of the Rocky Mountains' hydropower. Over 20 plants produce more than 300 MW of energy for the region. Two large pumped storage stations, Cabin Creek (324 MW) and Mount Elbert (230 MW), are also located in the Colorado Rockies (*Figure 6.19*).

The Rocky Mountains region has some of the highest potential for wind energy in the US (*see Figure 6.22*), though the area's terrain and lack of infrastructure has made tapping into this resource challenging. There are currently no large-scale wind power projects in the Southwestern Rocky Mountains.

Figure 6.18: A natural gas drilling rig in the North Park Basin, Colorado.

Figure 6.19: The Mount Elbert pumped storage power plant in Twin Lakes, Colorado. The plant generates power from water originally pumped from Twin Lakes and Turquoise Lake.

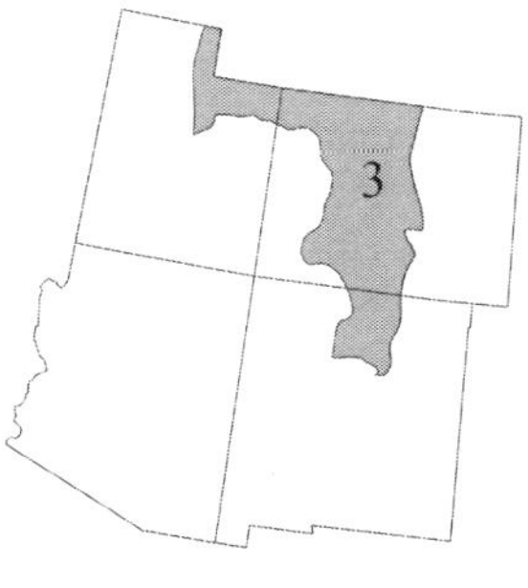

6 Energy

Region 4

> ***inland sea*** • *a shallow sea covering the central area of a continent during periods of high sea level.*

> ***chalk*** • *a soft, fine-grained, easily pulverized, white-to-grayish variety of limestone, composed of the shells of minute planktonic single-celled algae.*

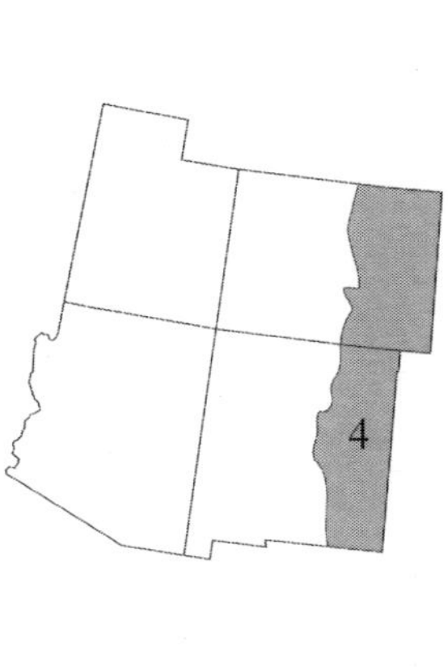

Energy in the Great Plains
Region 4

The Great Plains region is a broad expanse of flat land underlain by thick sequences of sedimentary rock and primarily covered in grassland and prairie. Ancient sedimentation patterns and tectonic activity have favored the placement of widespread fossil fuel resources in this region. Organic-rich sediments were deposited in **inland seas** that spread across much of the region, and Cenozoic swamps contributed plant matter to form thick beds of coal. The Great Plains' sedimentary basins contain vast oil, gas, and coal reserves that dominate energy production here (*see Figures 6.3 and 6.4*), but the area's topography and climate also make it favorable for large wind farms.

Oil and Gas

The Southwest is rich in fossil fuel resources, in part because of its history as an area of deposition. A variety of fine-grained organic-rich shales and coals, porous sandstones, and carbonate rocks—all excellent reservoirs and source rocks for fossil fuels—are found in the Great Plains' sedimentary basins. In this area, many vertically stacked reservoir rocks can be accessed through one well due to the large thicknesses of sedimentary rocks in these basins.

The Denver Basin is, in area, the largest of the basins in the Southwestern states, covering a large part of northeastern Colorado and the corners of Wyoming, Nebraska, and Kansas where they intersect Colorado. The Denver Basin contains thick sequences of Pennsylvanian through Paleocene rocks, as well as Cretaceous sediments from the Western Interior Seaway. Oil and gas have been produced from the basin since 1901, when oil was discovered in the Cretaceous Pierre Shale near Boulder (*Figure 6.20*). Just north of Denver, the Wattenberg Gas Field has been a major producer of natural gas since the 1970s, and has produced more than 113 billion cubic meters (4 trillion cubic feet) of natural gas. It is the ninth largest source of natural gas in the United States. Other well-known reservoir formations in the basin include the early Cretaceous Muddy Sandstone and the late Cretaceous Codell Sandstone.

Directly overlying the Codell Sandstone is the fine-grained Niobrara Formation. The Niobrara is formed of considerable **chalk** deposits, made up of the fine-grained carbonate skeletons of phytoplankton, and of shale weathered from mountains to the west. The Niobrara Formation was deposited in the deepest parts of the Western Interior Seaway, along its eastern margin; it is so widespread that it occurs in three of the four Southwestern regions in the Southwest (the Great Plains, Rockies, and Colorado Plateau) and in all the major Southwestern basins except the Permian Basin. The formation has been a major reservoir for oil and natural gas since the 1920s, and it has been drilled most intensively in the Denver Basin, particularly in Weld County. In the past decade, oil production rates in the Niobrara Formation have expanded enormously through the application of "unconventional" drilling, using horizontal drilling combined with high volume hydraulic fracturing. This method fractures rocks beneath the surface, releasing gas and oil trapped in source rocks that have very low permeability (also known as "tight" layers).

Figure 6.20: An oil pumpjack in the Pawnee National Grasslands, Colorado, part of the Denver Basin.

Hydraulic fracturing uses high volumes of water introduced at high pressure through horizontal wells along the source rock layer, to create thousands of tiny fractures (*Figure 6.21*). Most horizontal wells are drilled where the source rock is approximately 100–150 meters (330–490 feet) thick. The fractures are held open by small grains of sand carried by gel in the water, increasing its viscosity. A number of chemicals are added to the water to increase the recovery of fossil fuels, including a chemical to reduce friction as the mixture is introduced (thus the term "slickwater"). "Slickwater, high-volume hydraulic fracturing"—often shortened to "hydraulic fracturing" or simply "fracking"—has greatly increased the accessibility of available fossil fuel resources and the production rate of oil and gas. It has also been controversial, in part because of the use of large volumes of water, concerns about protection of shallow aquifers, the proper disposal of flow-back fluids, and associated environmental impacts. Unconventional drilling through low-permeability source rock layers has generated shale gas and shale oil booms around the US.

Another unconventional fossil fuel source in Colorado and Utah is Eocene-aged "oil shale" (as opposed to shale oil, *see box on p. 247*), in which the immature source rock is mined and heated to generate liquid hydrocarbons. This resource has been known and produced for many years elsewhere in the world, but has only recently been extracted in the Southwest because it is generally only economically viable to do so during times of high oil prices. Oil shale is more expensive to produce per unit of energy than oil and gas obtained through regular extraction, and has similar environmental consequences.

6 Energy

Region 4

dolostone • *a rock primarily composed of dolomite, a carbonate mineral.*

reef • *a feature lying beneath the surface of the water, which is a buildup of sediment or other material built by organisms, and which has positive relief from the sea floor.*

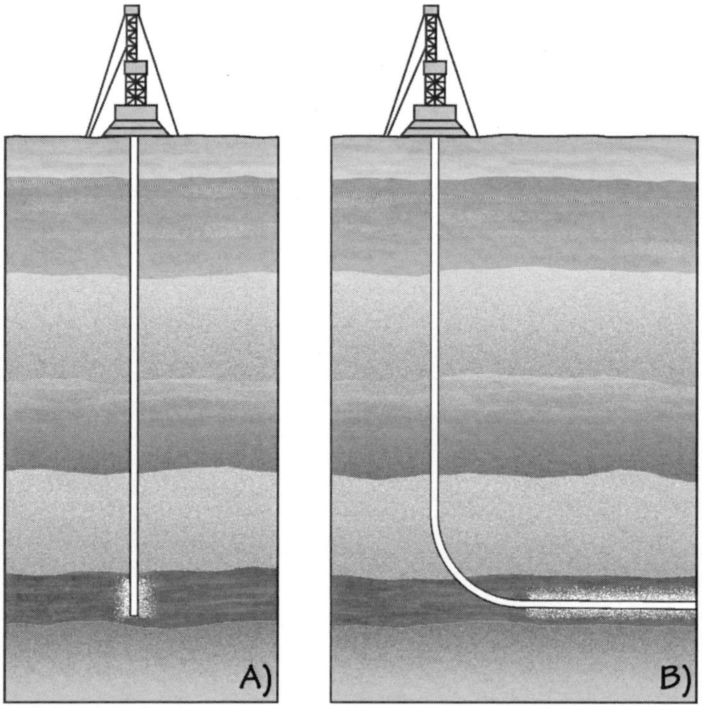

Figure 6.21: Oil wells (not to scale). A) A conventional vertical well. B) An unconventional horizontal well. Hydraulic fracturing may be carried out along horizontal wells running for 1.6 kilometers (1 mile) or more along layers with oil or gas trapped in pore spaces.

The Permian Basin, present partly in southeast New Mexico with a larger portion in western Texas, contains one of the world's thickest Permian rock sequences and accounts for nearly a fifth of US crude oil production. The "Greater" Permian Basin is composed of several sub-basins and other structural features: one of the two primary basins, the Delaware Basin, is present in southeast New Mexico (parts of Eddy and Lea Counties) and the western tip of Texas; the Midland Basin, located entirely in Texas, is the other. It contains a thick sequence of limestones and **dolostones** from **reef** and adjacent carbonate environments, together with evaporites, sandstones, and other reservoir structures that allowed hydrocarbons to accumulate. The Permian Basin's source rocks are largely organic-rich, shaly carbonates interbedded within the sequence. The first oil wells in the Permian Basin, in both New Mexico and Texas, were completed in the 1920s, and exploration in the 1940s led to significant production by the 1950s. In the past two decades, unconventional drilling in a number of carbonate-rich formations has increased oil production further. In New Mexico, new drilling has focused on Delaware Basin carbonates and a structure known as the Northwest Shelf, an area where shallow marine sediments accumulated just northwest of the Delaware Basin. Other formations of special interest include the early Permian Wolfcamp Formation, which occurs throughout the Permian Basin, as well as the more locally occurring early Permian Glorieta-Yeso, Abo-Yeso, and Bone Spring formations, and the late Permian Delaware Group.

See Chapter 3: Fossils to learn more about the ancient reefs of the Permian Basin.

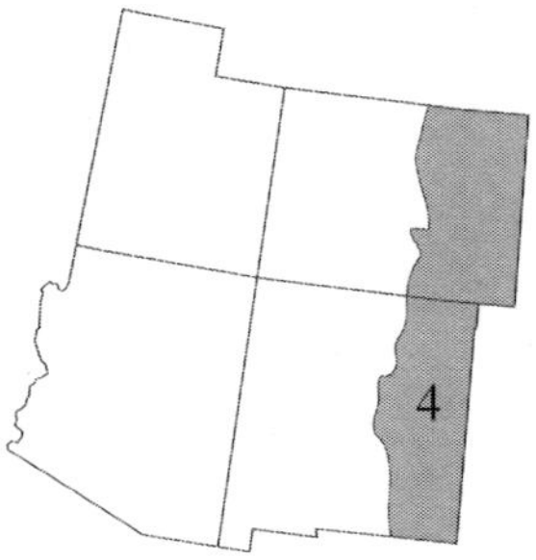

Coal

Significant amounts of coal are found in the Raton Basin, which lies along the boundary of Colorado and New Mexico. The basin contains a sequence of sediments that accumulated over a similar time interval to those of the Denver Basin. Cretaceous-aged coals in the Vermejo Formation formed along the Western Interior Seaway's deltas, and Cretaceous-Paleocene coals of the Raton Formation formed in swampy alluvial environments after the Western Interior Seaway retreated. Raton Basin coals have been mined since the 1870s. Coal mining declined substantially by the 1950s (Figure 6.22), but extraction of coalbed methane from the same formations began in the 1980s. For a time, coalbed methane from the Raton Basin became one of the largest sources of natural gas in the US, though it has since been eclipsed by the shale gas boom. Bituminous and **lignite** coals have historically been extracted from the Denver Basin, but coal mining ended there in 1979.

lignite • *a soft, brownish-black coal in which the alteration of plant matter has proceeded farther than in peat but not as far as in bituminous coal.*

Figure 6.22: Ovens in Cokedale, Colorado, used to produce a fuel called coke by baking bituminous coal. The coal mines and coke ovens of Cokedale ceased operation in 1946.

Alternative Energy

The Great Plains (in this case referring to the full area that runs from Texas to Montana and into Canada) has been called the "Saudi Arabia of Wind Energy," at least in terms of potential (*Figure 6.23*). Wind energy provides approximately a third of the renewable energy produced in the US, with hydroelectric representing approximately half; solar, geothermal, and biomass account for the remaining sixth. In contrast to hydroelectric, wind energy is growing rapidly: it grew tenfold on a national scale from 2004 to 2014, and wind farms on the Great Plains played a significant role in that growth. In the Southwest, the two Great Plains states are among the top 19 states for wind energy as a percentage of state electricity generation (Colorado 14%, New Mexico 6%). This is all the more remarkable considering the rate of local petroleum and coal extraction.

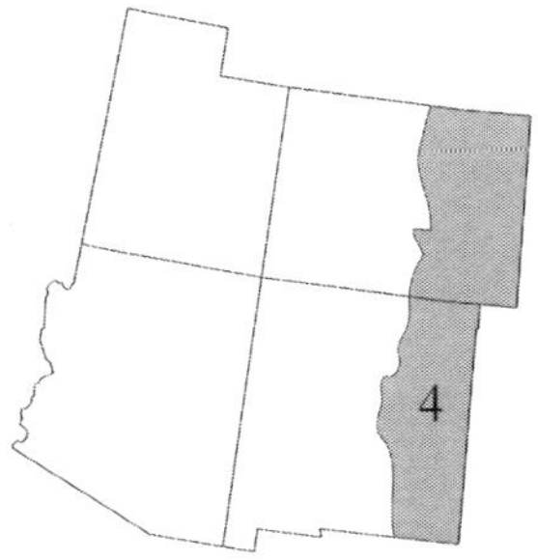

Region 4 - Facts

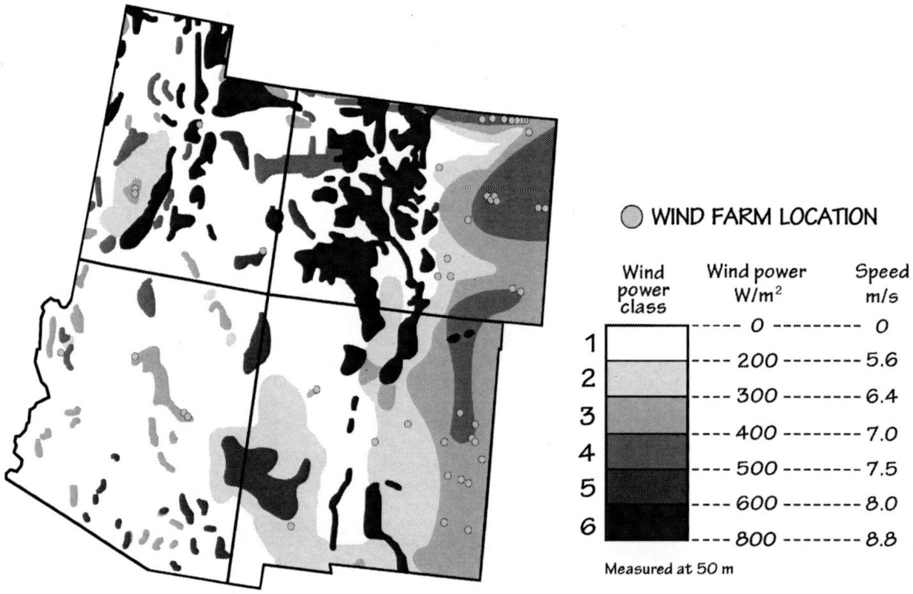

Figure 6.23: Wind energy potential in the Southwestern US, with locations of active wind farms. (See TFG website for a full-color version.)

Wind energy is strong and persistent on the high plains and the uplands of eastern Colorado and New Mexico. There are especially consistent high wind speeds in the upland area of northeastern Colorado, northeast of Denver, where over half-a-dozen wind farms of over 200 MW each have been built. The largest wind farm in the state, the Peetz Table Wind Energy Center, has a capacity of 430 MW. All together, the wind farms of Colorado's Great Plains make the state 10th in the nation for total wind energy. New Mexico's largest wind farm, the 250 MW capacity Roosevelt Wind Farm, is located on the Great Plains south of Clovis; the plant began operation in December 2015.

While the Great Plains also provide good opportunities for solar power capacity, this industry is still in the early stage of development in the region and only produces approximately 80 MW annually.

Energy Facts by State

Because of many local laws and guidelines, energy production and use is highly dictated by each state government. On the following pages, you will find a state by state assessment of energy use and production in the Southwestern US (from http://www.eia.gov/state/).

State Facts

Arizona

- Arizona's Palo Verde Nuclear Generating Station, rated at 3.937 net GW, is the largest nuclear power plant and the second largest power plant of any kind in the nation.

- Arizona's only operating coal mine, Kayenta, on the Navajo and Hopi reservations, supplies the 6.4 to 7.3 million metric tons (7 to 8 million tons) burned annually by the Navajo Generating Station's three 750 MW units.

- Approximately 25% of the energy consumed in Arizona homes is for air conditioning, which is more than four times the national average of 6%, according to EIA's Residential Energy Consumption Survey.

- Arizona, the 15th most populous state, ranked 44th in the nation in per capita energy consumption in 2013, partly because of the state's small industrial sector.

- Arizona ranked second in the nation in utility-scale electricity generation from solar energy in 2014.

- Arizona's Renewable Environmental Standard requires 15% of the state's electricity consumed in 2025 to come from renewable energy resources; in 2014, 8.9% of Arizona's net electricity generation came from renewable resources, primarily from the Glen Canyon and Hoover dams.

Arizona Energy Consumption Estimates, 2014

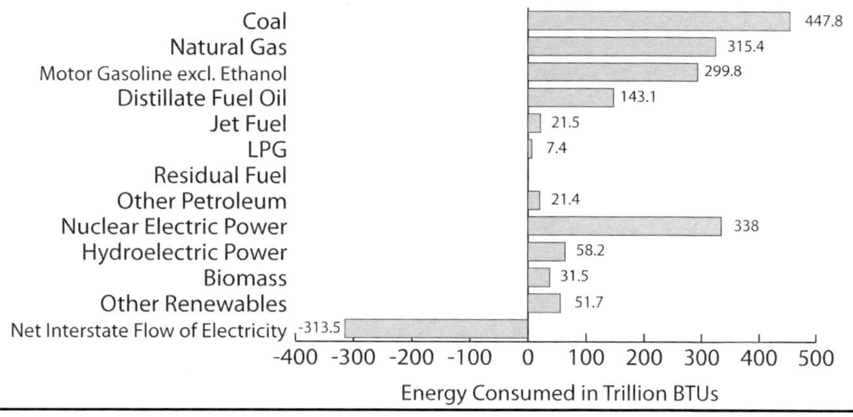

Arizona Energy Production Estimates, 2014

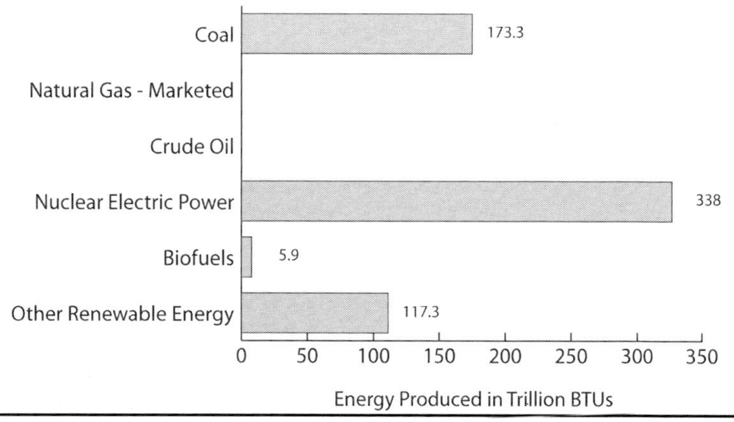

6 Energy

State Facts

Colorado

- Colorado's vast fossil fuel resources include the Niobrara Shale, with resource estimates running as high as two billion barrels of oil.
- Average household energy costs in Colorado ($1551 per year) are 23% less than the national average, primarily due to historically lower natural gas prices in the state, according to EIA's Residential Energy Consumption Survey.
- From 2004 to 2014, crude oil production in Colorado more than quadrupled; in the same period, marketed natural gas production rose 51%.
- In 2014, 60% of the electricity generated in Colorado came from coal, 22% from natural gas, and 18% from renewable energy resources.
- Colorado's Renewable Energy Standard requires investor-owned electric utilities to provide 30% of electricity sold from renewable energy sources by 2020, with 3% coming from distributed generation.
- In 2014, Colorado's grid-connected solar photovoltaic capacity of 430 MW was the ninth largest in the United States, and the state obtained nearly ten times as much net generation from solar power as it did just five years earlier in 2009.

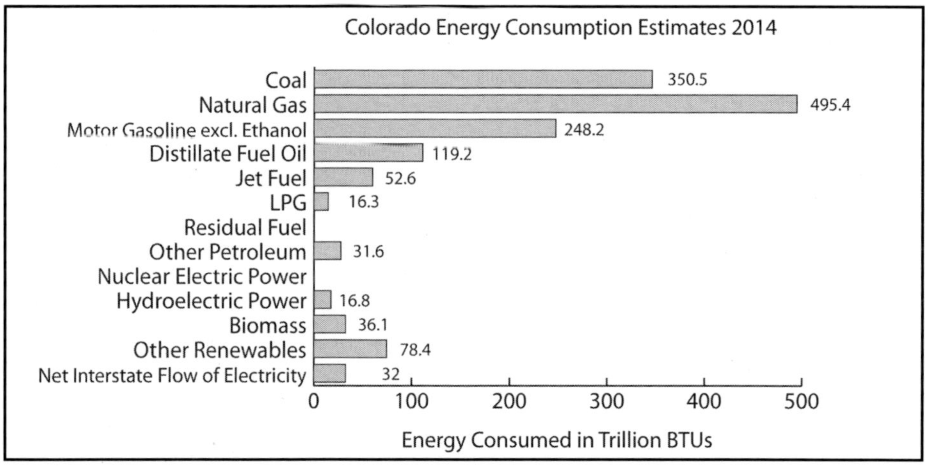

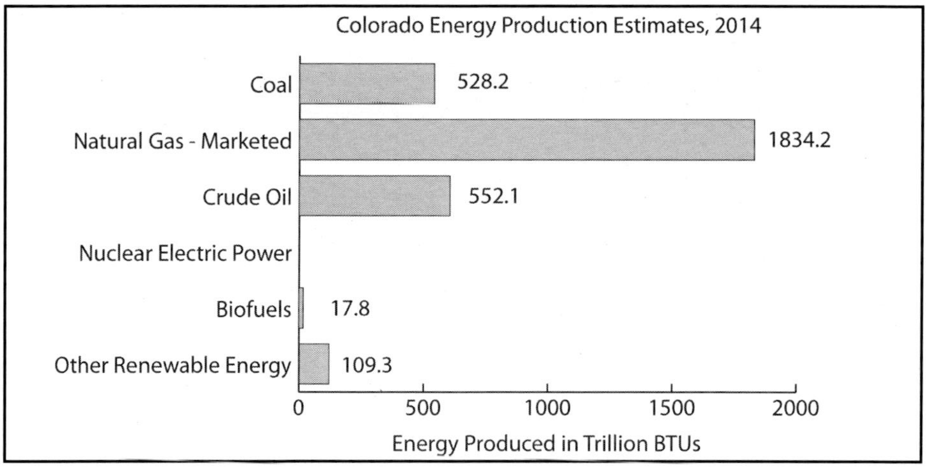

Energy 6

State Facts

New Mexico

- New Mexico has 26% of the nation's coalbed methane proved reserves, second only to Colorado in the United States.

- Excluding federal offshore areas, New Mexico ranked sixth in crude oil production in the nation in 2014.

- New Mexico's marketed production of natural gas accounted for 4.3% of U.S. marketed natural gas production in 2014, despite a decline in production of 30% from its 2001 peak.

- In 2014, New Mexico ranked sixth in the nation in utility-scale electricity generation from solar energy.

- New Mexico's Renewable Portfolio Standard requires that 20% of all electricity sold by investor-owned electric utilities, and 10% sold by cooperatives, come from renewable energy resources by 2020; in 2014, renewable energy supplied 9.3% of the electricity generated in the state.

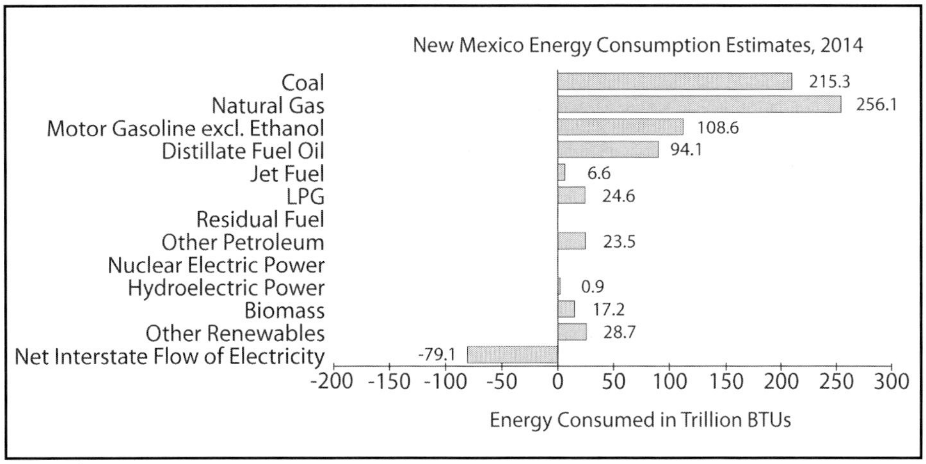

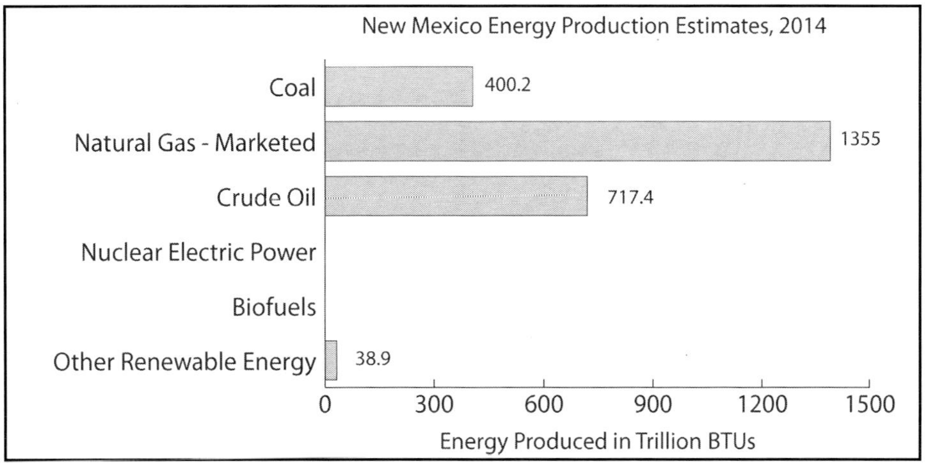

6 Energy

State Facts

Utah

- Utah's five refineries process crude oil primarily from Utah, Colorado, Wyoming, and Canada; the UNEV pipeline, opened in late 2011, is the first to connect Utah's refineries to Las Vegas, the largest city in Nevada.

- Utah produced 1.7% of U.S. coal in 2013 and shipped 27% of that production out of the state, of which nearly one-third was exported.

- In 2014, for the first time, coal produced only 76% of Utah's net electricity generation and natural gas produced 19%. State planners expect the natural gas share to continue rising as older coal units are shut down.

- Utah had the 10th lowest average electricity prices in the nation in 2014.

- Utah has a voluntary goal of using cost-effective eligible renewable energy resources to provide 20% of their 2025 adjusted retail electric sales; in 2014, 4.3% of net electricity generation came from renewable resources.

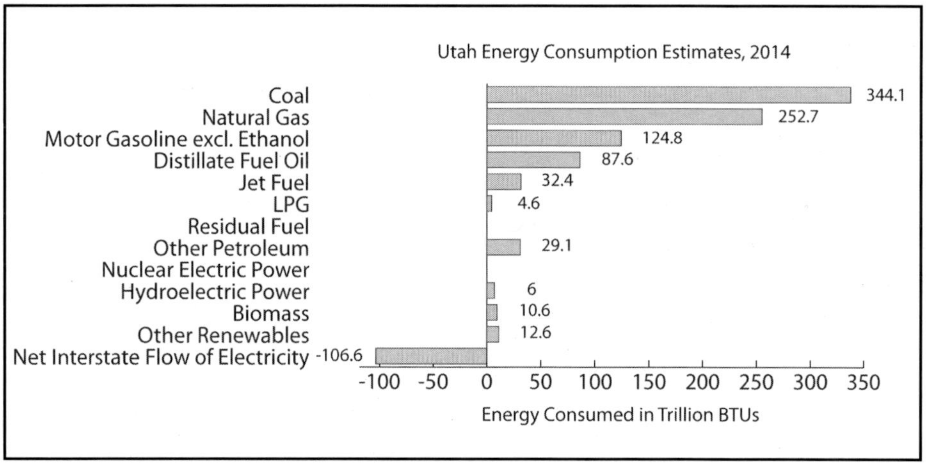

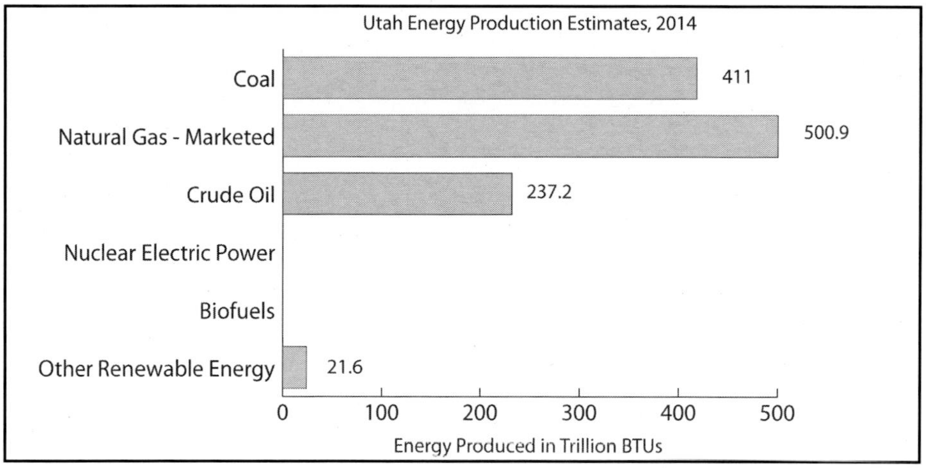

Energy 6

Energy and Climate Change

The Future of Energy in the US

Americans have come to rely on a diverse and abundant energy system, one that provides a continuous supply of energy with few interruptions. However, **climate change** is projected to play a big part in altering our supply, production, and demand for energy. Increases in temperatures will be accompanied by an increase in energy for cooling, while projected increases in the occurrence of **hurricanes**, floods, **tornados**, and other extreme **weather** events will continue to have a significant effect on the infrastructure of power grids and energy delivery systems. Drought and water shortages are already affecting energy production and supply. For example, in the Northeast, mild winter temperatures prior to the winter of 2013–2014 had decreased energy demands for heat, but they did not fully offset increased demands for cooling, and the regionally harsher winter of 2013–2014 saw increased demands for heating fuels. These types of disruptions affect us both locally and nationally, are diverse in nature, and will require equally diverse solutions.

See Chapter 8: Climate for more information about climate change and its effects on the Southwest.

Energy is a **commodity**, and supply and demand around the world will also affect the US energy system. As the global population grows, and industrialization of the world continues, demand for energy will increase even further as resources are depleted. These factors can significantly affect US energy costs through competition for imported and exported energy products. Reduction of our reliance on fossil fuels could have a huge positive impact on climate change. Unfortunately, there is no energy production system or source currently available that is truly **sustainable**. All forms of energy have negative impacts on the environment, as do many of the ways in which we use them.

Until we have a sustainable means of producing and delivering energy, we need to consider which means of energy production and transport make the least impact; we are faced with a sort of "energy triage." The answer to this problem will be multifaceted, depending in large part on which energy resources and delivery methods are available in each part of the US. The sources of energy that provide the least impact for the best price for people living in the Southwest are probably not the same as for those in other areas, such as the Midwest or Northeast.

Adaptation—changing our habits of energy use and delivery—can also make it easier for our existing energy infrastructure to adjust to the needs brought on by climate change. Investing in adaptation can pay off in the short term by reducing risks and vulnerabilities, thus minimizing future risks. Increasing sustainable energy practices (including harvesting and production) and improving infrastructure and delivery methods can go a long way toward not only decreasing the effects of climate change, but also our energy security.

Climate Change

climate change • *See global warming: the current increase in the average temperature worldwide, caused by the buildup of greenhouse gases in the atmosphere.*

hurricane • *a rapidly rotating storm system with heavy winds, a low-pressure center, and a spiral arrangement of thunderstorms.*

tornado • *a vertical funnel-shaped storm with a visible horizontal rotation.*

weather • *the measure of short-term conditions of the atmosphere such as temperature, wind speed, and humidity.*

commodity • *a good for which there is demand, but which is treated as equivalent across all markets, no matter who produces it.*

sustainable • *able to be maintained at a steady level without exhausting natural resources or causing severe ecological damage, as in a behavior or practice.*

efficiency • *the use of a relatively small amount of energy for a given task, purpose, or service; achieving a specific output with less energy input.*

Some of these changes are grounded in the development of new technologies for energy production and energy **efficiency**, while others may be related to changes in behavior. These changes in technology and behavior may go hand in hand; roughly 2% of electricity production now goes to data centers, for example—a use that did not exist in 1985. Additionally, the Internet is rapidly changing other ways we use energy, allowing us to telecommute and changing the way we shop.

In closing, some key points to keep in mind regarding the future of energy are:

Extreme weather events are affecting energy production and delivery facilities, causing supply disruptions of varying lengths and magnitudes and affecting other infrastructure that depends on energy supply. The frequency and intensity of extreme weather events are expected to increase.

1. Higher summer temperatures are likely to increase electricity use if homes are not adequately designed and constructed, causing higher summer peak loads, while warmer winters are likely to decrease energy demands for heating. Net energy use is projected to increase as rising demands for cooling outpace declining heating energy demands.

See Chapter 9: Earth Hazards to learn more about extreme weather events.

2. Both episodic and long-lasting changes in water availability will constrain different forms of energy production.

3. In the longer term, sea level rise will affect the coastal facilities and infrastructure on which many energy systems, markets, and consumers depend.

As we invest in new energy technologies, future energy systems will differ from those of the present in uncertain ways. Depending on the way in which our energy system changes, climate change will introduce both new risks and new opportunities.

Resources

General Books on Energy

Bird, K.J., 1989, North American fossil fuels, in: A. W. Bally and A. R. Palmer, eds.) *The Geology of North America: an Overview,* The Geology of North America, vol. A, Geological Society of America, Boulder, CO, pp. 555–574.

Duggan-Haas, D., R. M. Ross, and W. D. Allmon, 2013, *The Science Beneath the Surface: A Very Short Guide to the Marcellus Shale*, Paleontological Research Institution (Special Publication 43), Ithaca, NY, 252 pp.

Hinrichs, R., and M. H. Kleinbach, 2012, *Energy: Its Use and the Environment, 5th edition*, Thomson, Brooks/Cole, Belmont, CA, 640 pp.

Nye, D. E., 1998, *Consuming Power: a Social History of American Energies*, Massachusetts Institute of Technology Press, Cambridge, MA, 331 pp.

Richards, J., 2009, *Wind Energy*, Macmillan Library, South Yarra, Victoria, Canada, 32 pp. [For primary school age.]

Smil, V., 2006, *Energy: a Beginner's Guide*, Oneworld, Oxford, UK, 181 pp.

Smil, V., 2010, *Energy Myths and Realities: Bringing Science To the Energy Policy Debate*, AEI Press, Washington, DC, 213 pp.

Wohletz, K., and G. Heiken, 1992, *Volcanology and Geothermal Energy*, University of California Press, Berkeley, http://ark.cdlib.org/ark:/13030/ft6v19p151/.

General Websites on Energy

American Association of Petroleum Geology (AAPG), http://aapg.org.

American Wind Energy Association, Get the Facts, http://www.awea.org/Resources/Content.aspx?ItemNumber=5059.

Center for Climate and Energy Solutions, Renewable Energy, http://www.c2es.org/energy/source/renewables.

Climate Literacy & Energy Awareness Network (CLEAN). [A rich collection of resources for educators.] http://www.cleanet.org.

Coal Bed Methane, Montana State University Extension, http://waterquality.montana.edu/energy/cbm/.

Coalbed Methane Outreach Program (EPA), http://www.epa.gov/coalbed/faq.html.

Energy Literacy: Essential Principles and Fundamental Concepts for Energy Education, Energy.gov, http://www1.eere.energy.gov/education/energy_literacy.html.

Geotherm Energy...Power from the Depths. 1997, *Energy Efficiency and Renewable Energy Clearinghouse*, National Renewable Energy Laboratory (NREL) for the U.S. Department of Energy (DOE), 8 pp., http://www1.eere.energy.gov/library/pdfs/geothermal_energy_power_from_the_depths.pdf.

History of Energy Use in the United States, by Hobart King, Geology.com, http://geology.com/articles/history-of-energy-use/.

Renewable and Alternative Fuels, US Energy Information Administration, http://www.eia.gov/renewable/state/.

State-by-State CO_2 Emissions Data From Fossil Fuel Combustion, http://www.epa.gov/statelocalclimate/documents/pdf/CO2FFC_2011.pdf.

State-level Energy-related Carbon Dioxide Emissions, 2000–2011, Independent Statistics and Analysis, US Energy Information Administration (EIA), 2014, http://www.eia.gov/environment/emissions/state/analysis/pdf/stateanalysis.pdf.

US Department of Energy (DOE), http://energy.gov.

US Energy Information Administration (EIA). [A wealth of information on energy production and use in the United States.] http://www.eia.gov/.

US Energy Information Administration (EIA), by state, http://www.eia.gov/state/.

US Fuel Ethanol Plant Production Capacity (EIA), http://www.eia.gov/petroleum/ethanolcapacity/.

US Geological Survey Energy Resources Program, http://energy.usgs.gov/.

What is geothermal?, Geothermal Resources Council, http://www.geothermal.org/what.htmlhttp://www.geothermal.org/what.html.

Wind Energy Resource Atlas of the United States: Regional Summaries, http://rredc.nrel.gov/Wind/pubs/atlas/chp3.html.

6 Energy

Resources

Resources by State

Multistate Areas

Chidsey, T. C., Jr., 2016, Paleozoic shale-gas resources of the Colorado Plateau and eastern Great Basin, Utah—multiple frontier exploration opportunities, *Utah Geological Survey Bulletin* 136, 241 pp. [Technical] http://ugspub.nr.utah.gov/publications/bulletins/b-136/b-136.pdf.

Kirschbaum, M. A. and L. R. H. Biewick, 2000, A Summary of the Coal Deposits in the Colorado Plateau: Arizona, Colorado, New Mexico, and Utah, in: M. A. Kirschbaum, L. N. R. Roberts, and L. R. H. Biewick, eds., Geologic Assessment of Coal in the Colorado Plateau: Arizona, Colorado, New Mexico, and Utah, *US Geological Survey Professional Paper* 1625-B, 99 pp. [Technical] http://pubs.usgs.gov/pp/p1625b/Reports/Chapters/Chapter_B.pdf.

Nester, P., 2002, Is there oil in your backyard? The Southwest US, in: *The World of Oil: Oil in Your Backyard*, https://www.priweb.org/ed/pgws/backyard/sections/southrockies/southrockies1.html.

Southwest Energy Efficiency Project, http://www.swenergy.org/.

Thiel, A., 2013, *Climate Change Impacts on Hydropower in the Colorado River Basin*, Center for Water Policy, School of Freshwater Sciences, University of Wisconsin at Madison, 4 pp., http://uwm.edu/centerforwaterpolicy/wp-content/uploads/sites/170/2013/10/Colorado_Energy_Final.pdf.

What Every Westerner Should Know About Oil Shale, by P. Limerick, J. L. Hanson, and R. Rebhan, Center of the American West, University of Colorado at Boulder, http://centerwest.org/projects/energy/oil-shale/welcome-to-shale-country-3.

Arizona

Powering Arizona, http://arizonaexperience.org/innovate/powering-arizona.

Your Guide to Solar in Arizona, Arizona Solar Center, http://www.azsolarcenter.org/.

Colorado

Colorado Coal Field War Project, https://www.du.edu/ludlow/cfhist.html.

Colorado Energy Office, https://www.colorado.gov/energyoffice.

[Colorado] Energy Resources, Colorado Geological Survey, http://coloradogeologicalsurvey.org/energy-resources/.

Kirschbaum, M. A. and L. R. H., Biewick, A Summary of the Coal Deposits in the Colorado Plateau: Arizona, Colorado, New Mexico, and Utah, Chapter B of Geologic Assessment of Coal in the Colorado Plateau: Arizona, Colorado, New Mexico, and Utah, M. A. Kirschbaum, L. N. R. Roberts, and L. R. H. Biewick, eds., *US Geological Survey Professional Paper* 1625–B. [Technical] http://pubs.usgs.gov/pp/p1625b/Reports/Chapters/Chapter_B.pdf.

Scamehorn, H., 2002, *High Altitude Energy: a History of Fossil Fuels in Colorado*, University Press of Colorado, Boulder, 232 pp.

Weiner, C., 2014, *Oil and Gas Development in Colorado*, Colorado State University Extension, Fact Sheet 10.639, 4 pp., http://extension.colostate.edu/docs/pubs/consumer/10639.pdf.

New Mexico

Budzik, P. and J. Perrin, 2014, Six formations are responsible for surge in Permian Basin crude oil production, *Today in Energy*, July 9, 2014, US Energy Information Administration, https://www.eia.gov/todayinenergy/detail.cfm?id=17031.

Energy [of New Mexico], Natural Resource Programs, New Mexico Bureau of Geology & Mineral Resources, http://geoinfo.nmt.edu/resources/energy.html.

The Greatest Dam in the World: Building Hoover Dam, Teaching with Historic Places Lesson Plans, National Park Service, http://www.nps.gov/nr/twhp/wwwlps/lessons/140hooverdam/140hoover_dam.htm.

Utah

Magill, B., 2015, Tar Sands Mining Moves to Utah, *Scientific American*, http://www.scientificamerican.com/article/tar-sands-mining-moves-to-utah/.

Utah Geothermal Technologies Program, 2005, US Department of Energy, 2 pp., http://www.nrel.gov/docs/fy05osti/36552.pdf.

Utah State Rock: Coal, http://pioneer.utah.gov/research/utah_symbols/rock.html.

Utah Office of Energy, http://energy.utah.gov/.

Chapter 7: Soils of the Southwestern US

It's sometimes easy to take the soil beneath our feet for granted. Yet soil has always been with us—it is the foundation of our houses and roads, and from the soil comes our food, fiber, and paper. Soil is the interface between living earth and solid rock, between biology and geology. The engineer, the scientist, and the gardener may all look at the soil beneath them in different ways, but perhaps no one has a more integral relationship with soil than a farmer. The economic success of producing crops is intimately tied to the quality of the soil upon which those crops grow, and the most successful farmers are well versed in the science of their soil. Soils store and purify water, and they exchange gases with the **atmosphere**. They support agriculture and natural ecosystems and provide a grassy surface for our parks and fodder for our gardens. Everyone, everywhere, every day, depends upon the soil.

What is Soil?

Generally, **soil** refers to the top layer of earth—the loose surface of earth as distinguished from rock—where vegetation grows. The word is derived (through Old French) from the Latin *solum*, which means "floor" or "ground." Soil is one of the most important resources we have—the most basic resource upon which all terrestrial life depends. The Southwest has a wide variety of soils, and each type of soil has a story to tell of its origin.

Soils form from the top down, and typically reach a depth of about one meter (3.3 feet) at their more developed stages, although some can reach much deeper. Soils are composed of a mixture of two key ingredients. The first is plant litter, such as dead grasses, leaves, and fallen debris. Worms, bacteria, and fungi do the job of breaking these down into nutritious organic matter that helps soil to nourish future plant growth. The second important component of soil is sediment derived from the **weathering** of rock that is then transported by **wind**, water, or gravity. Both of these components influence the texture (*Figure 7.1*) and consistency of the soil, as well as the **minerals** available for consumption by plants.

All soils might seem alike, but there can be vast differences in soil properties even within small areas! A single acre may contain several different soil types, each with its own assets and drawbacks. Some types of soil are clayey or prone to flooding, while others are stable enough to be used as a foundation for buildings. The most identifiable physical properties of soils are texture, structure, and **color**, which provide the basis for distinguishing soil **horizons**. Texture refers to the percentage of sand, silt, and clay that makes up the soil. Soil textures have specific names, as indicated in *Figure 7.1*.

atmosphere • *a layer of gases surrounding a planet.*

weathering • *the breakdown of rocks by physical or chemical means.*

wind • *the movement of air from areas of high pressure to areas of low pressure.*

mineral • *a naturally occurring solid with a specific chemical composition and crystalline structure.*

color (soil) • *a physical property of soils, influenced by mineral content, the amount of organic material, and the amount of water it routinely holds.*

horizon • *a layer in the soil, usually parallel to the surface, which has physical characteristcs (usually color and texture) that are different from the layers above and below it.*

CHAPTER AUTHORS

Richard A. Kissel
Judith T. Parrish

7 Soils

Review

clay • the common name for a number of very fine-grained, earthy materials that become plastic (flow or change shape) when wet.

silt • fine granular sediment most commonly composed of quartz and feldspar crystals.

sand • rock material in the form of loose, rounded, or angular grains, and formed as a result of the weathering and decomposition of rocks.

loam • a soil containing equal amounts of clay, silt, and sand.

alluvial • a layer of river-deposited sediment.

biota • the organisms living in a given region, including plants, animals, fungi, protists, and bacteria.

parent material • the original geologic material from which soil formed.

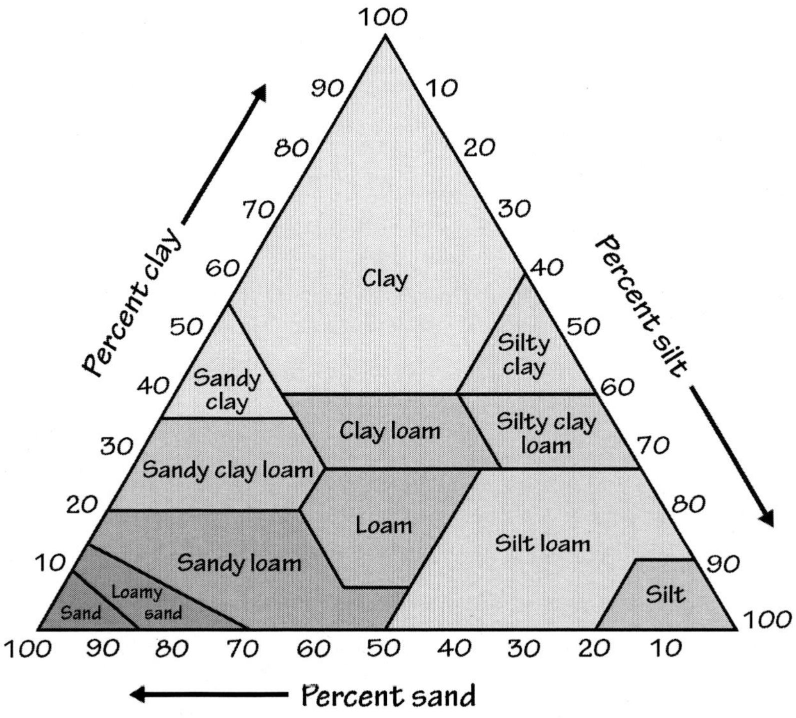

Figure 7.1: Soil texture triangle.

Generally, the best agricultural soils are those with about equal amounts of **clay**, **silt**, and **sand**. A soil of that type is called a **loam**. Soils that are mostly sand do not hold water very well and dry quickly, while soils with too much clay may never dry out. Soil structure refers to the way the soil forms clumps, known as **peds**. Peds are identified by the shape of the soil clods, which take the form of balls, blocks, columns, and plates. These structures are easiest to see in recently plowed fields, where the soil is often granular and loose or lumpy. Soil color is its most obvious physical property. The color is influenced by mineral content, the amount of organic material, and the amount of water it routinely holds. The colors are identified by a standard soil color chart called the Munsell chart.

Five main variables affect the characteristics of soil worldwide. In the Southwest, all soils are the products of subtle differences among these five factors:

1. *Parent material* is the original geologic material from which the soil formed. This can be bedrock, preexisting soils, or other materials such as **alluvium**, rock fragments, and windblown sediment.

2. *Climate* strongly determines the temperature regime, amount of moisture, and type of **biota** that interact with the **parent material**. This affects the extent of chemical and physical weathering on the soil-forming material. For example, if a particular **climate** lacks precipitation, mechanical weathering from wind or ice fracturing will predominate.

If, however, a climate has abundant precipitation, chemical **erosion** from water will be accelerated, resulting in substantial leaching.

3. *Topography*, or landscape, of the area is related to the relative position of the soil on the landscape. This includes the presence or absence of hills and the slopes between high and low areas. As the slope increases, water can carry larger sediment sizes, allowing for large sediment loads during major precipitation events. **Topography** also influences natural drainage. Gravity moves water down slopes to depressions or streams and pulls free water downward through the soil. Soils on hills tend to be dry, and soils in depressions and valleys are often wet or saturated. Areas with steep slopes that are susceptible to frequent erosion typically have very young soils, as they do not have long to develop before the ingredients are rearranged and the clock is reset. Flatter, more arid areas may have more time to develop, but they have significantly less plant life and will produce a very different soil than will a wetter environment. Slope also frequently determines the types of vegetation covering a soil—for example, different slopes on the same hill might receive varying amounts of sunlight during the growing season—which in turn can cause the characteristics of the soils to diverge if differing forms of vegetation dominate opposite slopes.

4. *Biota* or living organisms that live on or in the material affect soil development through their influence on the amount and distribution of organic matter in the soil. For example, plants contribute significantly to the formation of **humus**, and animals alter a soil's characteristics by leaving behind decayed remains and wastes. Decomposers like bacteria and fungi help to free up the nutrients locked away in these remains and wastes, and these freed nutrients are then recycled and used by new life forms within the same soil. In fact, more than 90% of the nutrients used by a forest in a given year are derived from the decomposition of old organic matter fallen to the forest floor. Animal burrows also create spaces in the soil horizons that allow for deeper penetration of air and water, which, in turn, aid plant development by helping to dissolve mineral nutrients into a form that plants can absorb and process. For its part, organic matter impacts the water-holding capacity of the soil, the soil's fertility, and root penetration.

5. *Time* is required for soils to develop while the four elements mentioned above interact. Older soils have deeper and thicker **subsoils** than do younger soils, but only if other soil forming factors remain constant. In northeastern Colorado, for example, it takes approximately 500 years to generate a new 2.5 centimeters (1 inch) of **topsoil** beneath the prairie grass—but it only takes a few years for erosion and weathering to destroy the same amount of unprotected topsoil.

Several types of **chemical reactions** are important for soil development; of these, acid-base reactions are some of the most important and complex. When carbon dioxide (CO_2) dissolves in water it forms weak carbonic acid. CO_2 found in soil water can come from the atmosphere, where it dissolves in rainwater.

Review

climate • *a description of the average temperature, range of temperature, humidity, precipitation, and other atmospheric/hydrospheric conditions a region experiences over a period of many years (usually more than 30).*

erosion • *the transport of weathered materials.*

topography • *the landscape of an area, including the presence or absence of hills and the slopes between high and low areas.*

humus • *a soil horizon containing organic matter.*

subsoil • *the layer of soil beneath the topsoil, composed of sand, silt and/or clay.*

topsoil • *the surface or upper layer of soil, as distinct from the subsoil, and usually containing organic matter.*

chemical reaction • *a process that involves changes in the structure and energy content of atoms, molecules, or ions but not their nuclei.*

Even more CO_2 usually comes from the soil itself, where it is produced by respiring organisms. The amount of CO_2 in soil gases can easily reach levels 10 times higher than the amount found in the atmosphere (over 4000 ppm in soil vs. 400 ppm in the atmosphere), making soil water potentially more acidic than rainwater. As this acidic water slowly reacts with fresh minerals, it buffers the soil's pH and keeps it in a range (6–8) preferred by many organisms. Acid-driven weathering breaks down the soil's primary **igneous** minerals, typically transforming them to **silica**-rich clays. As the soil's primary minerals are depleted, it loses the ability to buffer acidity, and the pH of highly weathered soil can drop to around 4. These weathered soils tend to be rich in **aluminum**, **iron**, and **titanium**.

In highly weathered settings, soil loses most of its nutrients, and the store of nutrients that remains is mostly found in organic matter. In weathered soils, only the top 25 centimeters (10 inches) or so may be very biologically active, and rooting depths are very shallow. If this thin layer is lost to erosion, the underlying mineral soil may be infertile and incapable of rapid recovery.

Soil Orders

Just as rocks are classified into different types based on how they formed (igneous, **metamorphic**, or **sedimentary**), their mineral composition, and other characteristics, soils also have their own classification scheme. Soil develops in horizons, or layers, whose formation is dependent on the available ingredients, environmental conditions, and the time it takes to mature. Since the organic and chemical processes that form soils first impact the top of the soil column and then work their way downward, horizontal layers of soil with different characteristics are formed, resulting in divergent colors, textures, and compositions.

A vertical cross-section of all the horizons or layers of soil present in a given area is referred to as a soil profile. Some horizons are completely absent in certain profiles while others are common to most. Each horizon corresponds to a stage in the weathering of rock and decay of plant matter, and each is found at a specific position beneath the surface *Figure 7.2*). The *O horizon* at the top of the profile contains partially decayed plant material and transitions down to the *A horizon*, which contains mineral matter with a mix of humus and is commonly referred to as topsoil. Below the A horizon lies the *B horizon* or subsoil, which contains mineral material that has leached from above. The *C horizon* at the base of the soil profile contains partially altered parent material.

Soils can also be categorized by their location (northern vs. southern soils), the type of vegetation growing on them (forest soils vs. desert soils), their topographic position (hilltop soils vs. valley soils), or other distinguishing features. The system used to classify soils based on their properties is called **soil taxonomy** *(Figure 7.3)*, and it was developed by the United States Department of Agriculture (USDA) with the help of soil scientists from across the country. It provides a convenient, uniform, and detailed classification of soils throughout the US (*Figure 7.4*), allowing for an easier understanding of how and why different regions have developed unique soils.

igneous rocks • *rocks derived from the cooling of magma underground or molten lava on the Earth's surface.*

silica • *a chemical compound also known as silicon dioxide (SiO_2).*

aluminum • *a metallic chemical element (Al), and the most abundant metal in the Earth's crust.*

iron • *a metallic chemical element (Fe).*

titanium • *a metallic chemical element (Ti) that is important because of its lightweight nature, strength and resistance to corrosion.*

metamorphic rocks • *rocks formed by the recrystallization and realignment of minerals in pre-existing sedimentary, igneous, and metamorphic rocks when exposed to high enough temperature and/or pressure.*

sedimentary rock • *rock formed through the accumulation and consolidation of grains of broken rock, crystals, skeletal fragments, and organic matter.*

Soil Orders

soil taxonomy • the system used to classify soils based on their properties.

soil orders • the twelve major units of soil taxonomy, which are defined by diagnostic horizons, composition, soil structures, and other characteristics.

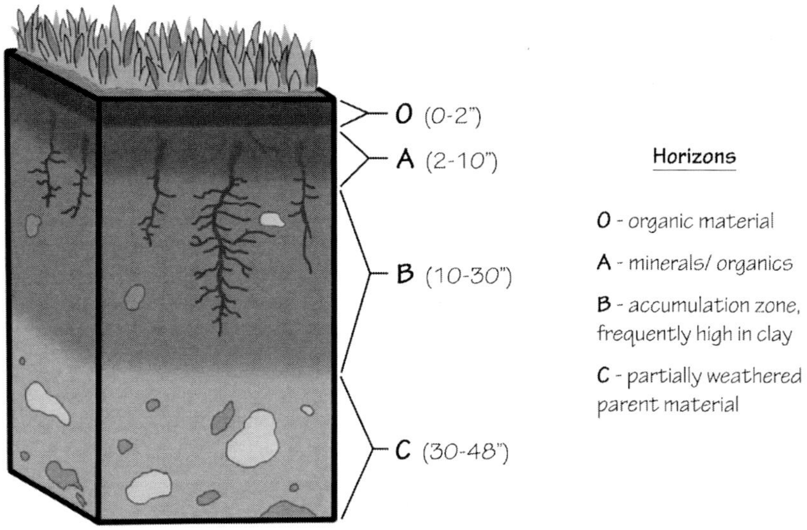

Figure 7.2: A typical soil profile shows the transition from the parent material (horizon C) to the highly developed or changed horizons (O through B). Not every soil profile will have all the horizons present.

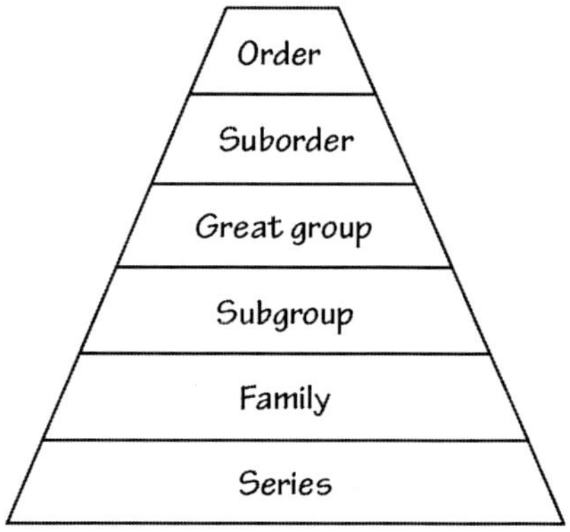

Figure 7.3: Soil taxonomy.

In soil taxonomy, all soils are arranged into one of 12 major units, or **soil orders**. These 12 orders are defined by diagnostic horizons, composition, soil structures, and other characteristics. Soil orders depend mainly on climate, parent material, and the organisms within the soil. These orders are further broken down into 64 suborders based on properties that influence soil development and plant growth, with the most important property being how wet the soil is throughout the year. The suborders are, in turn, separated into great groups (300+) and subgroups (2400+). Similar soils within a subgroup are grouped into even more selective families (7500+), and similar soils within families are grouped together into the most exclusive category of all: a series. There are more than 19,000 soil series described in the United States, with more being defined every year.

7 Soils

Soil Orders

Figure 7.4: Dominant soil orders of the United States. (See TFG website for a full-color version.)

The 12 Soil Orders

Name	Description	Controlling Factors	Percentage of global ice-free land surface	Percentage of US ice-free land surface
Alfisols	Highly fertile and productive agricultural soils in which clays often accumulate below the surface. Found in humid and subhumid climates.	climate and organisms	~10%	~14%
Andisols	Often formed in volcanic materials, these highly productive soils possess very high water- and nutrient-holding capabilities. Commonly found in cool areas with moderate to high levels of precipitation.	parent material	~1%	~2%
Aridisols	Soils formed in very dry (arid) climates. The lack of moisture restricts weathering and leaching, resulting in both the accumulation of salts and limited subsurface development. Commonly found in deserts.	climate	~12%	~8%

7 Soils

Soil Orders

Entisols	Soils of relatively recent origin with little or no horizon development. Commonly found in areas where erosion or deposition rates outpace rates of soil development, such as flood-plains, mountains, and badland areas.	time and topography	~16%	~12%
Gelisols	Weakly weathered soils formed in areas that contain permafrost within the soil profile.	climate	~9%	~9%
Histosols	Organic-rich soils found along lake coastal areas where poor drainage creates conditions of slow decomposition and peat (or muck) accumulates.	topography	~1%	~2%
Inceptisols	Soils that exhibit only moderate weathering and development. Often found on steep (relatively young) topography and overlying erosion-resistant bedrock.	time and climate	~17%	~10%
Mollisols	Agricultural soils made highly productive due to a very fertile, organic-rich surface layer.	climate and organisms	~7%	~22%

Oxisols	Very old, extremely leached and weathered soils with a subsurface accumulation of iron and aluminum oxides. Commonly found in humid, tropical environments.	climate and time	~8%	~.02%
Spodosols	Acidic soils in which aluminum and iron oxides accumulate below the surface. They typically form under pine vegetation and sandy parent material.	parent material, climate, and organisms	~4%	~4%
Ultisols	Soils with subsurface clay accumulations that possess low native fertility and are often red hued (due to the presence of iron oxides). Found in humid tropical and subtropical climates.	climate, time, and organisms	~8%	~9%
Vertisols	Clayey soils with high shrink/swell capacity. During dry periods, these soils shrink and develop wide cracks; during wet periods, they swell with moisture.	parent material	~2%	~2%

7 Soils

Soil Orders

calcium carbonate • a chemical compound with the formula $CaCO_3$, commonly found in rocks in the mineral forms calcite and aragonite, as well as the shells and skeletons of marine organisms.

floodplain • the land around a river that is prone to flooding.

hyperthermic • soils in which the mean annual temperature 50 centimeters (20 inches) below the surface is at least 22°C (71.6°F), with 5 or more degrees variation between winter and summer.

Dominant Soils of the Southwest

Six soil orders are found in the Southwest, with the greatest diversity found within the Basin and Range.

Alfisols are partially leached soils with a high degree of fertility, in which clays often accumulate below the surface. They tend to develop in cooler, more forested environments, and commonly form a band separating humid areas from more arid areas. In the Southwest, they occur primarily at higher elevations (*Figure 7.5*).

Aridisols are very dry soils that form in arid environments, such as the Basin and Range region. Water content is very low or even nonexistent for most of the year, leading to limited leaching. The lack of leaching means that there is abundant **calcium carbonate** in these soils, making them quite alkaline. This soil type is unsuitable for plants that are not adapted to store water or to survive extreme drought. Because of the Southwest's dry climate, Aridisols are the most abundant soil type there (*Figure 7.6*).

Entisols are soils of recent origin with poorly developed horizons, typically formed near **floodplains** and on recent erosional surfaces. These soils are found throughout the Southwest in association with river courses, rapid erosion, and wind-transported sediment (*Figure 7.7*).

Inceptisols are soils with poorly developed horizons that are associated with steep slopes and erosion-resistant parent material. Inceptisols are located in cool to very warm humid and subhumid regions. They are uncommon in the Southwest thanks to the area's generally dry climate, and are concentrated at high elevations, especially in the Rocky Mountains (*Figure 7.8*).

Mollisols are the dominant soils of grasslands. Their thick, black A horizon makes these soils extremely productive and valuable to agriculture. These soils are extensive in the Great Plains, and are also found in bands along parts of the Colorado Plateau (*Figure 7.9*).

Vertisols are dark soils, rich in swelling clays. They dramatically change in volume depending on the moisture level of the soil, forming deeply cracked surfaces during dry periods and swelling again in the wet season to seal all the cracks. As a result, they are very difficult soils to build roads or other structures on. Vertisols are rare in the Southwest, and are present only in a few areas of Arizona and New Mexico (*Figure 7.10*).

Although these soil orders occur in other parts of the country, the Southwest's arid climate allows for a secondary way of classifying soils—by the temperature zone in which they occur. Exposure to extreme temperatures means that many Southwestern soils are considered to be **hyperthermic**, and may not seem to have a lot in common with their equivalents in other regions. In addition, many of the area's soils are thin because the requisite conditions for the formation of deep soils are not present.

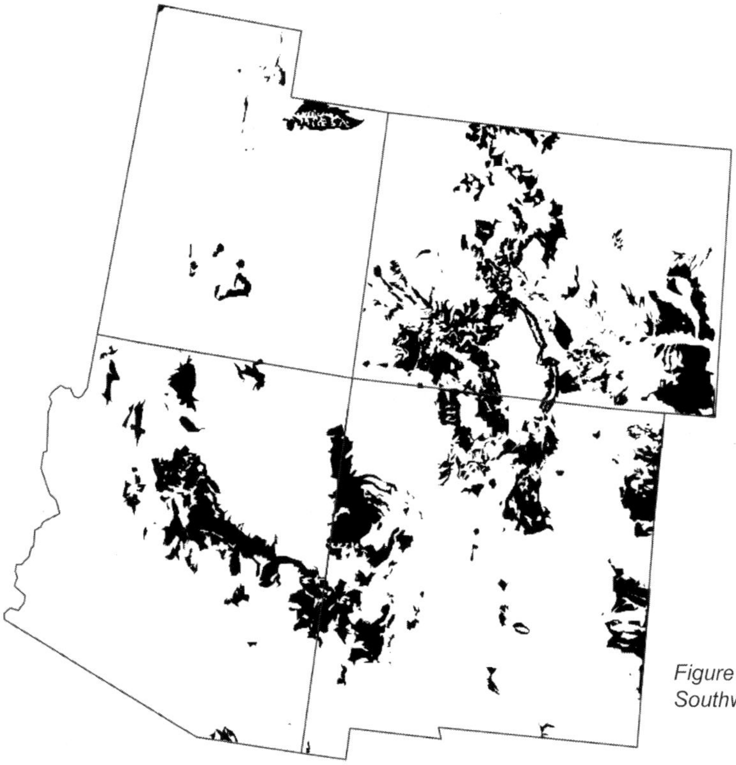

Figure 7.5: Alfisols of the Southwest.

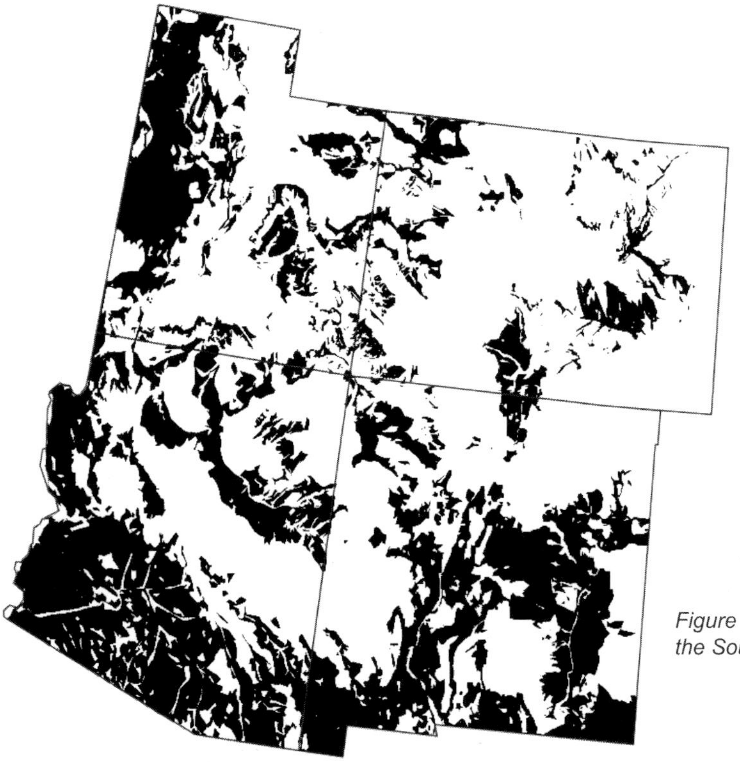

Figure 7.6: Aridisols of the Southwest.

7 Soils

Soil Orders

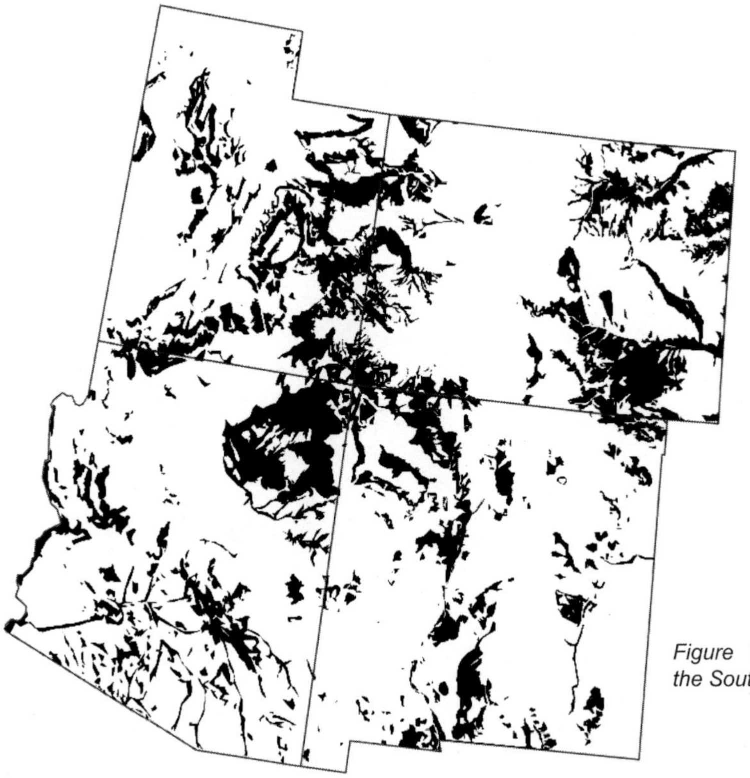

Figure 7.7: Entisols of the Southwest.

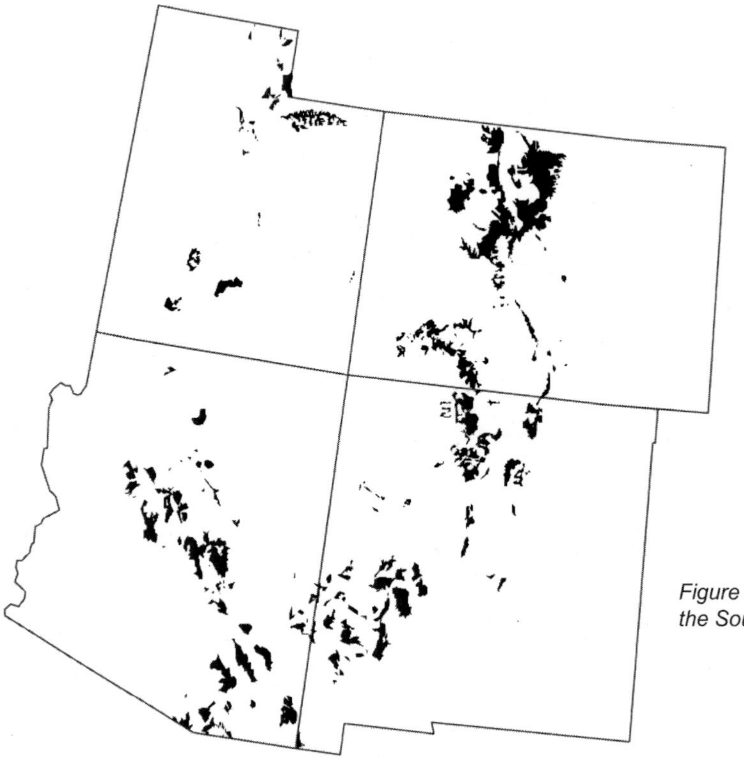

Figure 7.8: Inceptisols of the Southwest.

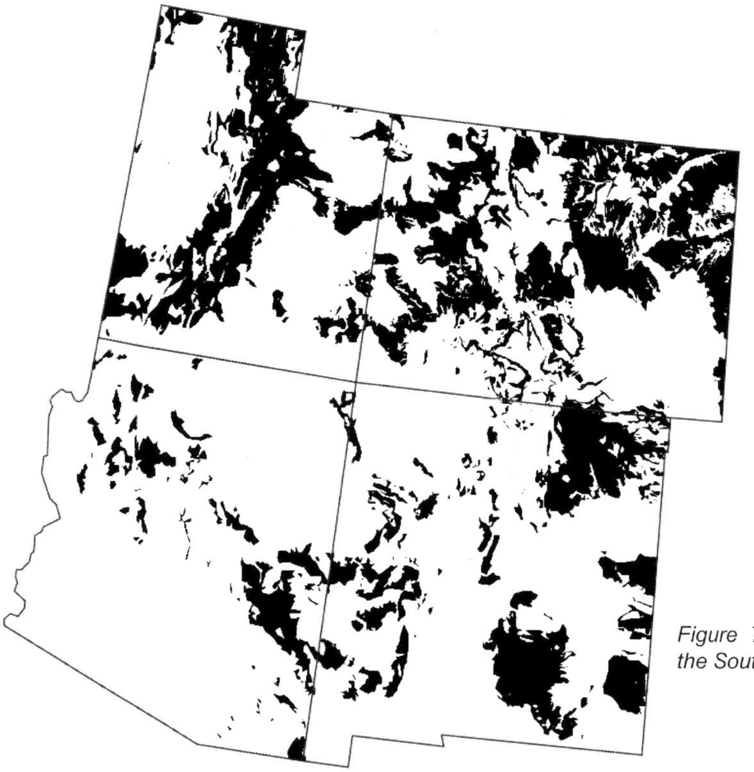

Figure 7.9: Mollisols of the Southwest.

Figure 7.10: Vertisols of the Southwest.

7 Soils

Soil Orders

hydrography • *The science of the physical aspects of water on the surface of the Earth.*

Geology of the Southwest: Parent Material

The Southwest is home to a wide variety of parent materials—the minerals and organic matter from which its soils are derived (*Figure 7.11*). Mineral material determines a soil's overall fertility and the vegetation it supports.

Weathered sedimentary rock is a ubiquitous parent material throughout many areas of the Southwest. The western half of Utah, for example, is part of a much larger **hydrographic** area known as the Great Basin, which contains **sandstone**, **shale**, and **limestone** that accumulated over millions of years. The eastern halves of Colorado and New Mexico are also home to an abundant supply of weathered sedimentary rock—sandstone, siltstone, limestone, and shale are among the most common bedrocks in this portion of the Southwest. Much of this rock was laid down during the **Paleozoic,** when shallow **inland seas** repeatedly flooded the landscape.

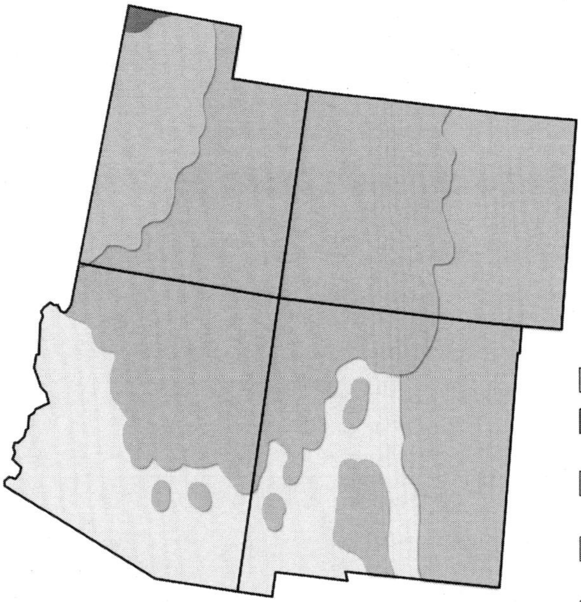

Figure 7.11: Physiographic and regolith map of the Southwest. (See TFG website for full-color version.)

sandstone • *sedimentary rock formed by cementing together grains of sand.*

shale • *a dark, fine-grained, laminated sedimentary rock formed by the compression of successive layers of silt- and clay-rich sediment.*

limestone • *a sedimentary rock composed of calcium carbonate ($CaCO_3$).*

The majority of the Southwest contains extremely varied parent material generated during the tectonic events that led to the **uplift** of the Rocky Mountains. **Orogenic** events during the **Mesozoic**, eruptions of ash and **lava** from **volcanic** mountain ranges to the west, and inundation by the late **Cretaceous** Western Interior Seaway all contributed to the formation of the igneous, sedimentary, and metamorphic rock types found throughout the area. Soils in the Rocky Mountain and Colorado Plateau regions are derived largely from the erosion of these materials, while soils in the Basin and Range are mostly composed of **gravel**, sand, and other debris eroded from the desert and mountainous landscape.

Soils 7

Soils of the Colorado Plateau
Region 1

The Colorado Plateau is a high region, with an average elevation of 1500 meters (5000 feet). Slopes are relatively gentle across the region, except in mountain ranges, such as the Henry Mountains of Utah, which owe their existence to **magmatic intrusion**. The geology of the Colorado Plateau comprises mostly Mesozoic-aged sedimentary rocks.

See Chapter 2: Rocks to learn more about mountains formed by magmatic intrusion.

The most common soils on the Colorado Plateau are Aridisols, which occur in the region's mid-elevations. These dry, coarse soils formed from the weathering of limestones and **carbonate** parent material deposited in ancient seas. Due to a lack of precipitation that would leach out soluble minerals, Aridisols contain high concentrations of **gypsum**, carbonates, and salt, which sometimes solidify into **caliche**—a hard, light-colored layer **cemented** together by **lime**. They support drought-resistant vegetation, especially sagebrush steppe ecosystems (*Figure 7.12*). Irrigation can convert these soils into useable cropland, but they are also managed for livestock grazing.

Figure 7.12: Aridisol soils support the growth of this sagebrush steppe near Utah's Henry Mountains.

Paleozoic • *a geologic time interval that extends from 541 to 252 million years ago.*

inland sea • *a shallow sea covering the central area of a continent during periods of high sea level.*

uplift • *upward movement of the crust due to compression, subduction, or mountain building.*

orogeny • *a mountain-building event generally caused by colliding plates and compression of the edge of the continents.*

Mesozoic • *a geologic time period that spans from 252 to 66 million years ago.*

lava • *molten rock located on the Earth's surface.*

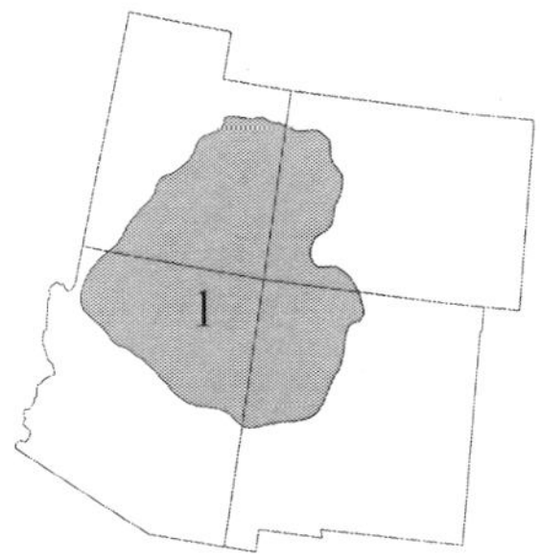

7 Soils

Region 1

conifer • a woody plant bearing cones that contain its seeds.

windward • upwind; facing into the prevailing winds, and thus subject to orographic precipitation.

Alfisols are relatively uncommon on the Colorado Plateau. These soils generally form in forested areas as a result of weathering processes that leach minerals from the surface layer into a clay-rich subsoil, where nutrients are retained. Many occur beneath **conifer** forests in the region's higher elevations, especially along the Mogollon Rim in Arizona (*Figure 7.13*) and in the Zuni Mountains in western New Mexico, the southern Chuska Mountains in eastern Arizona, and the Henry Mountains in south-central Utah.

See Chapter 4: Topography to learn about the physiography of the Colorado Plateau.

Figure 7.13: Pale-colored, leached Alfisols are exposed on this path in Coconino National Forest, Arizona.

Mollisols constitute the most fertile soils on the Colorado Plateau, formed where organic matter accumulates beneath prairie grasses and in poorly drained forests (*Figure 7.14*). These soils are rich in humus, which stores mineral nutrients and contributes to the soil's high moisture and nutrient content. In the central parts of the Plateau, they are scattered at mid-elevations on the **windward** sides of mountain ranges where moisture is slightly more abundant. Mollisols are particularly rich in western Colorado, allowing for the cultivation of fruit orchards and other crops. They are also found in a wide band stretching through central Utah along the Plateau's western margin, where they form some of the state's most productive and important agricultural soils.

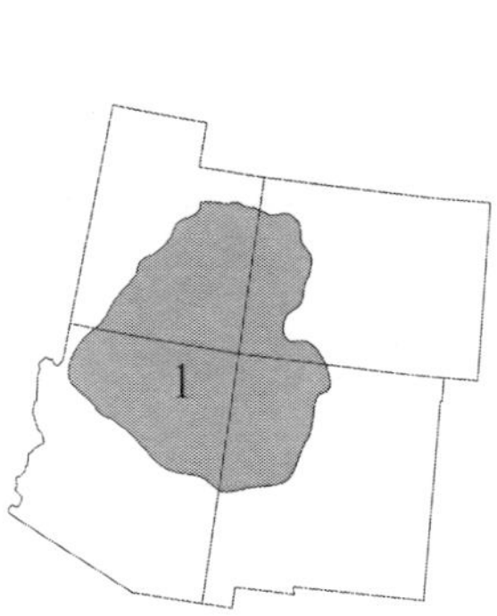

Figure 7.14: A farm in Paonia, Colorado irrigates this field of rich Mollisols with a system of channels.

Entisols, young soils lacking in horizons, are found where erosion and deposition occur faster than the rate of soil formation. In the Colorado Plateau they are common on recent erosional surfaces, especially along the Colorado River and other river courses (*Figure 7.15*). They also occur as shallow soils on the bedrock of arid uplands. Inceptisols are rare here; these weakly developed soils can be found scattered throughout the region in mountainous areas and on steep slopes.

Figure 7.15: An accumulation of sandy Entisols marks a dry riverbed near Palisade, Colorado.

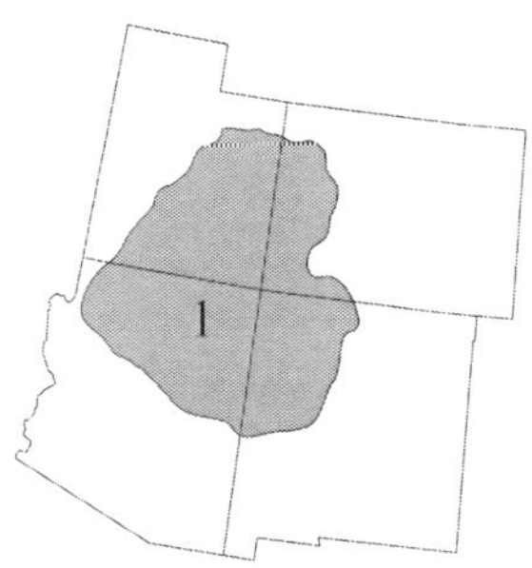

7 Soils

Region 2

crust • the uppermost, rigid outer layer of the Earth.

Neogene • the geologic time period extending from 23 to 2.6 million years ago.

fault • a fracture in the Earth's crust in which the rock on one side of the fracture moves measurably in relation to the rock on the other side.

salt • a mineral composed primarily of sodium chloride (NaCl).

playa lake • ephemeral or dry lakebed that occasionally contains only a thin layer of quickly evaporating water.

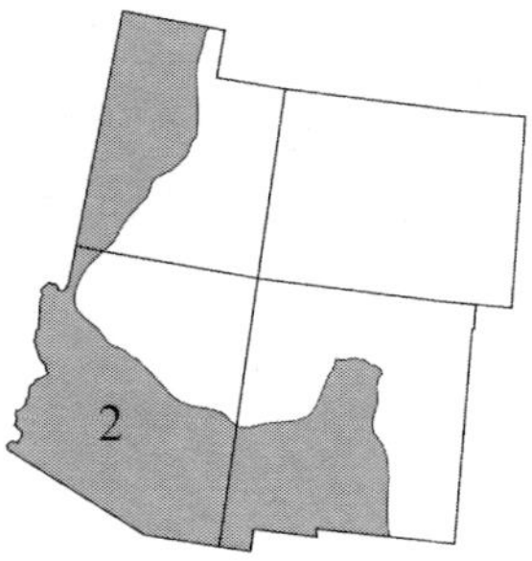

Soils of the Basin and Range
Region 2

The Basin and Range is a unique topographical region consisting mostly of north- and south-running mountain ranges and valleys. These features formed as a result of extension of the continental **crust** during the **Neogene**. This extension, or stretching, caused the crust to break along straight, nearly parallel **faults**, along which steep mountains rose and flat-bottomed valleys dropped. In the Southwest, valley bottoms can be close to sea level, particularly in southwestern Arizona (in Death Valley, California, the Basin and Range actually has valleys below sea level). The mountains, on the other hand, can top 3000 meters (10,000 feet)! The region's low valleys are commonly occupied by dry, alkaline lakes with very little vegetation, but in the slightly cooler, higher valleys of southern Arizona and southwestern New Mexico, the soil is more hospitable to plant growth.

See Chapter 4: Topography for more information about the formation of the Basin and Range.

With the exception of high mountaintops, a very dry climate dominates the Basin and Range. Aridisols are by far the most common soils, occupying the low- and mid-elevation parts of the basins and adjacent mountain ranges. These soils tend to contain a high amount of calcium carbonate, and are moderately to strongly alkaline; they support drought-resistant plants including sagebrush, Joshua tree, and yucca. While many Aridisols are beyond the practicality of common agricultural and economic practices, not all have been left undeveloped. Many of the region's Aridisols belong to the suborder Argids, which are rich in clay; with irrigation, these can be productive soils (*Figure 7.16*). In northwestern Utah, however, the dominant suborder is Salids—these fine-grained and **salt**-heavy soils are commonly found in depressions and **playa lakes**, including the extensive Great Salt Lake Desert.

Figure 7.16: The famous chile pepper farms of southern New Mexico produce their crops in Aridisol soils.

Entisols are the next most common type of soil in the Basin and Range. These soils, young and lacking in horizons, are found scattered along floodplains and streams where alluvial sediments are deposited, and river valleys where unconsolidated sediment has been transported from the steep surrounding slopes. In Utah, several areas of highly sandy sediment have been designated as dune land. One of the largest dune fields in the state is Little Sahara, whose sand dunes result from silicate-rich deposits left by the Sevier River during the **Pleistocene** (*Figure 7.17*). Strong winds carried the sediment from the river **delta** to its current location, where it covers an area of about 570 square kilometers (270 square miles).

Pleistocene • *a subset of the Quaternary, lasting from 2.5 million to about 11,700 years ago.*

delta • *a typically wedge-shaped deposit formed as sediment is eroded from mountains and transported by streams across lower elevations.*

Figure 7.17: Sand Mountain, a 210- to 245-meter (700- to 800-foot) wall of Entisol sand at Little Sahara in Utah, is a popular location for ATV riding.

Other soil types that appear in the Basin and Range are uncommon. Soils weathered from the region's mountains are characteristically rocky, shallow, and gravelly. Dry Alfisols occur near the tops of the highest mountains in southern Arizona and New Mexico, and along the Mogollon Rim at the edge of the Colorado Plateau. Freely drained Inceptisols, associated with steep slopes and resistant parent material, occur scattered across mid-elevations in southern Arizona and southwestern New Mexico. These soils are commonly associated with the newly formed soils of mountainous terrain and do not lend themselves well to agriculture due to their poor development. Arid Mollisols, which are fertile but can become dusty and dry during drought conditions, are present at mid-elevations that surround mountains where moisture is slightly higher.

Region 2

Mesozoic • a geologic time period that spans from 252 to 66 million years ago.

In the Basin and Range, patches of soil have been heavily influenced by existing sedimentary rock material lain down during the uplift of the Rockies and the deposition of **Mesozoic** sediments. The erosion of exposed Cretaceous marine shales produces Vertisols, soils that experience drastic fluctuations in volume when exposed to water. The only Vertisols found in the Southwest occur in the Basin and Range, in southern Arizona and New Mexico (*Figure 7.18*). During dry periods, these clayey soils shrink and form wide cracks at the surface; the cracks seal shut again when moisture enters the soil. Because Vertisols shrink and swell so readily, it is extremely difficult—and even dangerous—to build structures or roads on top of them. Expansion of the clay minerals can cause foundations to crack and roads to buckle; millions of dollars are spent every year on repairing damage done by expansive soils. The action of shrinking and swelling within the soil also prevents the formation of distinct horizons.

See Chapter 9: Earth Hazards to learn more about the dangers associated with expansive soils.

Figure 7.18: Cracked Vertisol soils near Ash Fork, Arizona.

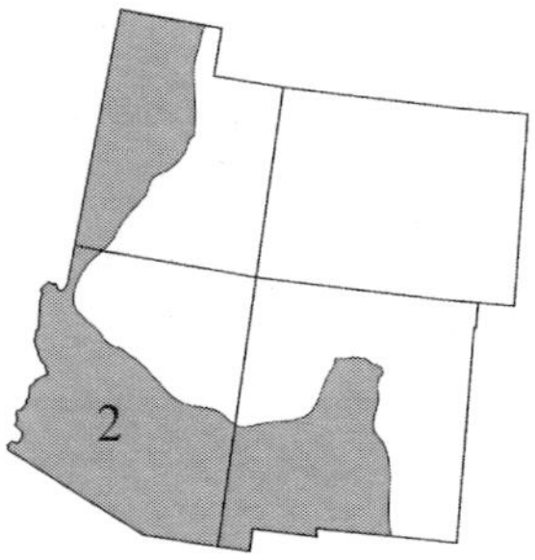

Soils of the Rocky Mountains
Region 3

The Rocky Mountains of the Southwest encompass the mountains of central and northwestern Colorado and northeasternmost Utah. The region took form as a result of several orogenic events starting in the Paleozoic, and the present-day Rockies owe their current topography to events that occurred in the **Paleogene**. The Rockies actually consist of many mountain ranges with high-elevation valleys in between. These valleys are higher than the average elevation of the Colorado Plateau, generally reaching heights of around 1700 to more than 2100 meters (5500 to more than 7000 feet). The Rocky Mountains are home to many of the highest peaks in the continental US, with 54 peaks topping 14,000 feet (approximately 4270 meters)! At the highest elevations, bare rock predominates, but soils have formed below the **tree** line.

See Chapter 4: Topography for more about the highest points in the Southwest.

Paleogene • *the geologic time period extending from 66 to 23 million years ago.*

tree • *any woody perennial plant with a central trunk.*

permeability • *a capacity for fluids and gas to move through fractures within a rock, or the spaces between its grains.*

Not surprisingly, the soils of the Rocky Mountains are quite different from those found in surrounding areas of the Southwest. The region's higher elevation means that water is more abundant, both because temperatures are cooler, which reduces evaporation, and because rainfall and runoff are higher, particularly on the windward sides of the mountains. For this reason, Aridisols, which are widespread elsewhere in the Southwest, are uncommon in the Rocky Mountains, although clay- and carbonate-rich soils are scattered throughout the mountains of northwestern Colorado and northeastern Utah. Alfisols, in contrast, are common at higher elevations and support forested areas with a short, cool growing season. Entisols are uncommon, but do occur in the region's valleys.

Inceptisols are found on reasonably steep slopes and involve parent rock material that is quite resistant to weathering, so they are frequently associated with mountain formations, and often overlie erosion-resistant bedrock. While Inceptisols represent a level of soil development one step above that of Entisols, they are still very poorly developed. The Rocky Mountains of northern Colorado, especially the Medicine Bow, Front, and Park ranges, host a high concentration of these soils (*Figure 7.19*). The thin, rocky nature of Inceptisol soil prevents significant water retention, placing lower limits on plant cover.

Mollisols are common throughout lower elevations of the Rockies. These loamy soils are well drained and **permeable**, containing ample organic matter and a high nutrient content. Valleys that contain these highly productive dark soils naturally support forest, grass, and shrubland. They are used to raise a variety of crops, including hay, wheat, and barley, and also as rangeland for cattle.

Figure 7.19: The Nokhu Crags, part of Colorado's Never Summer Mountains, erode to form weathered Inceptisols.

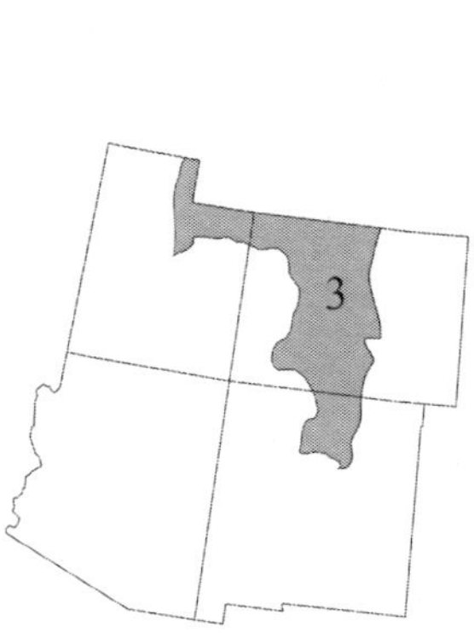

Soils of the Great Plains
Region 4

The Great Plains, a broad plateau that is home to intermediate and short grasslands, stretches for 3200 kilometers (2000 miles) from the Canadian interior south to the Mexican border. In the Southwest, this relatively flat region runs along the eastern edge of Colorado and New Mexico. Here, the Great Plains tends to be at a higher elevation than it is farther east; the area of lowest elevation runs through southeastern New Mexico, in the broad valley occupied by the Pecos River. Most of the region's soils are derived from Mesozoic- and **Cenozoic**-aged sedimentary rock that mantles the area.

Mollisols are the most common type of soil in the Great Plains. The region is dominated by dry Mollisols belonging to the suborder Ustolls, which form in semi-arid conditions. These soils can become even more dusty and dry during drought conditions, limiting crop yields and leading to damaging dust storms such as those that occurred during the Dust Bowl of the 1930s. Decreased precipitation and lower soil fertility provides for a localized agricultural economy based heavily in rangeland livestock (*Figure 7.20*), and crops here often require irrigation from local **aquifers** or various surface water impoundments. Despite these limitations, fruits, nuts, vegetables, and wheat are all grown in the Great Plains' Mollisols.

Figure 7.20: Cows graze in New Mexico rangeland underlain by dusty Mollisols.

Cenozoic • *the geologic time period spanning from 66 million years ago to the present.*

aquifer • *a water-bearing formation of gravel, permeable rock, or sand that is capable of providing water, in usable quantities, to springs or wells.*

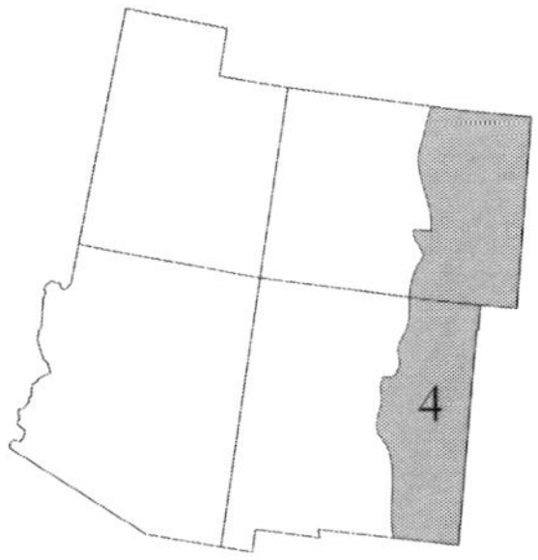

Entisols are also common throughout the Great Plains, especially at lower elevations and along river courses where they are generated by slope erosion. Sandy deposits of wet Entisols can be found all along the tributaries and banks of Colorado's Platte River, and the Pecos River in New Mexico frequently deposits alluvial materials on its floodplains. Layers of sandy soil in the northeastern part of Colorado provide a productive foundation for rangeland and pasture.

Alfisols occur mostly in the high-elevation areas of the Great Plains, especially in southeastern Colorado, where open forests of juniper and other conifers grow.

State Soils

Just as many states have official state flowers, birds, and **fossils**, they also have official soils. State soils are most often determined by a vote of soil scientists in the state, and, absent any political wrangling, usually represent the most productive soils and those that most closely resemble everyone's favorite soil: loam. As mentioned earlier, loam soils are almost equal parts sand, silt, and clay.

Arizona

The state soil of Arizona is the Casa Grande series, an Aridisol. This soil is known to cover roughly 110,000 **hectares** (275,000 acres) but may actually cover nearly one million hectares (several million acres) of the state. This dry soil series forms in alluvium derived from weathered and eroded **granite**, **rhyolite**, **andesite**, **quartzite**, limestone, and **basalt**. With irrigation, these soils become productive and are used to raise vegetables, grain, and cotton.

Colorado

The Seitz series is a group of Alfisols that cover approximately 140,000 hectares (350,000 acres) of primarily mountainous terrain throughout southwestern and central Colorado. These soils support spruce and fir populations at higher altitudes, and are also home to an understory of grasses, forbs, and shrubs.

New Mexico

In New Mexico, Aridisols called Penistaja soils cover more than 400,000 hectares (one million acres). Named for a Navajo word meaning "forced to sit," Penistaja soils are formed in mixed alluvium and **aeolian** material derived from sandstone and shale. They support productive rangeland and are commonly used to graze livestock.

Utah

Covering more than 80,000 hectares (200,000 acres) of southeastern Utah, Mivida soils form in warm, semiarid climates. These coarse and loamy Aridisols are home to many unique plants, including Wyoming big sagebrush, Indian ricegrass, galleta, and blue grama.

fossil • *preserved evidence of ancient life.*

hectare • *a metric unit of area defined as 10,000 square meters.*

granite • *a common and widely occurring type of igneous rock.*

rhyolitic • *a felsic volcanic rock high in abundance of quartz and feldspar.*

andesite • *a fine-grained extrusive volcanic rock, with a silica content intermediate between that of basalt and dacite.*

quartzite • *a hard metamorphic rock that was originally sandstone.*

basalt • *an extrusive igneous rock, and the most common rock type on the surface of the Earth.*

aeolian • *pertaining to, caused by, or carried by the wind.*

Resources

General Books and Articles on Soils

Lindbo, D. L. and J. Mannes, 2008, *Soil!: Get the Inside Scoop*, Soil Science Society of America, Madison, WI, 32 pp.

Lindbo, D. L., 2012, *Know Soil, Know Life*, Soil Science Society of America, Madison, WI, 206 pp.

Logan, W. B., 1995, *Dirt: the Ecstatic Skin of the Earth*, Riverhead Books, New York, 202 pp.

O'Meara, C., 2013, Do you know dirt about soil? Here's a three-step primer, *Denver Post*, March 27, 2013, http://www.denverpost.com/2013/03/27/do-you-know-dirt-about-soil-heres-a-three-step-primer/.

Soil Survey Staff, 2014, *Keys to Soil Taxonomy, 12th edition*, US Department of Agriculture, Natural Resources Conservation Service, Washington, DC, 362 pp., http://www.nrcs.usda.gov/ wps/PA_NRCSConsumption/download?cid=stelprdb1252094&ext=pdf.

Soil Survey Staff, 2014, Illustrated Guide To Soil Taxonomy, US Department of Agriculture, Natural Resources Conservation Service, National Soil Survey Center, Lincoln, NE, 498 pp., http://www. nrcs.usda.gov/wps/PA_NRCSConsumption/download?cid=stelprdb1247203&ext=pdf.

General Websites on Soils

K–12 Soil Science Teacher Resources, Soil Science Society of America, http://www.soils4teachers.org/.

Soil Survey Reports (by state and county/parish), The Cooperative Soil Survey, http://soils.missouri.edu/survey/selectstate.asp.

Soil Sustains Life, Soil Science Society of America, https://www.soils.org.

Soils, USDA Natural Resources Conservation Service. [Soil surveys by state, technical publications, soil data, and much more.] http://www.nrcs.usda.gov/wps/portal/nrcs/site/soils/home/.

Soils Tutorial, The Cooperative Soil Survey, http://soils.missouri.edu/tutorial/index.asp.

The Twelve Soil Orders Soil Taxonomy, University of Idaho College of Agricultural and Life Sciences, http://www.cals.uidaho.edu/soilorders/.

Books and Websites on Soils of Specific Parts of the Southwest

Boettinger, J. L., 2009, *Soils of Utah*, Soil STATSGO2 Soil Database, US Department of Agriculture, pp. 46–48, http://extension.usu.edu/utahrangelands/files/uploads/RRU_Section_Six.pdf.

Hendricks, D. M., ed., 1986, *Arizona Soils*, College of Agriculture, University of Arizona, Tucson, AZ, 244 pp., http://www.library.arizona.edu/exhibits/swetc/azso/.

Chapter 8:
Climate of the Southwestern US

Climate is a description of the average temperature, range of temperatures, humidity, precipitation, and other **atmospheric**/hydrospheric conditions a region experiences over a period of many years. These factors interact with and are influenced by other parts of the Earth **system**, including geology, geography, insolation, currents, and living things.

Because it is founded on statistics, climate can be a difficult concept to grasp, yet concrete examples can be illuminating. Terms like "desert," "rain forest," and "tundra" describe climates, and we have gained a general understanding of their meaning. Climate can also encompass the cyclical variations a region experiences; a region with a small temperature variation between winter and summer—San Francisco, for example—has a different climate from one that has a large variation, such as Buffalo. Scientists have settled on 30 years as the shortest amount of time over which climate can be defined, but it can of course also define time periods millions of years in length.

You cannot go outside and observe climate. **Weather**, on the other hand, can be observed instantly—it is 57 degrees and raining *right now*. Weather varies with the time of day, the season, multi-year cycles, etc., while climate encompasses those variations. Our choice of clothing in the morning is based on the weather, while the wardrobe in our closet is a reflection of climate. Due to the area's great regional variety, from the arid zones of the Basin and Range to the higher, wetter parts of the Rocky Mountains, residents of the Southwest have a diverse wardrobe. While the entire Southwest experiences seasonal variation, southernmost Arizona and New Mexico experience much less variation due to the warmth of their winters.

Past Climates

Climate, like other parts of the Earth system, is not static but changes over time, on both human and **geologic time scales**. Latitude, for example, has a very direct effect on climate, so as the continents shift over geologic time, the climates on them also shift. Furthermore, the conditions on Earth as a whole have varied through time, altering what kinds of climates are possible. Throughout most of its long history, the Southwest has been tropical or temperate, but it has also ranged from very wet to very dry.

Ancient climates are reconstructed through many methods. Written records and **tree** rings go back hundreds of years, **glacial** ice cores hundreds of thousands of years, and **fossils** and rocks that indicate different climates go back hundreds of millions of years. These clues, coupled with modeling and a knowledge of physics and chemistry, help climatologists put together an increasingly detailed history of the Earth's climate, and of that of the Southwest. Unfortunately, we do not have as clear an understanding of climate for the earliest part of Earth

atmosphere • *a layer of gases surrounding a planet.*

system • *a set of connected things or parts forming a complex whole.*

weather • *the measure of short-term conditions of the atmosphere such as temperature, wind speed, and humidity.*

geologic time scale • *a standard timeline used to describe the age of rocks and fossils, and the events that formed them.*

glacier • *a body of dense ice on land that does not melt away annually and has sufficient mass to move under its own weight.*

fossil • *preserved evidence of ancient life.*

CHAPTER AUTHORS

Ingrid H. H. Zabel
Judith T. Parrish
Andrielle N. Swaby

Past

helium • a gaseous chemical element (He), which is the second most abundant and second lightest element in the universe.

volcanism • the eruption of molten rock onto the surface of the crust.

meteorite • a stony or metallic mass of matter that has fallen to the Earth's surface from outer space.

salt • a mineral composed primarily of sodium chloride (NaCl).

mineral • a naturally occurring solid with a specific chemical composition and crystalline structure.

energy • the power derived from the use of physical or chemical resources.

greenhouse gas • a gas in the atmosphere that absorbs and emits heat.

iron • a metallic chemical element (Fe).

history as we do for the later parts, because the oldest rocks are much more difficult to find. However, we can still say something about the climate of the ancient Earth, in large part due to our knowledge of atmospheric chemistry.

Ancient Atmosphere

Not long after the Earth first formed, more than 4.5 billion years ago, its atmosphere was composed mostly of hydrogen and **helium**. **Volcanic** activity and collisions with **meteorites** and comets added water vapor, carbon dioxide (CO_2), and nitrogen to the atmosphere. As the Earth cooled enough for liquid water to form, the vapor formed clouds from which the rain poured forth in such a deluge as the planet will never experience again. These torrential rains were constant for *millions* of years, absorbing **salt** and other **minerals** from the earth as the rainwater coursed to the lowest areas, forming Earth's oceans and seas.

At this time, the sun produced significantly less **energy** than it does today, so one might expect that once the oceans formed, they would continue to cool and eventually freeze. Yet temperatures stabilized, perhaps because there was a greater concentration of potent **greenhouse gases** in the atmosphere and less land surface to reflect light, so temperatures remained high enough for liquid water to exist. Indirectly, the ocean was responsible for the final ingredient of the modern atmosphere because it was home to the first life on Earth. Photosynthetic bacteria appeared perhaps as early as 3.5 billion years ago, but abundant **iron** and organic matter quickly absorbed the oxygen they produced. After hundreds of millions of years, these sinks were filled, and free oxygen could finally build up in the atmosphere. With this addition, the modern atmosphere was complete, though the relative amounts of the gases composing it would, and still continue to, shift. *The composition of the atmosphere and the huge volume of water on Earth are two of the most important factors affecting climate.*

Much of the light from the sun passes unimpeded through the atmosphere and hits the Earth. Approximately 70% of that light is absorbed and retransmitted from the surface as heat. The transmitted heat, which has a longer wavelength than light, is trapped by gases in the atmosphere including water vapor, carbon dioxide, and methane. The similarity between this process and that which warms a greenhouse earned these "greenhouse gases" their moniker.

While the atmosphere was forming about 3.7 billion years ago, the surface of the Earth was cooling to form a solid **crust** of rock (although there are indications that this process may have started as early as 4.4 billion years ago). Regardless of precisely when this took place, it represented the beginning of tectonic processes that have continued ever since. Molten rock from the **mantle** constantly wells up from deep fissures and solidifies into relatively **dense** rock, while more buoyant rock floats higher on the **magma** and is pushed around on the slow conveyor belts of mantle-formed rock (*Figure 8.1*). Denser rock forms oceanic **plates** that are lower and covered in water, and lighter rock forms continental plates, though part or all of a continental plate may be submerged under a shallow sea. The motion of these plates, the rearrangement of the continents, and the amount and types of minerals exposed to the atmosphere play a huge role in the climate. Not only do the continents and oceans move through different climate zones, but the continents also affect climate based on their size, and the **weathering** of rock on the continents plays a large role in the composition of the atmosphere. For example, rock that is enriched in organic matter will release abundant amounts of carbon dioxide as it weathers, while rock rich in **feldspar** and **mica** will take up carbon dioxide.

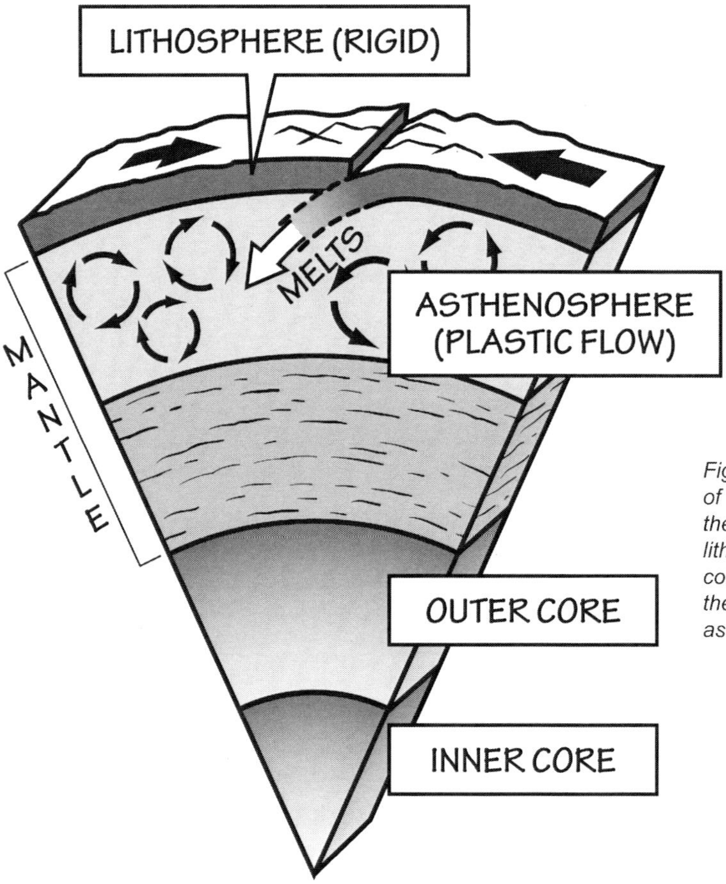

Figure 8.1: The layers of the Earth include the rigid crust of the lithosphere, which is constantly moving over the plastically flowing asthenosphere.

Nearly one billion years ago, the Earth began fluctuating between warm and cool periods lasting roughly 150 million years each. During cool periods, there is usually persistent ice at the poles, while during warm periods there is little or no glaciation anywhere on Earth. Today, we are still in a cool period—although the world has been cooler than it is at present, it has been far hotter for much of its history (*Figure 8.2*). Through the shifting global climate and the movement of the continents, what is now the Southwest has at times been at the bottom of a shallow sea, a coastal plain with swamps and rivers, and even inundated by monsoonal rains or clouded by intense dust storms.

crust • *the uppermost, rigid outer layer of the Earth.*

mantle • *the layer of the Earth between the crust and core.*

density • *a physical property of minerals, describing the mineral's mass per volume.*

magma • *molten rock located below the surface of the Earth.*

plates • *large, rigid pieces of the Earth's crust and upper mantle, which move and interact with one another at their boundaries.*

weathering • *the breakdown of rocks by physical or chemical means.*

feldspar • *an extremely common group of rock-forming minerals found in igneous, metamorphic and sedimentary rocks.*

mica • *a large group of sheetlike silicate minerals.*

8 Climate

Past

Huronian glaciation • *a glaciation beginning about 2.4 billion years ago, that covered the entire surface of the Earth in ice for as long as 300 million years.*

plate tectonics • *the process by which the plates of the Earth's crust move and interact with one another at their boundaries.*

ice sheet • *a mass of glacial ice that covers part of a continent and has an area greater than 50,000 square kilometers (19,000 square miles).*

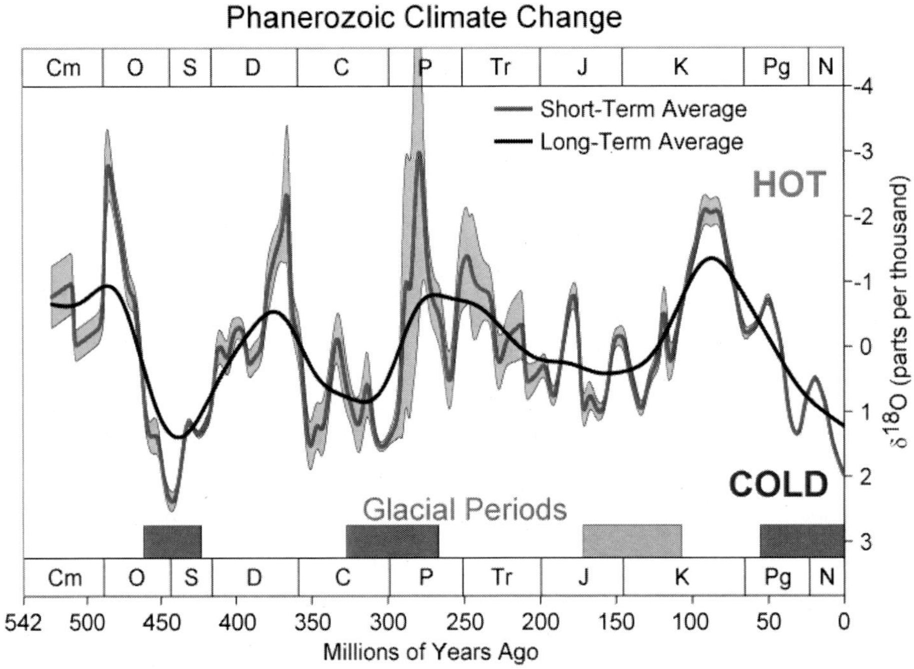

Figure 8.2: Changing global climate throughout the last 542 million years. These data were compiled using the ratios of stable oxygen isotopes found in ice cores and the carbonate skeletons of fossil organisms. (See TFG website for ful-color version.)

Snowball Earth

There is evidence suggesting that the entire surface of the planet has been covered in ice several times, a hypothesis called Snowball Earth (*Figure 8.3*). Glacial deposits discovered near Lake Huron and elsewhere show that starting about 2.4 billion years ago the entire surface of the Earth may have been covered in ice for as long as 300 million years, an event known in North America as the **Huronian glaciation**. At that time the continental plates made up less than half as much of the Earth's surface as they do today and were unified as the continent Arctica. It may have been early life's production of oxygen that reacted with and lowered the amount of the greenhouse gas methane in the atmosphere, tipping the Earth toward a series of cooling feedbacks and causing ice to spread from pole to pole.

An ice-covered planet would remain frozen because almost all of the sun's energy would be reflected back into space, but this did not happen on Earth because of **plate tectonics**—the Snowball Earth cycle was eventually disrupted by volcanic activity. While the Earth was covered in ice, volcanoes continued to erupt, dumping carbon dioxide and methane into the atmosphere. While these gases are usually removed from the atmosphere by organisms and the weathering of rocks, this was not possible through miles of ice! After millions of years, the concentrations of methane and CO_2 increased to the point that greenhouse warming began to melt the **ice sheets**. Once the melting started, more of the sun's energy was absorbed by the surface, and warming feedbacks began. Because the oceans had been covered, nutrients derived from volcanic

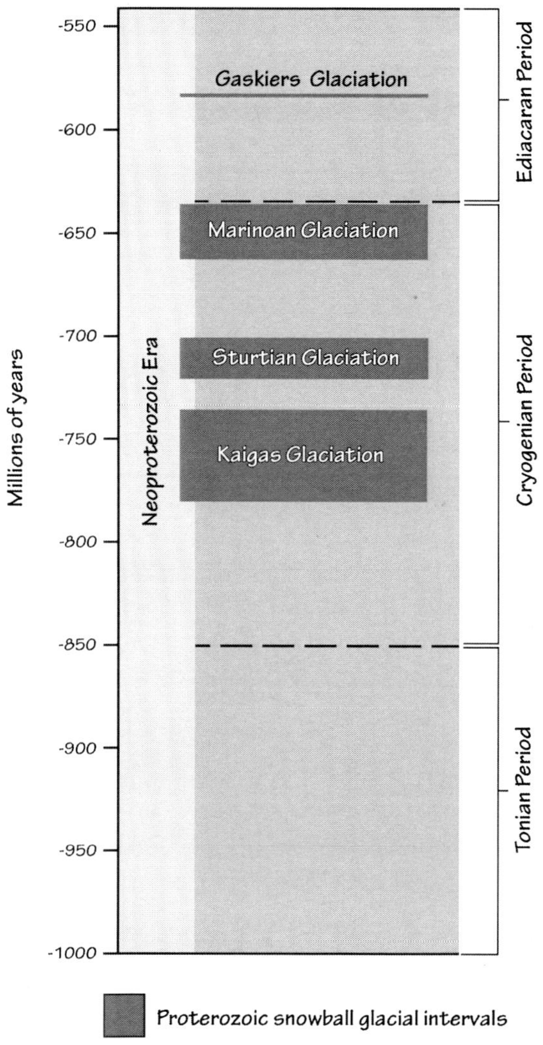

Figure 8.3: Snowball Earth periods during the Proterozoic.

gases and chemical changes in the rocks accumulated in the waters. Once they were re-exposed to light, a population explosion of **cyanobacteria** produced more and more oxygen, which was capable of combining with freshly thawed carbon sources to make more carbon dioxide, further enhancing the warming.

For the next 1.5 billion years, the Southwest, free of ice, drifted around the surface of the Earth. A new supercontinent—**Rodinia**—formed, and the part that is now North America was stable, creating what is known as a **craton**, or continental interior relatively free of the folding and **faulting** that characterizes continental margins that are subjected to mountain building and other plate tectonic processes. Since the Southwest was part of that craton, it was probably underwater for most of this time. About 850 million years ago, during the **Cryogenian**, the Earth entered a 200-million-year **ice age**, during which there were two more Snowball Earth cycles. Although the part of Rodinia that would eventually become North America was located near the equator, the fact that North America was at such a low latitude, yet had glaciers, is strong evidence

Past

cyanobacteria • *a group of bacteria, also called "blue-green algae," that obtain their energy through photosynthesis.*

Rodinia • *a supercontinent that contained most or all of Earth's landmass, between 1.1 billion and 750 million years ago, during the Precambrian.*

craton • *the old, underlying portion of a continent that is geologically stable relative to surrounding areas.*

fault • *a fracture in the Earth's crust in which the rock on one side of the fracture moves measurably in relation to the rock on the other side.*

Cryogenian • *a geologic period lasting from 850 to 635 million years ago, during the Precambrian.*

ice age • *a period of global cooling of the Earth's surface and atmosphere, resulting in the presence or expansion of ice sheets and alpine glaciers.*

tillite • *glacial till that has been compacted and lithified into solid rock.*

Proterozoic • *a geologic time interval that extends from 2.5 billion to 541 million years ago.*

Paleozoic • *a geologic time interval that extends from 541 to 252 million years ago.*

mass extinction • *the extinction of a large percentage of the Earth's species over a relatively short span of geologic time.*

sandstone • *sedimentary rock formed by cementing together grains of sand.*

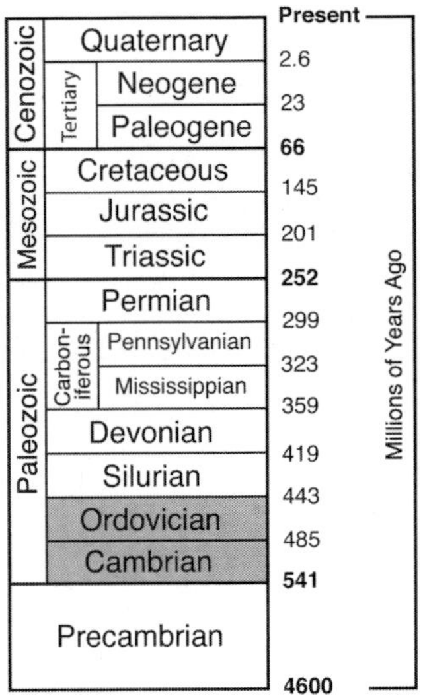

that the Earth really did freeze over completely. **Tillites**—rocks composed of ancient glacial sediment—found in Utah provide direct evidence for **Proterozoic** glaciation.

See Chapter 2: Rocks for more about the Proterozoic rocks of Utah.

Life and Climate

With the start of the **Paleozoic** era, climates across the world were warm, and North America was located in the low and warmer latitudes of the Southern Hemisphere. As the **Cambrian** progressed, North America moved northward, and what would become much of the Southwestern US was located near the Tropic of Capricorn (*Figure 8.4*). Shallow seas invaded the continent, ultimately covering the whole area until the late **Carboniferous**. During this time, the only exposed areas were islands in western Colorado and parts of New Mexico. Although there is a rich marine fossil record from the areas between these islands, we have no record of what kinds of plants colonized the land after land plants evolved in the late **Ordovician** and **Silurian**.

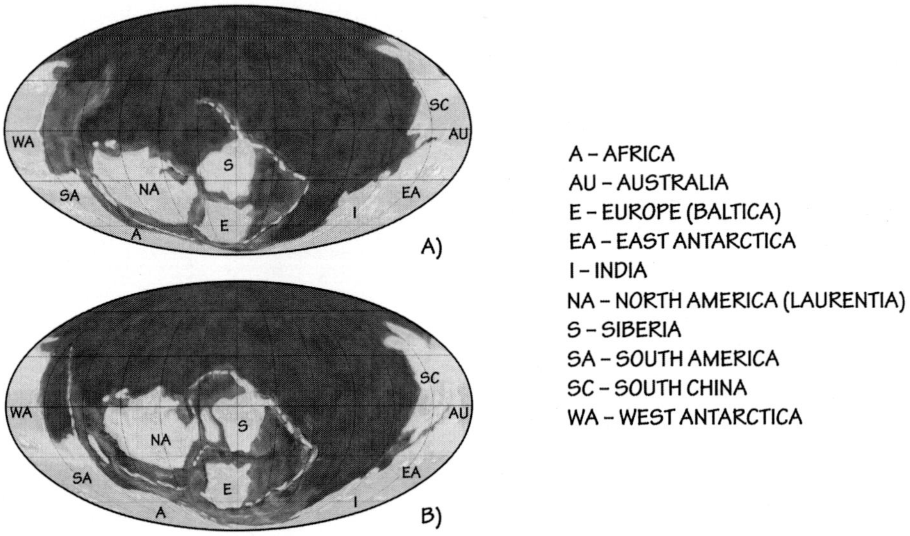

Figure 8.4: The location of the continents during the A) early and B) late Cambrian. Note the position of North America relative to the equator.

In the late Ordovician (about 460 to 430 million years ago), the Earth fell into another brief but intense ice age. Glaciers covered most of the world's southern landmasses, which were located over the South Pole. This led to global cooling, which was associated with the first of five major **mass extinctions** that have occurred over the last half-billion years. Although global sea level dropped during this event, North America's position near the equator kept its climate relatively warm. Ordovician deposits across the Southwest indicate warm, shallow seas rich in invertebrate life; shelly **sandstones** in Utah represent vast tidal flats.

Past

In the Silurian and **Devonian** (430 to 300 million years ago), North America moved north across the equator, and the cycle of warming and cooling was repeated yet again. Glaciation in the Southern Hemisphere occurred during the late Devonian, while the supercontinent **Gondwana** was located over the South Pole, and intensified during the early Carboniferous. At the same time—while the Southwest was still submerged—the oceans between Gondwana and North America began to close (*Figure 8.5*). In the early Carboniferous, ice capped the South Pole and began to expand northward. Much of the Southwest became an archipelago of warm shallow seaways and **uplifted** islands, with terrestrial swampy forests and shallow sea floors populated by **bivalves**, **brachiopods**, **arthropods**, corals, and fish.

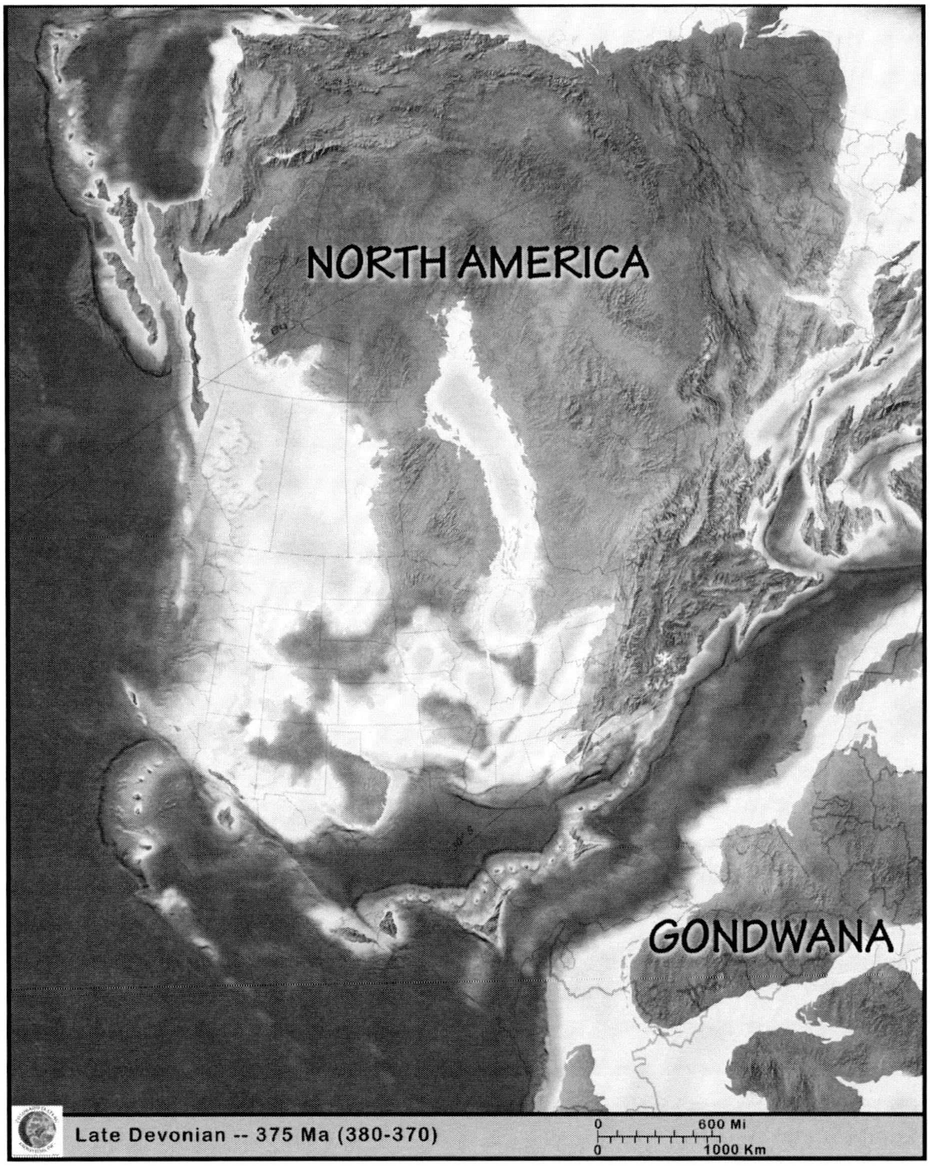

Figure 8.5: By the late Devonian (375 million years ago), the oceans between Gondwana and Euramerica had begun to close.

Gondwana • *the supercontinent of the Southern Hemisphere, composed of Africa, Australia, India, and South America.*

uplift • *upward movement of the crust due to compression, subduction, or mountain building.*

bivalve • *a marine or freshwater invertebrate animal characterized by right and left calcareous shells (valves) joined by a hinge.*

brachiopod • *a marine invertebrate animal characterized by upper and lower calcareous shell valves joined by a hinge, and a crown of tentacles (lophophore) used for feeding and respiration.*

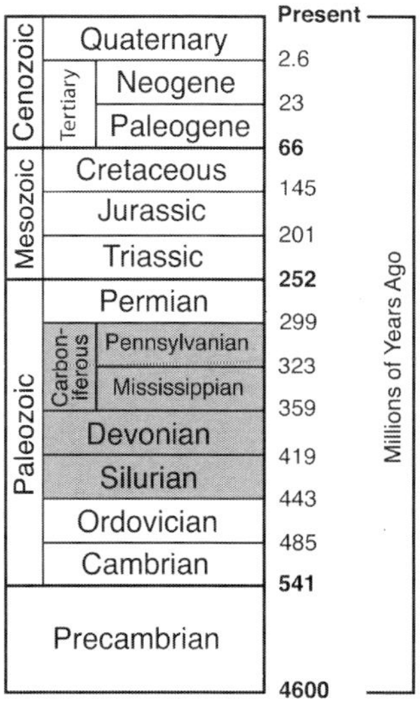

8 Climate

Past

Pangaea • *supercontinent, meaning "all Earth," which formed over 300 million years ago and lasted for almost 150 million years.*

gypsum • *a soft sulfate mineral that is widely mined for its use as fertilizer and as a constituent of plaster.*

synapsid • *a group of tetrapod vertebrates posessing one opening in the skull behind each orbit (eye hole), and a bony arch beneath.*

horsetail • *see sphenopsid: a terrestrial plant belonging to the family Equisetaceae, characterized by hollow, jointed stems with reduced, unbranched leaves at the nodes.*

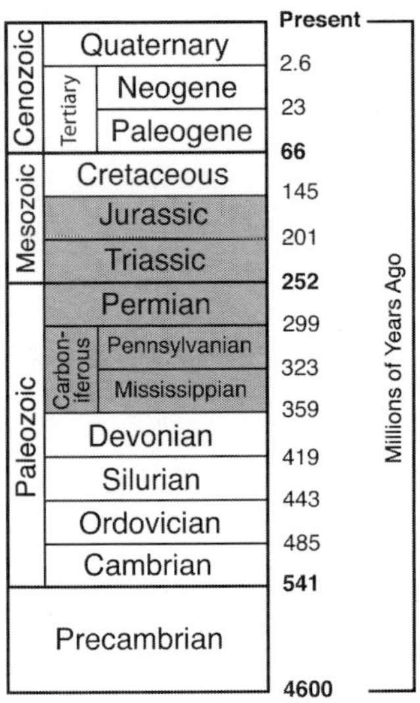

By the late Carboniferous, North America had collided with Gondwana, leading to the formation of **Pangaea**—a supercontinent composed of nearly all the landmass on Earth. Although the mountain building that occurred during this event was mostly far to the east, the Southwest was influenced by both fluctuating sea levels and a few significant tectonic changes. The climate remained warm, despite large southern ice sheets, but it had grown much drier. In the late Carboniferous, thick salt deposits accumulated in the northwestern Four Corners area as the seas evaporated. Where the land was exposed, deposits of dust (**loess**) accumulated and were blown across much of the Southwest. In southern New Mexico and Arizona, shallow marine deposits, laid down when the ice in Gondwana retreated and sea level rose, alternate with layers of dust blown in when the ice in Gondwana advanced and sea level fell. Loess is often, though not exclusively, associated with dry areas around glaciers. One controversial hypothesis proposes that an area of western Colorado—one of the islands that dotted the early Carboniferous sea—was, in fact, glaciated.

During the **Permian**, shallow marine waters gave way to lowland coastal areas across portions of the Southwest. Extensive Permian deposits throughout the Southwest are home to a host of fossils, including terrestrial amphibians, reptiles, and **synapsids**. The climate was drier than that of the Carboniferous, and mudflats with salt and **gypsum** formed across the Southwestern states. **Sand** dunes started to become widespread. A shift in plant type—from water-loving ferns and **horsetails** to those better adapted to drier conditions—further suggests a change in climate during the Permian. A large, low-latitude desert formed along Pangaea's western margin, generating extensive dune deposits.

See Chapter 2: Rocks to learn more about the cross-bedded Coconino Sandstone and other Permian dune deposits.

By the end of the Permian, the southern ice sheets had disappeared. As the Triassic period began, the Southwest moved north from the equator. The world warmed, and would stay warm through the Mesozoic. The continued growth of Pangaea created an intense monsoonal climate, similar to that of Asia today, that affected large parts of the continent. As Pangaea reached its greatest size during the early Triassic, the monsoon's intensity increased, and the vast dune deserts of the late Permian were replaced by rivers and floodplains. Soils associated with these floodplains testify to the extreme seasonality of rainfall during that time. The monsoon's intensity waned by the early Jurassic, and the rivers and floodplains were replaced by even larger deserts. The Southwest's Triassic-Jurassic dune deposits are some of the most extensive in the world, and the dune field that existed during the Jurassic may be the largest in Earth history. These deposits, including the Navajo Sandstone, are responsible for spectacular scenery in the national parks and recreation areas of northernmost Arizona and southern Utah. Despite the area's arid climate, the dunes were surprisingly full of life, particularly in southeastern Utah. Here, oases with large trees, large colonies of burrowing animals, and reptile **trackways** punctuated the otherwise dry and sandy landscape. These oases were fed by groundwater that originated in the higher country of what is now western Colorado. Later in the Jurassic, the climate became more moderate; dune fields were replaced by rivers and floodplains populated by a rich **dinosaur** fauna (exemplified by the Morrison Formation) and large trees along rivers, streams, and grasslands.

Pangaea began to break up during the Jurassic, **rifting** apart into continents that would drift toward their modern-day positions (*Figure 8.6*). The supercontinent was split by spreading along the mid-Atlantic ridge, initiating the formation of the Atlantic Ocean. As a result of displacement due to continental rifting and sea-floor spreading, sea level throughout the **Cretaceous** was much higher than it is today. Global temperatures during the Cretaceous were very warm, as much as 10°C (18°F) above those at present. There was likely little or no glacial ice anywhere on Earth, and temperatures were highest in lower latitudes. Shallow seaways spread over many of the continents, including South America, Africa, Eurasia, and North America. In the middle Cretaceous, oceans covered most of the Southwest, with the exception of parts of Arizona and New Mexico. By the start of the late Cretaceous, this **inland sea**, called the Western Interior Seaway, divided North America in two (*Figure 8.7*); the water was rich with **mosasaurs**, giant clams, and other marine life. In the late Cretaceous, however, sea level dropped and the western Southwest became a broad coastal plain that hosted lush forests, abundant dinosaurs, and large swamps. By the end of the Cretaceous, uplift to the west was great enough that the resulting hills shed large amounts of sand and **gravel** in an easterly direction, pushing the shoreline eastward until sediment (combined with a worldwide drop in sea level) filled the area formerly occupied by the Western Interior Seaway. As the continents moved closer to their modern positions, the Southwest experienced a hot and humid tropical climate. At the close of the Mesozoic, global climate—though warmer than today—was cooler than at the start of the era.

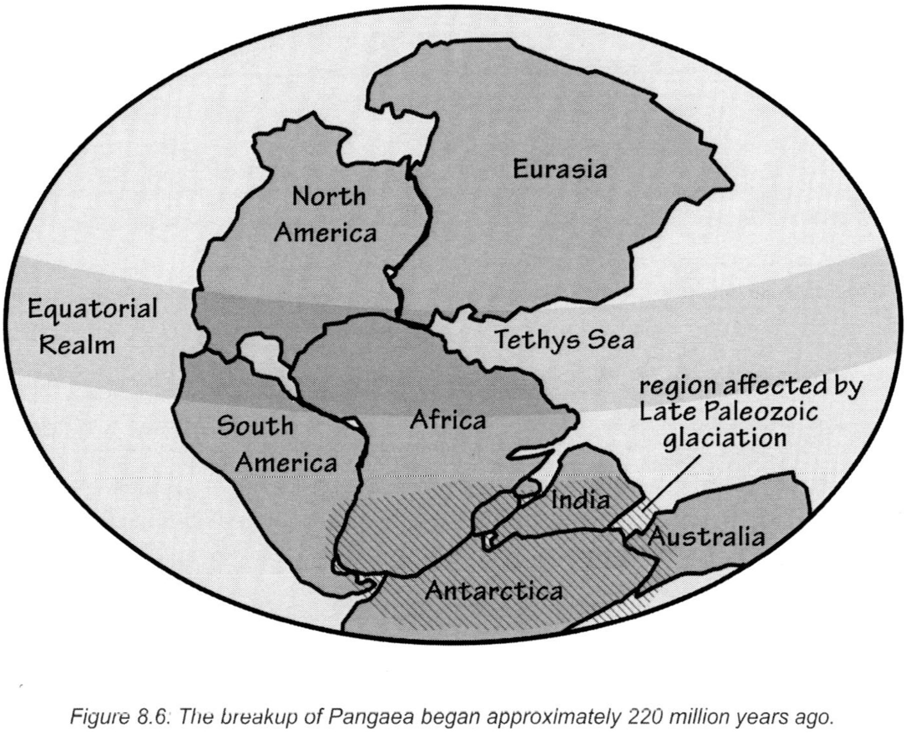

Figure 8.6: The breakup of Pangaea began approximately 220 million years ago.

rift • *a break or crack in the crust that can be caused by tensional stress as a landmass breaks apart into separate plates.*

inland sea • *a shallow sea covering the central area of a continent during periods of high sea level.*

mosasaur • *an extinct, carnivorous, marine vertebrate reptile characterized by a streamlined body for swimming, a powerful fluked tail, and reduced, paddle-like limbs.*

gravel • *unconsolidated, semi-rounded rock fragments larger than 2 millimeters (0.08 inches) and smaller than 75 millimeters (3 inches).*

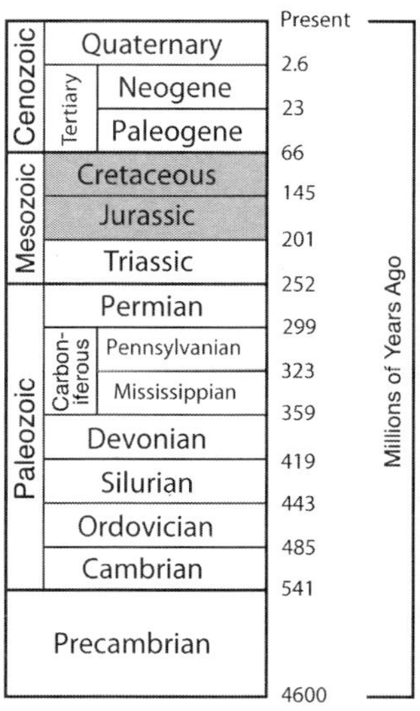

8 Climate

Past

bolide • an extraterrestrial object of any composition that forms a large crater upon impact with the Earth.

Eocene • a geologic time period extending from 56 to 33 million years ago.

Paleocene • a geologic time interval spanning from about 66 to 56 million years ago.

extinction • the end of species or other taxonomic groups, marked by death of the last living individual.

Oligocene • a geologic time interval spanning from about 34 to 23 million years ago.

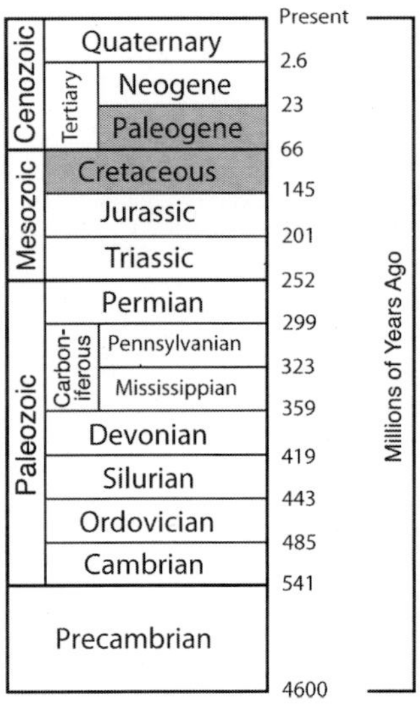

Figure 8.7: The Western Interior Seaway.

At the very end of the Cretaceous, the Gulf Coast experienced an enormous disruption when a large asteroid or **bolide** collided with Earth in what is now the northern Yucatán Peninsula in Mexico. The impact vaporized both water and rock, blocking out sunlight for weeks to years, which led to a collapse of photosynthesis and food webs on land and in the oceans. The event devastated the Southwest, shifting a densely forested landscape to one primarily covered with fast-growing herbs and ferns.

After this event, the climate may have cooled briefly, but it soon rebounded to a warmer state, and the world reached one of its warmest episodes during the **Eocene**. Right at the boundary between the **Paleocene** and Eocene epochs (approximately 56 million years ago), temperatures spiked upward in what geologists call the Paleocene-Eocene Thermal Maximum. During this event, which lasted perhaps only approximately 10,000 years, the atmosphere and ocean warmed by as much as 8°C (14°F) in as little as 4000 years, and deep oceans became acidic, with low levels of dissolved oxygen. The causes of this event remain unclear, but may have involved the sudden release of methane from sediments on the seafloor. The resulting greenhouse effect persisted for 100,000 years. The abrupt climatic change was associated with major migrations, the **extinction** of plants and animals on land, and a mass extinction in the deep sea. The Southwest's climate was warm and wet, with strong volcanic activity, and large mammals roamed the forested landscape. Large lakes covered parts of northern Utah and Colorado.

In the late Eocene, the Earth began to cool, and global temperatures fell sharply at the boundary between the Eocene and **Oligocene** epochs (approximately 35 million years ago), due in part to the separation of South America's southern tip from Antarctica. This allowed for the formation of the Antarctic Circumpolar Current, which insulated Antarctica from warm ocean water coming from lower latitudes and led to the formation of the continent's glaciers. The continents approached their modern configuration, and India began to collide with Asia to form the Himalayas. Global temperatures fell further in the late **Miocene** thanks to the formation of the Himalayas—this event had a significant impact on global climate, as weathering of the newly exposed rock began to serve as a sink to take up atmospheric CO_2. With the reduction of this greenhouse gas, temperatures cooled worldwide, and this cooling has continued more-or-less to the present day. Volcanic activity intensified in the Southwest, and the Basin and Range region began to form, leading to the **topography** that is seen in those areas today (i.e., low valleys alternating with high mountain ranges). While most of the evidence for cooling at the Eocene-Oligocene boundary comes from the deep sea, fossil mammals in the Rocky Mountains show clear evidence of a change from forests to grasslands, which is associated with global cooling.

See Chapter 4: Topography to learn more about the volcanism and tectonics that shaped the Basin and Range.

Silicate and *carbonate rocks* both weather chemically in reactions that involve CO_2 and water, typically creating *clays*, bicarbonate, and calcium ions. Silica weathering occurs relatively slowly, taking place on a large scale in the weathering and *erosion* of mountain ranges, and it may have an impact on atmospheric carbon dioxide levels on time scales of tens or hundreds of millions of years. On the other hand, carbonate rocks weather (in this case, dissolve) quickly, relative to silicates. In both cases, the products of weathering often end up in seawater, where they may be used in the *calcium carbonate* skeletons of marine organisms or taken up during photosynthesis. Skeletal material and organic matter often sink to the sea floor and become buried, effectively removing carbon from the global carbon cycle (and thereby the atmosphere) for many millions of years.

Miocene • *a geological time unit extending from 23 to 5 million years ago.*

topography • *the landscape of an area, including the presence or absence of hills and the slopes between high and low areas.*

silica • *a chemical compound also known as silicon dioxide (SiO_2).*

carbonate rocks • *rocks formed by accumulation of calcium carbonate, often made of the skeletons of aquatic organisms.*

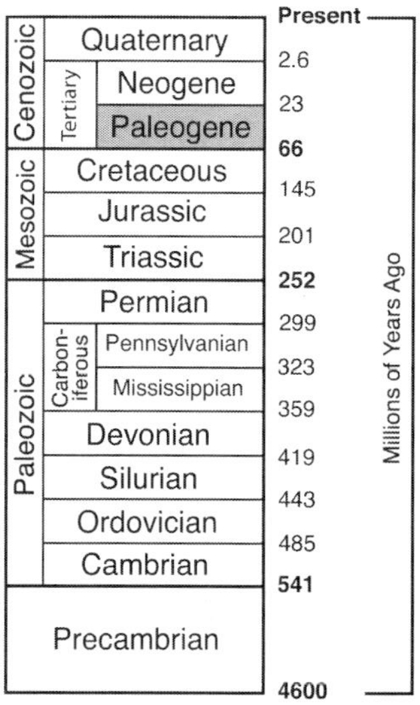

8 Climate

Past-Present

Approximately 3.5 million years ago, glacial ice began to form over the Arctic Ocean and on the northern parts of North America and Eurasia. Surprisingly, a major contributing factor to this event was a geological change that occurred half a world away. The Central American Isthmus, which today makes up most of Panama and Costa Rica, rose out of the ocean at approximately this time, formed by undersea volcanoes. The new dry-land isthmus blocked the warm ocean currents that had been flowing east-to-west from the Atlantic to the Pacific for more than 100 million years, diverting them into the Gulf of Mexico and ultimately into the western Atlantic Gulf Stream. The strengthened Gulf Stream carried more warm, moist air with it into the northern Atlantic, which caused increased snowfall in high latitudes, leading to accelerating cooling. These changes in ocean circulation throughout the Caribbean and Gulf of Mexico also affected nutrient supplies in the coastal ocean, which may have contributed to an increase in the extinction of marine animals (including everything from mollusks and corals to whales and dugongs) during the late **Pliocene**.

__Pliocene__ • a geologic time interval extending from roughly 5 to 2.5 million years ago.

__ice cap__ • an ice field that lies over the tops of mountains.

__Pleistocene__ • a subset of the Quaternary, lasting from 2.5 million to about 11,700 years ago.

__pluvial lake__ • a landlocked basin that fills with rainwater or meltwater during times of glaciation.

__heat__ • a form of energy transferred from one body to another as a result of a difference in temperature or a change in phase.

Era	Period		Millions of Years Ago (start)
			Present
Cenozoic	Quaternary		2.6
Cenozoic	Tertiary	Neogene	23
Cenozoic	Tertiary	Paleogene	66
Mesozoic	Cretaceous		145
Mesozoic	Jurassic		201
Mesozoic	Triassic		252
Paleozoic	Permian		299
Paleozoic	Carboniferous	Pennsylvanian	323
Paleozoic	Carboniferous	Mississippian	359
Paleozoic	Devonian		419
Paleozoic	Silurian		443
Paleozoic	Ordovician		485
Paleozoic	Cambrian		541
	Precambrian		4600

Eventually, a sheet of sea ice formed over the Arctic, and ice sheets spread over northern Asia, Europe, and North America, signaling the start of the most recent ice age. Since a mere 800,000 years ago, a type of equilibrium has been reached between warming and cooling, with the **ice caps** growing and retreating primarily due to the influence of astronomical forces (i.e., the combined gravitational effects of the Earth, Sun, moon, and planets). The large ice sheets in the Northern Hemisphere did not extend into the Southwest, even at their largest. However, large glaciers were found at higher elevations, and temperatures were cool. Fossil mammals adapted to colder temperatures are found in the **Pleistocene** of Colorado. In southern New Mexico, Pleistocene fossil mammals are found that now live at higher elevations in the mountains of northern New Mexico, indicating cooler temperatures and more available moisture in the area during the late Pleistocene. Large lakes formed in low areas, and the Southwest's most striking ice age feature was Lake Bonneville—a massive **pluvial lake** that covered much of Utah (*Figure 8.8*). Its remnant exists today as the Great Salt Lake. The last glacial advance of the modern ice age peaked some 18,000 years ago, and today nearly all the glaciers in the Southwest are gone, while the climate is now in an arid state.

See Chapter 1: Geologic History to learn more about the formation of glacial lakes during the last ice age.

Present Climate of the Southwest

The location of the Southwest and the topographical extremes across this area strongly influence its weather. The Southwest experiences nearly every variety of extreme weather; **heat** and cold waves, droughts, floods, blizzards, and even **tornados** are all considerations for residents of the Southwestern states.

See Chapter 9: Earth Hazards for more information on extreme weather in the Southwest.

Present

tornado • *a vertical funnel-shaped storm with a visible horizontal rotation.*

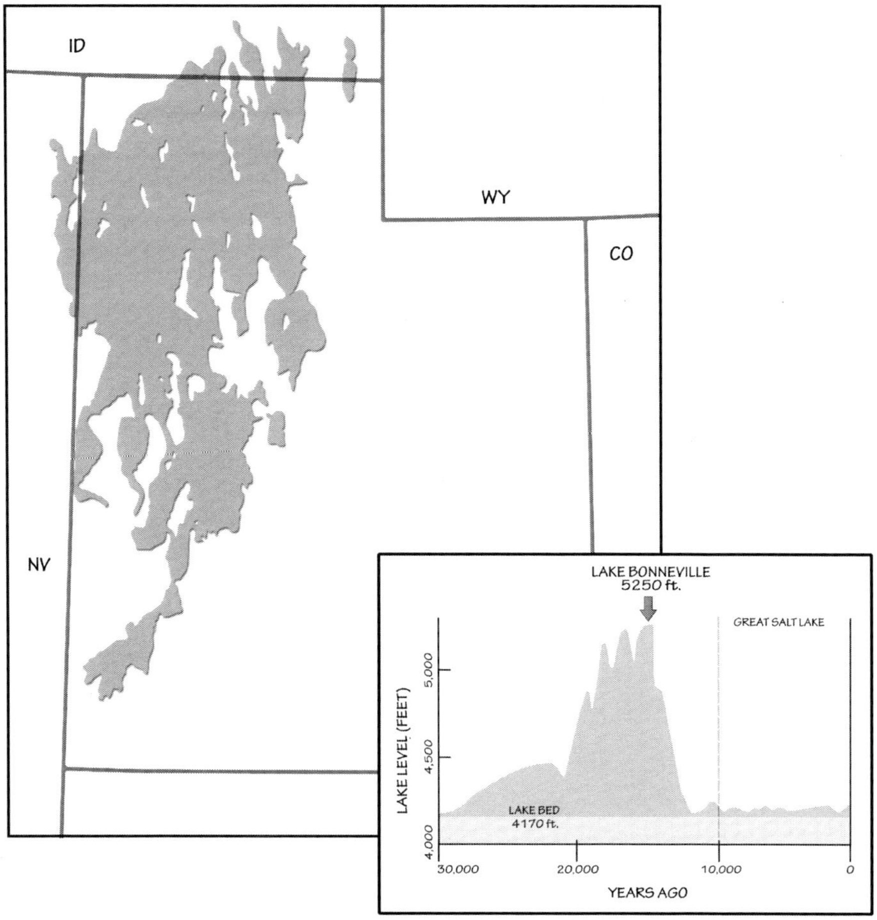

Figure 8.8: Lake Bonneville's maximal extent during the Pleistocene. Inset: Graph of the lake's changing level.

Although much of the Southwest falls within the category of an arid zone, using a single label to describe the Southwest's climate would belie its diversity. The main features that influence the area's climate are latitude, regional topography, and a low atmospheric moisture content that leads to quick evaporation. For example, parts of the Colorado Rockies experience cool annual temperatures and over 8 meters (25 feet) of snowfall every year, while the dry deserts in southwestern Arizona receive only about 8 centimeters (3 inches) of precipitation a year and can experience as much as a 15°C (60°F) degree temperature difference between night and day.

Average temperatures found in the Southwest tend to decrease northward, which is largely the influence of latitude and elevation. Lower latitudes receive more heat from the sun over the course of a year: for each degree increase in latitude, there is approximately a 1°C (2°F) decrease in temperature. Higher elevations (such as those found in the Rockies and on the Colorado Plateau) are also cooler, with approximately a 1.5°C (3°F) decrease in mean annual temperature for each 300-meter (1000-foot) increase in elevation.

8 Climate

Present

The warmest temperatures in the Southwest are found in Arizona and New Mexico, while the coolest are found in Utah and Colorado (*Figure 8.9*). The Southwest's overall average high temperature of 19.2°C (66.6°F) and average low of 2.8°C (37.0°F) are indicative of a varied climate, one much less uniform than that found in many other parts of the United States. By comparison, the average high and low temperatures for the entire United States are 17°C (63°F) and 5°C (41°F), respectively.

Another factor besides latitude and elevation that influences temperature in the Southwest is its arid climate. The lack of moisture in the air allows heat trapped in the earth during daylight hours to rapidly radiate away, leading to cool evenings. Thus, each Southwestern state experiences both extreme highs and lows. In New Mexico, for example, the average difference between the daily high and low temperatures ranges from 14° to 19°C (25° to 35°F). Record high temperatures for the Southwest range from 53°C (128°F) in Arizona to 47°C (117°F) in Utah, while record low temperatures range from −56°C (−69°F) in Utah to −40°C (−40°F) in Arizona.

Average Annual Temperatures			
	Overall (°C [°F])	**Low (°C [°F])**	**High (°C [°F])**
Arizona	16.0 (60.8)	7.6 (45.7)	24.5 (76.1)
New Mexico	11.9 (53.2)	3.0 (37.4)	20.3 (68.5)
Utah	9.1 (48.4)	1.5 (34.7)	16.7 (62.1)
Colorado	7.5 (45.5)	−0.9 (30.4)	15.4 (59.7)

The average amount of precipitation for the United States is 85.6 centimeters (33.7 inches). In the Southwest, average precipitation ranges from only 34 centimeters (13.4 inches) in Utah to 39.9 centimeters (15.7 inches) in Colorado, which is indicative of the area's general aridity (*Figure 8.10*). Elevation does, however, play a key role in precipitation received throughout the Southwest. In New Mexico, for example, average annual precipitation ranges from less than 25 centimeters (10 inches) within the Great Plains and Basin and Range regions to more than 50 centimeters (20 inches) at the higher elevations to the northwest. Arizona's highest elevations receive an average of 65 to 76 centimeters (25 to 30 inches), with lower areas in the state's southwestern portion averaging less than 8 centimeters (3 inches). In Utah, areas below 1200 meters (4000 feet) receive less than 25 centimeters (10 inches) per year, while higher elevations in the Wasatch Mountains receive more than 100 centimeters (40 inches).

Across New Mexico, Arizona, and Utah, summer rains originate from moisture brought into the area from the Gulf of Mexico. Warm, moist air from the south occasionally but infrequently moves into Colorado during the summer. During the winter, moisture travels from the west, as storms from the Pacific Ocean move east. Pacific storms lose most of their moisture as they pass over the Rocky Mountains, so much of the Southwest's winter precipitation falls as snow within the area's mountainous regions.

The Köppen Climate Map

Wladimir Köppen developed a commonly used system of climate categorization based on the kinds of vegetation areas sustain. He defined 12 climate types, many of which are familiar: rainforest, monsoon, tropical savanna, humid subtropical, humid continental, oceanic, Mediterranean, steppe, subarctic, tundra, polar ice cap, and desert. Updated by Rudolf Geiger, it has been refined to five groups, each with two to four subgroups.

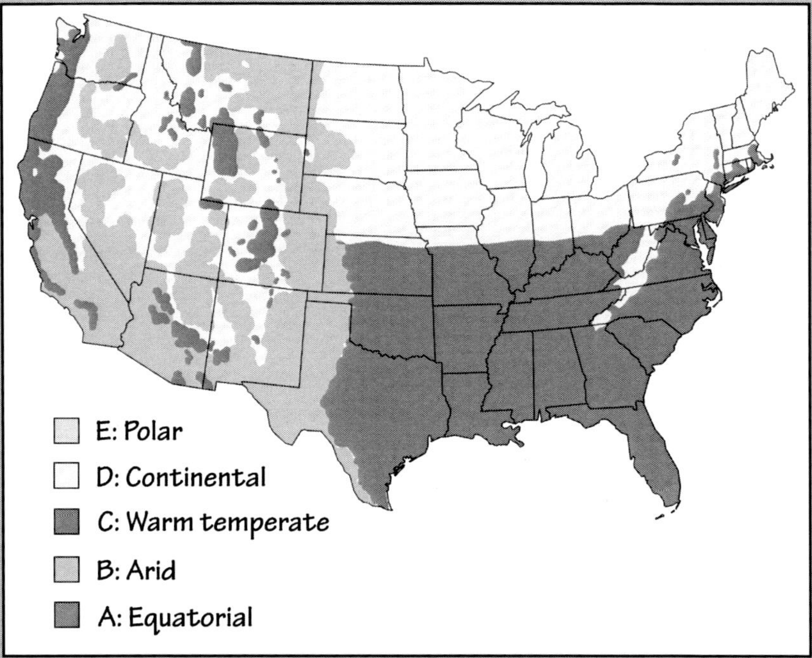

(See TFG website for full-color version.)

In a broad sense, the Southwest's climate is mostly dry and hot, with much of the region characterized as arid (represented by "B" in the *Köppen system*). Such conditions are common throughout the Great Plains, Colorado Plateau, and Basin and Range. Cold continental conditions (represented by "D") dominate the higher altitudes, especially within the Rocky Mountains. Scattered pockets of drier, Mediterranean temperatures (represented by "C") can also be found.

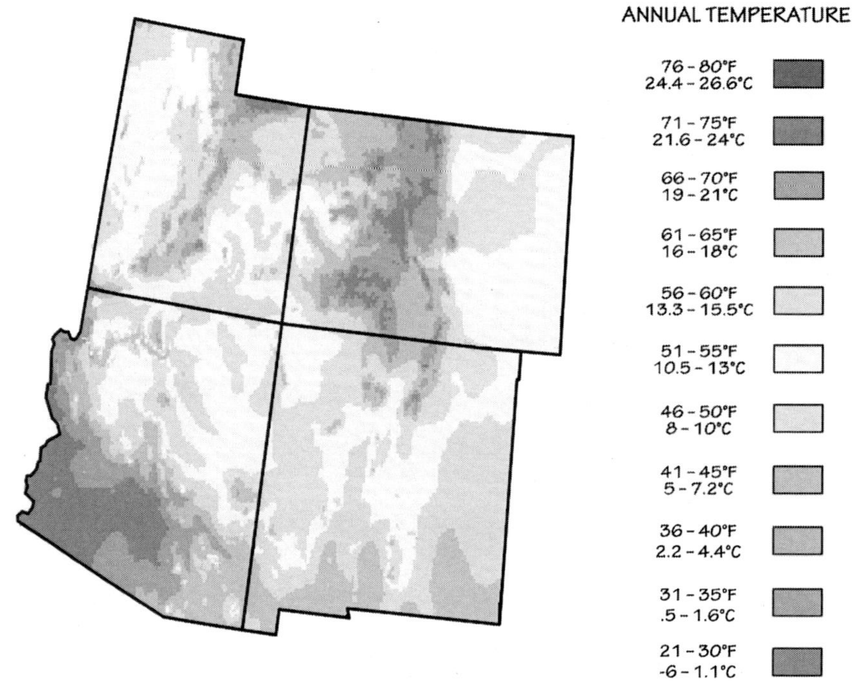

Figure 8.9: Mean annual temperature for the Southwestern states. (See TFG website for full-color version.)

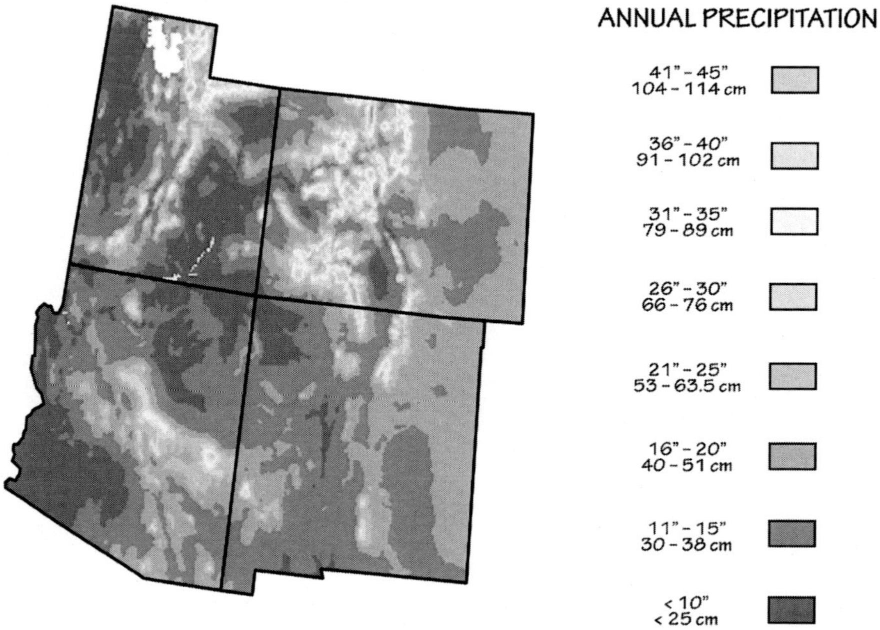

Figure 8.10: Mean annual precipitation for the Southwestern states. (See TFG website for full-color version.)

Present

A strong temperature difference at different heights creates instability—the warmer the air near the surface is relative to the air above it, the more potential energy it has to move up. The Great Plains receive warm, moist air moving north from the Gulf of Mexico, and cold, dry air moving in from the Rocky Mountains and the northern US. Where these air masses meet, vigorous mixing causes thunderstorms. Because warm air can hold more moisture than cool air can, **convective** mixing with cool air forces moisture to condense out of warm air as vapor (clouds) and precipitation. This movement of air in different directions is also the reason for the high incidence of powerful tornados that occur along "Tornado Alley" in the Great Plains (*Figure 8.11*).

See Chapter 9: Earth Hazards to learn more about tornados in the Southwest.

convection • *the rise of buoyant material and the sinking of denser material.*

wind • *the movement of air from areas of high pressure to areas of low pressure.*

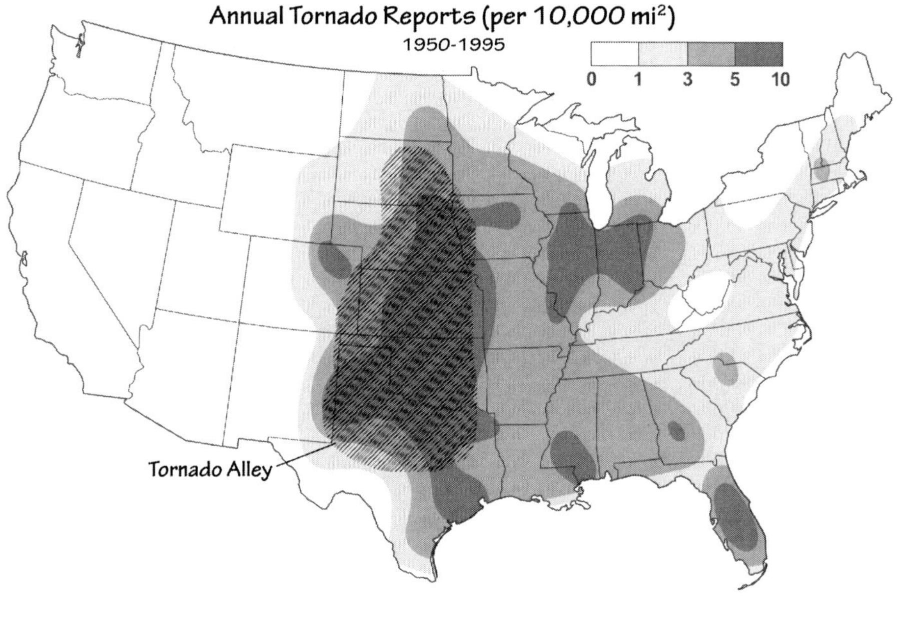

Figure 8.11: Frequency of tornados in the continental US. "Tornado Alley" is an area of the central US known for its violent tornados.(See TFG website for a full-color version.)

Colorado has a generally cool and continental climate, with low humidity. The climate of the eastern plains is fairly uniform, with hot, **windy** summers and prevalent thunderstorms. The state's highest temperatures occur in the northeastern plains, where they can exceed 46°C (115°F). Moving westward, Colorado's foothills and mountainous areas experience an overall cooler climate and higher levels of precipitation. Here, the state's varied topography leads to wide changes in climactic conditions that occur across short distances. For example, the difference in annual mean temperature between Pikes Peak (4302 meters [14,114 feet]) and Las Animas (1188 meters [3898 feet]), only 145 kilometers (90 miles) to the southeast, is equivalent to that between Iceland and southern Florida! Precipitation also varies widely—Cumbres in the San Juan Mountains receives nearly 7.6 meters (300 inches) of snowfall annually, while Manassa, less than 50 kilometers (30 miles) away in the San Luis Valley, receives only about 63 centimeters (25 inches) of snow a year.

8 Climate

Present-Future

hurricane • *a rapidly rotating storm system with heavy winds, a low-pressure center, and a spiral arrangement of thunderstorms.*

interglacial • *a period of geologic time between two successive glacial stages.*

anthropogenic • *caused or created by human activity.*

Arizona's climate is influenced by three main topographical areas: the high Colorado Plateau (about 1520–2130 meters [5000–7000 feet] in elevation), the rugged mountains to the west (2740–3660 meters [9000–12,000 feet] high), and the low southwestern mountains with desert valleys (as low as 30 meters [100 feet] above sea level). While the state is generally arid, its high western mountains experience more precipitation each year than the desert southwest and the high northeastern plateau do. The desert also experiences higher temperature extremes, especially between day and night, with a daily change of as much as 15°C (60°F) during the driest parts of the year.

See Chapter 4: Topography for a list of the highest and lowest elevations by state.

In New Mexico, climate is characterized by arid, semiarid, or continental conditions, with light precipitation, low humidity, and abundant sunshine. As in Arizona, the desert experiences a high temperature range daily; the state's mountainous areas, however, have climate characteristics that more closely follow those found in the Colorado Rockies. Summer rains fall almost entirely during brief but intense thunderstorms on the Great Plains, although the occasional **hurricane** in the Gulf of Mexico may push heavier precipitation inland. Winter is the driest season in New Mexico, because precipitation from eastward-travelling Pacific storms is left behind in the western mountains of Arizona and Utah.

Utah's distance from both the Pacific Ocean and the Gulf of Mexico prevents heavy precipitation, and much of the state is typically sunny year-round, with light to moderate winds, although changes in atmospheric pressure during the late fall and winter can lead to an accumulation of haze. Light precipitation travels eastward over the Sierra Nevada and Cascade mountains after dropping heavy snowfall in areas of high elevation. Because high mountains to the west and north act as a barrier to cold Arctic air masses, most areas of Utah rarely experience temperatures below freezing or prolonged periods of extreme cold.

Future Climate of the Southwest

By using techniques that help to reconstruct past climates, and by tracking trends in the present, we can predict how current climates might change. Overall, the world is warming, yet, as we are still in an ice age, eventually the current **interglacial** period should end, allowing glaciers to advance towards the equator again (although likely not for about 100,000 years). However, because the Earth is already getting warmer, the effects of **anthropogenic** warming are

The Earth's orbit, tilt, and wobble alter its position with respect to the Sun, affecting the global climate. These changes in the Earth's movement are cyclical, and the changes in Earth's climate associated with them are known as *Milankovitch cycles.*

amplified through feedback. Some scientists worry that, if not curbed, human activity could actually disrupt the cycle and knock the planet entirely out of the interglacial period, melting all the ice on Earth.

Causes of Change

While astronomical and tectonic forces will continue to cause climatic shifts, they act so slowly that they will be overshadowed in the near term by human-induced effects. In 1956, NOAA established the Mauna Loa Observatory (MLO) in Hawai'i to measure a variety of atmospheric parameters, including carbon dioxide (CO_2) concentration. The CO_2 record extends from 1958 to present, and it shows the influence of both natural and anthropogenic processes (*Figure 8.12*). The zigzag pattern is the result of seasonal photosynthesis in the Northern Hemisphere. In spring and summer, the growth and increased photosynthetic activity of plants draws CO_2 out of the atmosphere. Conversely, it accumulates in the atmosphere during fall and winter when plants are dormant. The overall upward trend is caused by human activity. Industrialization, **fossil fuel** combustion, and deforestation all contribute CO_2 to the atmosphere, adding it at a rate much faster than natural processes can remove it. Analyses of ancient atmosphere samples preserved in glacial ice cores show CO_2 levels to have been 180 parts per million (ppm) at the height of the last ice age and 280 ppm at its end. The amount of CO_2 in the atmosphere has been increasing at a rapid rate since the start of the industrial revolution, and it has accelerated since the end of World War II. In May 2013, measurements at MLO reached 400 ppm CO_2 for the first time.

fossil fuels • *fuel for human use that is made from the remains of ancient biomass.*

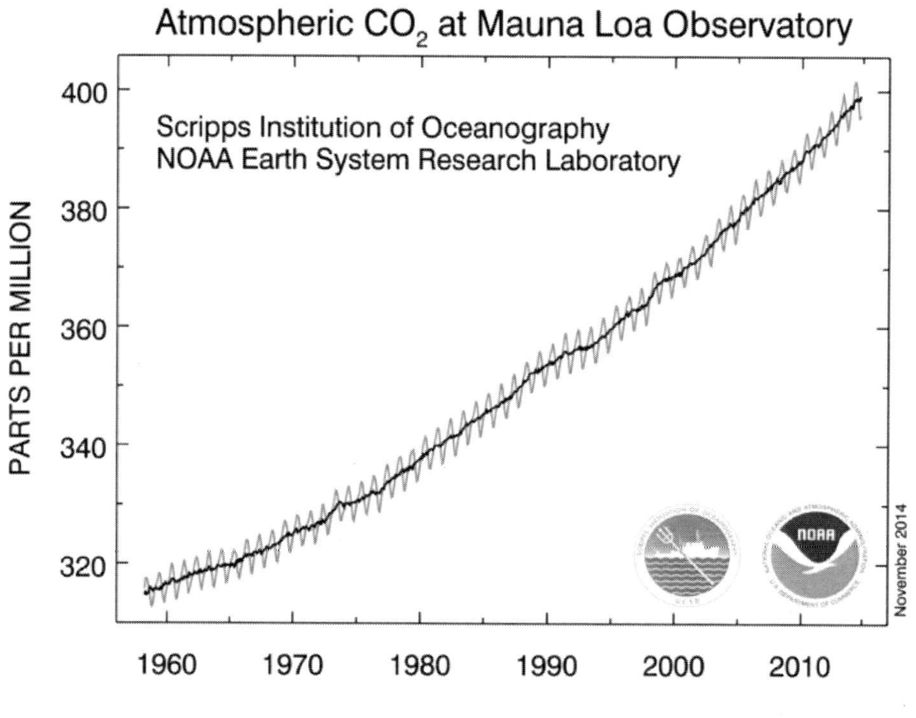

Figure 8.12: Measured concentration of atmospheric carbon dioxide (1958 to present) at MLO.

8 Climate

Future

permafrost • *a layer of soil below the surface that remains frozen all year round. Its thickness can range from tens of centimeters to a few meters.*

aerosol • *tiny solid or liquid particles in the air.*

climate change • *See global warming: the current increase in the average temperature worldwide, caused by the buildup of greenhouse gases in the atmosphere.*

petroleum • *a naturally occurring, flammable liquid found in geologic formations beneath the Earth's surface.*

fuel • *a material substance possessing internal potential energy that can be transferred to the surroundings for specific uses.*

natural gas • *a hydrocarbon gas mixture composed primarily of methane (CH4), but also small quantites of hydrocarbons such as ethane and propane.*

While some atmospheric carbon dioxide is necessary to keep Earth warm enough to be a habitable planet, the unprecedentedly rapid input of CO_2 to the atmosphere by human beings is cause for concern. Everything we know about atmospheric physics and chemistry tells us that increased CO_2 leads to a warmer planet. Multiple paleoclimate data sets verify this conclusion, and modern measurements confirm that we are living in an increasingly warmer world. The increasing heat is causing glaciers and sea ice around the globe to melt, and as the ground and ocean they covered is exposed, these darker surfaces absorb and re-radiate increasing amounts of heat.

As **permafrost** in high latitudes melts, carbon in the soil becomes free to enter the atmosphere and, worse, buried organic material can be converted by bacteria into the even more potent greenhouse gas methane. Less directly, higher temperatures lead to more frequent and severe droughts, which, in turn, lead to more wildfires that release carbon and **aerosols** into the atmosphere. Aerosols can have a cooling effect since they reflect away radiation from the sun, but they can also pose a public health hazard.

Water is extremely good at absorbing heat: water vapor is actually the most effective greenhouse gas. Higher temperatures increase evaporation and allow the air to retain more water. While water vapor feedback is the most significant reinforcer of climate warming, water tends to move out of the atmosphere in a matter of weeks—other greenhouse gases, such as carbon dioxide and methane, linger in the atmosphere for years.

The Southwest contributes significantly to **climate change**. The population of any industrialized and particularly wealthy country produces pollution; the majority of these emissions come from the use of **petroleum**. The more than 16 million residents of the Southwest use carbon-rich fossil **fuels** to provide electricity for lighting, cooling, and appliances, to fuel their transportation and industry, and to make the products they use. Burning those fossil fuels releases carbon into the atmosphere, which warms the Earth. Of the Southwestern states, Arizona emits the most greenhouse gases, releasing 94 million metric tons of carbon dioxide per year. Although this pales in comparison to emissions from the nation's highest CO_2 producer—Texas, which releases nearly 656 million metric tons of CO_2 per year—Arizona's greenhouse gas emissions are rising rapidly compared with the nation as a whole. In the last decade, the United States has *decreased* the total amount of energy-related carbon dioxide emissions by almost ten percent, yet Arizona's emissions have *increased* by 9% thanks to a growing population that relies heavily on oil and **natural gas** for energy. Emissions from Colorado and Utah have also increased over the past decade, growing 7% and 2%, respectively.

See Chapter 6: Energy for more information about fossil fuel extraction in the Southwest.

On the other hand, Southwestern states are making changes to reduce human impact on the climate. New Mexico has reduced its CO_2 emissions by more than four metric tons in the last decade. The cities of Aspen and Lafayette, Colorado, as well as the state of New Mexico, were early adopters of the 2030 Challenge, an effort to reduce fossil fuel use in buildings so that both new and renovated buildings would qualify as carbon neutral by the year 2030. Additionally, states are beginning to step up their use and production of **renewable energy**. As of 2015, Arizona ranks 30th in the nation for renewable energy production, much of which it produces from hydroelectricity and **biomass**.

renewable energy • *energy obtained from sources that are virtually inexhaustible (defined in terms of comparison to the lifetime of the Sun) and replenish naturally over small time scales relative to human life spans.*

biomass • *organic material from one or more organisms.*

Trends and Predictions

Studies show that the Southwest's climate is changing right now, and that change has accelerated in the latter part of the 20th century. These changes include the following:

- The number of days with temperatures above 35°C (95°F) and nights above 24°C (75°F) has been steadily increasing since 1970, and the warming is projected to continue (*Figures 8.13 and 8.14*).
- The onset of stream flows from melting snow in Colorado has shifted two weeks earlier due to warming spring temperatures. Flows in late summer are correspondingly reduced, leading to extra pressure on the state's water supplies.
- Streamflow totals for the last decade in the Great Basin, Rio Grande, and Colorado River were between 5% and 37% lower than their 20th-century averages.
- Since 1980, tree mortality in forests and woodlands across the Southwest has been higher and more extensive than at any time during the previous 90-year record; this is attributed to higher temperatures, drought, and the eruption of bark beetles that are able to survive through warmer winter weather.
- Increased heat in the Pacific Ocean has altered the weather patterns of Pacific storms, decreasing snowfall in the mountains of western Utah and Arizona.
- In the last decade, the Southwest's frost-free season has increased by approximately 7% compared to the average season length for the 20th century.
- The seasonality and transmission frequency of insect-borne diseases and other infectious diseases prevalent in the Southwest, including plague, valley fever, and Hanta, are influenced by warming trends.

heat wave • *a period of excessively hot weather that may also accompany high humidity.*

Recent warming within the Southwest has been among the most rapid in the United States, and models predict that the area's climate will continue to warm. The average annual temperature in most of the Southwest is predicted to rise 2.2° to 5.5°C (4° to 10°F) by 2100. Summer **heat waves** will become hotter and longer, while winter cold snaps will occur less often. These increased temperatures lead to a whole host of other effects, including a decrease in snowpack, declines in river flow, drier soils from more evaporation, and the increased likelihood of drought and fires. In winter, rising temperatures have increased the amount of frost-free days—today, most of the Southwest

Future

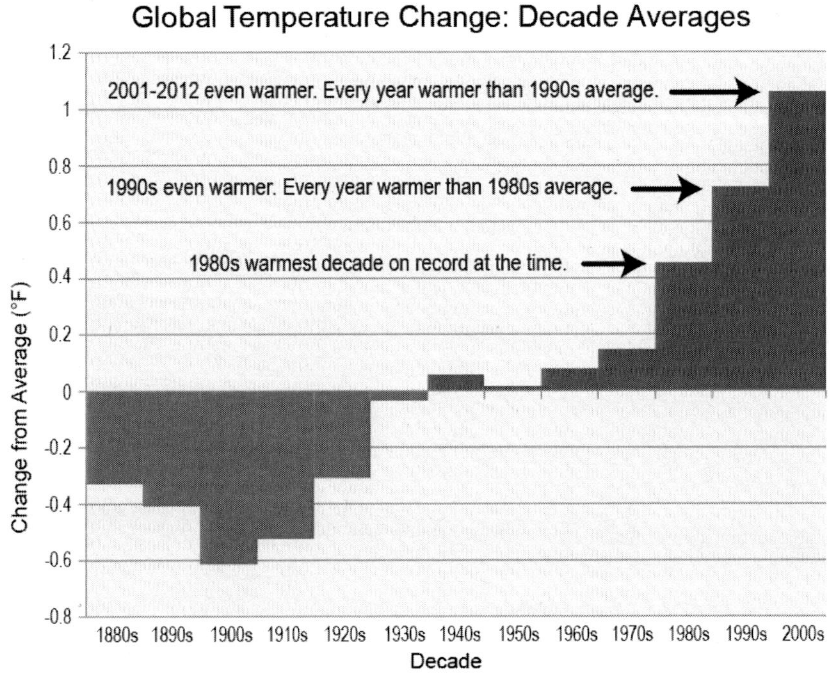

Figure 8.13: Global temperature change since the 1880s. The Earth's average surface temperature has progressively risen over the last five decades.

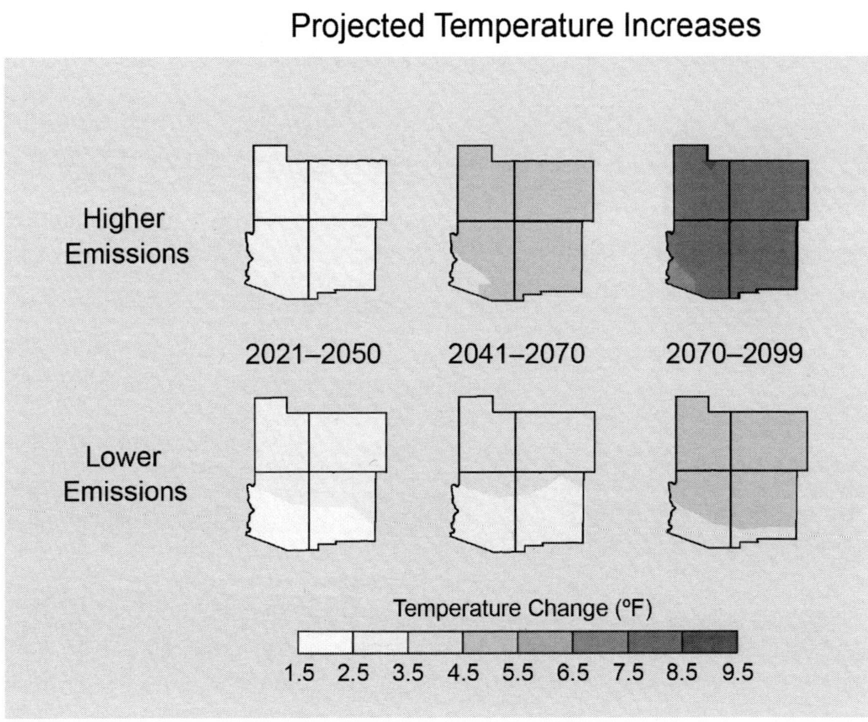

Figure 8.14: Projected temperature increases for the Southwestern states over the next century, as compared to the average for 1971–1999. The "higher emissions" scenario assumes emissions continue to rise, while the "lower emissions" scenario assumes a substantial reduction in emissions. In both cases, temperatures will continue to rise. (See TFG website for a full-color version.)

experiences about 17 fewer freezing days than it did over the last century. By 2070, one can expect up to 38 more days of freeze-free weather each year (*Figure 8.15*). These warmer temperatures and increased precipitation have helped bring on longer growing seasons. While changes in the growing season can have a positive effect on some crops (such as melons and sweet potatoes), altered flowering patterns due to more frost-free days can lead to early bud bursts, damaging **perennial** crops such as nuts and stone fruits.

perennial • *continuous; year-round or occurring on a yearly basis.*

Warmer temperatures also make it easier for insect pests to overwinter and produce more generations. Bark beetles, which normally die in cold weather, have been able to survive through the winter and reproduce, increasing tree mortality. For example, high winter temperatures between 2000 and 2003 correlated to bark beetle outbreaks that devastated pinyon pine throughout the Southwest, leading to nearly 90% mortality at some sites in Colorado and Arizona. As of 2010, bark beetles in Arizona and New Mexico have affected more than twice the forest area burned by wildfires in those states.

Water supply is an important issue in the Southwest, and communities will need to adapt to changes in precipitation, snowmelt, and runoff as the climate changes. Agriculture accounts for more than half of the Southwest's water use, so any major reduction in the availability of water resources will create a serious strain on ecosystems and populations. Drier days and higher temperatures will amplify evaporation, increasing the desertification of already arid areas and affecting natural ecosystems as well as increasing pressure on the water supply for agriculture and cities (*Figure 8.16*). An increased frost-free season length also leads to increased water demands for agriculture and heat stress on plants. Cattle ranches throughout the Southwestern states rely on rain-fed grazing

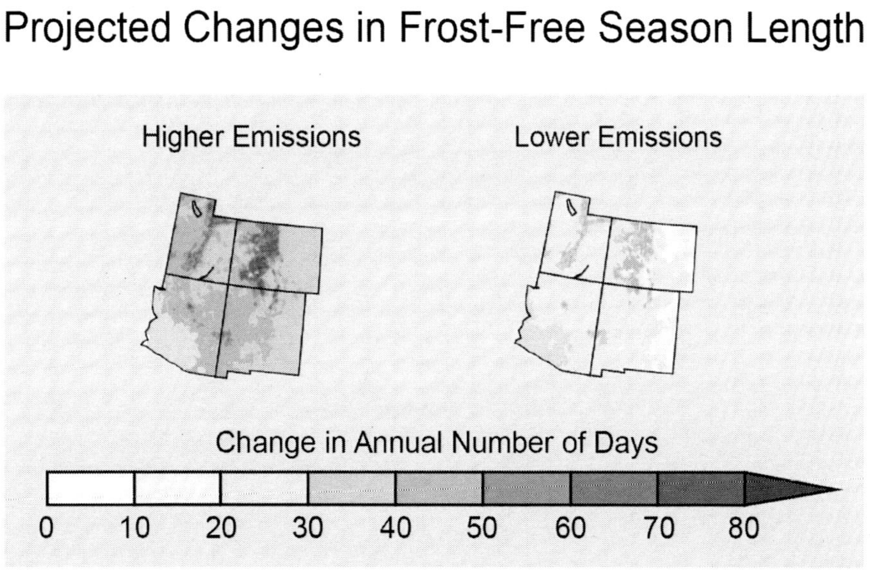

Figure 8.15: Projected frost-free days for the Southwestern states over the next century, as compared to the average for 1971–2000. The "higher emissions" scenario assumes emissions continue to rise, while the "lower emissions" scenario assumes a substantial reduction in emissions. Gray areas are projected to experience more than 10 frost-free years. (See TFG website for a full-color version.)

Future

forage, making them extremely susceptible to climate change and drought. In addition, temperature increases and recent drought lead to earlier spring snowmelt and decreased snow cover on the lower slopes of high mountains, bringing about more rapid runoff and increased flooding. These changes to rain and snow-pack are already stressing water sources and affecting agriculture. Precipitation has become more variable from year to year, and heavy downpours across the US have increased in the last 20 years. Because higher temperatures mean greater evaporation and warmer air can hold more water, precipitation will occur in greater amounts at a time, but less frequently. Although there has so far been little regional change in the Southwest's annual precipitation, the area's average precipitation is expected to decrease in the south and remain stable or increase in the north. Most models predict a decrease in winter and spring precipitation by the middle of the century, and more frequent precipitation extremes during the last half of the century.

See Chapter 9: Earth Hazards for more about the effects of climate change on the environment.

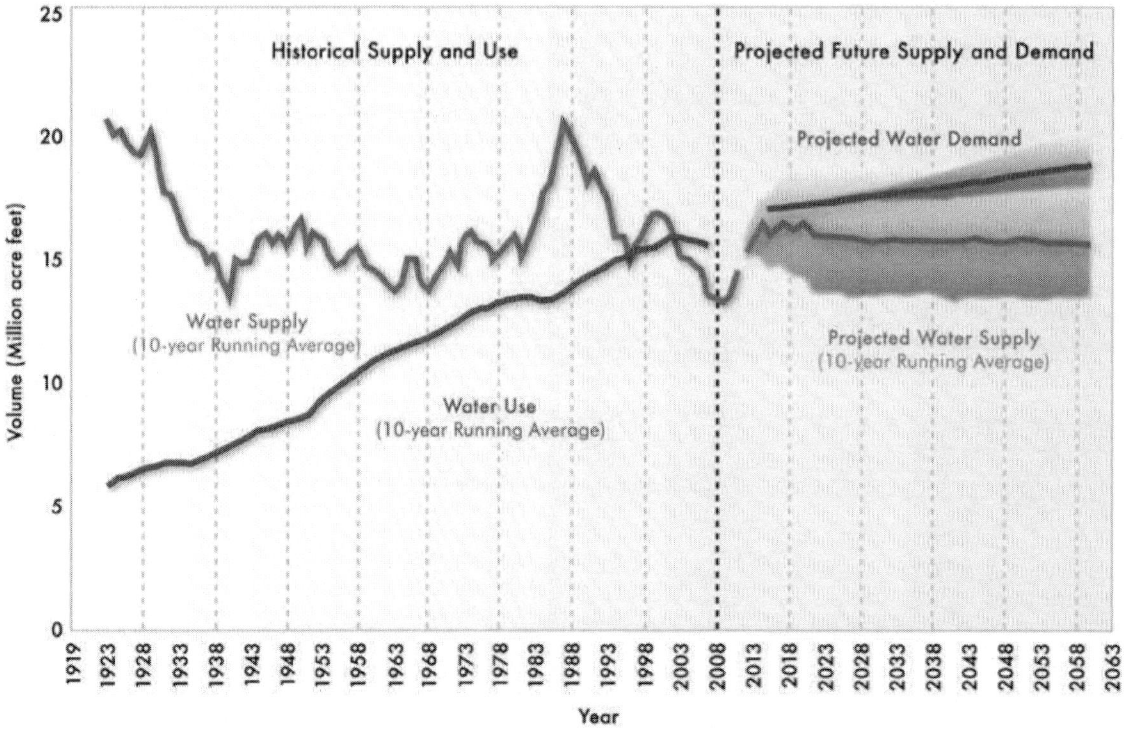

Figure 8.16: Projected 21st-century supply-demand imbalance for the use of water from the Colorado River. The Colorado drains roughly 15% of the continental United States, and is relied upon for municipal and agricultural use by over 35 million people in seven states. (See TFG website for a full-color version.)

The causes of specific weather events such as tornados and severe thunderstorms are incredibly complex, although climate change has enhanced some correlated factors, such as increased wind speed and an unstable atmosphere. Higher atmospheric moisture content has also been correlated with an increased incidence of tornados and winter storms. However, although climate change is predicted to enhance the intensity of severe weather, there is currently no way to calculate what effect climate change will have on the frequency of specific storm events—for example, we might see more powerful tornados, but we do not know if we will see *more* of them.

All over the Southwestern US, residents and communities have begun to adapt to climate change, and to plan for future changes that are expected to come.

Resources

Books and Articles on Climate

Allmon, W. D., T. A. Smrecak, and R. M. Ross, 2010, *Climate Change—Past, Present & Future: a Very Short Guide*, Paleontological Research Institution, Ithaca, NY, 200 pp.

Committee on the Importance of Deep-Time Geologic Records for Understanding Climate Change Impacts, 2011, *Understanding Earth's Deep Past Lessons for our Climate Future*, National Academies Press, Washington, DC, http://www.nap.edu/download.php?record_id=13111.

Karl, T. R., J. M. Melillo, and T. C. Peterson, eds., 2009, *Global Climate Change Impacts in the United States*, Cambridge University Press, Cambridge, NY, 188 pp., http://downloads.globalchange.gov/usimpacts/pdfs/climate-impacts-report.pdf.

Kottek, M., J. Grieser, C. Beck, B. Rudolf, and F. Rubel, 2006, World map of Köppen-Geiger climate classification updated, *Meterologische Zeitschrift*, 15: 259–263, http://koeppen-geiger.vu-wien.ac.at/.

Melillo, J. M., T. C. Richmond, and G. W. Yohe, eds., 2014, *Climate Change Impacts in the United States: The Third National Climate Assessment*, US Global Change Research Program, 841 pp., http://www.globalchange.gov/nca3-downloads-materials.

Retallack, G. J. 2007. Cenozoic paleoclimate on land in North America, *The Journal of Geology*, 115: 271–294, http://blogs.uoregon.edu/gregr/files/2013/07/cenozoicnorthamerica-gmt02r.pdf.

Rosenzweig, C., A. Iglesius, X. B. Yang, P.R. Epstein, and E. Chivian, 2001, Climate change and extreme weather events—implications for food production, plant diseases, and pests, *Global Change and Human Health*, 2(2): 90–104, and *NASA Publications*, Paper 24, http://digitalcommons.unl.edu/cgi/viewcontent.cgi?article=1023&context=nasapub.

Ruddiman, W. F., 2014, *Earth's Climate: Past and Future, 3rd edition*, W. H. Freeman, New York, 445 pp.

Websites on Climate

Climate, National Oceanic and Atmospheric Administration, http://www.noaa.gov/climate.html.

Climate has Changed Throughout Earth's History, National Park Service, http://nature.nps.gov/geology/nationalfossilday/climate_change_earth_history.cfm.

Climate Literacy & Energy Awareness Network (CLEAN). [A rich collection of resources for educators.] http://www.cleanet.org.

Envisioning Climate Change Using a Global Climate Model, by B. Youngman, M. Chandler, L. Sohl, M. Hafen, T. Ledley, S. Ackerman, and S. Kluge, SERC Earth Exploration Toolkit, http://serc.carleton.edu/eet/envisioningclimatechange/index.html.

Global Climate Change: Vital Signs of the Planet, NASA. [Information about global climate change, including spectacular satellite images.] http://pmm.nasa.gov/education/websites/global-climate-change-vital-signs-planet.

Global Greenhouse Gas Reference Network, Global Monitoring Division, National Oceanographic and Atmospheric Administration Earth System Research Laboratory. [Data and visualizations.] http://www.esrl.noaa.gov/gmd/ccgg/data-products.html.

Global Weather, Jetstream, Online School for Weather, National Weather Service, http://www.srh.noaa.gov/jetstream/global/global_intro.htm.

Intergovernmental Panel on Climate Change, Fifth Assessment Report (AR5), http://www.ipcc.ch/.

JetStream, Online School for Weather, National Weather Service, National Oceanographic and Atmospheric Administration, http://www.srh.noaa.gov/jetstream/index.htm.

National Climate Assessment. [Reports summarizing impacts of climate change.] http://nca2014.globalchange.gov.

National Weather Service, National Oceanographic and Atmospheric Administration, http://www.weather.gov.

NOAA's El Nino Portal, National Oceanographic and Atmospheric Administration, http://www.elnino.noaa.gov/.

North America During the Last 150,000 Years, compiled by J. Adams, http://www.esd.ornl.gov/projects/qen/nercNORTHAMERICA.html.

Paleomap Project. [Maps and information about Earth's tectonic and climate history.] http://scotese.com/.
Regional Climate Trends and Scenarios for the U.S. National Climate Assessment, National Oceanographic and Atmospheric Administration, http://www.nesdis.noaa.gov/technical_reports/142_Climate_Scenarios.html.
US Map of Köppen-Geiger Climate Classification, http://koeppen-geiger.vu-wien.ac.at/pics/KG_USA.gif.
Weather Base. [Weather and climate data by country, state, and city.] http://www.weatherbase.com.
Weatherunderground Maps. [A variety of types of weather maps, including surface, temperature, moisture, wind, cloud cover, precipitation.] http://www.wunderground.com/maps.
Wind Energy Resource Atlas of the United States, by D. L. Elliott, C. G. Holladay, W. R. Barchet, H. P. Foote, and W. F. Sandusky, 1986, National Renewable Energy Laboratory, Golden, CO, http://rredc.nrel.gov/wind/pubs/atlas/atlas_index.html.

Websites on State- or Region-Specific Climate Resources

Climate Assessment for the Southwest (CLIMAS), The University of Arizona, http://www.climas.arizona.edu/.
Climate Change and Drought in the American Southwest, Climate Institute, http://climate.org/effects-of-21st-century-climate-change-on-drought-disk-in-the-american-southwest/.
Climate of Arizona, Western Regional Climate Center, Desert Research Institute, http://www.wrcc.dri.edu/narratives/ARIZONA.htm.
Climate of Colorado, Western Regional Climate Center, Desert Research Institute, http://www.wrcc.dri.edu/narratives/COLORADO.htm.
Climate of New Mexico, Western Regional Climate Center, Desert Research Institute, http://www.wrcc.dri.edu/narratives/NEWMEXICO.htm.
Climate of Utah, Western Regional Climate Center, Desert Research Institute, http://www.wrcc.dri.edu/narratives/UTAH.htm.
Colorado Plateau Carbon Connections, collaboration among Northern Arizona University, Biological Sciences Curriculum Study, and Oregon Public Broadcasting. [High school curriculum on climate change.] http://cpcc.bscs.org/index.html.
Extreme Weather and Climate Change: The Southwest, 2011, Climate Central, http://www.climatecentral.org/library/climopedia/extreme-weather-and-climate-change-the-southwest.
Fischetti, M., 2015, U.S. droughts will be the worst in 1,000 years: the southwest and central Great Plains will dry out even more than previously thought, *Scientific American*, http://www.scientificamerican.com/article/u-s-droughts-will-be-the-worst-in-1-000-years1/.
Garfin, G., A. Jardine, R. Merideth, M. Black, and S. LeRoy, eds., 2013, Assessment of Climate Change in the Southwest United States: a Report Prepared for the National Climate Assessment, Southwest Climate Alliance, Island Press, Washington, DC, 506 pp., http://swccar.org/sites/all/themes/files/SW-NCA-color-FINALweb.pdf.
Gordon, E. and D. Ojima, 2015, *Colorado Climate Change Vulnerability Study: a Report Submitted to the Colorado Energy Office*, University of Colorado at Boulder, and Colorado State University, Fort Collins, 175 pp., http://wwa.colorado.edu/publications/reports/co_vulnerability_report_2015_final.pdf.
Hereford, R., R. H. Webb, and S. Graham, 2002, Precipitation History of the Colorado Plateau Region, 1900–2000, *USGS Fact Sheet* 119-02, 4 pp., http://pubs.usgs.gov/fs/2002/fs119-02/fs119-02.pdf.
Kunkel, K. E, L. E. Stevens, S. E. Stevens, L. Sun, E. Janssen, D. Wuebbles, K. T. Redmond, and J.G. Dobson, 2013, Regional Climate Trends and Scenarios for the U.S. National Climate Assessment. Part 5. Climate of the Southwest U.S., *NOAA Technical Report* NESDIS 142-5, 79 pp., https://scenarios.globalchange.gov/regions/southwest.
NOAA, United States Climate Page. [Clickable state map of the US to get city-by-city climate data, 1961–1990.] http://www.esrl.noaa.gov/psd/data/usclimate/states.fast.html.
The Paleontology Portal. [North American fossil record and geologic and climate histories, by state.] http://paleoportal.org/.
Our Changing Climate: Southwest, National Climate Assessment, http://nca2014.globalchange.gov/report/regions/southwest.
Ray, A. J., J. J. Barsugli, and K. B. Averyt, 2008, *Climate Change in Colorado: a Synthesis to Support Water Resources Management and Adaptation*, A Report for the Colorado Water Conservation Board, University of Colorado at Boulder, CO, 52 pp., http://wwa.colorado.edu/publications/reports/WWA_ClimateChangeColoradoReport_2008.pdf.

8 Climate

Resources

Southwest Climate Change Network, The University of Arizona, http://www.southwestclimatechange.org/.

Southwest, Regional Impacts, Climate Nexus: Changing the Conversation, http://climatenexus.org/learn/regional-impacts/southwest.

Weatherbase. [Monthly averages and forecasts for cities for each state.] http://www.weatherbase.com/weather/state.php3?c=US&s=&countryname=United-States.

Chapter 9: Earth Hazards of the Southwestern US

Natural hazards or earth hazards are events or processes that have significant impacts on human beings and the environment. Extreme **weather** conditions or geologic activity can cause substantial short-term or long-term changes to our environment. These changes can influence many aspects of the world around us, including crops, homes, infrastructure, and the **atmosphere**. The 4.6-billion-year-old Earth has experienced many naturally generated hazards, while other events are byproducts of human activities, created during mineral and **energy** extraction or in construction practices that modify the landscape.

The Southwest, like any other part of the US, has numerous hazards—based largely on its geography—that directly infringe upon people's property and safety. Dangerously hot weather and drought are commonplace in the Southwest's arid environment. Weather hazards such as tornados, thunderstorms, and winter storms frequently occur over the Great Plains, thanks to the unobstructed movement of air masses over areas of low **topographic relief**. The Rocky Mountains are susceptible to extreme winter weather such as heavy snow, blizzards, and high **winds**. Flooding can occur in areas of low elevation and along large rivers. Geological hazards, including avalanches, **earthquakes**, **landslides**, and rockfalls, also occur throughout the Southwest, especially in areas with rugged, mountainous terrain.

weather • *the measure of short-term conditions of the atmosphere such as temperature, wind speed, and humidity.*

atmosphere • *a layer of gases surrounding a planet.*

energy • *the power derived from the use of physical or chemical resources.*

topography • *the landscape of an area, including the presence or absence of hills and the slopes between high and low areas.*

relief • *the change in elevation over a distance.*

wind • *the movement of air from areas of high pressure to areas of low pressure.*

Landslides

The term "landslide" refers to a wide range of **mass wasting** events that result in rock, **soil**, or fill moving downhill under the influence of gravity (*Figure 9.1*). These events occur when friction between the earth material (i.e., rock and soil) and the slope is overcome, allowing the earth material to fail and move downslope. Landslides may be triggered by high rainfall, earthquakes, **erosion**, deforestation, groundwater pumping, or **volcanic** eruptions. They range in size from the simple raveling of a stream embankment to the collapse of an entire mountainside that involves tens of thousands of cubic meters (yards) of material. In the Rocky Mountains, every year at least one road will be temporarily closed as the result of an avalanche, earth movement, or rockfall event. Mass wasting events can also dam streams and rivers, creating lakes. If such dams fail, a flood will result somewhere downstream.

Landslides are common in mountainous parts of the Southwest thanks to a combination of steep terrain, poorly consolidated sediments, and melting snowpack that leads to soil saturation (*Figure 9.2*). They often occur in high valleys with little vegetative cover. In years that are particularly wet or rainy,

CHAPTER AUTHORS

Luke P. McCann
Andrielle N. Swaby

9 Earth Hazards

Landslides

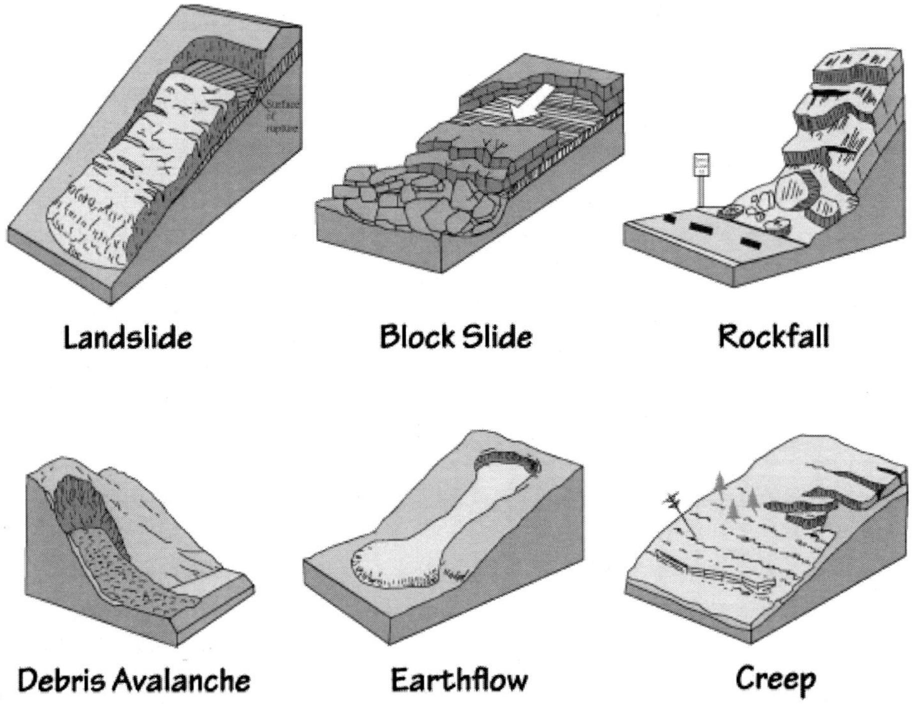

Figure 9.1: Common types of landslides.

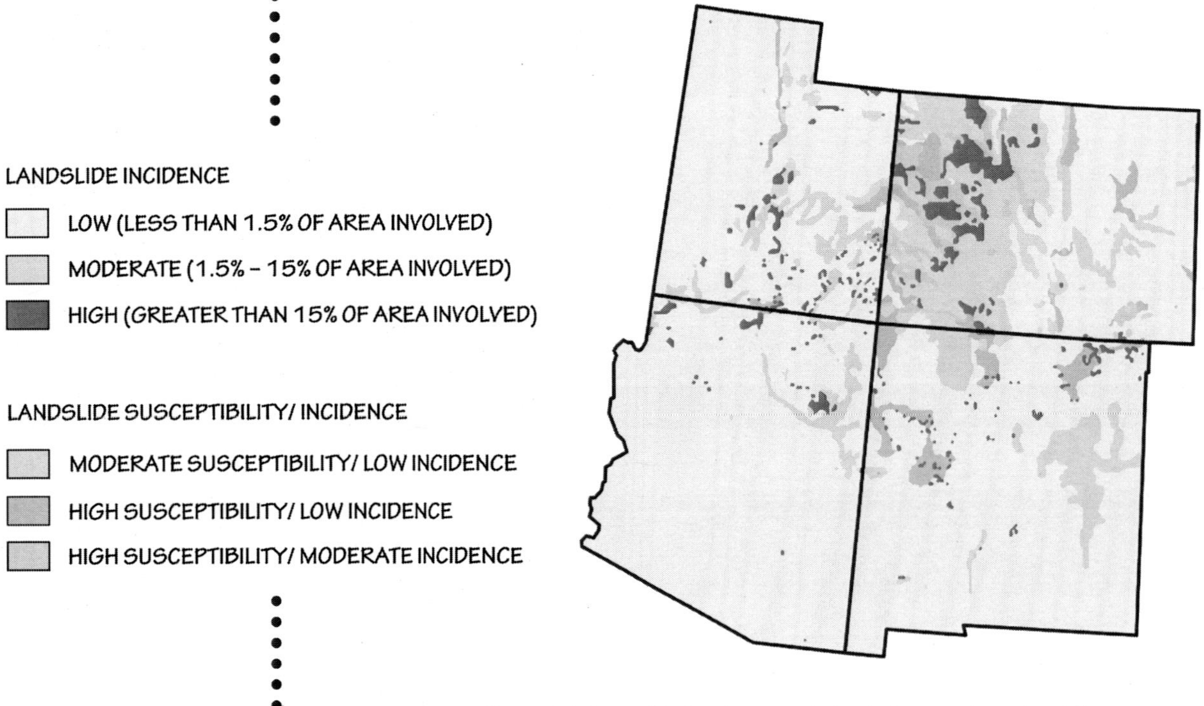

Figure 9.2: Landslide incidence and risk in the Southwestern US.

landslide incidence increases as unstable soils on saturated slopes break free of the rock. Some very fast landslides can reach speeds exceeding 32 kilometers per hour (20 miles per hour). Although many slides in the Rockies are small, or take place in remote and inaccessible locations, people and property are impacted each year by these events. In the winter, many of the same mountainous areas that are prone to landslides during the year are subject to avalanches—rapid flows of snow, ice, and rock. Avalanches occur when the strength of the snow is overcome, or when a weak layer in the snow fails. These snow failures can result from storms, warming weather, sunny slopes, earthquakes, and people moving over the snow. Hundreds of avalanches occur every winter in the mountains of Colorado and Utah.

Utah has seen some of the largest landslides in US history. In April 1983, a massive landslide dammed the Spanish Fork River, destroying roads and flooding the town of Thistle with more than 80 million cubic meters of water that backed up behind the naturally formed dam (*Figure 9.3*). This dam eventually created a lake 60 meters (200 feet) deep and 5 kilometers (3 miles) long. Thistle was almost completely destroyed, and the nearby railroad and highways had to be rebuilt on higher ground. While these transportation routes were closed, communities in eastern and southeastern Utah were completely cut off from the rest of the state for up to eight months. Direct and indirect costs of the Thistle landslide have been estimated to be as high as $950 million (adjusted for inflation); the state of Utah and the United States Geological Survey have categorized this landslide as the costliest in the nation. More recently, in April 2013, a massive landslide at Utah's Bingham Canyon Mine (also known as the Kennecott Copper Mine) displaced almost 70 million cubic meters (2.5 billion cubic feet) of dirt and rock from the side of the pit. This was the largest non-volcanic landslide in the history of North America; luckily, thanks to an early warning system, no injuries occurred. Two years later, the mine is still cleaning up debris from the slide. Massive landslides in Utah aren't just restricted to recent history, either. In 2014, scientists in Dixie National Park discovered the remnants of the largest known landslide anywhere on earth. This major prehistoric slide occurred 21 million years ago and stretched over 2700 meters (1700 miles)—an area the size of Rhode Island. Geologists studying the site have concluded that it originated when a volcanic field collapsed, and took place over an extremely short period of time, during which the friction of moving blocks pulverized and even melted the surrounding rocks.

Mudflows or earthflows are fluid, surging flows of debris that have been fully or partially **liquefied** by the addition of water. They can be triggered by heavy rainfall, snowmelt, or high levels of ground water flowing through cracked bedrock. Higher temperatures, thick melting snowpack, and an increase in spring rainstorms are thought to have generated the 2012 mudflow in Mesa County, Colorado, in which a slide five kilometers (three miles) long and 1.2 kilometers (¾ of a mile) wide claimed the lives of three men as well as triggered a small earthquake (*Figure 9.4*). The Grand Mesa area, where the slide occurred, is prone to landslides due to a soft underlying layer of claystone that erodes easily from runoff and snowmelt.

Landslides

liquefaction • *a process by which water-saturated unconsolidated sediment temporarily loses strength and behaves as a fluid when vibrated.*

9 Earth Hazards

Landslides

debris flow • *a dangerous mixture of water, mud, rocks, trees, and other debris that can move quickly down valleys.*

tree • *any woody perennial plant with a central trunk.*

fault • *a fracture in the Earth's crust in which the rock on one side of the fracture moves measurably in relation to the rock on the other side.*

joint • *a surface or plane of fracture within a rock.*

Figure 9.3: In spring 1983, a major landslide near Thistle, Utah created this dam and the resulting "Lake Thistle," inundating the town.

Figure 9.4: The catastrophic 2012 mudslide in Mesa County, Colorado, triggered by melting snow and unusually heavy rainfall, rushed down a mountain near the town of Collbran into the West Salt Creek Valley.

Landslides

Debris flows are a dangerous mixture of water, mud, rocks, **trees**, and other debris that moves quickly down valleys. The flows can result from sudden rainstorms or snowmelt that creates flash floods. In Chalk Cliffs, Colorado, one or more small debris flows occur every year after periods of intense rainfall. Though less hazardous than debris flows that occur in populated areas, these deposits have blocked roads and diverted streams (*Figure 9.5*). Debris flows can also occur in otherwise stable landscapes after the occurrence of large wildfires, which can destabilize the ground due to the removal of vegetation and desiccation of the soil. Heavy rainfall following the fire can then cause the burned slopes to fail. The Sandia and Manzano Mountain areas in central New Mexico have been studied extensively regarding their susceptibility to post-wildfire debris flows.

Figure 9.5: Remnants of a 2002 debris flow near Buena Vista, Colorado, which blocked Chaffee County Road 306 in 11 places and trapped several motorists.

In the Rocky Mountains, where the bedrock contains many discontinuities (folded bedding planes, **faults**, **joints**, and **cleavage**) resulting from several episodes of mountain building (the **Antler**, **Laramide**, and **Sevier orogenies**), rock slides and rockfalls are common, especially along transportation routes running through the mountains. US Highway 6 in Colorado, State Route 9 (the Zion-Mount Carmel Highway) in Utah, and I-70 in west-central Colorado are often impacted by rockfalls, leading to frequent road closures (*Figure 9.6*). Stretches of highway can remain closed for periods of several months. Rockfalls can also have fatal consequences in populated areas where buildings have been constructed in high-hazard zones (*Figure 9.7*)

Not all mass wasting events are rapid—slow land movement, known as soil **creep**, is generally not hazardous, but can impact structures over a long period

cleavage • *a physical property of minerals that occurs when it breaks in a characteristic way along a specific plane of weakness.*

Antler Orogeny • *a period of Mississippian mountain building that deformed rocks in a belt extending from the California–Nevada border northward into Idaho.*

Laramide Orogeny • *a period of mountain building that began in the Late Cretaceous, and is responsible for the formation of the Rocky Mountains.*

Sevier Orogeny • *a mountain-building event resulting from subduction along the western edge of North America, occurring mainly during the Cretaceous.*

orogeny • *a mountain-building event generally caused by colliding plates and compression of the edge of the continents.*

creep • *the slow movement or deformation of a material under the influence of pressure or stress.*

Figure 9.6: Boulders weighing as much as 60 metric tons (66 tons) are blasted and removed from I-70 at Glenwood Canyon, Colorado, after a major rockfall closed the highway. The rocks punched holes in elevated sections of the roadway; luckily, no one was injured.

Figure 9.7: This house in Rockville, Utah, was demolished in a 2013 rockfall that destroyed the house, garage, and car and killed two residents. A motorist who witnessed the event estimated that it lasted only 10 seconds. Multiple nearby houses are also located in this high-hazard zone.

Landslides

of time. **Slumps** and creep are common problems in parts of the Southwest with a wetter **climate** and/or the presence of unstable slopes, especially in the Great Plains and on the Colorado Plateau. Many areas in the Southwest contain expansive soils generated from **clay**-rich **parent materials**, especially **volcanic ash** or debris. Certain clay **minerals** can absorb water and swell up to twice their original volume. The pressures exerted through expansion of the minerals in the soil can easily exceed 22 metric tons per square meter (5 tons per square foot)—a force capable of causing significant damage to highways and buildings. An estimated $9 billion of damage to infrastructure built on expansive clays occurs each year in the United States, making swelling soils one of the costliest hazards. In addition, when the clay dries and contracts, the particles settle slightly in the downhill direction. This process can result in soil creep, a slow movement of land that causes fences and telephone poles to lean downhill, while trees adjust by bending uphill (*Figure 9.8*). Human development can exacerbate this process when homes are built along steep embankments, disturbing vegetation that would otherwise stabilize the slope or adding water to the land in the form of yard irrigation or septic systems.

See Chapter 7: Soils to learn more about the types and locations of expansive soils found in the Southwest.

Expansive soils can be found all over the US, and every state in the Southwest has bedrock units or soil layers that are possible sources (*Figure 9.9*). Clay minerals that expand and contract when hydrated and dehydrated due to their

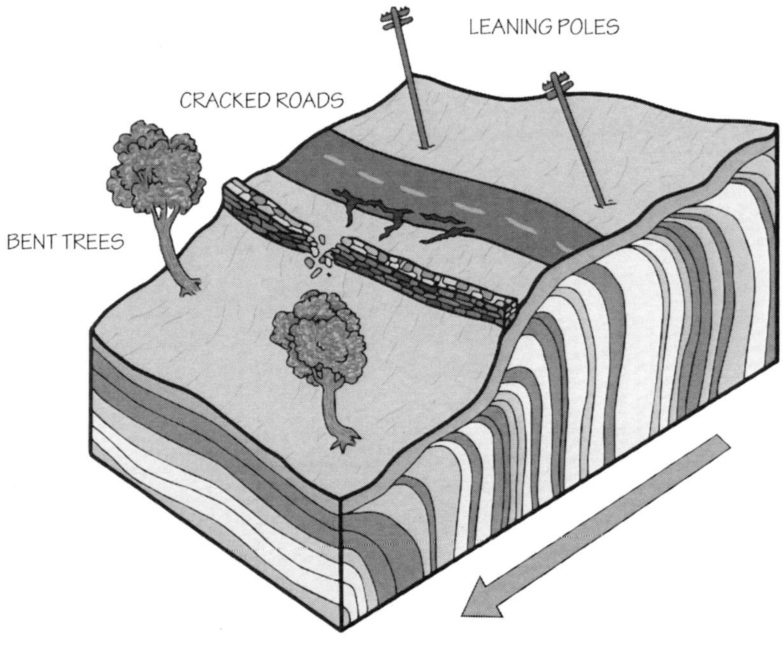

Figure 9.8: Some influences of soil creep on surface topography.

slump *• a slow-moving landslide in which loosely consolidated rock or soil layers move a short distance down a slope.*

climate *• a description of the average temperature, range of temperature, humidity, precipitation, and other atmospheric/hydrospheric conditions a region experiences over a period of many years (usually more than 30).*

clay *• the common name for a number of very fine-grained, earthy materials that become plastic (flow or change shape) when wet.*

parent material *• the original geologic material from which soil formed.*

volcanic ash *• fine, unconsolidated pyroclastic grains under 2 millimeters (0.08 inches) in diameter.*

mineral *• a naturally occurring solid with a specific chemical composition and crystalline structure.*

bentonite • *a clay, formed from decomposed volcanic ash, with a high content of the mineral montmorillonite.*

alluvial • *a layer of river-deposited sediment.*

layered molecular structure are generically referred to as smectite; soils that tend to form deep cracks during drought are often indicative of the presence of smectite. The Colorado Plateau and Great Plains regions have the highest risk of damage caused by swelling soil. Here, clays are typically composed of montmorillinite or **bentonite**, which have a very high shrink/swell potential. In the Basin and Range, the clay-rich beds of the Pantano Formation are prone to expansion, as are old **alluvial** fan surfaces along river **terraces**.

See Chapter 4: Topography for more on physiographic regions, including the Basin and Range.

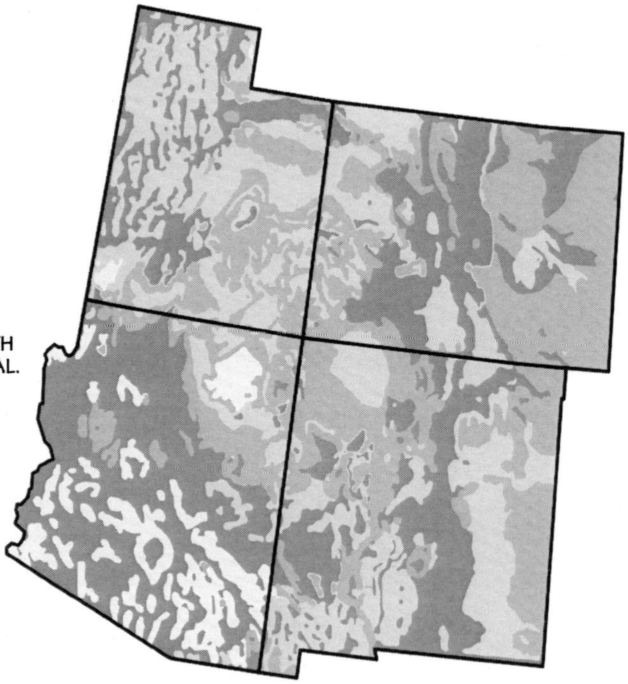

Figure 9.9: Approximate distribution of expansive soils in the Southwestern US. This map is based on the distribution of types of bedrock, which are the origin of soils produced in place. (Where substantial fractions of the soil have been transported by wind, water, or ice, the map will not be as accurate.) (See TFG website for a full-color version.)

terrace • *a flat or gently sloped embankment or ridge occurring on a hillside, and often along the margin of (or slightly above) a body of water, representing a previous water level.*

silt • *fine granular sediment most commonly composed of weathered grains of quartz and feldspar, of grain diameter 1/250 to 1/16 millimeters.*

topsoil • *the surface or upper layer of soil, as distinct from the subsoil, and usually containing organic matter.*

Significant or repeated changes in moisture, which can occur from human use or in concert with other geologic hazards such as earthquakes, floods, or landslides, greatly increase the hazard potential of expansive soils. Because precipitation is infrequent in much of the Southwest, low-moisture soils also have a high potential for hydrocompaction, where dry **silt** and clay particles lose their cohesion upon wetting. This process causes the soil to collapse, settling lower. If hydrocompaction occurs over deeper layers that have been severely dried due to prolonged drought or receding groundwater levels, the settling **topsoil** may fall into and expose giant underground fissures, called desiccation cracks (*Figure 9.10*). These fissures can be up to a meter (3 feet) wide, 3 meters (9 feet) deep, and as much as 300 meters (1000 feet) long.

lime • *an inorganic white or grayish-white compound made by roasting limestone (calcium carbonate, $CaCO_3$) until all the carbon dioxide (CO_2) is driven off.*

Figure 9.10: This giant desiccation crack appeared in a road in Graham County, southeastern Arizona. Such cracks typically occur as a network of polygonal fissures.

Slumping occurs when expansive minerals are present on steeper slopes, and involves the downward movement of a larger block of material along a surface that fails when the weight of the saturated soils can no longer be supported (*Figure 9.11*). Slumping is common near roads and highways, thanks to the presence of steeper hills, roadcuts, and construction. On steep, high slopes, slumping often precedes earthflows and mudflows that develop farther downslope as water is added to the slump while it mixes the moving material.

The key to reducing expansive soil hazards is to keep the water content of the soil constant—in the dry Southwest, the best option is to utilize proper drainage methods, prevent the infiltration of surface water, and use moisture protection barriers around houses and other structures. There are also chemical stabilizers, including **lime**, potassium, and ionic agents, that can increase the clay's structural stability. Damage to life and property from larger mass-wasting events can be reduced by avoiding landslide hazard areas or by restricting access to known landslide zones. Hazard reduction is possible by avoiding construction on steep slopes or by stabilizing the slopes. There are two main ways to accomplish stabilization: 1) preventing water from entering the landslide zone through runoff, flooding, or irrigation and 2) stabilizing the slope by placing natural or manmade materials at the toe (bottom) of the landslide zone or by removing mass from the top of the slope.

9 Earth Hazards

Earthquakes

crust • the uppermost, rigid outer layer of the Earth.

seismic waves • the shock waves or vibrations radiating in all directions from the center of an earthquake or other tectonic event.

compression • flattening or squeezing as a result of forces acting on an object from all or most directions.

uplift • upward movement of the crust due to compression, subduction, or mountain building.

magnitude (earthquake) • a logarithmic scale used to measure the seismic energy released by an earthquake. Magnitudes range from 1 to 10, with M3 earthquakes classed as minor and earthquakes of M8 or greater being classified as great.

Figure 9.11: A hillslide slump in Rocky Flats National Wildlife Refuge, Colorado. The slump occurred on a steep, manmade embankment after a severe storm saturated the soil.

Earthquakes

Earthquakes occur less frequently in the Southwestern US than they do in some other regions, but modest-sized earthquakes nonetheless represent potential hazards for the Southwestern states. Earthquakes occur when a critical amount of stress is applied to the Earth's **crust** and the crust responds by moving. According to the elastic rebound theory, rocks can bend elastically up to a point, until they finally break. The rocks then snap apart, releasing energy in the form of **seismic waves** (*Figure 9.12*). The plane defined by the rupture is known as a fault, and the surrounding rock layers become offset along it.

Many earthquakes, including most of those that occur in the Southwestern US, arise along pre-existing faults. In cases such as these, stress may accumulate from lateral **compressive** pressure, as the rocks are temporarily locked in position by friction and other constraints, until sufficient strain energy has built up to cause sudden slippage along the fault (i.e., an earthquake). Earthquakes have many different effects on the rocks in which they occur, including breaking and movement along faults, **uplift**, and displacement.

There are two common ways to measure the size of earthquakes: **magnitude** and intensity. Magnitude (M) is the measure of the energy released by the earthquake, whereas the intensity is what people actually experience. The first scale used to measure magnitude was the Richter scale (abbreviated

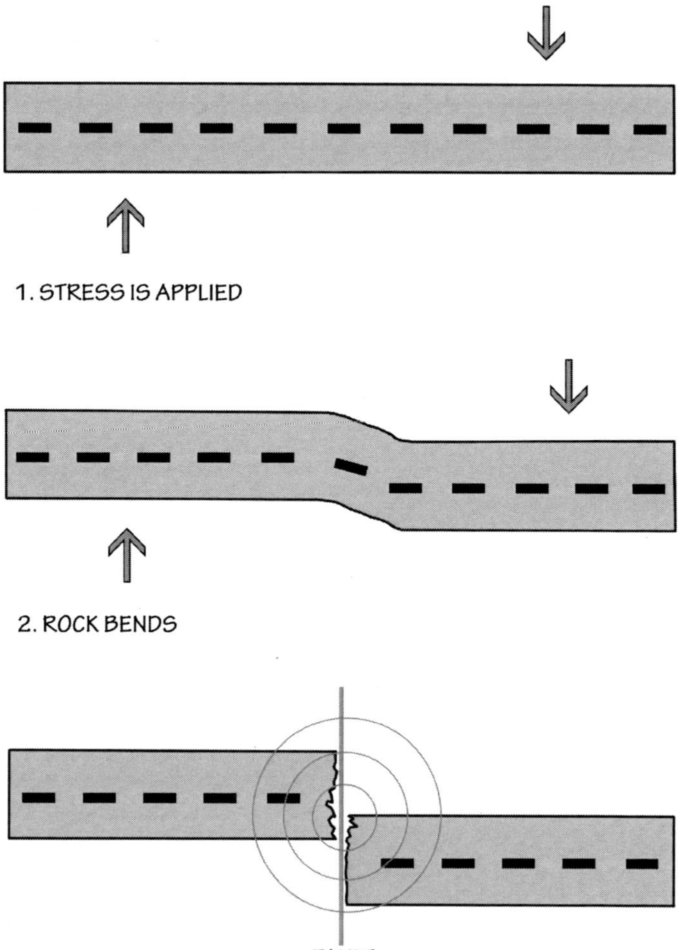

Figure 9.12: Elastic rebound.

M_L), which measures the amplitude of a seismic wave at a defined distance from the source of the earthquake. The Richter scale was designed to classify earthquakes at a local scale, but it does not do a very good job of describing the energy released by very large earthquakes. Geologists therefore developed another measurement, the Moment Magnitude scale (abbreviated M_w), which was introduced in 1979. The Moment Magnitude estimates the total energy released by an earthquake along an entire fault surface.

Both the Richter and Moment Magnitude scales are logarithmic, meaning that an M9.0 earthquake has 10 times the amplitude, and releases 32 times the energy, of an M8.0 earthquake. Accordingly, an M9.0 earthquake would have 100 times the amplitude and 1024 times the energy of an M7.0 earthquake. Both scales may appear to reach maximum values of 10 (since the largest recorded

Earthquakes

intensity • *a subjective measurement that classifies the amount of shaking and damage done by an earthquake in a particular area.*

plates • *large, rigid pieces of the Earth's crust and upper mantle, which move and interact with one another at their boundaries.*

active plate margin • *the boundary between two plates of the Earth's crust that are colliding, pulling apart, or moving past each other.*

seismic belt • *a narrow geographic zone along which most earthquakes occur.*

seismic zone • *a regional zone that encompasses areas prone to seismic hazards, such as earthquakes or landslides.*

earthquakes are slightly greater than 9), but technically there is no upper limit. The United States Geological Survey (USGS) describes earthquakes as *minor* (M3.0–3.9), *light* (M4.0–4.9), *moderate* (M5.0–5.9), *strong* (M6.0–6.9), *major* (M7.0–7.9), and *great* (M8.0 and higher). The largest recorded earthquake in US history was the 1964 Alaskan earthquake, which had an M_w of 9.2. By comparison, the largest recorded earthquake in the Southwest occurred in 1934 in Kosmo, Utah (M6.6). Notable earthquakes that have occurred just outside the Southwestern states, such as the 1887 Sonoran earthquake (M7.4) and the 1940 Imperial Valley earthquake (M7.1), have also caused extensive shaking and property damage, especially in Arizona.

Notable Earthquakes of the Southwestern States

Date	Location	M_w
03-12-1934	Kosmo, UT	6.6
11-08-1882	Denver, CO	6.6
11-15-1906	Socorro, NM	6.5
07-21-1959	AZ-UT border	5.6
09-02-1992	Springdale, UT	5.6
08-09-1967	Denver, CO	5.3
08-22-2011	Trinidad, CO	5.3
06-29-2014	NM-AZ border	5.2
01-23-1966	Dulce, NM	5.1

The magnitude of an earthquake does not tell us how much damage it causes. The amount of shaking and damage is known as the earthquake's **intensity**, and it can be measured by the Modified Mercalli Intensity (MMI) scale. This scale uses the Roman numerals I–XII to describe the effects of the earthquake in a particular location. For example, near the epicenter of a small earthquake, or at a location far from a large earthquake, the intensity may be described with an MMI of II: *"Felt only by a few persons at rest, especially on the upper floors of buildings. Delicately suspended objects may swing."* Unlike the Moment Magnitude scale, the MMI scale is a subjective gauge, and the USGS has attempted to improve the accuracy of MMI shake maps by soliciting data from the public. *Figure 9.13* shows the intensities felt in surrounding areas after the 1934 earthquake near Kosmo, Utah, which is the largest earthquake known to have occurred in the state.

Large earthquakes are relatively uncommon in the Southwestern US, due to the area's distance from current **plate** boundaries—the Southwest is located in the center of a tectonic plate rather than at an **active plate margin**. All earthquakes that occur in the Southwestern US are therefore referred to as "intraplate" earthquakes, and they are largely related to faults that localize earthquakes in particular areas, along linear **seismic belts** or **zones**. Many of the largest earthquakes in the Southwest, especially those in the Rocky Mountains, stem from activity along the *Intermountain Seismic Belt* (*Figure 9.14*). This linear zone of earthquake activity extends 1290 kilometers (800 miles) from northwestern Montana southward along the Idaho-Wyoming border, through

Earth Hazards 9

Earthquakes

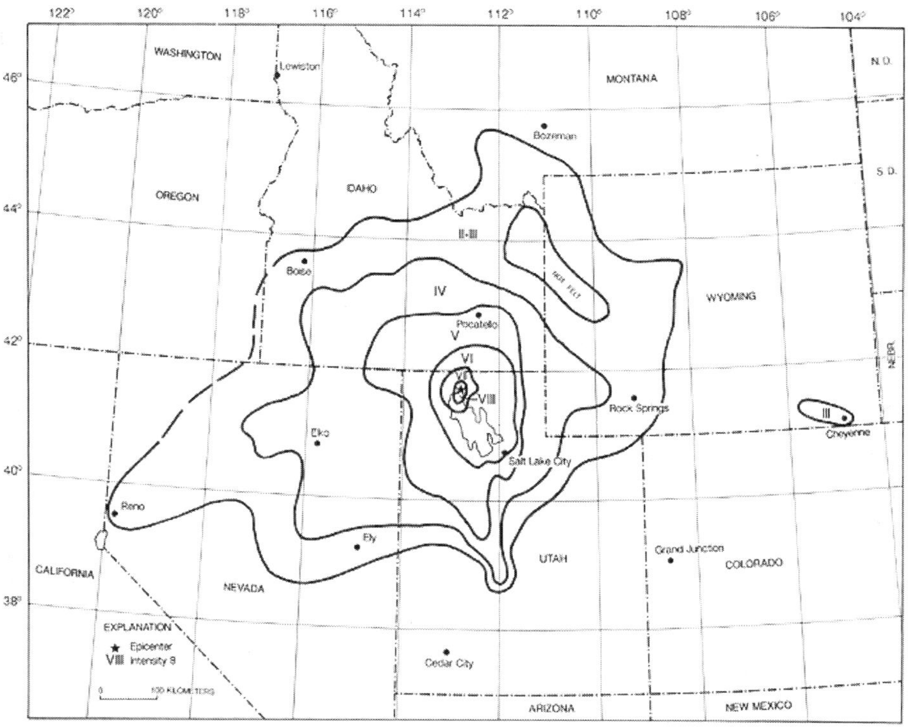

Figure 9.13: Intensity map of the 1934 Kosmo earthquake.

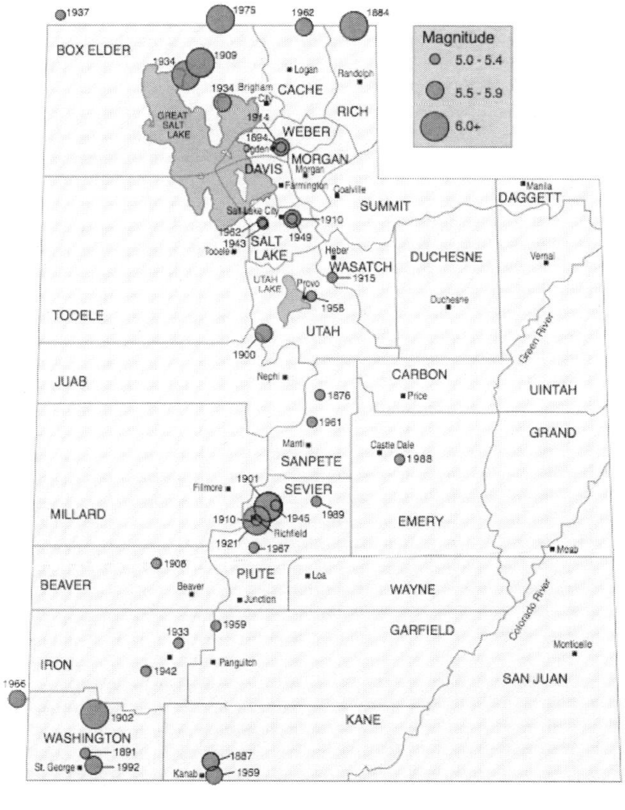

Figure 9.14: Since 1850, there have been 35 earthquakes in Utah with a magnitude of 5 or higher. All have occurred in or near the Intermountain Seismic Belt. (See TFG website for a full-color version.)

fault scarp • an escarpment directly beside a fault line, where the ground on one side of the fault has moved vertically with respect to the other side, creating step-like topography.

Quaternary • a geologic time period that extends from 2.6 million years ago to the present.

Utah, and into southern Nevada. Many active fault lines occur along this belt; the largest and most active is the Wasatch Fault, which marks the eastern edge of Basin and Range extension (*Figure 9.15*). Geologic studies indicate that the Wasatch Fault has experienced 19 or more surface-faulting earthquakes in the last 600 years. Some of these prehistoric earthquakes displaced the land surface by as much as 3 meters (10 feet) in a 30- to 65-kilometer (20- to 40-mile) radius, while others formed **fault scarps** over 6 meters (20 feet) high. Because the Wasatch Front is such a desirable place to live—about 80% of Utah's population resides along this mountain range, known for its spectacular views—the area is designated as having the greatest earthquake risk in the interior western US. Scientists estimate that the Wasatch Range has a 1-in-7 chance of being hit by a M7.0 earthquake sometime in the next 50 years.

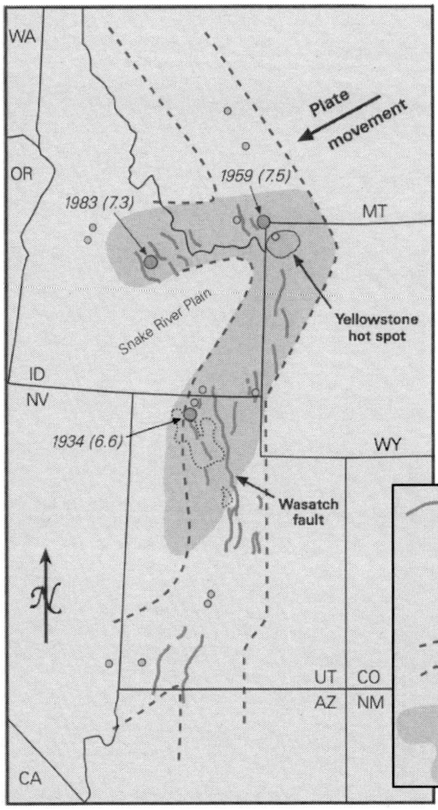

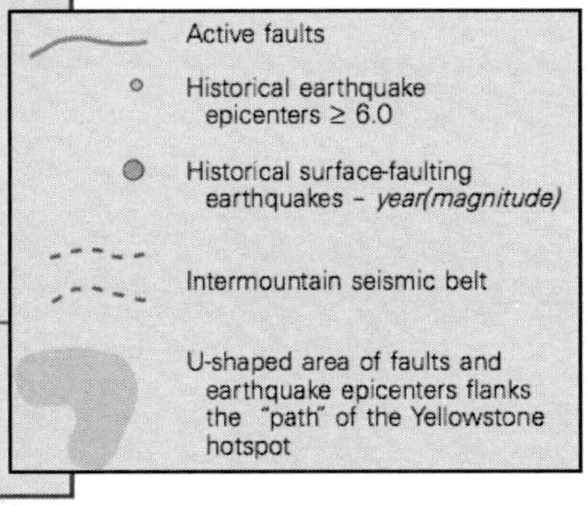

Figure 9.15: The Wasatch Fault is one of many active faults in the Intermountain Seismic Belt, which runs from Montana through Utah. The largest earthquakes in the Southwest, both historical and prehistoric, have occurred in this seismic belt. (See TFG website for a full-color version.)

The *Northern Arizona Seismic Belt* is an offshoot of the Intermountain Seismic Belt that extends south into Arizona along the Colorado Plateau. Faults in this zone are **Quaternary** in age, and are thought in part to have formed due to stress between the Basin and Range and edge of the Colorado Plateau. Swarms of tiny earthquakes occur along these fault lines in Arizona every year, most too small to be felt. Major quakes can and do occur, though, and geologists are keeping a close eye on the Anderson Mesa Fault near Flagstaff, which could conceivably produce an earthquake between M5.0 and M6.5.

Earthquakes can also occur through human causes, or "induced seismicity." These events are specifically linked to the high-pressure injection of wastewater from **oil** and gas extraction operations into the ground. The pressure of the water increases the likelihood that a rupture might occur along an otherwise locked fault. In early 2016, the US Geological Survey released a list of states considered to be at the highest risk for manmade earthquakes. Colorado and New Mexico rank fourth and fifth respectively due to the presence of the Raton Basin, an important source of coalbed methane and **natural gas**.

See Chapter 6: Energy to learn more about the process of extracting coalbed methane.

Networks of seismograph stations have improved geologists' ability to detect and accurately locate earthquake hazards (*Figure 9.16*), and specific fault zones are being studied throughout the Southwest. This information on earthquake risk can lead to better designs for high-risk infrastructure like dams, high-rise buildings, and **power** plants—and it can also be used to inform the public of potential hazards to lives and property. The hazards associated with earthquakes are mainly related to collapsing buildings and other structures, as well as fire related to broken gas lines and other utilities (and broken water lines that prevent fire-fighting).

oil *• See petroleum: a naturally occurring, flammable liquid found in geologic formations beneath the Earth's surface and consisting primarily of hydrocarbons*

natural gas *• a hydrocarbon gas mixture composed primarily of methane (CH4), but also small quantites of hydrocarbons such as ethane and propane.*

power *• the rate at which energy is transferred, usually measured in watts or, less frequently, horsepower.*

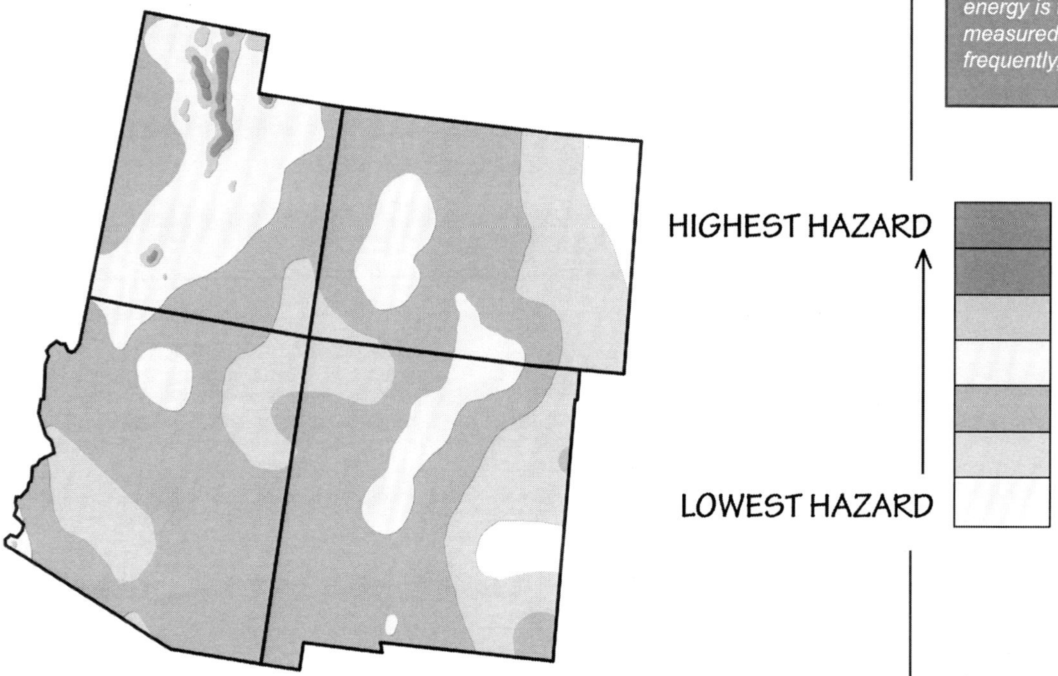

Figure 9.16: Seismic hazard map of the Southwestern US, based on 2014 data. (See TFG website for a full-color version.)

9 Earth Hazards

Karst

karst topography • a kind of landscape defined by bedrock that has been weathered by dissolution in water, forming features like sinkholes, caves, and cliffs.

sedimentary rock • rock formed through the accumulation and consolidation of grains of broken rock, crystals, skeletal fragments, and organic matter.

carbonate rocks • rocks formed by accumulation of calcium carbonate, often made of the skeletons of aquatic organisms.

gypsum • a soft sulfate mineral that is widely mined for its use as fertilizer and as a constituent of plaster.

dolomite • a carbonate mineral, consisting of calcium magnesium carbonate ($CaMg(CO_3)_2$).

porosity • the percentage of openings in a body of rock such as pores, joints, channels, and other cavities, in which gases or liquids may be trapped or migrate through.

Karst and Sinkholes

Karst topography forms in areas where the underlying bedrock is composed of material that can be slowly dissolved by water. Examples of this type of **sedimentary rock** include **carbonate rocks** such as limestone, halite, **gypsum**, **dolomite**, and anhydrite. Carbonate rocks may develop karst and other dissolution features due to the effects of circulating groundwater that has been made slightly acidic through the presence of dissolved carbon dioxide (which creates carbonic acid that reacts with the rock, dissolving it). Sinkholes and caverns can form, creating potential hazards (i.e., the land surface could subside or collapse into the underground openings). This may principally occur in areas where cavities filled with water are emptied through groundwater withdrawal or other natural processes, resulting in the cavities being filled with air and reducing support for the overlying rock. Many parts of the Southwest are underlain by karst and soluble carbonate bedrock (see *Figure 9.19*), especially Arizona's Colorado Plateau and New Mexico's Basin and Range. Because karst terrain is very **porous** and **fractures** easily, groundwater pollution can also be a serious problem. Contaminants that might otherwise be filtered through underlying sedimentary rock are quickly transported into **aquifers** by runoff. The hazards of pollution are increased by rampant industrial, agricultural, and residential development over karst features.

The Colorado Plateau of northern Arizona contains extensive surface limestone and subsurface gypsum/**salt** deposits. As these beds dissolve beneath the surface through the movement of groundwater, sinkholes form through the collapse of overlying layers. Karst features such as open caverns also commonly form at the surface. The mountains of southeastern Arizona also contain limestone layers that have dissolved to form caverns such as Colossal Cave near Tuscon—these features are less extensive than those on the Plateau and collapse at the surface is uncommon. In New Mexico, karst is concentrated in the northern Sacramento Mountains and the Guadalupe Mountains, where a large number of impressive caverns (including Carlsbad Cavern) have formed in **Permian reef** limestone (*Figure 9.17*). Although karst collapse is less prevalent in New Mexico than in many other parts of the United States, it is still an environmental issue of concern. In Colorado, the highest karst and sinkhole hazards are located in the Roaring Fork and Eagle river valleys, where hundreds to thousands of meters (yards) of subsidence has already occurred via subsurface dissolution and deformation of **evaporite** rocks. Colorado's sinkholes also form in arid and easily eroded soils, creating a landform known as "pseudokarst."

See Chapter 4: Topography to learn about the unique way in which sulfuric acid has generated karst in New Mexico.

Sinkholes are funnel-shaped depressions in the land surface formed by the dissolution of near-surface rocks or by the collapse of underground channels and caverns (*Figures 9.18 and 9.20*). Sinkholes can form by several different mechanisms, but all require dissolution of rock beneath the surface (*Figure 9.21*). Manmade sinkholes can also occur through the collapse of mine shafts

Karst

Figure 9.17: Carlsbad Cavern in New Mexico is an extensive cave system formed in the Permian limestone of the Capitan Formation. Uniquely, this and many other karst caves of New Mexico were formed through dissolution by sulfuric acid rather than the more common carbonic acid.

Figure 9.18: Hole-in-the-ground sinkhole, Millard County, Utah.

fracture • a physical property of minerals, formed when a mineral crystal breaks.

aquifer • a water-bearing formation of gravel, permeable rock, or sand that is capable of providing water, in usable quantities, to springs or wells.

salt • a mineral composed primarily of sodium chloride (NaCl).

Permian • the geologic time period lasting from 299 to 252 million years ago.

reef • a feature lying beneath the surface of the water, which is a buildup of sediment or other material built by organisms, and which has positive relief from the sea floor.

evaporite • a sedimentary rock created by the precipitation of minerals directly from seawater, including gypsum, calcite, dolomite, and halite.

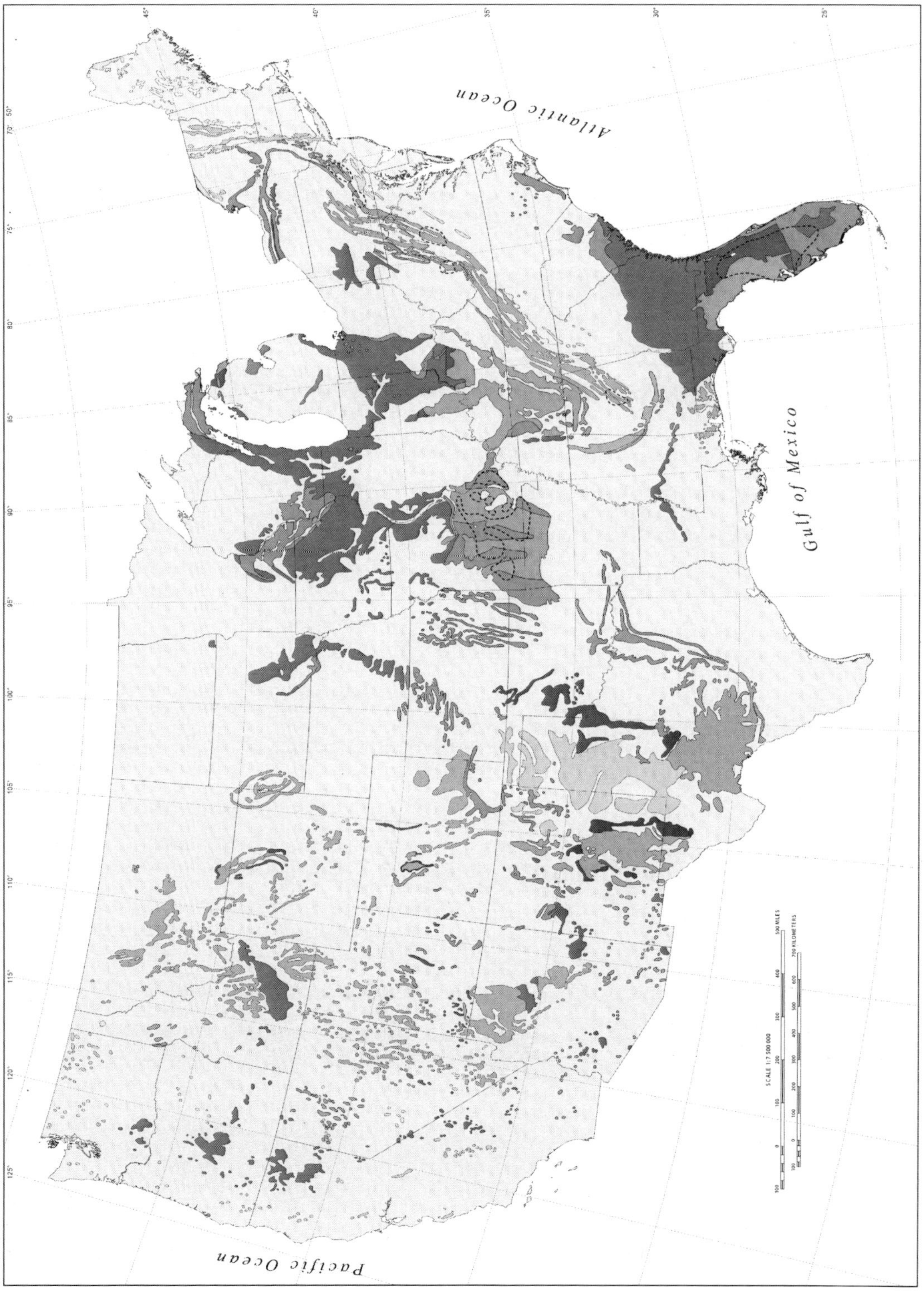

Figure 9.19: Areas of karst in the continental US, associated with carbonate and evaporate rocks. See Key on facing page. (See TFG website for full-color version.)

Earth Hazards 9

Karst

FISSURES, TUBES, AND CAVES OVER 1,000 FT (300 M) LONG; 50 FT (15 M) TO OVER 250 FT (75 M) VERTICAL EXTENT

- In metamorphosed limestone, dolostone, and marble
- In moderately to steeply dipping beds of carbonate rock
- In gently dipping to flat-lying beds of carbonate rock
- In gently dipping to flat-lying beds of carbonate rock beneath an overburden of noncarbonate material 10 ft (3 m) to 200 ft (60 m) thick
- In moderately to steeply dipping beds of gypsum
- In gently dipping to flat-lying beds of gypsum

FISSURES, TUBES, AND CAVES GENERALLY LESS THAN 1,000 FT (300 M) LONG; 50 FT (15 M) OR LESS VERTICAL EXTENT

- In metamorphosed limestone, dolostone, and marble
- In crystalline, highly siliceous intensely folded cabonate rock
- In moderately to steeply dipping carbonate rock
- In gently dipping to flat-lying carbonate rock
- In gently dipping to flat-lying beds of carbonate rock beneath an overburden of noncarbonate material 10 ft (3 m) to 200 ft (60 m) thick
- In moderately to steeply dipping beds of gypsum
- In gently dipping to flat-lying beds of gypsum
- In gently dipping to flat-lying beds of gypsum beneath and overburden of nongypsiferous material 10 ft (3 m) to 200 ft (60 m) thick
- In carbonate zones in highly calcic granite (Alaska only)
- In moderately to steeply dipping beds of carbonate rock with a thin cover of glacial till and frost-derived residual soil (Alaska only)

FISSURES, TUBES, AND CAVES GENERALLY ABSENT; WHERE PRESENT IN SMALL ISOLATED AREAS, LESS THAN 50 FT (15 M) LONG; LESS THAN 10 FT (3 M) VERTICAL EXTENT

- In crystalline, highly siliceous intensely folded carbonate rock
- In moderately to steeply dipping beds of carbonate rock
- In gently dipping to flat-lying beds of carbonate rock

FEATURES ANALOGOUS TO KARST

- Fissures and voids present to a depth of 250 ft (75 m) or more in areas of subsidence from piping in thick unconsolidated material
- Fissures and voids present to a depth of 50 ft (15 m) in a of subsidence from piping in thick, unconsolidated mate
- Fissures, tubes, and tunnels present to a depth of 250 ft (75 m) or more in lava
- Fissures, tubes, and tunnels present to a depth of 50 ft (15 m) in lava
- Areas in which extensive historical susidence has occurred

9 Earth Hazards

Karst

Figure 9.20: Aerial view of large aligned sinkholes in the Permian Kaibab Formation, southeast of Winslow, Arizona.

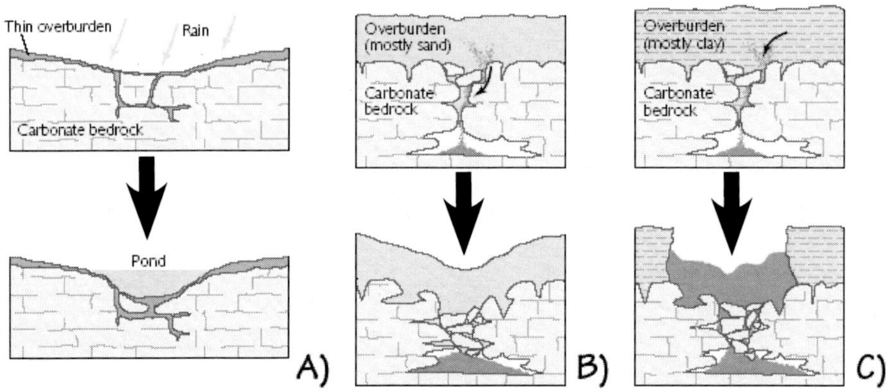

Figure 9.21: Three mechanisms of sinkhole formation.

A) Dissolution: Rain and surface water percolate through carbonate bedrock, dissolving a hole from the top down.

B) Cover-subsidence: Carbonate bedrock dissolves beneath a permeable overlying layer such as sand. As the sand falls into the hole below, slow downward erosion leads to a depression.

C) Cover-collapse: Carbonate bedrock dissolves beneath an overlying layer made largely of clay. The clay collapses from beneath into the cavity below, abruptly forming a dramatic sinkhole when the surface is breached. This type of sinkhole causes the most catastrophic damage, as it is not easily detected before it forms.

and tunnels, or the removal of groundwater and oil (*Figure 9.22*). Sinkhole formation commonly damages roads, buildings, and utilities, and it is a problem in all four Southwestern states.

Sinkholes may be very small or large enough to swallow even **hectares** (acres) of land, along with any structures that had been built upon the surface. The early stages of sinkhole development may be indicated by signs of mass wasting such as "pistol-grip"-shaped trees, cracked building foundations, and leaning fence posts. Structural damage from sinkholes can be mitigated, but usually only at significant cost. It is therefore often far more prudent to avoid building in such locations altogether. Evaluating sinkhole risk commonly involves foundation testing by drilling or remote sensing (e.g., measuring electrical resistivity) prior to construction.

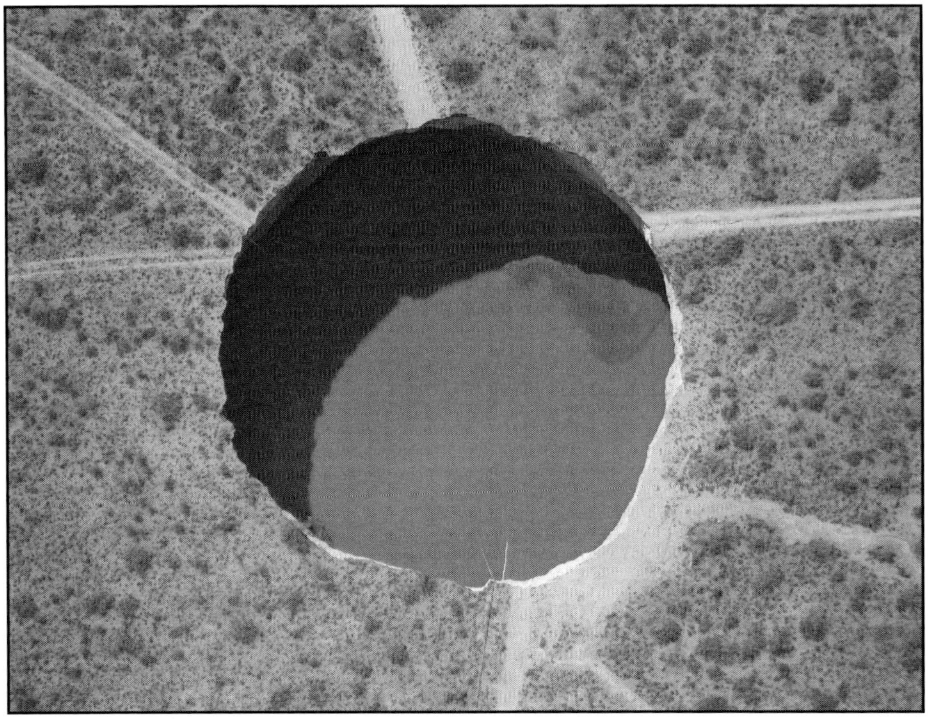

Figure 9.22: This sinkhole near Carlsbad, New Mexico formed after improper solution mining practices created a large cavity beneath the desert. Three days after its collapse, the sinkhole was 200 meters (670 feet) long and 140 meters (450 feet) wide, and there are major concerns that it could expand into the town of Carlsbad.

Asbestos

The name **asbestos** is used to describe a variety of fibrous minerals, including chrysotile and crocidolite, which occur naturally as bundles of fibers that can be separated into durable filaments. The fibers are non-conductive and naturally resistant to chemicals and **heat**. During the early twentieth century, asbestos was commonly mined for its numerous industrial applications such as weather

hectare • *a metric unit of area defined as 10,000 square meters.*

asbestos • *a fibrous silicate mineral that is resistant to heat, flames, and chemical action.*

heat • *a form of energy transferred from one body to another as a result of a difference in temperature or a change in phase.*

Asbestos

insulation and fireproofing for buildings. However, when asbestos-laden dust is inhaled, the microscopic mineral fibers are capable of piercing and damaging the cells in which they come into contact. This can cause lung irritation and lead to serious health problems, including cancer.

While many people worry about the asbestos insulation hazards found in older buildings, few consider the hazards associated with the minerals' natural occurrence. Natural asbestos sources can be found throughout the Southwest, and it has been mined in both Utah and Arizona (*Figure 9.23*), though these mines are no longer in operation thanks to recent limitations placed on the minerals' use. Remediation attempts on abandoned mines include blocking off access to contaminated areas and burying contaminated soil that has been found near surface water sources.

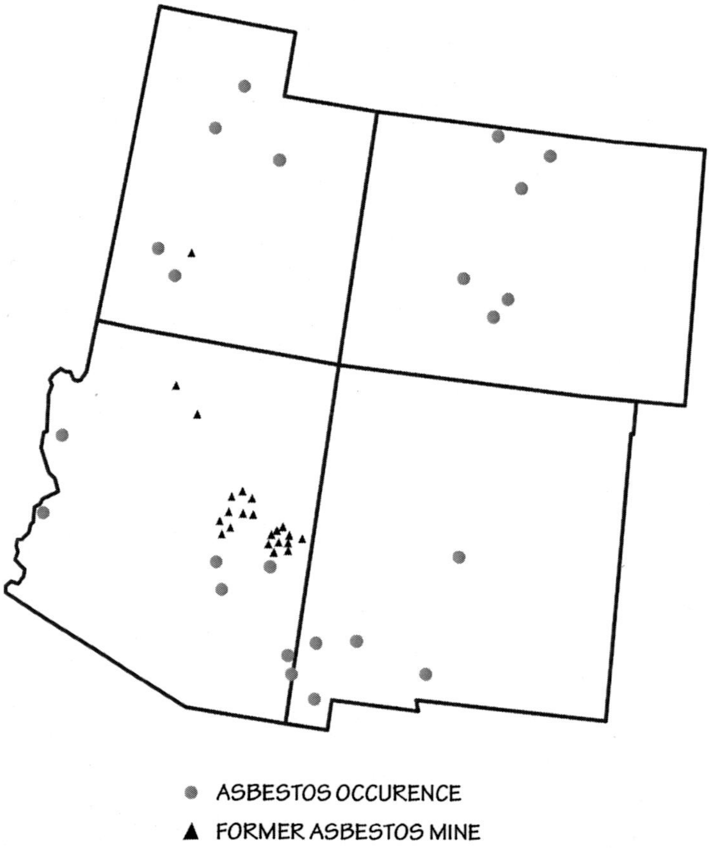

Figure 9.23: Presence of significant asbestos sources and former mines in the Southwestern states.

Natural events such as landslides can release previously trapped asbestos minerals, which can then be transported across great distances by the wind or even carried by surface water running over an asbestos site. Asbestos crystals can then make their way into streams and lakes, spreading contamination over large areas. People within the vicinity of exposed sources are at risk from windblown particles and mud particles collected on their shoes, clothing, and vehicles. The particles can then be carried into homes.

Earth Hazards 9

Radon

Radon is a naturally occurring **radioactive**, colorless, odorless gas. It is the leading cause of lung cancer in American non-smokers, and the second leading cause of lung cancer overall. It can collect in homes, buildings, and even in the water supply. Radon gas is formed naturally when uranium-238 undergoes radioactive decay, producing energy and several radioactive products such as radon-222 and **thorium**-232. (The thorium later decays to emit energy and radon-220.) Radon is more commonly found where uranium is relatively abundant in bedrock at the surface, often in **granite**, **shale**, and limestone. The EPA has produced a map of the US showing geographic variation in radon concentrations, divided into three levels of risk: high, medium, and low (*Figure 9.24*).

Although radon is more or less universally present, high levels of radon are associated with areas containing uranium-rich bedrock. Most rocks have a small amount of uranium, but certain rocks tend to have higher concentrations of the radioactive element, such as light-**colored** volcanic rocks, granites, dark shales, sedimentary rocks with **phosphates**, and **metamorphic rocks**. Radon concentrations are generally high in the Southwest's mountainous areas, as uranium is relatively concentrated in the granites, black shales, and metamorphic rocks of the Rocky Mountains (*Figure 9.25*). The sediments eroded from those areas also carry a high radon hazard potential, leading to moderate radon presence throughout the Southwest.

Radon is chemically inert, meaning that it does not react or combine with elements in the ground, and it can move up through rocks and soil into the atmosphere. It is dangerous primarily when it accumulates indoors, creating a health hazard similar to that of secondhand smoke. Radon gas finds its way through cracks in basement foundations, sump pump wells, dirt floor crawlspaces, and basement floor drains. It can also be found in well and municipal water. Since radon is more easily released from warm water than from cold water, one of the greatest forms of exposure likely occurs while showering in water with high radon levels.

Radon cannot be detected by sight or smell, so there is no way that the body can sense its presence. Fortunately, with proper monitoring and mitigation (reduction) techniques, radon gas can be easily reduced to low levels. One technique that is often used in homes involves sealing cracks in the basement floor, covering drains, and installing ventilation systems. A well-ventilated space will prevent the radon from accumulating and will reduce the risk of exposure. Most states have licensed radon mitigation specialists who are trained in the proper testing and mitigation of radon levels in buildings. The EPA has also published a homebuyer's guide designed to help citizens make informed decisions about radon gas. For radon in water, filtration systems can be installed to mitigate exposure in the home.

radioactivity • *the emission of radiation by an unstable atom.*

thorium • *a radioactive rare earth element.*

granite • *a common and widely occurring type of igneous rock.*

shale • *a dark, fine-grained, laminated sedimentary rock formed by the compression of successive layers of silt- and clay-rich sediment.*

color (mineral) • *a physical property determined by the presence and intensity of certain elements within the mineral.*

phosphate • *an inorganic salt of phosphoric acid, and a nutrient vital to biological life.*

metamorphic rocks • *rocks formed by the recrystallization and realignment of minerals in pre-existing sedimentary, igneous, and metamorphic rocks when exposed to high enough temperature and/or pressure.*

9 Earth Hazards

Radon

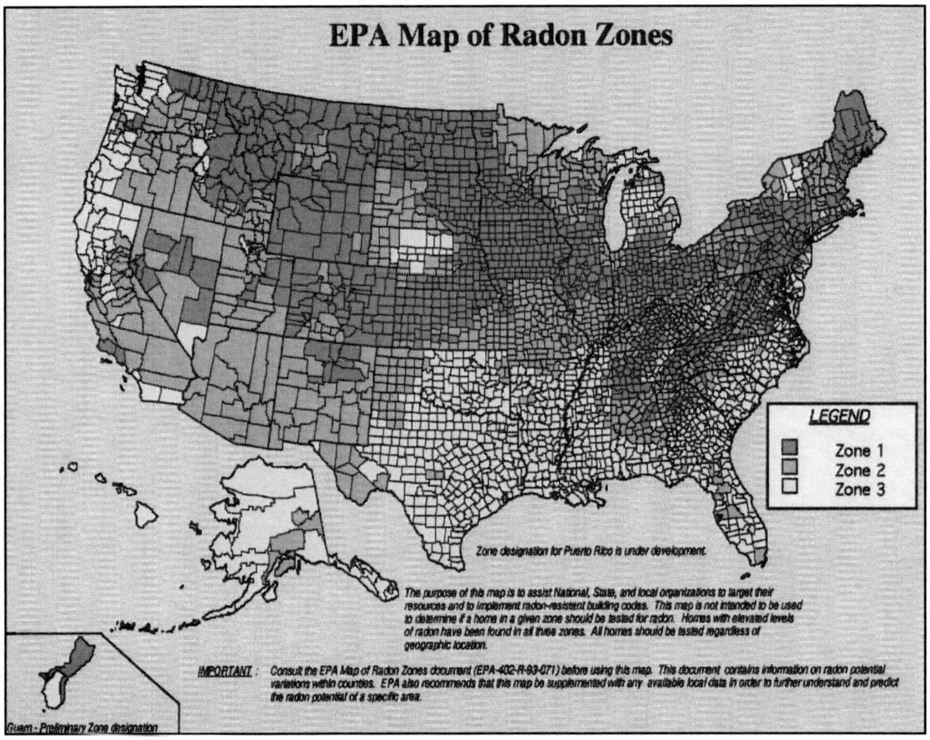

Figure 9.24: Radon zone map of the US. (Note: Zone 1 contains the highest radon levels.) (See TFG website for full-color version.)

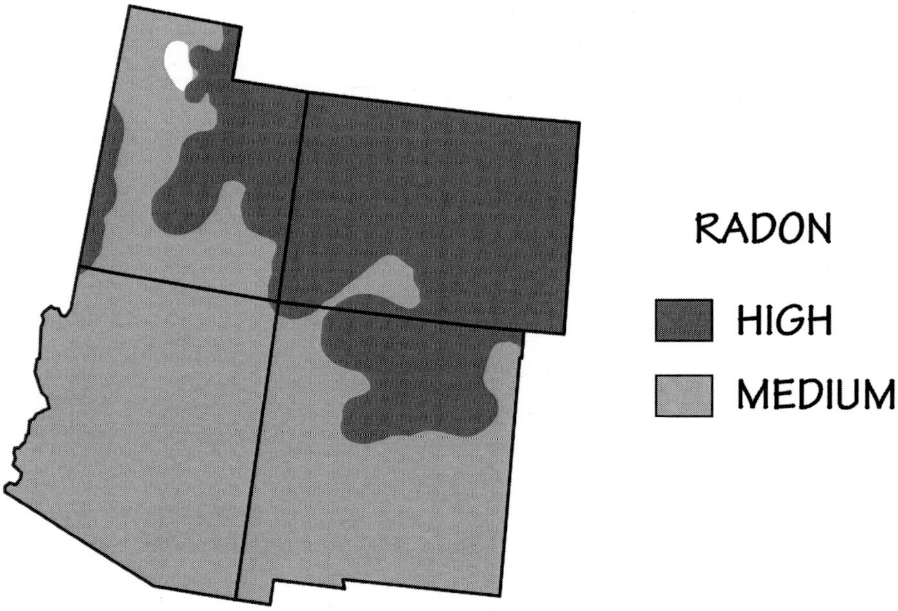

Figure 9.25: Radon risk levels at the surface in the Southwestern US. (See TFG website for a full-color version.)

Floods

Floods

Although the Southwest has an overall arid climate, there are several large rivers that flow through the area, including the Colorado River and Rio Grande. Many of the Southwest's largest floods have occurred along the Colorado River and its tributaries (*Figure 9.26*). Along **floodplains**, the soil is fertile thanks to nutrients deposited by the rivers, and nearby water allows for easy irrigation. These factors encourage development on flood-prone areas throughout the Southwest. In the Great Plains, a large proportion of farmland—a significant industry in the Southwest—is located on floodplains along rivers that flow through the region. Before humans settled along rivers, floods were often beneficial events: a flood would wash away nutrient-depleted soil and then deposit fresh minerals and other nutrients to help support future plants. People now face the dilemma of whether or not to build in areas that are potentially subject to flooding. Water control structures such as dams are engineered to protect infrastructure and lives, but nature is not always so easily controlled.

floodplain • *the land around a river that is prone to flooding.*

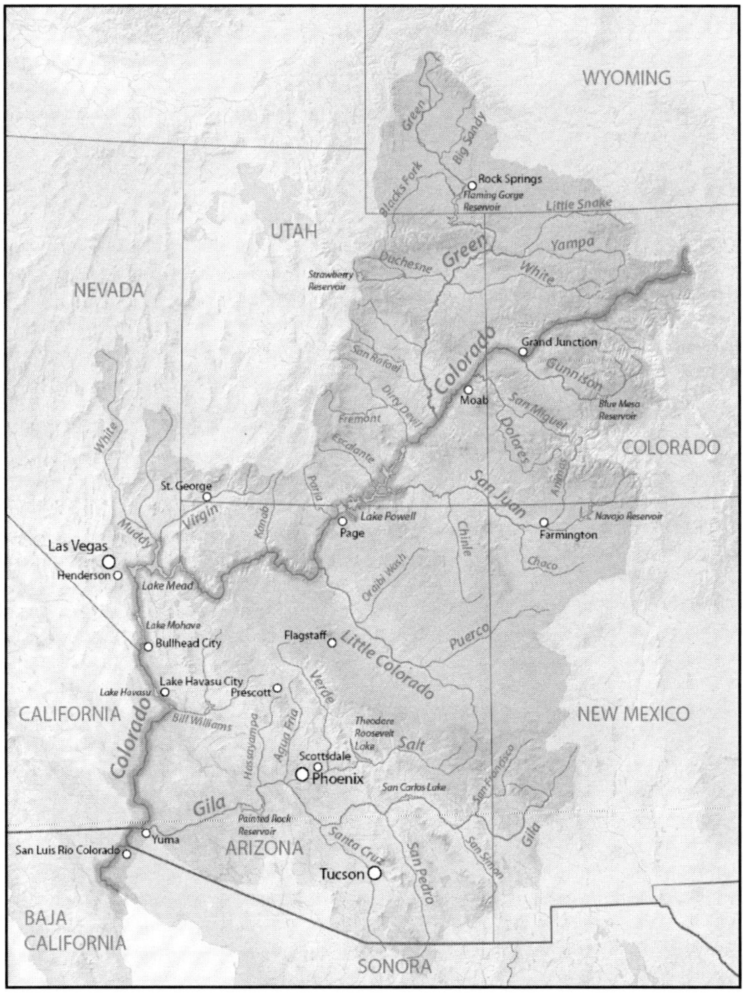

Figure 9.26: The Colorado River and its tributaries in the Southwestern states. (See TFG website for full-color version.)

Floods

Floods are controlled by the rate of precipitation, run-off, stream flow, and shape of the land surface. They may occur as water overflows the banks of a standing water body (such as a lake) or flowing water (such as a stream), or when rainwater accumulates in an area that normally has neither standing nor flowing water. Areas near rivers, tributaries, creeks, and streams are likely to experience flooding during periods of heavy rainfall.

See Chapter 8: Climate to learn more about precipitation in the Southwestern states.

system • a set of connected things or parts forming a complex whole.

Flooding can occur at any time of the year and is caused when more water enters a stream/river channel than the channel can contain. This situation can develop when water is unable to soak into the ground and instead runs off into a river channel. Runoff can occur if the ground is already saturated (full of water) or if the ground is too dry, hard, or frozen. The slope of a river (i.e., the topography of the land) can also contribute to flooding. If rivers have a steep slope, water can quickly move through the channel and continue downstream. If rivers have a shallow slope, water moves slowly through the river channel and remains in the area instead of moving downstream. Large floods typically result from unusually rapid regional melting of snow in the spring or from major weather **systems** that bring heavy rainfall over a large region. Flash floods—rapid flooding of low-lying areas—are often associated with heavy rain, which can quickly waterlog soil and lead to mudslides on steep terrain, resulting in damage to roads and property. In areas of lower elevation, flash floods can be produced when slow-moving or multiple thunderstorms occur over the same area. When storms move more quickly through an area, flash flooding is less likely. Although flash floods may be of only a short duration, they can cause major damage—they have been known to wash coffins out of graveyards, destroy structures, and demolish manmade dams.

In the Southwest, arid air travelling from the western mountains draws in moisture from the south where there are no mountains to block the moisture, a phenomenon known as a monsoon climate. Warm, moist air has a concentration of energy that may be released in sudden, violent thunderstorms, generating downpours that lead to flash floods. Monsoon floods occur in every Southwestern state, and can reach heights of 9 meters (30 feet) or more, moving rocks and trees, sweeping away vehicles, and destroying buildings (*Figure 9.27*). Flash floods in the Southwest also tend to be especially deadly and destructive due to the area's many canyons, which funnel water to great speeds and depths. In September 2015, extreme rainfall generated by Pacific Hurricane Linda flooded Keyhole Canyon in Zion National Park, Utah. In only 15 minutes, the Virgin River's flow increased from 1.5 cubic meters (55 cubic feet) per second to 74.5 cubic meters (2630 cubic feet) per second. Seven hikers were swept away and killed. Near Hildale, Utah, rainfall from the same event caused major flash floods that swept away vehicles, killing 13 people, as well as destroying water lines, bridges, and power infrastructure for the town (*Figure 9.28*).

Floodplains are areas adjacent to rivers and streams that occasionally flood but are normally dry, sometimes for many years. When storms produce more runoff than a stream can carry in its channel, waters rise and inundate adjacent lowlands, leaving behind layers of settled sediment. Significant damage and

Figure 9.27: In August 2006, runoff from heavy rains sent a wall of water into the town of Hatch, New Mexico. No one was injured, but damages exceeded $4 million. The summer of 2006 was a record monsoon season in New Mexico, with a total of 91 flash flood events.

Figure 9.28: The remains of a vehicle swept away by the Hildale flash flood in September 2015. Ten of the vehicle's eleven occupants were killed.

sometimes loss of human life can occur when buildings and other human infrastructure are built on floodplains, under the assumption that future floods may never occur or will only occur in the distant future. Floods can occur at any time, but major floods are more frequent in spring and fall after periods of heavy or sustained rains when stream levels rise rapidly. For example, rapid runoff from distant storms in the Rocky Mountains has had devastating effects, both in the mountains and where streams spread over broad areas of more open land. These floods have damaged structures, property, and put lives in peril. For example, in September 2013, torrential rains over Colorado's Front Range resulted in catastrophic flooding along the South Platte River and related tributaries. Up to 510 millimeters (20 inches) of rain fell over a three-day period; water levels of the river reached as high as 2.7 meters (8.8 feet) above flood level and affected 17 counties (*Figure 9.29*).

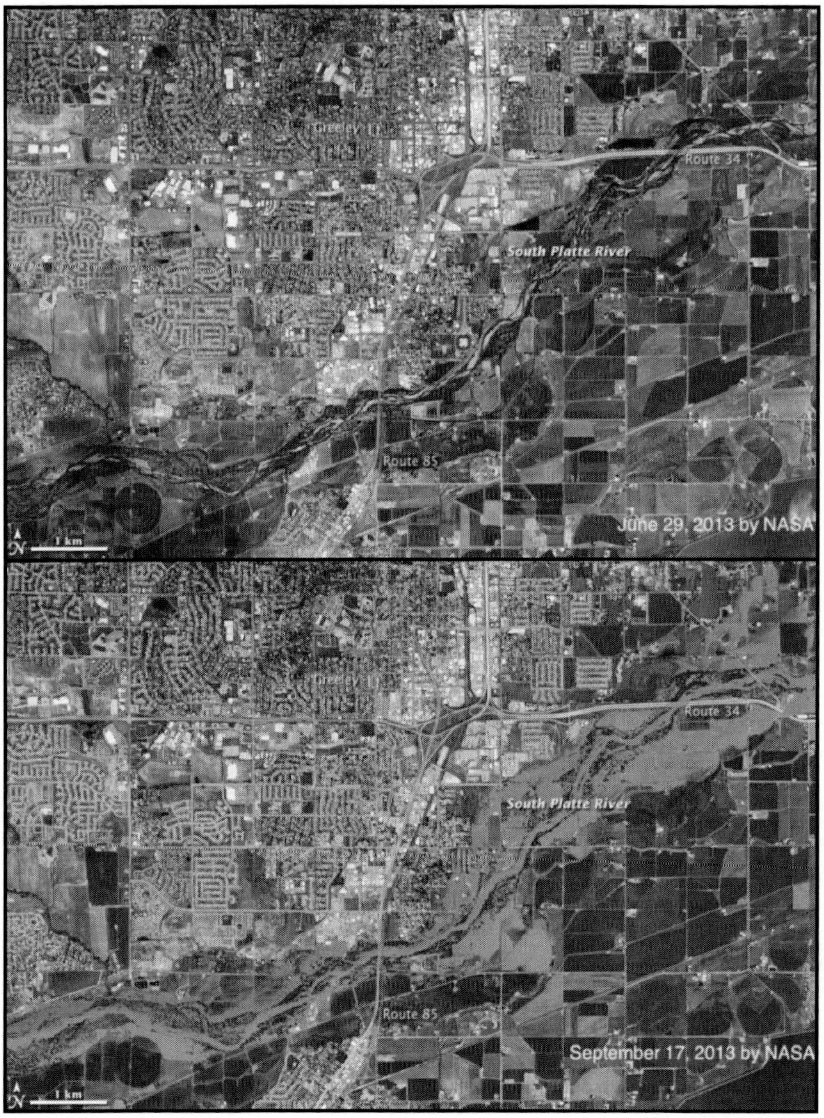

Figure 9.29: Before (top) and after (bottom) images of the South Platte River flood near Greely, Colorado, in September 2013.

While floods are always considered a hazard to life and property, they present a compound threat when they trigger mudslides or contribute to the conditions that cause expansive soils and karst topography. While there is no way to completely avoid the destructive impacts of flooding, good community planning and informed decision-making can greatly reduce the safety concerns and economic impacts of these events. Flood control is part of the mission statements of many government agencies, including the National Resource Conservation Service (NRCS), US Corps of Engineers (USCE), and US Geological Survey (USGS). These agencies and others maintain gauges on most large rivers and streams in the Southwest from which flow data are gathered. Using historical records and flow data collected over a long period of time, hydrogeologists can apply statistics to calculate the frequency and **recurrence intervals** of flows of different magnitude. These data have been used by the USGS to produce special topographic maps showing flood-prone areas. The Federal Emergency Management Agency (FEMA) provides guidelines for communities that are planning mitigation strategies designed to minimize the impacts of natural hazards such as flooding.

recurrence interval • *the time elapsed between major events, such as floods.*

tornado • *a vertical funnel-shaped storm with a visible horizontal rotation.*

cold front • *the boundary between the warm air and the cold air moving into a region.*

Weather Hazards

Weather is the measure of short-term atmospheric conditions such as temperature, wind speed, and humidity. The Southwest is an active location for atmospheric events such as thunderstorms and **tornados**. It also experiences a variety of other weather hazards, including high temperatures and drought.

Storms and Tornados

Several types of severe storms present challenges to people living in the Southwest. Summer brings severe thunderstorms associated with **cold fronts**. Fall and spring can bring ice storms, while winter brings snow and, in some cases, blizzard conditions. In March 2016, for example, a major blizzard dumped 60 centimeters (2 feet) of snow on the Denver metropolitan area and Colorado's Front Range, knocking out power, shutting down the Denver International Airport, and closing schools. A second event in April 2016—dubbed Winter Storm Vexo—inundated the Southwest with more heavy snowfall, from 1.3 meters (51 inches) in Pinecliffe, Colorado to 28 centimeters (11 inches) near Questa, New Mexico and 18 centimeters (7 inches) in Bellemont, Arizona.

Rainstorms occur where colder air from higher latitudes abruptly meets warmer air. Severe thunderstorms are a common occurrence for people living in the eastern Southwest because the conditions over the Great Plains are perfect for the development of severe weather. The region's flat, open fields are warmed by the summer sun, which sits high in the sky during this time of year. This results in large temperature differences when cold air masses move across the country. At the boundary between warmer and cooler air, buoyant warm air rises, and then cools because air pressure decreases with increasing height in the atmosphere. As the air cools, it becomes saturated with water vapor, condensation occurs, and clouds begin to form. Because liquid water droplets in the clouds must be very small to remain suspended in the air, a significant amount of condensation causes small water droplets to come together, eventually becoming too large to remain suspended. Sufficient

wind shear • *when wind speed and/or direction changes with increasing height in the atmosphere.*

moisture and energy can lead to dramatic rainstorms. Because warm air has a lower pressure relative to cold air, and the movement of air from areas of high pressure to areas of low pressure generates wind, the significant difference in air pressure associated with these boundaries and rainstorms also generates strong winds. Hail is also a possible occurrence during storms as a result of moisture high in the atmosphere that condenses and forms rain droplets. If the wind is strong enough to keep the droplets suspended, and cold enough to freeze them, they may become hailstones. If the wind continues to persist and keeps the hail suspended in the clouds long enough, they can even grow as large as golf balls. Once they reach a mass that is too great for the wind to keep them suspended, they fall to the Earth, where they can do considerable harm upon impact. Anyone caught in a significant hailstorm can expect some bruises or stinging sensations. If the hail is large enough, property can be damaged; car windshields, sunroofs, and canopies are especially susceptible.

With freak and intense thunderstorms comes the added risk of lightning strikes. Friction in the atmosphere from a chaotic storm can produce a buildup of static electricity and an unbalanced electrical charge. When the imbalance is great enough, the accumulated energy will discharge itself in the form of a lightning bolt. This discharge can be heard as the sound of thunder. The intense beam of energy can scorch or kill any life that is unlucky enough to be at the point of contact. If a lightning strike occurs in an arid, vegetated area, the resulting fires may develop into full-blown forest fires.

Some severe thunderstorms, called supercells, have the potential to develop into tornados that can cause serious property damage and endanger lives. These storm events are associated with **wind shear**, which occurs when the wind's speed or direction changes with increasing height in the atmosphere. Wind shear can happen when a cold front moves rapidly into an area with very warm air. There, the condensing water droplets mix with the cooler, drier air in the upper atmosphere to cause a downdraft. At the frontal boundary, warm, moist air rapidly rises as cooler, dry air descends; in the meantime, the pressure differences between the warm and cold air masses cause strong winds. Clouds with a visible horizontal rotation can form, appearing to roll like waves crashing on the shore of a beach. This horizontal motion can tilt, lifting the rotating cloud vertically, and the rolling cloud will form a tornado. Most tornados will last a few seconds to several minutes. During that time, many tornado-prone areas will use tornado sirens to alert residents of the danger. A smaller tornado might generate flying debris that can cause injury or damage to buildings, while larger tornados can cause buildings and houses to be completely broken apart. Tornados are classified by their ranking on the En-hanced Fujita scale, or EF scale. These classifications are estimates of wind speeds based on the type of damage that is observed following the storm.

The word tornado has its roots in the Spanish word *tonar*, which means to turn.

"Tornado Alley" is the nickname for an area, extending from Texas to Minnesota, that experiences a high number of exceptionally strong tornados due to its flatter topography and high incidence of severe thunderstorms. The Great Plains of Colorado and New Mexico are part of Tornado Alley, leading to more tornados

Measuring Tornado Intensity

Tornado intensity is measured on the Fujita scale, or simply F-scale, based on the amount of damage that a tornado can cause. The scale ranges from F0 to F5. The scale was modified recently to more accurately reflect specific wind speeds; this newer scale is known as the "Enhanced Fujita scale" and is labeled EF0 to EF5.

EF Scale	Estimated Wind Speed (kph)	Estimated Wind Speed (mph)
EF0	104–137	65–85
EF1	138–177	86–110
EF2	178–217	111–135
EF3	218–266	136–165
EF4	267–322	166–200
EF5	> 322	> 200

in this part of the Southwest (*Figure 9.30*). From 1991 to 2010, for example, an annual average of 53 and 11 tornados occurred in Colorado and New Mexico, respectively (*Figure 9.31*). To the west, fewer tornado strikes occur, with an annual average of five and three striking Arizona and Utah, respectively. The boundaries of Tornado Alley vary in application, depending on whether the frequency, intensity, or number of events per location are used to determine its borders.

Dust Storms

In arid climates, even under non-drought conditions, dust storms are a hazard. Dust storms occur when winds hold dust aloft, sometimes briefly over a local area, and sometimes over broad regions for days. They can be hazardous to health and, because they drastically reduce visibility, dangerous to motor vehicle and airline traffic.

Among the most spectacular dust storms are those known as haboobs (or monsoonal dust storms), which occur when strong thunderstorm downdrafts blow loose sediments up from the desert, sending dust up to over 1000 meters (3300 feet) into the sky. Large haboobs can be as much as 100 kilometers (62 miles) across, and travel at speeds of 50 to 100 kilometers per hour (about 30 to 60 miles per hour) for over an hour. These storms occur in the summer, across southernmost New Mexico and Arizona (*Figure 9.32*), as well as in California and Texas.

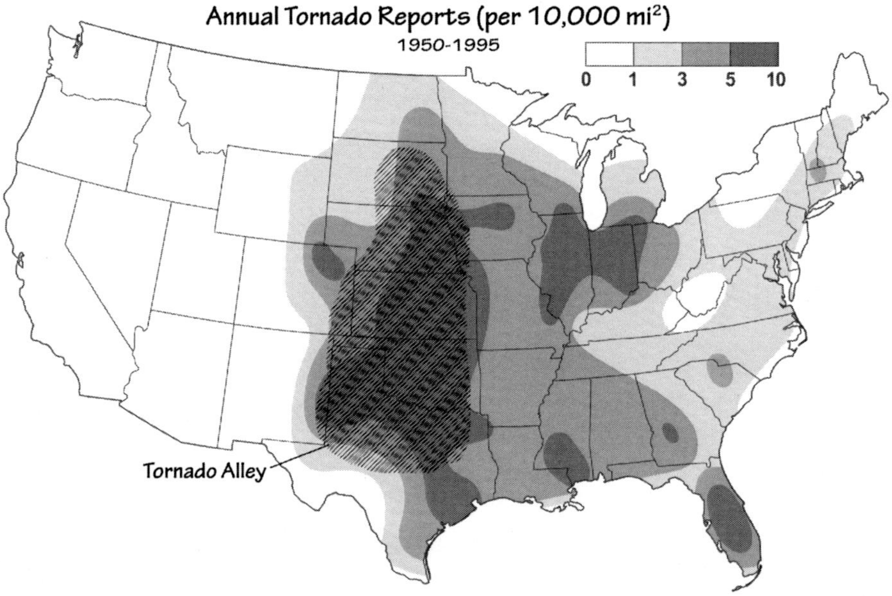

Figure 9.30: Annual tornado reports per 29,500 square kilometers (10,000 square miles) in the continental US, between 1950 and 1995. (See TFG website for full-color version.)

Figure 9.31: A tornado touches down over the hills near Roswell, New Mexico.

Figure 9.32: A haboob blows into the village of Ahwatukee, part of Phoenix, Arizona, in August 2003.

heat wave • a period of excessively hot weather that may also accompany high humidity.

In addition to the inhalation of silt and clay dust, other health hazards associated with dust storms include fungi, bacteria, pollutants, and heavy metals. These materials can irritate the lungs and trigger asthma attacks, allergic reactions, and other illnesses. One fungus, *Coccidioides*, causes "valley fever," which causes cold- and flu-like symptoms and sometimes rashes. Though most recover without treatment, it can have serious consequences and even lead to death for some people with weak immune systems.

Extreme Temperature and Drought

Extreme temperatures can create dangerous conditions for people and may lead to property damage. Summer temperatures in the arid Southwest can reach dangerously high levels, and temperatures around or above 38°C (100°F) are not uncommon. High heat can lead to a series of health complications if not properly dealt with—heat exhaustion, heat stroke, and dehydration can all result from exposure to extreme temperatures. Since the human body can only survive a few days (typically three) in the desert without water, a stranded and unlucky hiker or camper can easily die of dehydration if a suitable water supply cannot be reached in time. **Heat waves** are periods of excessively hot weather that may also accompany high humidity. Temperatures of just 3°C (6°F) to 6°C (11°F) above normal are enough to reclassify a warm period as a heat wave. Under these conditions, the mechanism of sweating does little to cool people down because the humidity prevents sweat from evaporating and cooling off the skin. Heat waves have different impacts on rural and urban settings. In

Drought

heat island effect • a phenomenon in which cities experience higher temperatures than do surrounding rural communities.

rural settings, agriculture and livestock can be greatly affected. Heat stress recommendations are issued to help farmers protect their animals, particularly pigs and poultry, which, unlike cattle, do not have sweat glands.

The impacts of heat waves on urban settings include a combination of the natural conditions of excessive heat and the social conditions of living in a densely populated space. Cities contain a considerable amount of pavement, which absorbs and gives off more heat than vegetation-covered land does. Air conditioning units that cool down the inside of buildings produce heat that is released outside. Pollution from cars and industry also serve to elevate the outdoor temperatures in cities. This phenomenon, in which cities experience higher temperatures than surrounding rural communities do, is known as the **heat island effect**. Other social conditions can increase the hazards associated with heat waves in urban areas. People who are in poor health, live in apartment buildings with no air conditioning, or are unable to leave their houses are at greatest risk of death during heat waves. In the summer of 2015, a record-setting heat wave occurred across the desert Southwest, scorching Arizona and leaving Phoenix with a new daily record of 46°C (115°F). Temperatures hovered above 38°C (100°F) even later than 10 pm. The heat wave contributed to severe drought, amplified heat-based health emergencies, and caused a heavy spike in electricity usage (related to increased air conditioning use) that generated a record-breaking demand on the power grid. Other times when heat waves have affected the Southwestern states include the heat wave of June-July 2013, which scorched the Southwest and Great Plains, baking Utah and setting records for extreme heat across New Mexico (*Figure 9.33*); and the heat wave of June-July 2012, which broke previous record highs across Colorado.

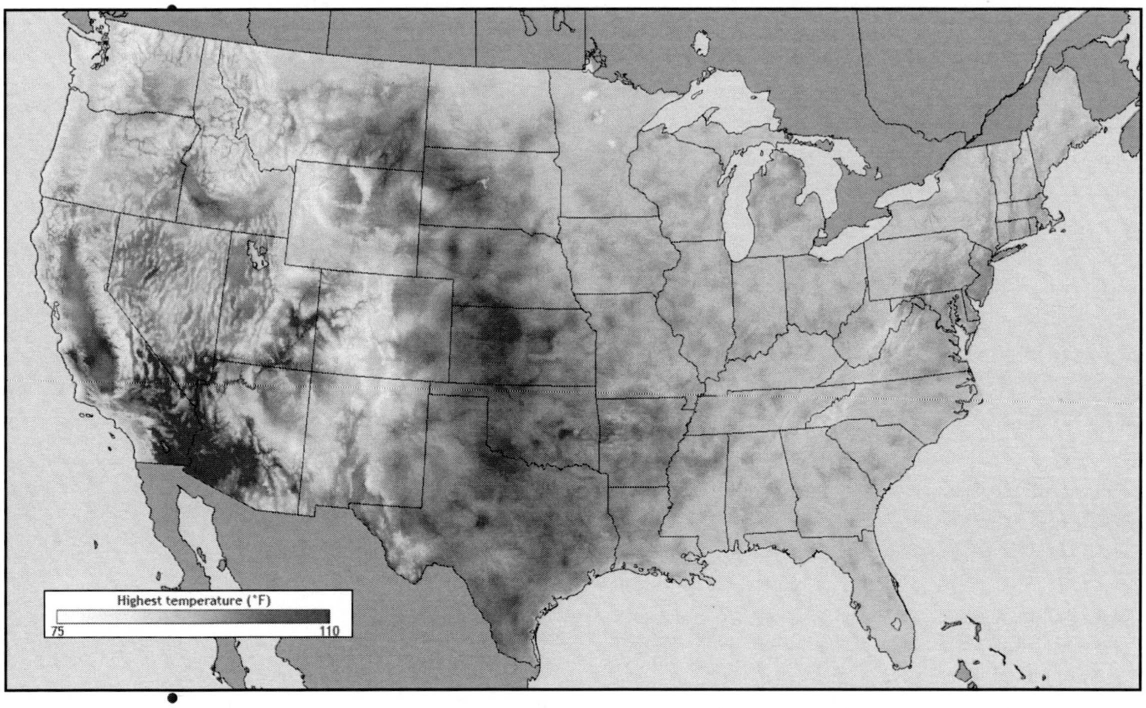

Figure 9.33: Air temperatures across the continental US during July 2013. (See TFG website for full-color version.)

Drought

While high temperatures can be directly dangerous, a larger scale hazard arises when these temperatures are coupled with lack of precipitation in an extended drought period. The Southwest has experienced both short-term and even decade-long periods of drought. Unlike other hazards, drought sets in slowly and takes time to be recognized. Agricultural areas can be seriously affected by a lack of rainfall and insufficient water supplies. Even higher-altitude forests show signs of stress since the combination of heat and long-term lack of precipitation deprives the land of one of its key resources. Lack of precipitation does not simply mean a lack of rain—it also means less seasonal snowfall in the mountains. Relatively little mountain snow in the winter translates into a lack of water for crop irrigation and household use in desert portions of the Southwest. Change in the flow rate of the Colorado River, which originates in the Rocky Mountains, serves as an excellent diagnostic for the effects of drought. This river is crucial for the irrigation of crops, and it feeds manmade reservoirs such as Lake Powell that supply drinking water to much of the region.

See Chapter 8: Climate to learn how global warming has influenced changes in the Colorado River's flow rate.

Many significant droughts have occurred in the Southwestern states—one notable instance of catastrophic drought in the Southwest was the Dust Bowl of the 1930s. Severe drought led to a drying of much of the topsoil, which was crucial to the agriculture of the area. High winds stripped the land of this topsoil, making crop growth impossible. This, in turn, led to the collapse of the farming industry, which was one of the main factors contributing to the Great Depression. More recently, severe drought conditions in 2002 forced Denver, Colorado to impose mandatory limits regarding water use; in addition, from 2011–2014, New Mexico was struck by its worst drought since the Dust Bowl. As of May 2016, much of the Southwest, especially Arizona, is experiencing conditions of moderate drought or abnormal dryness. Compiled tree-ring records over the past several thousand years shows that there have been past "megadroughts" that have been worse, and lasted longer, than recent ones. Models suggest that the likelihood of such droughts is expected to increase due to the effects and continuing patterns of climate change. Recent research using both models and data suggests that the climate of the Southwestern US has become and will remain drier, as subtropical dry zones move north.

Careful planning for seasonal drought, as well as extended drought, is the most effective way to reduce the chance of storage depletion in the Southwest. Conservation must be implemented as a series of progressive steps to be taken as water becomes scarcer. Out of necessity, the Southwest actually implements some of the most effective water management strategies in the United States. Still, no amount of planning can eliminate the long-term threat of drought, especially in an area dominated by deserts and under threat of the influence of changing climate.

Climate Change

global warming • *the current increase in the average temperature worldwide, caused by the buildup of greenhouse gases in the atmosphere.*

Climate Change

It is important to understand that most of the extreme climate change in Earth's history occurred before humans existed. That being said, the rapid release of carbon dioxide into the atmosphere from human activity is currently causing a **global warming** event. The seemingly slight increase in the average annual temperatures in the Southwest over the past 25 years has been accompanied by more frequent heat waves, shorter winters, and an increased likelihood of drought and wildfires.

Although wildfires can occur during any season, summer fires are the most common, since increased dryness contributes to fire risk. Today these most often start due to human activities, such as a poorly extinguished campfire, but they can also occur by natural ignition from lightning. Hundreds of square kilometers (miles) of forest have been lost to wildfires despite our best efforts to prevent, control, and extinguish them. Rural towns and summer homes, along with the people who inhabit them, can be suddenly caught in the blaze. Not only do these fires spread quickly, but human attempts to extinguish the blaze are hindered by the lack of available water to fight the fire. The Wallow Fire, which raged from May to July 2011, was the largest fire in Arizona's history; it consumed 2180 square kilometers (840 square miles) of land, destroyed 17 structures, and caused the evacuation of over 6000 people. In 2012, one of the Southwest's worst wildfire years, 1041 fires burned across Colorado, destroying 90,875 hectares (224,559 acres) of land, while over 1000 fires in Utah scorched more than 171,000 hectares (422,000 acres). And fires don't have to be large to be destructive—the most destructive fire in Colorado history, 2013's Black Forest Fire, burned only 5780 hectares (14,280 acres) but destroyed 511 homes and led to two fatalities (*Figure 9.34*).

Figure 9.34: The remains of a home destroyed by Colorado's Black Forest Fire on June 12, 2013.

Climate Change

Water supply is also a critical issue for the Southwestern states. Here, most water is obtained from precipitation, snowmelt, and runoff, which will dramatically decrease in quantity as temperature and aridity rise. In addition, parts of Colorado and New Mexico obtain agricultural and drinking water from the Ogalalla aquifer, an underground layer of water-bearing **permeable** rock. Part of the High Plains aquifer system, this underground reservoir supplies vast quantities of groundwater to the Great Plains. As drought intensifies and temperature rises, the amount of water drawn from the aquifer (especially for agricultural irrigation) has increased, while the rate at which the aquifer recharges has decreased. The aquifer's average water level has dropped by about 4 meters (13 feet) since 1950, and in some areas of heavy use, the decrease is as high as 76 meters (250 feet) (*Figure 9.35*). However, the aquifer only replenishes at a rate no greater than 150 millimeters (6 inches) per year. Some estimates indicate that at its current rate of use, the entire Ogalalla aquifer could be depleted by as early as 2028, threatening human lives, our food supply, and the entire Great Plains ecosystem.

permeability *• a capacity for fluids and gas to move through fractures within a rock, or the spaces between its grains.*

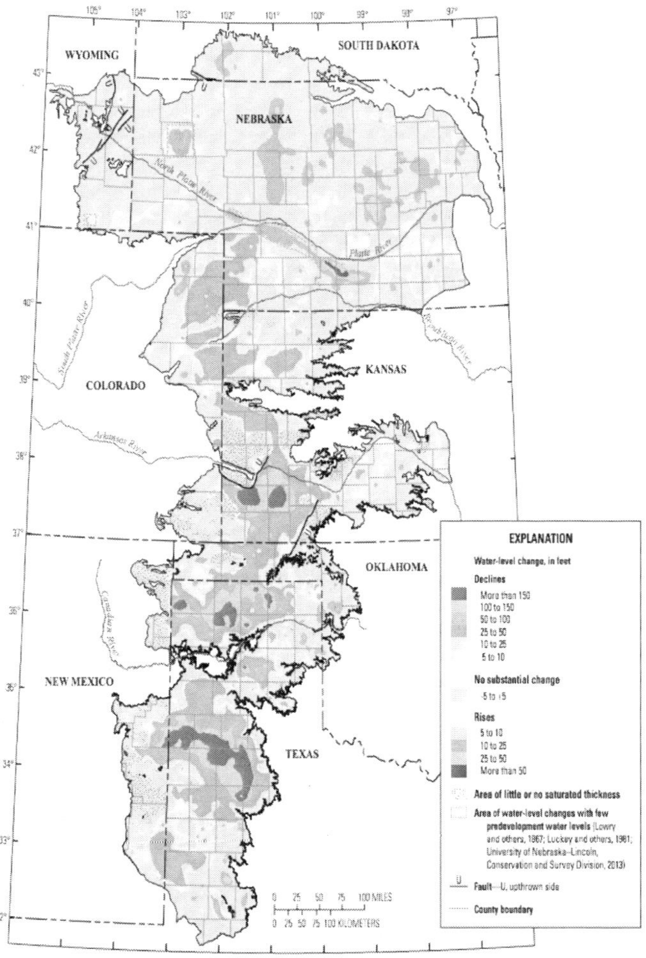

Figure 9.35: *Water level change in the Ogalalla aquifer between 1950 and 2013. (See TFG website for full-color version.)*

Climate Change

water table • *the upper surface of groundwater.*

hurricane • *a rapidly rotating storm system with heavy winds, a low-pressure center, and a spiral arrangement of thunderstorms.*

In rural desert and semi-desert areas that are not served by well-planned regional or municipal systems, most people are dependent upon streams and wells. Streams often run dry, especially in the summer. The **water table** (the level of underground water) then migrates deeper, forcing people to extend wells deeper into the ground. Unfortunately, this is only a temporary solution.

In most of these areas, water is being withdrawn much more quickly than it is naturally replenished. Another hazard arising from excessive pumping of groundwater in the Southwest is land subsidence and subsidence-related earth fissures. In sum, lack of water reserves can lead to a cycle of economic disasters as well as the displacement of populations and businesses. The preservation and storing of water in large aquifers (water banking) for future use is an important technique to help adapt to drought.

Increasing temperatures also allow certain pests, such as ticks and mosquitoes, to live longer, thereby increasing the risk of contracting the diseases they carry. In addition, organisms that damage ecosystems, such as the bark beetle, are better able to survive warmer winters, thrive, and multiply. In recent decades, bark beetles are estimated to have affected more than twice the forest area burnt by wildfires in New Mexico and Arizona.

Another concern regarding hazards exacerbated by climate change in the Southwest is whether or not there has been or will be an increase in the number or severity of storms, such as **hurricanes** and tornados. According to NASA, the present data is inconclusive in terms of whether hurricanes are already more severe, but there is a greater than 66% chance that global warming will cause more intense hurricanes in the 21st century. Since climate is a measure of weather averaged over decades, it might take many years to determine that a change has occurred with respect to these types of storms. Scientists are certain that the conditions necessary to form such storms are becoming more favorable due to global warming. The Union of Concerned Scientists has created an infographic that demonstrates the relative strength of the evidence that various hazards are increasing as a result of climate change (*Figure 9.36*).

See Chapter 8: Climate for more on the effects of climate change in the Southwest.

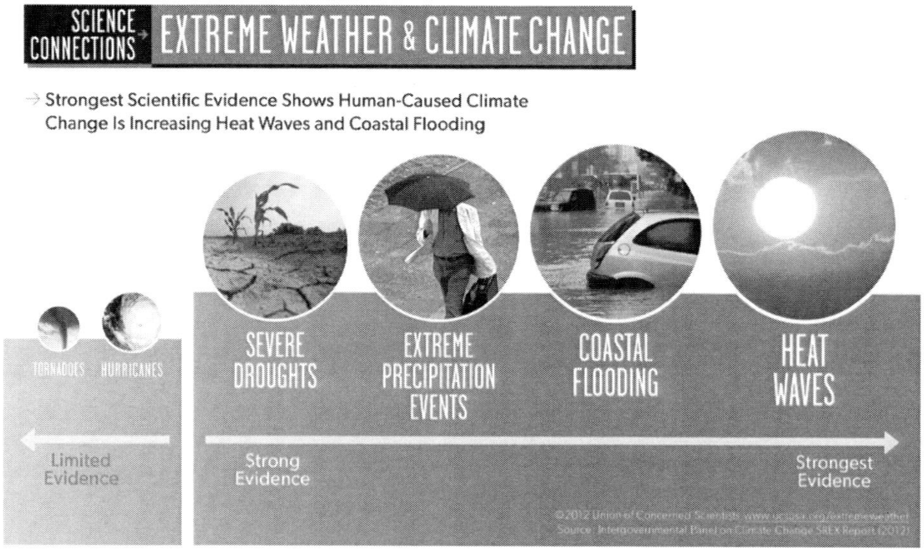

Figure 9.36: The strength of evidence supporting an increase in different types of extreme weather events caused by climate change.

9 Earth Hazards

Resources

Resources

General Resources on Earth Hazards

Gunn, A. M., 2001, *The Impact of Geology on the United States: a Reference Guide to Benefits and Hazards*, Greenwood Press, Westport, CT, 296 pp.
Macdougall, J. D., 2011, *Why Geology Matters: Decoding the Past, Anticipating the Future*, University of California Press, Berkeley, 285 pp.
NASA Earth Observatory Natural Hazards map. [Monthly images of Earth hazards occurring globally.] http://earthobservatory.nasa.gov/NaturalHazards/.

General Resources on Earth Hazards in the Southwest

Environmental Geology and Geologic Hazards in Arizona, The Arizona Geological Survey, http://www.azgs.az.gov/hazards.shtml.
Geologic Hazards, Colorado Geological Survey, http://coloradogeologicalsurvey.org/geologic-hazards/.
Geologic Hazards, New Mexico Bureau of Geology and Mineral Resources, http://geoinfo.nmt.edu/geoscience/hazards/home.html.
Harris, R. C., and P. A. Pearthree, 2002, *A Homebuyer's Guide to Geologic Hazards in Arizona*, Down-To-Earth 13, Arizona Geological Survey, Tucson, AZ, 36 pp., http://www.azgs.az.gov/HomeOwners-OCR/home_buyers_guide_2002_full.pdf.
Hazards, Ready Colorado, Colorado Homeland Security Emergency Management, https://www.readycolorado.com/about-ready-colorado.
Highland, B., 2013, *Colorado Natural Hazard Events: an Overview of When and Where Nature Impacted People in Colorado*, Highland & Associates, Morrison, CO, 68 pp.
King, B., 2008, Drought, *Utah Natural Hazards Handbook*, Utah Division of Water Resources, pp. 95–108, http://www.utah.gov/beready/documents/HazardsHandbookDraft8.pdf.
Natural Hazards Center, University of Colorado at Boulder, https://hazards.colorado.edu/.
Natural Hazards in Arizona, Arizona Geological Survey. [Interactive map] http://data.azgs.az.gov/hazard-viewer/.
Utah Natural Hazards Handbook, Utah Division of Homeland Security, 2008, 128 pp., http://www.utah.gov/beready/documents/HazardsHandbookDraft8.pdf.

This "General Resources on Earth hazards" section may contain additional information on each of the specific topics in the lists below.

Resources on Climate- and Weather-related Earth Hazards

Floods

Cook, T., 2013, Releasing a flood of controversy on the Colorado River, *Earth Magazine*, http://www.earthmagazine.org/article/releasing-flood-controversy-colorado-river.
Desert Safety Tips and Desert Survival. [This has a section on flash floods.] http://www.desertusa.com/thingstodo/du_safetytips.html.
Preparing for the Flash Flood Season, What is the Wash?, Las Vegas Wash Coordination Committee, https://www.lvwash.org/html/what_visit_flood.html.
Rosen, J., 2014, Colorado River researchers find signs of ancient, devastating floods, *Los Angeles Times*, June 19, 2014, http://www.latimes.com/science/sciencenow/la-sci-sn-colorado-floods-20140618-story.html.

Droughts and High Temperatures

Climate Change and Drought in the American Southwest, Climate Institute, http://climate.org/effects-of-21st-century-climate-change-on-drought-disk-in-the-american-southwest/.

Drought Monitoring, Climate Assessment for the Southwest, The University of Arizona, http://www.climas.arizona.edu/sw-climate/drought-monitoring.

Fischetti, M., 2015, U.S. droughts will be the worst in 1,000 years: the Southwest and central Great Plains will dry out even more than previously thought, *Scientific American*, http://www.scientificamerican.com/article/u-s-droughts-will-be-the-worst-in-1-000-years1/.

Harris, R. C., Giant desiccation cracks in Arizona, Arizona Geological Survey, *Arizona Geology*, 34(2): 1–3, http://www.azgs.az.gov/Hazards_ocr/Swelling_Shrinking%20soils/Desiccation%20Cracks%20in%20AZ-2004.pdf.

Dust, Dust Storms, and Haboobs

Dust Storm and Haboob Safety Tips, National Weather Service, http://www.nws.noaa.gov/airquality/dust_storm.shtml.

Environmental Mediation of Valley Fever, Department of Geography, US Geological Survey, and University of Arizona, http://www.casa.arizona.edu/~peter/valleyfever/web/index.html.

Hinkley, T. K., *Mineral Dusts in the Southwestern U.S.*, US Geological Survey, http://geochange.er.usgs.gov/sw/impacts/geology/dust/.

Thomas, M., The mysterious fungus infecting the American Southwest, *The Atlantic*, August 8, 2014, http://www.theatlantic.com/health/archive/2014/08/the-mysterious-fungus-infecting-the-american-southwest/375191/.

Tornados

Edwards, R., *Tornado Safety*, Storm Prediction Center, National Weather Service, http://www.spc.noaa.gov/faq/tornado/safety.html.

Historical Records and Trends [in Tornados in the US], National Centers for Environmental Information, National Oceanic and Atmospheric Administration, https://www.ncdc.noaa.gov/climate-information/extreme-events/us-tornado-climatology/trends.

The Most "Important" US Tornadoes by State, Tornado Project Online, http://www.tornadoproject.com/alltorns/worststs.htm.

Tornado History Project. [Searchable database of reported US tornadoes since 1950.] http://www.tornadohistoryproject.com/.

Vaughan, K. and B. Hubbard, Colorado ranks near top nationally in tornado activity, *Rocky Mountain PBS News*, June 13, 2013, http://inewsnetwork.org/2013/06/19/colorado-ranks-near-top-nationally-in-tornado-activity/.

Resources on Landscape-related Earth Hazards

Expansive Soils

Harris, R. C., and P. A. Pearthree, 2002, Problem soils, in: *A Homebuyer's Guide to Geologic Hazards in Arizona*, Down-To-Earth 13, Arizona Geological Survey, Tucson, AZ, pp.11–15, http://www.azgs.az.gov/HomeOwners-OCR/HG4_problemsoils.pdf.

King, H., *Expansive Soil and Expansive Clay: the Hidden Force Behind Basement and Foundation Problems*, Geologic Hazards, Geology.com, http://geology.com/articles/expansive-soil.shtml.

Noe, D. C., C. L. Jochim, and W. P. Rogers, 1997, *A Guide to Swelling Soils For Colorado Homebuyers and Homeowners, 2nd edition*, Colorado Geological Survey Special Publication 43, 52 pp., http://vonscolorado.com/wp-content/uploads/2014/04/A_Guide_to_Swelling_Soils_Pub43.pdf.

Landslides

Burt, C. B., 2014, *Worst Landslides in U.S. History*, Weather Underground, March 25, 2014, https://www.wunderground.com/blog/weatherhistorian/worst-landslides-in-us-history.

Gootee, B. F., 2006, *The Marcus Landslide: an Ancient Landslide in the McDowell Mountains*, The Arizona Geological Survey, http://www.azgs.az.gov/MarcusLandslide_2008.shtml.

Highland, L., Landslide types and processes, 2004, *US Geological Survey Fact Sheet* 2004-3072, 4 pp., http://pubs.usgs.gov/fs/2004/3072/pdf/fs2004-3072.pdf.

Highland, L. M., & Bobrowsky, P., 2008, The landslide handbook—A guide to understanding landslides, *US Geological Survey Circular* 1325, 129 pp.

Resources

Landslide Hazards Program, US Geological Survey, http://landslides.usgs.gov/.

Landslides in the United States Since 2007 (map), National Geographic, http://news.nationalgeographic.com/news/2014/04/140422-surveying-american-landslides-interactive/.

Wald, L., *Reconstruction of an Avalanche: The West Salt Creek Rock Avalanche, Colorado, May 25, 2014*, Landslide Hazards Program, USGS, http://landslides.usgs.gov/research/featured/2015-westsaltcreek/.

Karst, Sinkholes, and Fissures

Arizona's Earth Fissure Center, The Arizona Geological Survey, http://www.azgs.az.gov/EFC.shtml.

Hill, C. A., 2003, Caves and karst in New Mexico, New Mexico Bureau of Geology and Mineral Resources, *New Mexico Earth Matters*, 3(1): 1–4, https://geoinfo.nmt.edu/publications/periodicals/earthmatters/3/n1/em_v3_n1.pdf.

Lindberg, P. A., 2009, *Devil's Kitchen Sinkhole—Sedona, AZ: a Potential Geohazard*, The Arizona Geological Survey, http://www.azgs.az.gov/arizona_geology/winter09/article_devilskitchen.html.

Resoures on Tectonics-related Earth Hazards

Earthquakes

A Brief History of Earthquake Activity in Arizona, AZ Central, http://www.azcentral.com/story/news/arizona/2014/06/30/arizona-earthquake-timeline/11787839/.

Earthquakes [in Colorado], Colorado Geological Survey, http://coloradogeologicalsurvey.org/geologic-hazards/earthquakes/.

Fellows, L. D., 2000, Earthquake hazard in Arizona, Arizona Geological Survey, *Arizona Geology*, 30(1): 1–4, http://www.azgs.az.gov/Hazards_ocr/earthquakes/Earthquake%20Hazard%20in%20AZ-2000.pdf.

Haneberg, B., 2016, *Do We Have Earthquakes in New Mexico?*, New Mexico Bureau of Geology and Mineral Resources, http://tremor.nmt.edu/faq/here.html.

Incorporated Research Institutions for Seismology (IRIS) Education and Public Outreach, http://www.iris.edu/hq/programs.epo.

Induced Earthquakes, Earthquake Hazards Program, US Geological Survey, http://earthquake.usgs.gov/research/induced/.

IRIS Seismic Monitor, Incorporated Research Institutions for Seismology (IRIS), http://www.iris.edu/seismon/.

Putting Down Roots in Earthquake Country: Your Handbook for Earthquakes in Utah, 2008, Utah Seismic Safety Commission, 33 pp., http://www.utah.gov/beready/documents/roots_earthquake_low.pdf.

Stover, C. W., and J. L. Coffman, 1993, Seismicity of the United States, 1568–1989 (revised), *US Geological Survey Professional Paper* 1527, 418 pp, http://pubs.usgs.gov/pp/1527/report.pdf.

Eldgredge, S. N., 1996, The Wasatch Fault, *Utah Geological Survey Public Information Series* 40, 17 pp., http://files.geology.utah.gov/online/pdf/pi-40.pdf.

Today in Earthquake History, Earthquake Hazards Program, US Geological Survey. [Content abridged from Stover & Coffman, 1993.] http://earthquake.usgs.gov/learn/today.

U of U Seismic Stations, Earthquake Center, University of Utah, http://quake.utah.edu/earthquake-center/quake-map.

US Earthquake Information by State/Territory, USGS Earthquake Hazards Program, http://earthquake.usgs.gov/earthquakes/states/.

US Earthquake Monitor, US Geological Survey, http://earthquake.usgs.gov/earthquakes/map/.

Utah Faults, Utah Geological Survey, http://geology.utah.gov/hazards/earthquakes-faults/utah-faults/.

Volcanoes

Duffield, W., 1997, *Volcanoes of Northern Arizona: Sleeping Giants of the Grand Canyon Region*, Grand Canyon Association, Grand Canyon, AZ, 68 pp.

Frequently Asked Questions about New Mexico Volcanoes, New Mexico Bureau of Geology and Mineral Resources, https://geoinfo.nmt.edu/faq/volcanoes/.

Hagadorn, J., 2015, *Colorado's Volcanoes, Just the Facts*, Front Porch Stapleton [News], https://frontporchstapleton.com/article/colorados-volcanoes/.

Pfeiffer, T., *Volcanoes of Canada and USA (Mainland)*, Volcano Discovery. [Volcanos active within the past 10,000 years.] https://www.volcanodiscovery.com/north-america.html.

The Volcanoes of New Mexico, Online Exhibits, New Mexico Museum of Natural History and Science, http://www.nmnaturalhistory.org/online-exhibits-geoscience/volcanoes-new-mexico.
U.S. Volcanoes and Current Activity Alerts, Volcano Hazards Program, US Geological Survey, http://volcanoes.usgs.gov/index.html.

Resources on Materials-related Earth Hazards

Radon

Radon, Colorado Department of Public Health and Environment, https://www.colorado.gov/pacific/cdphe/categories/services-and-information/environment/radon.
Radon Fact Sheet, Air Check Inc., 2009, http://www.radon.com/radon/radon_facts.html.
Radon: Health Risks, Environmental Protection Agency, 2013, http://www.epa.gov/radon/healthrisks.html.
Radon Information, Environmental Protection Agency, http://www.epa.gov/radon/index.html.
Radon levels for [any US state], http://**-radon.info/ [in which ** = two-letter state code, such as AZ, CO, NM, or UT].
Radon (Rn), United States Environmental Protection Agency (EPA). [Includes state radon maps with county-level data.] http://www.epa.gov/radon/whereyoulive.html.

Asbestos

Harris, R. C., 2004, Asbestos in Arizona, Arizona Geological Survey, *Arizona Geology*, 34(1): 1–4, http://www.azgs.az.gov/Mineral%20Scans/Asbestos%20in%20AZ%2004.pdf.
Van Gosen, B. S., 2008, Reported historic asbestos mines, historic asbestos Prospects, and natural asbestos occurrences in the southwestern United States (Arizona, Nevada, and Utah), *US Geological Survey Open-File Report* 2008-1095, 1 plate, http://pubs.usgs.gov/of/2008/1095.

Earth Hazards Teaching Resources

Investigating Speed and Acceleration Using Tornado Tubes, Hamline University Graduate School of Education MnSTEP Teaching Activity Collection, http://serc.carleton.edu/sp/mnstep/activities/27202.html.
Karst Formation, City of Austin Youth Education resources, http://austintexas.gov/sites/default/files/files/Watershed/youth_education/karst_lesson_high_school.pdf.
Radke, J., *Impact of Natural Disasters on the Earth*, bHamline University Graduate School of Education MnSTEP Teaching Activity Collection, http://serc.carleton.edu/sp/mnstep/activities/19789.html.
Radon Activities from the Alabama Radon Program, Alabama and Auburn Universities Extension, http://www.aces.edu/fcs/hndh/radon/alradon.php.
Teaching Quantitative Concepts in Floods and Flooding, SERC Resources for Undergraduate Students and Faculty, http://serc.carleton.edu/quantskills/methods/quantlit/floods.html.

Chapter 10: Real and Virtual Fieldwork:

"Why Does This Place Look the Way it Does?"

All the major topics in *The Teacher-Friendly Guides™* were built upon observations of the natural world, and these observations are the clues that scientists use to reconstruct the history of the Earth. Shelly fossils along the Himalayas tell of ancient sea floors that have been uplifted into mountains. Ripple marks that have since turned to stone tell of ancient shorelines. And scratches along the bedrock in Central Park tell of massive glaciers that—some 20,000 years ago—created a skyline much different than the one of steel and glass found in New York today. A number of forces and processes have made seas, forests, deserts, and the life those ecosystems hosted appear and disappear from the landscape over the course of geologic time. Many of these changes left behind hints that we can interpret today when we tell the story of a place. That massive glaciers once advanced as far south as New York is not a conclusion derived from mathematical modeling in a lab; it is instead evidenced by not only those scratches, but also by a host of observed glacial deposits that litter not only New York, but much of northern North America.

> *The story of a place is written in its landscape, rocks, fossils, and biota; fieldwork investigations help scientists—and students and teachers—tell that story.*

Introducing students to the practice of fieldwork can be a tremendous experience. Its central role in the education of geoscientists makes fieldwork a "signature pedagogy" in the preparation of professionals within the field, and fieldwork warrants a larger place in the K-12 curriculum. For these reasons, real and virtual fieldwork practices are well suited for addressing both *The Next Generation Science Standards* and *The Common Core Learning Standards.* Fieldwork as a topic is also fundamentally different from the other chapter topics in this guide. Therefore, this chapter is somewhat different in structure and is significantly longer than the other chapters in the Guide. The chapter begins by laying out some of the rationale for engaging in real and virtual fieldwork, and it then addresses some of the nuts-and-bolts issues for planning, carrying out, and documenting fieldwork with your students.

Exploring local natural history through inquiry-based approaches emphasizes critical thinking. And by conducting such investigations, students have taken a tremendous leap: they are not merely learning about science; they are doing science! But getting students into the field can be difficult. An alternative is for the educator to visit the field on his or her own time, returning to the classroom with a series of images and specimens that permit a Virtual Field Experience

CHAPTER AUTHORS

Don Duggan-Haas
Richard A. Kissel

(VFE). Virtual fieldwork offers the opportunity to explore an area without leaving the classroom, and it allows multiple "visits" to a site. VFEs can also enhance and extend the experience when actual fieldwork is possible. The Earth is a system, after all, and any one site—virtual or real—can display a host of natural phenomena, from simple erosion and deposition to the principles of superposition and faunal succession to the formation of ripple marks or mud cracks. By adding to a VFE year after year, you can also document changes within the environment, such as changes to a stream's course, the succession of an ecosystem, or the nature of human disturbance. Ideally, virtual fieldwork in the classroom captures the active experience of a scientist examining an area: It provides opportunities to actively explore, discover, ask questions, and make observations that help to answer those questions, ultimately allowing students to develop educated responses to the question *"Why does this place look the way it does?"*

Commonalities of Virtual and Actual Fieldwork

This chapter addresses both actual and virtual fieldwork and the many connections between them. The process of making VFEs, at least in the ways we lay out here, involves doing actual fieldwork. Much of the work of making a VFE involves simply following good fieldwork practices in combination with a heightened attention to sharing the experience with students or other learners. While VFEs can be used in place of actual fieldwork, they can also be used to both prepare for and reflect upon actual fieldwork. Engaging students as partners in the creation of VFEs is an opportunity for teaching through inquiry while also building a resource that is useful to people outside of the school, as well as to future students. What follows addresses all of these possibilities.

NASA scientists routinely conduct actual fieldwork remotely.

We also draw attention to the distinction between field*work* and field *trips*. We strive to engage learners in *figuring things out*, while field trips—whether actual or virtual—are too often characterized by trip leaders *pointing things out*. Building in the opportunity for genuine discovery is challenging but promises to yield longer-term engagement and understanding.

Just Go (and Don't Stop)

The minimum requirement for conducting fieldwork is your own sweet self. This chapter discusses a wide range of tools and approaches, but doing fieldwork of any (safe) sort that doesn't damage the site is a key objective. The tools and approaches discussed in this chapter will extend your senses and help you to capture the experience in ways that will make it easier to share with students. Work within your comfort zone (but perhaps at its edge) and at a pace appropriate to what life allows, and gradually build your virtual representation of the local environment over the course of years, increasing student participation in the process as time goes by. Use the local landscape to nurture skills within your students that will allow them to read any type of landscape. Through this process, your students can teach members of your community about the story of your site while also creating and extending resources that can teach other learners around the country about where you live. Building a deep understanding of place through VFE development and then comparing your local environment with VFEs created by other teachers and students is an excellent way to use the local environment to understand the global environment.

Whether the fieldwork is real or virtual, it can either involve a single visit or be extended over many, many visits. Scientists may reach points where they have figured out particular pieces of the puzzle when understanding the nature of a site, but they never fully understand all aspects of a place's story. Fieldwork, therefore, is something that is never "finished." Whether it is the second or seven-hundredth visit to a site, there is always more to discover. This is part of what makes science fascinating! It connects to the idea that while fieldwork may focus primarily upon a single topic, researchers (whether K-12 students, educators, or professional scientists) who develop a deep understanding of the story of a place must understand the roles of geology, ecology, climatology, anthropology, and more. Of course, this type of understanding will not come from a single class period of fieldwork, or even a single course infused with fieldwork, but the appreciation of this systems idea can be planted and nurtured.

Start local

In choosing a field site, whether it is local or distant or for actual or virtual fieldwork, it should be interesting from an Earth systems science perspective. Fortunately, if you know how to look, *every* site is interesting from an Earth system science perspective. Over the grand course of Earth history, the story of any location is a fascinating one that involves myriad changes. The work of telling the story of any environment is a form of rich inquiry. While it would also be fascinating to find a place that hasn't changed, no such place exists on the surface of Planet Earth!

While VFEs provide the opportunity to study distant or otherwise difficult to access locations, we suggest starting close to home or school, at a location that students are already familiar with or have access to. What is outside your classroom door has more immediate relevance to the lives of your students than anywhere else on Earth. Nearly every unit in an Earth or environmental science course, and most of the units in a biology course, play out in some meaningful way in the local environment, and the local environment can extend the boundaries of the classroom tremendously with little or no cost. Things are only understood in comparison to something else, so comparing sites to one another can deepen one's understanding of both or even of all sites—but it is still best to start with the local.

Students can use real or virtual field sites to study how all the major topics in their Earth or environmental science curriculum are manifest in the "real world." In an ideal situation, the classroom is immediately adjacent to a safe, accessible field site, and there is flexibility within the school schedule that allows for in-depth study of the site in ways that cut across disciplinary boundaries.

Just Go

Unfortunately, it's not always practical to repeatedly visit an actual field site with 30 students throughout the year or semester. Through virtual fieldwork, students can come to see how the rock types and flora and fauna outside their classroom tell part of the story of that place.

In order to create VFEs, authors must closely study their field sites with an eye toward doing fieldwork with students. VFEs are a stepping-stone to bringing students into the field, even if the field is "only" the schoolyard. VFEs can be used to prepare students for the field and/or to process the fieldwork after visiting the actual site. Ideally, students will participate in the creation and extension of VFEs, but we recognize that getting to this point may take years.

Connecting to Earth Science Bigger Ideas, the Next Generation Science Standards, and the Common Core

Fieldwork investigations have the potential to be extended indefinitely in time and can involve the integration of a wide range of science and non-science disciplines. *"Why does this place look the way it does?"* is a bottomless question, meaning that it can be productively investigated for a very, very long time. Field scientists, be they professionals or fifth graders, will never fully answer this driving question absolutely or at every scale.

The act of VFE creation is a valuable type of professional development (PD) that creates useful evidence of having done the PD. Through the creation and continued use of virtual fieldwork, a teacher can become a true expert on his or her local environment—perhaps the preeminent expert. The process of VFE creation and use can also create evidence of inquiry teaching aligned to relevant standards. The VFE you create or augment can serve as a key piece of a professional portfolio.

The ultimate goal of our instruction is to build understanding of the Earth system and the ways in which science is used to build that understanding. We bring focus through the use of a small set of bigger ideas and overarching questions. These are discussed in detail in the Big Ideas Chapter and are also summarized below.

Overarching questions:

- How do we know what we know?
- How does what we know inform our decision making?

Earth system science bigger ideas:

- The Earth is a system of systems.

Connecting

- The flow of energy drives the cycling of matter.
- Life, including human life, influences and is influenced by the environment.
- Physical and chemical principles are unchanging and drive both gradual and rapid changes in the Earth system.
- To understand (deep) time and the scale of space, models and maps are necessary.

Fieldwork should provide the opportunity to explore, describe, and build understanding of these questions and ideas. These ideas and questions map onto the *Next Generation Science Standards'* Disciplinary Core Ideas, Crosscutting Concepts, and Science and Engineering Practices. The Crosscutting Concepts and Scientific and Engineering Practices are shown in *Table 11.1*. As you read through the rest of this chapter, and as you and your students carry out fieldwork, revisit these lists of concepts and practices frequently in order to draw attention to how they connect to the work of reading the landscape.

Table 11.1: NGSS's Scientific and Engineering Practices and Crosscutting Concepts. As you and your students engage in fieldwork, consider how the practices and concepts are being used to make sense of the environment. See the Big Ideas Chapter for a more in-depth discussion.

Scientific and Engineering Practices	Crosscutting Concepts
1. Asking questions and defining problems	1. Patterns
2. Developing and using models	2. Cause and effect
3. Planning and carrying out observations	3. Scale, proportion, and quantity
4. Analyzing and interpreting data	4. Systems and system models
5. Using mathematics and computational thinking	5. Energy and matter
6. Constructing explanations and designing solutions	6. Structure and function
7. Engaging in argument from evidence	7. Stability and change
8. Obtaining, evaluating, and communicating information	8. Interdependence of science, engineering, and technology
	9. Influence of engineering, technology, and science on society and the natural world

Fieldwork Challenges and Benefits

Of course, VFEs also allow for some kind of "fieldwork" experience when actual fieldwork is difficult or impossible to carry out. The reasons that actual fieldwork is difficult are fairly obvious:

- **Fieldwork is logistically challenging.** It's hard to fit into a typical class period, or even a double lab period. To go off site requires permission slips, busing, and figuring out how to deal with behavior outside the normal classroom setting.

- **It costs money.** Field trip budgets have been slashed, and weren't even very common at the secondary level before budget cuts.

- **Many teachers have only limited experience doing field science themselves.** Earth science has more teachers teaching out of field than any other science discipline, and fieldwork is not a component of many Earth, biology, or environmental science teacher certification programs. It is intimidating to lead fieldwork if you haven't been through it yourself.

- **Fieldwork poses safety and behavior concerns different from those in the classroom.** Falling off a cliff has different consequences than falling off a chair.

- **Teaching in the field employs a different set of skills than teaching in the classroom.** The logistics of moving groups of students from place to place and focusing their attention on the goals of the fieldwork takes careful planning, especially if multiple classes are involved.

These issues shouldn't preclude fieldwork, but they undeniably complicate it. These challenges are not insignificant, but the rewards of doing fieldwork are worth the trouble. Field trips are among the most memorable and most valued school experiences.

Fieldwork 101: Gathering Information and Creating Your Own VFE

What follows are recommendations. These recommendations are intended to help prepare you for fieldwork, but they are just guidelines, not steadfast rules. Bringing the field to the classroom at any scale is better than not bringing the field to the classroom at all. The careful attention to detail described here will prove extremely helpful, but avoid being discouraged if your first trip to the field isn't as productive as you had initially imagined. Scientists of all disciplines continually refine their methods and procedures, leading to more productive and "better" results over time. With time and more fieldwork, your confidence will grow. Get into the field, be safe, and do your best to capture the experience in a way that allows you to best reproduce it for your students!

Fieldwork 10

Fieldwork 101

Before visiting the site: understand the natural history of the region

In order to make sense of a local site, it's helpful to understand the geologic history of the larger region before your visit. Did inland seas once flood the area? Have mountain-building events shaped the landscape and its rocks? Was it glaciated? Since the reasons that a place looks the way it does are dependent upon more than the geology, you want to pay attention to this concept as well. That being said, since the geology is the base upon which the landscape is built, starting there makes good sense. *The Teacher-Friendly Guides*™ are an excellent source for discovering the history of a region, as well as that history's effect on the rocks, fossils, and other features of the area.

Questions to Keep in Mind

When visiting or examining any area, the ultimate question to answer is: *Why does this place look the way it does?* But to help understand such an overarching concern, it is important to have certain other questions in mind. These questions will guide exploration, and they will help ensure that important information is recorded during your visit:

- What kind(s) of rock(s) are found in the area? How do you know?
- In what environment did these rocks probably form?
- What is the arrangement of the rocks?
- Are fossils preserved in the rocks? If so, what can they tell you about past environments?
- What has happened to this area to make it look the way it does today? (That is, what has happened to the area since the rocks formed?) Why do you think so? (What is the evidence for your claim?)

We have put together a set of questions that build upon the fundamentals listed above and that can be asked of any site. This is a key idea—that there are questions that can be asked productively about any environment. Recognizing that idea is a key step toward being able to take the lessons of one field trip and applying them to the "reading" of any landscape. These questions are included in the graphic organizer in *Figure 11.1*, and as a checklist in the section entitled Back in the Classroom.

10 Fieldwork

Fieldwork 101

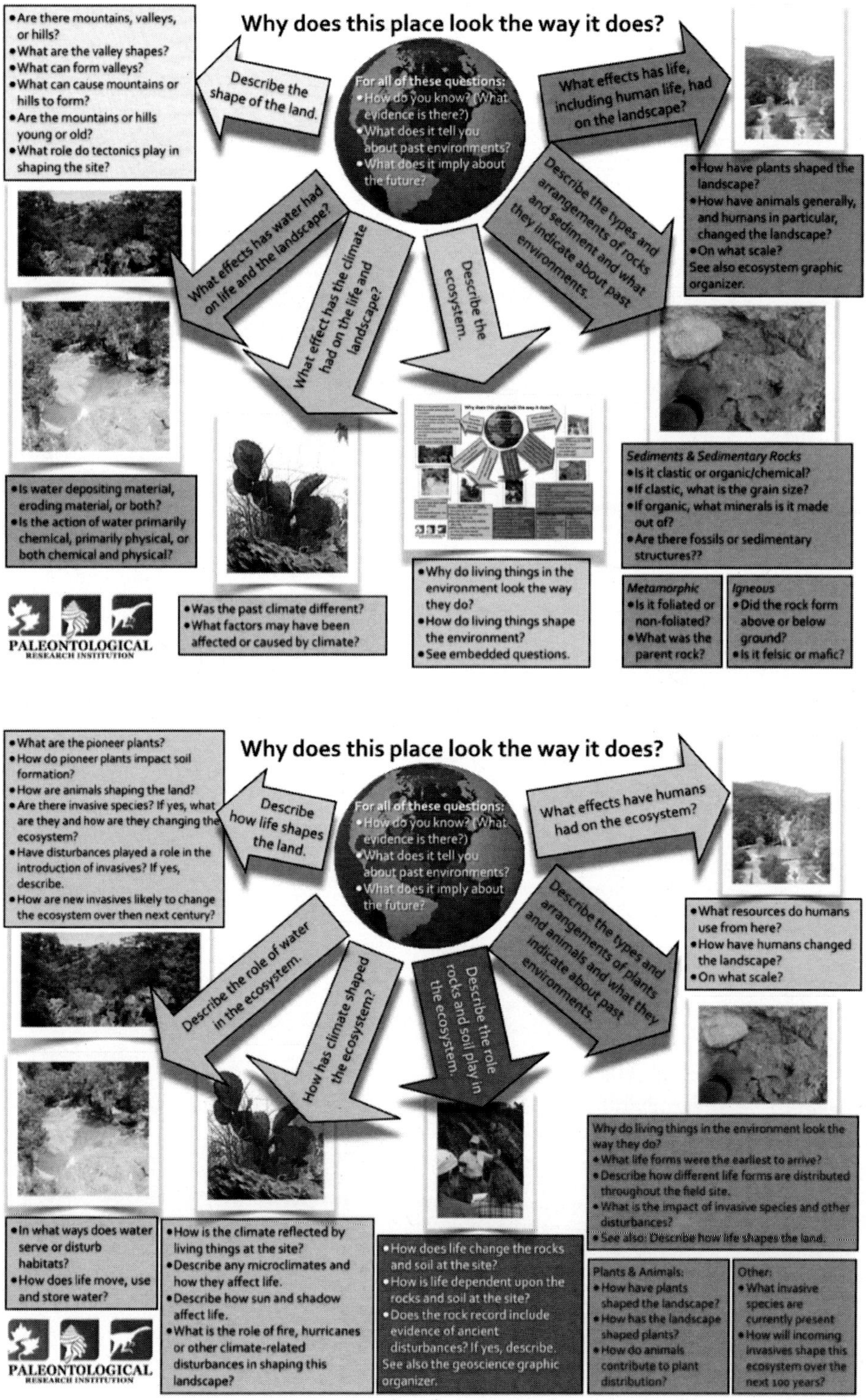

Figure 11.1: This pair of graphic organizers shows various paths of inquiry that stem from the question: Why does this place look the way it does? *The top graphic focuses upon the geosciences, and the bottom focuses upon the environmental sciences. The questions within the diagrams are also included as printable checklists in the section "Back in the Classroom."*

Safety and Logistics in the Field

At the Site

Considerations are different for an adult or a group of adults in the field than they are for taking students into the field, but certain measures related to safety are universal. At any field site, safety is the first priority. No photograph, measurement, or fossil is worth the risk of personal injury or death. To ensure safe and productive fieldwork, keep the following thoughts in mind:

- Always carry a small, standard first-aid kit.
- Wearing the proper clothing is very important. Long pants are recommended, as are sturdy boots, which will help prevent twisted ankles as you scurry over uneven or loose surfaces.
- While walking through a valley or next to any outcrop, always be on the lookout for rock falls. Remember, slopes with no vegetation tend to produce more falls.
- If more than one individual is climbing an outcrop, do not climb single file. Rocks dislodged from one climber can quickly tumble down the outcrop and hit the next climber.
- When using your rock hammer, protective eyewear should always be worn. If your hammer possesses a sharp pick opposite the flat surface, always use the flat surface when striking. And if you are working with others, notify all in the vicinity before striking any surface with your hammer.
- Never use one hammer to strike another. Metal chips can be broken off and thrown at high speeds.

Sunscreen, insect repellent, flashlights, food, and water should be considered in relation to environmental conditions and length of the field excursion. Please note that this chapter is written with shorter excursions in mind where substantial supplies will not generally be required. The next section offers more detail on the materials to take with you into the field.

Give appropriate consideration to group management. We suggest taking individual classes into the field for short trips before attempting either longer fieldwork excursions or trips with multiple classes. Managing larger groups or longer trips requires attention to logistics that will not be addressed in depth here. Whether the group is large or small, consider the benefits of a buddy system and measures to keep track of where everyone is—both children and adults. If groups are spread out on the trail, the lead group should stop at trail crossings to make sure everyone follows the intended trail. Younger students should not be left unsupervised for *any* length of time. Schedules and rendezvous points are important for longer trips and larger groups. All teachers and chaperones should have one another's cell phone numbers.

10 Fieldwork

Needs

Things You Might Use in the Field

The Essentials and Near Essentials

As noted above, the essential materials for going in the field (besides yourselves) are clothing (especially footwear) that is suited to the weather and trail conditions and a first-aid kit appropriate to the situation. You will likely also want tools or devices to extend your senses, to preserve your observations, to collect materials (where safe and legal), to take photographs, and to store data, all of which will allow for continued observation and analysis after you return from the field. If your fieldwork is on the school grounds, or adjacent to it, you perhaps won't need anything different than what is needed on a typical class day, at least for the initial visit.

To extend your senses, start with simple things like magnifying loupes and rulers and potentially move on to include more sophisticated tools like probeware (to measure pH, temperature, and dissolved oxygen) or field microscopes. Since tools are used for both extending your senses and for capturing and preserving your observations, the most obvious tools for preserving one's observations are notebooks, pencils, cameras, GPS units, smartphones, and tablets.

As varied as field science is, a few items should be in every scientist's gear whether you are investigating rocks, observing streams, or documenting ecology. Even though processes and concepts are universal, each place is also unique, a product of its position on the Earth, its geological and ecological history, and the local human impacts. Making sense of why a place looks the way it does must take that context into account. Further, good science depends upon repeatability of observations: if another scientist (or your next class!) wants to analyze or build upon your observations, he or she must be able to know precisely where your study took place and how you made your observations. It is thus critical to locate the position of your studies on a map as precisely as possible. With modern GPS technology, it has never been easier to record a location to within a few meters, though you can certainly follow good science practices even if you don't have this capability. *Table 11.2* lists equipment and materials that are useful in the field.

Maps and Notebooks

Large-scale maps provide a way to see your field site in the context of other features in the area. At a closer scale they also provide a way to show the position of several sites relative to each other. At still higher resolution, maps provide the medium to store and display spatial information from one site. You will therefore probably want maps at all of these scales.

Large- and medium-scale maps for providing context can be found online. Google Maps and Google Earth are two of the best known interactive sources. If students need help understanding maps and scale, a helpful exercise is to create a "Powers of Ten" map of your schoolyard, starting with an overhead shot of the school yard that students recognize, then zooming out—making each of the new images increase in dimension by ten times—until one can see the site from the perspective of the whole Earth. A video tutorial, inspired

Table 11.2: Materials to take in the field. (Items in bold are highly recommended.)

<table>
<tr><th>For Safety and Comfort</th><th>For Extending the Senses</th><th>For Preserving and Extending Observations</th></tr>
<tr><td rowspan="3">☐ Yourself
☐ Appropriate footwear
☐ First aid supplies
☐ Water
☐ Sunscreen
☐ Insect repellent
☐ Food
☐ Safety goggles
☐ Flashlight

Common sense should be your guide when determining what is needed for a particular visit to the field. Trips that last a class period and are adjacent to the school may require nothing beyond materials for a typical class—a notebook and a pencil.</td><td>☐ Ruler or scale card
☐ Measuring tape or meter stick
☐ Magnifying loupe or hand lens (about 10× magnification)
☐ Water test kit
☐ Compass
☐ Clinometer
☐ Field microscope
☐ Field guides</td><td>☐ Notebook
☐ Pencil
☐ Materials for collecting
o Baggies
o Specimen labels
o Sharpies
☐ Rock hammer
☐ Camera</td></tr>
<tr><td colspan="2">For Both Extending the Senses and Preserving Observations</td></tr>
<tr><td colspan="2">☐ Maps
☐ Camera (possibly with video)
☐ Probeware and interface (like the Vernier LabQuest)
☐ Digital field microscope
☐ GPS unit, smartphone, or tablet
☐ Apps used in the field might include:
o GPS
o Google Earth or other virtual globe
o Skitch (or other image-annotating app) for adding notes to photos. Skitch also includes a map annotation function.
o Photosynth or other panorama app
o Video (the YouTube Capture app allows for basic video editing on your smartphone or tablet)
o Other specialized photography apps
o Audio recorder
o Notes
o Photo management software, such as Web Albums</td></tr>
</table>

by the classic film, is available at http://www.virtualfieldwork.org. It is simple to add your field site to the same Google Earth file containing the Powers of Ten centered on your school. This can help students better understand the location of the field site in relation to the school.

Field scientists typically show information about their field site: the location of observations (such as photographs and specimen collection) and also the scientific data (such as rock type, position of faults, areas of bedrock exposure, water quality information, and much more). For these purposes you may want to have a paper copy of a map you can bring into the field upon which you can make notes. Commonly topographic maps are used as base maps, in part because the contours can help you locate yourself on the map (if it's not completely flat) and partly because the topography itself is often relevant to Earth and the environmental data being collected. If your field area is larger than about 100 meters (330 feet) on a side, you can create a topographic map tailored to your needs using online software (http://www.gpsvisualizer.com). USGS topographic maps of the entire US are available as free downloads at http://www.usgs.gov/pubprod/. You may wish to download the local map and take an excerpt of the area surrounding your site.

10 Fieldwork

Needs

Positions of samples, photographs, and observations can be located using GPS. In this case, you can make notes about your GPS locations, and plot the locations on a computer later, or make use of an app like Skitch that allows you to annotate digital maps in the field. Photos taken with smartphones, tablets, and GPS-enabled cameras will include location data with pictures. Those familiar with Geographic Information Systems (GIS) can make elaborate maps using your own sets of coordinates and data. While GPS and GIS technology are now standard in most types of fieldwork, they are not essential for doing good fieldwork. Standard, intuitive tools for measuring are, however, quite helpful. A compass (either traditional or digital) can be helpful in orienting your field site in space, and a ruler and protractor can be helpful when drawing the field site in correct proportions (e.g., the position of samples along a transect or the angle of bedding or faults). Bring a clipboard so that you have a flat surface to write upon in the field—pencils and a good eraser are the best writing implements for drawing and annotating your map.

It is possible in principle to capture all your data electronically, but most field scientists still use a notebook even if they have access to the latest technology. Certain information can be captured very simply in the field with a pencil and paper while it may prove challenging with digital technology, such as when making annotated sketches of the field site and taking written notes. Normally pencil is used, in part because it doesn't smear if it gets wet, but also because it's erasable; while not essential, field scientists who know they may have to work in wet conditions will purchase notebooks with waterproof paper (Rite-in-the-Rain notebooks). An audio recorder (smartphone or standalone digital recorder) is handy when writing a lot of text is impractical, though it does create transcription work at the end of the day. Remember that it is considered a form of "best practice" to make sure that each entry includes the date, time, and locality.

Documentation and Specimen Collection

Photographs

Once at a field site it is easy to immediately begin taking photographs without recording notes to accompany them—a problem experienced by professional and amateur scientists alike. But the lack of proper documentation is perhaps the most common mistake made in the field, especially with digital photography, where it is easy to take tens or even hundreds of photographs at a single site. Also, before you begin photographing it is advisable to first explore the entire location and develop a plan for how you will communicate the site to your students back in the classroom. This plan will guide your photography, and the recorded notes will ensure that every image makes sense long after you've visited the site. Proper documentation includes the following steps:

- Note the location and orientation of the photographs you take. Recording this information on a map is very helpful.

- In each photograph, it is important to have a sense of scale. For smaller structures (like ripple marks or fossils) or close-ups of an outcrop or rock, it is important to show scale by using a common object, such as a penny, rock hammer, an unsharpened pencil, or (ideally) a clearly marked ruler. For larger structures, a really great scale is a person, so feel free to step into the picture! The importance of a scale cannot be overstated, as the proper identification of geologic features in photographs often depends on knowing the feature's size.

- In addition to showing scale within photographs, be sure to pay attention to different scales across the set of photographs you take. That is, include photographs across a wide range of scales, from the smallest fossil or mineral crystal to panoramic shots of the landscape. Maps and virtual globe software, such as Google Earth, can extend scales from the local landscape to a global perspective.

Drawings

Although photographs are key, simple sketches or drawings are also useful for documenting a field site. In fact, subtle changes in rock layers, for example, may not be visible in photographs, so to capture such features, drawing may be required. Drawing also forces you (or your students) to observe closely. It will be helpful to use either a Rite in the Rain notebook or a large, clear plastic bag to hold your notebook in case of rain. When drawing, keep in mind that you should document the same type of information that is documented in photographs (location, orientation, and scale). Drawing also requires close study in a way snapping a photograph does not. Louis Agassiz once said that "...a pencil is one of the best of eyes." While drawing, you have to think about the relationship of the elements you are representing, their scale, and their arrangement.

Annotating Photographs

The use of smartphones and tablets in the field allows for a hybrid of photographs and drawings. Many apps allow for captioning photos in the field, and some allow you to draw and write text on photos as you take them. Skitch is one such app, and it also allows for the taking of notes on the maps themselves. Photos taken on smartphones and tablets are also (typically) geo-referenced. This means that they can easily and quickly be included in a Google Earth or other GIS program in the precise location where the image was taken. If you are unable to annotate photographs in the field, or you wish to add more detail than is practical on your electronic device while you are at the field site, the "old fashioned" technique is to take a picture, then make a simple notebook sketch containing labels of key features. Later you can annotate a digital or printed version of the photograph using your field notes. If the conditions are poor for note taking either digitally or manually, it may be more practical to record audio notes that you can later match to your picture.

Using Field Guides

Select field guides appropriate to the focus of your work and consider whether or not you wish to bring others. The appropriate field guide might be something as simple as a single sheet with line drawings of the fossils common at your field site, a few pages containing a dichotomous key of common rock types, or a collection of field guides on fossils, birds, mammals, butterflies, rocks, flowering plants, and more. While scientists will come to know by sight the kinds of specimens commonly found at their site, they do not typically set out to memorize them, and uncommon things are sometimes found that send even experts back to their field guides.

Collecting Specimens

Rocks and fossils often provide significant clues for interpreting past environments. Layers of basalt indicate past volcanism, for example, whereas shales bearing trilobite and other fossils indicate deposition in a shallow sea. Collecting specimens from a site provides a wonderful opportunity to take a piece of the field into the classroom, allowing you to engage students in hands-on learning. Collecting specimens also permits further study away from a site where time and field conditions can impose certain limitations. You can and are encouraged to identify rocks, minerals, fossil types, and flora and fauna in the field. So, what do you need to know about collecting specimens?

- **You first need to confirm that collecting specimens at the site you are visiting is legal**. Typically, collecting is not allowed in parks, so be sure to check.
- Just as you made decisions about photography based on how you plan to communicate the site to students, collect specimens that will help tell the story of the site back in the classroom. If rock types change from area to area, either vertically or horizontally, then specimens of each type are ideal.
- Before collecting a specimen, take a photograph of it in situ, both close up as well as from a distance. Don't forget to include an object for scale in the photograph!
- Document the location from which the specimen is collected, preferably on a map of the area. Labeling the specimen with a number that corresponds to a number on your map is an effective technique.
- Specimens should be broken directly from the outcrop so the exact source is known. Eroded rocks scattered about on the floor of the site may have originated from multiple locations.
- The weathered surface of rocks often carries a different appearance than a "fresh" break. Ideally, collected specimens

possess one weathered surface but are otherwise not weathered. Rocks broken directly from outcrops will ensure fresh surfaces.

- As specimens are collected, place each in a separate resealable bag, noting on the bag with permanent marker each specimen's location as indicated on your map. Include a specimen label within the bag, including the information shown in *Figure 11.2*.

PALEONTOLOGICAL RESEARCH INSTITUTION

ReaL Earth Inquiry Specimen Label

Location rock was collected:

Kind of rock or fossil:

Geological period or age of rock:

Collector:

Date collected:

Figure 11.2: This specimen label, printed six to a page, is available for download at http://virtualfieldwork.org/Assessments_and_Student_Materials.html.

Back in the Classroom: Virtual Field Experiences (VFEs)

Following your trip to a field site, perhaps the most critical step after returning to your lab or classroom is to examine all of your photographs, illustrations, specimens, and notes associated with each. Sometimes even the most diligent geologist forgets to record notes that, in hindsight, are critical. It is therefore recommended that one makes sure that his or her notes are legible and complete. Recopy your notes. Such an activity will not only ensure legibility for the future, but it will help indicate any gaps in your note taking. If gaps exist, then it is easiest to fill them in when your memory of the site is fresh.

Once your materials from the site visit are in order, it is time to develop an activity that will allow your students to experience the site much like you did—but in the classroom. VFEs allow you to compile this information in a way that is easy to share with others who wish to learn about the site. Ideally, VFEs provide opportunities for open-ended exploration, just as actual fieldwork does. Scientists in the field are not limited to a single possible way to operate, nor do they have a guide explaining what they see at every turn. In the field, one might pick up a rock and take a closer look, or pull out a magnifying glass and look at a cliff face. Exploration drives inquiry in the field, and inquiry and exploration are key goals of VFEs.

The concept of VFEs can take on multiple forms. For example, kits containing maps, printed photographs, and specimens (with notes on the map indicating where the specimens were collected or where the photographs were taken) can be produced. Or, your digital photographs can be embedded within a PowerPoint or Prezi presentation, a website, or a Google Earth tour with placemarks containing photos, video, or other data in the exact locations where the specimens were collected. Maps can also be overlain. Historic maps can be included, and Google Earth has historical imagery included for much of the world. Many VFEs incorporate more than one technological platform.

Keep in mind that these electronic presentations may take on a very linear, directed feel. In that respect, be careful that your VFE does not turn into a Virtual Field Trip. Virtual Field Trips have become increasingly common at many levels of education, but these experiences are typically guided tours rather than opportunities for inquiry. An online search will yield many examples of these tours, as will a search of the Digital Library of Earth System Education (DLESE). Such resources clearly have value, but they are passive experiences for students. VFEs, in contrast, should stress the importance of inquiry; learning for understanding involves students figuring things out. The act of making new, or extending existing, VFEs may be the simplest way to bring inquiry to the use of VFEs.

In considering VFEs as a recurring practice, initial experiences are perhaps more guided than the later experiences; allow a gradual transfer of responsibility from teacher to student. But VFEs ideally offer the same opportunities for exploration as those provided at an actual field site, with occasional moments of discovery that lead to new questions about the site. By asking such questions and then seeking answers, students are doing science. And it is perfectly reasonable to virtually visit a site several times for further data collection, or even to study different concepts at the same site. Scientists, of course, do exactly the same thing.

Prezi and PowerPoint VFE Templates

This section discusses templates intended to simplify VFE production in addition to providing general information on VFE development and use. There are templates in both Prezi and PowerPoint formats, each with a version of the graphic organizer shown in *Figure 10.1* as its centerpiece. Questions in the graphic organizers and in the rest of the templates are written generically, so they may be applied to any site. The templates serve as starting tools that are useful for creating an "entry level" VFE. They are available at http://virtualfieldwork.org/Template.html. The template includes graphic organizers for both Earth and environmental science, with the environmental science organizer embedded within the geoscience organizer.

How are teachers using virtual fieldwork?

VFEs might be used as a single, in-class exercise, or they can be explored across an entire year. We hope that teachers who use and develop VFEs will eventually use them across the entire curriculum, but it makes sense to start smaller. There is no single correct approach to using VFEs in the classroom. Here are some examples of ways teachers are using virtual fieldwork:

- Students in a rural community are using Google Earth to create Powers of Ten tours centered on their homes (based on the Eames' classic film). This helps students to internalize the abstraction that is central to making maps and to build deeper understandings of scale.
- Students are making geologic maps of the local bedrock.
- Students are creating an interpretive guide for a county forest.
- Students are exploring lakes, dams, streams, outcrops, quarries, waterfalls, and more.

For more VFEs, see our growing database at http://virtualfieldwork.org/.

What do I need to consider as I begin to build my VFE?

Considerations fall into four categories:

- **Logistical:** What do I have the attitude, time, resources, and skills to do? (Attitude is listed first as it is the most important factor.)
- **Pedagogical:** How do I bring the scientific content together with technologies in a way that best builds enduring understandings of bigger ideas and overarching questions, as well as of the smaller scale ideas and questions I deem important?
- **Technological:** What hardware and software do I need to assemble the materials for the VFE and to make it accessible to my students? This may include traditional scientific tools, like a rock hammer or a compass, as well as the computer technologies discussed in this chapter and on our website.
- **Content:** What scientific knowledge, ideas, processes, and practices do I want my students to understand and be able to do at the end of the experience?

Of course, these categories overlap and interplay substantially—teachers of Earth science use Google Earth in different ways than other Google Earth users do.

Most of the remainder of this chapter is a set of checklists to help you address these different considerations when outlining your VFE design. Take it with you into the field as you collect pictures and other kinds of data for your VFE; use it to identify issues you think are most important for the development of your VFE. Most of the items in the checklists are there to start you thinking about how to address a particular issue. Content is listed last for the sake of readability, as the checklists for the content section are longer than they are for the other categories.

> **The framework for understanding how to effectively blend technology, pedagogy, and content knowledge is known by its acronym TPACK.**

Table 11.3: A checklist of cross category issues. Many of the questions in the checklist relate to more than one of the categories identified above. Because of this overlap, only the cross-category issues and content sections are of significant length.

Have I considered this?	**Question:**	Logistical	Pedagogical	Technical	Content
	Do I have appropriate safety and first aid equipment and materials?	√		√	
	What content do I want to address?	√	√	√	√
	Do I have connections in mind to at least a couple of the bigger ideas and overarching questions? • The Earth is a system of systems. • The flow of energy drives the cycling of matter. • Life, including human life, influences and is influenced by the environment. • Physical and chemical principles are unchanging and drive both gradual and rapid changes in the Earth system. • To understand (deep) time and the scale of space, models and maps are necessary. • How do we know what we know? • How does what we know inform our decision-making?		√		√
	How much time do I realistically have to spend on VFE creation?	√			
	How much class time do I want to dedicate to VFEs?	√	√	√	√
	Am I okay with the trade-off between some expected frustration and the pedagogical payback?	√	√	√	√
	Can I productively engage students in VFE development? *Or is that something to aspire to for next year?*	√	√	√	√
	How does the technology I have serve the goals I wish to meet?	√		√	
	Do I have enough batteries for my powered equipment?	√		√	
	Is the site accessible to me? *This includes legal, safety and proximity considerations.*	√	√		
	Are my students familiar with the site? If not, is it accessible to *all* of my students? *If the answer to both questions is no, select another site.*	√	√		
	Are the required pedagogical, technological, and content skills and knowledge needed to create the VFE within my reach? *Ideally, select challenges that are just within (or just beyond) your reach so that you grow professionally.*	√	√	√	√
	Do I have the hardware (including field equipment) and software needed for VFE creation? *The bare essentials are an Internet-connected computer, a digital camera, and either PowerPoint or Google Earth.*	√	√	√	

Logistical

We hope that VFE development is used to expand teachers' skills and knowledge. Performing fieldwork for the first time can be overwhelming, but remember that science is a process, and not even professional scientists capture all that they need in one visit. With practice, and the proper attitude, you will become more and more comfortable when visiting the field.

Pedagogical

While most pedagogical questions also address other categories as noted above, there are issues that deserve explicit attention here.

- Does the data you are collecting go toward answering why this place looks the way it does? *Or is there a good reason to introduce distracting information?*

- If the site is especially striking or unusual, have you considered how to get yourself and your students beyond the "novelty space" of the location? Crudely summarized, novelty space is the idea that you can't figure out what's going on at a field site if you're either awed by its beauty or freaked out by its perceived dangers. This is one of several reasons for choosing a site that is already familiar to the students.

Technological

Most technological issues are also logistical; these are addressed in the table above.

Content

Why does this place look the way it does? The driving question of our work can serve as an entry into any major topic in Earth or environmental science curricula. It also brings relevance to the science since we want to start with sites near the school that are already somewhat familiar to the students. We want students to look at the familiar with new eyes, and to become skilled at reading their local landscape. Ultimately, we want the skills built by reading the local landscape (being able to tell the story of why a place looks the way it does) to be transferable to *any* landscape.

What scientific content do you want your students to better understand through their work in the VFE? How does this fit into the larger goals of the course? Can you draw, and help your students to draw, connections to bigger ideas and overarching questions? What topics in Earth science can be addressed by doing fieldwork?

Below are questions taken from the geoscience and environmental science graphic organizers. Most teachers will likely use one sheet or the other, but not both. Your VFE likely won't address all of the questions (on either sheet), but you should be able to strategically select what you minimally wish to address.

Understandings will be made much deeper in schools where teachers in more than one subject or grade level engage their students in studying the local environment.

VFEs

For the Geosciences:

For all of the following questions:
- How do you know? (What evidence is there?)
- What does it tell you about past environments?
- What does it imply about the future?

☐ **Describe the shape of the land.**
- Are there mountains, valleys, or hills?
- What are the valley shapes?
- What can cause valleys to form?
- What can cause mountains or hills to form?
- Are the mountains or hills young or old?
- What roles does tectonics play in shaping the site?

☐ **What effects has water had on the landscape?**
- Is water depositing material, eroding material, or both?
- Is the action of water primarily chemical, primarily physical, or both chemical and physical?

☐ **What effect has the climate had on the landscape?**
- Was the past climate different?
- What factors may have been affected or caused by climate?
- How has fire played a role in shaping the environment?

☐ **Describe the ecosystem.**
- See the ecosystem graphic organizer and checklist.

☐ **What does the arrangement of the rocks and soils indicate about past conditions?**
- Do the rocks seem to form a sequence?
- Where would you find the oldest rocks? The youngest rocks?
- Does the rock record include evidence of ancient disturbances? If yes, describe.
- Are there different kinds of rocks at different outcrops?

☐ **What types of rock and soils are there and what do they indicate about past conditions?**

Sediments and Sedimentary Rocks
- Is the sample clastic or organic /chemical?
- If clastic, what is the grain size?
- If organic, what minerals is it made out of?
- Are there fossils?

Metamorphic
- Is the rock foliated or non-foliated?
- What was the parent rock?

Igneous
- Did the rock form above or below ground?
- Is it felsic or mafic?

☐ **What effects has life, including human life, had on the landscape?**
- How have plants shaped the landscape?
- How have animals generally, and humans in particular, changed the landscape?
- On what scale?

For the Environmental Sciences:

For all of the following questions:

- How do you know? (What evidence is there?)
- What does it tell you about past environments?
- What does it imply about the future?

☐ **Describe how life shapes the land.**

- What are the pioneer plants?
- How do pioneer plants impact soil formation?
- How are animals shaping the land?
- Are there invasive species? If yes, what are they, and how are they changing the ecosystem?
- Have disturbances played a role in the introduction of invasives? If yes, describe.
- How are new invasives likely to change the ecosystem over the next century?

☐ **Describe the role of water in the ecosystem.**

- In what ways does water serve or disturb habitats?
- How does life move, use, and store water?

☐ **How has climate shaped the ecosystem?**

- How is the climate reflected by living things at the site?
- Describe any microclimates and how they affect life.
- Describe how sun and shadow affect life.
- What roles do fire, hurricanes, or other climate-related disturbances play in shaping this landscape?

☐ **Describe the role rocks and soil play in the ecosystem.**

- How does life change the rocks and soil at the site?
- How is life dependent upon the rocks and soil at the site?
- Does the rock record include evidence of ancient disturbances? If yes, describe.
- See also the geoscience questions.

☐ **Describe the types and arrangements of plants and animals and what they indicate about present and past environments.**

- Why do living things in the environment look the way they do?
- What life forms were the earliest to arrive?
- Describe how different life forms are distributed throughout the field site.
- What is the impact of invasive species and other disturbances?
- See also the **Describe how life shapes the land** section.

Plants

- How have plants shaped the landscape?
- How has the landscape affected the plants?

Animals

- How do animals contribute to plant distribution?
- How has the landscape affected the animals?

Other biota

☐ **What effects have humans had on the landscape?**

- What resources do humans use from here?
- How have humans changed the landscape?
- On what scale?

Closing Thoughts

This chapter was written to help get you started in the creation of VFEs and, in a broader sense, to help you learn more about fieldwork. But how do you know when to stop? It may be more productive to think of VFEs or activities involving actual fieldwork as undertakings that are becoming ready for use rather than as finished products. Here is a nice quote from Wendell Berry's essay "Faustian Economics" that relates to this concept:

> *It is the artists, not the scientists, who have dealt unremittingly with the problem of limits. A painting, however large, must finally be bounded by a frame or a wall. A composer or playwright must reckon, at a minimum, with the capacity of an audience to sit still and pay attention. A story, once begun, must end somewhere within the limits of the writer's and the reader's memory. And of course the arts characteristically impose limits that are artificial: the five acts of a play, or the fourteen lines of a sonnet. Within these limits artists achieve elaborations of pattern, of sustaining relationships of parts with one another and with the whole, that may be astonishingly complex. And probably most of us can name a painting, a piece of music, a poem or play or story that still grows in meaning and remains fresh after many years of familiarity.*

Fieldwork 10

Resources

Resources

Field Geology Teaching Practices

Greene, J. P., Kisida, B., & Bowen, D. H., 2014, The Educational Value of Field Trips, *Education Next*, 14(1), 78–86.

Issigonis , M., 2006, Field trips as an aid to teaching Earth science courses, *The Earth Scientist*, 22(3): 14–16.

Johnson, J.K., and Reynolds, S.J., 2005, Concept sketches—using student- and instructor-generated annotated sketches for learning, teaching, and assessment in geology courses. *Journal of Geoscience Education*, 53: 85–95.

Orion, N., and Hofstein, A., 1994, Factors that influence learning during a scientific field trip in a natural environment, *Journal of Research in Science Teaching,* 31: 1097–1119.

Russell, H.R., 1998, *Ten-Minute Field Trips, 3rd edition*, National Science Teachers Association: Alexandria, VA. [Focused on elementary and junior high; chapter on Earth science on pp. 113–137.]

Shulman, L. S., 2005, Signature pedagogies in the professions, *Daedalus*, 134(3): 52–59.

Whitmeyer, S.J., Pyle, E.J., Mogk, D.W. (eds.), 2009, *Field Geology Education: Historical Perspectives and Modern Approaches*, GSA Special Papers 461. [29 articles, focused on undergraduate education.] http://specialpapers.gsapubs.org/content/461.toc.

Extraordinary Science Field Trips, *NSTA Reports*, Summer 2013, 25(1): 1–2. http://www.nsta.org/docs/NSTAReports201307.pdf.

My Geologic Address: Locating Oneself in Geologic Time and Process, by Ault, K., SERC InTeGrate workshop "Teaching the Methods of Geoscience" activities, http://serc.carleton.edu/integrate/workshops/methods2012/activities/ault.html.

Teaching in the Field, National Association of Geoscience Teachers. [Set of resources for teaching field geology.] http://nagt.org/nagt/teaching_resources/field/index.html.

Guides to Fieldwork

(Mostly focused on post secondary education, but useful as references)

Coe, A., T. Argles, D. Rothery, and R. Spicer, 2010, *Geological Field Techniques,* Wiley Blackwell, Chichester. [This is a current standard.]

Compton, R. R., 1962, *Manual of Field Geology*, John Wiley & Sons, New York. [An old classic.]

Compton, R., 1985, *Geology in the Field,* Wiley, New York. [An updated version.]

Lambert, D., 2006, *The Field Guide to Geology*. The Diagram Group.

Lisle, R., P. Brabham, and J. Barnes, 2011, *Basic Geological Mapping,* Wiley-Blackwell, Hoboken, NJ.

Maley, T. S., 2005, *Field Geology Illustrated, 2nd edition*, Mineral Land Publications, Boise, ID.

Mathur, S. M., 2004, *Guide to Field Geology*, Prentice Hall of India, New Delhi.

Spencer, E., 2006, *Geologic Maps : a Practical Guide to the Preparation and Interpretation of Geologic Maps*, Waveland Press, Long Grove, IL.

Walker, J. & H. Cohen, 2009, *The Geoscience Handbook: AGI Data Sheets,* American Geological Institute, Alexandria, VA.

Walker, J. D., and H. A. Cohen, 2007, *The Geoscience Handbook: AGI Data Sheets, 4th edition,* American Geological Institute.

How to Read a Geologic Map, Wisconsin Geological and Natural History Survey, http://wgnhs.uwex.edu/wisconsin-geology/bedrock-geology/read-geologic map/.

Appendix: The Teacher-Friendly Guides™, Virtual Fieldwork, and the NGSS's Three-Dimensional Science

The Next Generation Science Standards contain a set of learning goals that define and describe the ideas and practices that we need in order to think scientifically. The NGSS are not a curriculum. They tell teachers not how to teach, but rather, are tools to show what to teach. They also help families know what children are expected to learn, and help schools and teachers know what to assess. So, how do you teach in ways that align with NGSS, if NGSS itself doesn't tell you? The strategies, tools and resources associated with the ReaL Earth Inquiry project, like this *Teacher-Friendly Guide™*, are intended to offer a partial answer to that question.

"ReaL Earth Inquiry" is the project name of the NSF grant (0733303) to the Paleontological Research Institution to develop teacher resources such as Teacher-Friendly Guides™ to regional Earth science and Virtual Fieldwork Experiences. "ReaL" refers to Regional and Local.

Acronyms frequently used in *The Next Genertion Science Standards* (NGSS):

PE: Performance Expectation
DCI: Disciplinary Core Idea
CC: Crosscutting Concept
SEP: Scientific and Engineering Practice
PS: Physical Sciences
LS: Life Sciences
ESS: Earth & Space Sciences
ETS: Engineering, Technology, and the Applications of Science

The vision of NGSS differs in a number of important ways from current common practice in schools and classrooms across the country. Teaching about local and regional Earth and environmental science can and has worked well for many teachers under more traditional standards, but by attending to the three dimensions of the NGSS (see below), we believe it can work even better. Deep understandings of why your local environment looks the way it does requires understanding the local environment from multiple disciplinary perspectives, and understanding the connections amongst these different disciplinary ideas. That is, to understand your local environment, a systems perspective is needed. Scientifically accurate meaningful understanding can and does come out of single lessons, single units, and single courses, but these understandings become richer, deeper, and more durable if they are connected across courses. The NGSS vision includes recognition that building a deep understanding of big ideas is both very important and a process that takes years of coordinated effort. Fortunately, the many processes that shape the local environment are part and parcel of existing curricula, and especially for Earth science, biology, and environmental science courses, nearly every unit has central aspects that play out on a human scale just outside the school door. A coordinated approach to the study of the local environment across units within a single course and across grade levels

CHAPTER AUTHOR

Don Duggan-Haas

and courses can be a fairly subtle change in each teacher's daily routines, but it has the potential for big returns in terms of the depth of student understanding. This deeper understanding pertains not only to the local environment and the way course topics are represented within it, but also to systems more generally, to the nature and importance of scale, and to much, much more.

NGSS builds upon the earlier work in the National Science Education Standards (NSES), but brings more of a systems approach not only to its representation of science, but to the standards themselves. NSES defined science not just as a body of ideas, but an evolving body of ideas extended by inquiry. NGSS continues this work by clarifying inquiry and the sciences as a set of relationships amongst three dimensions: Disciplinary Core Ideas (DCIs), Scientific and Engineering Practices and Crosscutting Concepts." Each of the three dimensions is judged to be of roughly equal importance and they are seen as interdependent. To truly, deeply, understand science and how scientific understandings develop, learners must not only understand each dimension, but how the dimensions are related to one another—the whole is greater than the sum of the parts. By coming to understand these interconnections, teachers and students will also come to better understand the nature of both scientific inquiry and of complex systems.

A Perspective on Science Education Priorities

The bulk of the NGSS is a series of Standards, each a page or two in length, with "Performance Expectations" (PEs) at the top of the first page, followed "Foundation Boxes" and "Connection Boxes" supporting the PEs. It's tempting to jump into the discussion of NGSS by starting there. It's also tempting to start with the Disciplinary Core Ideas (DCIs), especially for those who specialize in a particular scientific discipline. But readers shouldn't do either of those things. Appendix K of NGSS notes, "The goal is not to teach the PEs, but rather to prepare students to be able to perform them by the end of the grade band course sequence." It's important to understand the basic three-dimensional structure of the NGSS before looking at the PEs or DCIs. We will give them both their due, but we won't start with either of them.

If you have a degree in a particular science, and this is the science that makes up the bulk of your teaching load, it's natural to go straight for your area of expertise in the NGSS, to see how that's addressed. But don't do that, or, if you already have, try to imagine that you haven't. Before considering the concepts and practices essential to being literate in your discipline, consider what you think everyone needs to know about science disciplines *outside your area of specialization*, and consider the ideas that are broadly applicable across all the sciences. That is, think about the fundamentals of science.

Imagine having magical powers that allowed you to make every American understand six or eight profound scientific ideas – ideas that, if everyone understood them, would help people make the world a better place because they would make better decisions. Imagine again that this power could also be used to give everyone a small set of well-developed scientific skills. What should these ideas and skills be? Ponder what these ideas and skills are before reading further, perhaps going so far as to put them down on paper. Ask your colleagues, and your former students the same question. What are the most important ideas and skills for everyone to understand or be able to do related to science?

The profound scientific ideas you thought of are likely to be something like NGSS's Crosscutting Concepts, and the scientific skills are likely to be something like the Scientific and Engineering Practices (*Table A.1*). In reviewing the NGSS, teachers at the secondary and college levels who specialize in a particular subject are often naturally drawn first to the Disciplinary Core Ideas for their discipline, and when they find a favorite topic that is not addressed to what they consider an appropriate depth, they are upset that NGSS is not providing the content necessary to prepare their students for the future. But, decades of educational practice teaching science courses with thousand-page textbooks and scores of key ideas has not yielded a scientifically literate populace. It is essential to focus on smaller sets of truly big ideas (see also the *Big Ideas* chapter) and work across grade-levels to build understandings over time. This may mean, however, that your favorite topics are no longer explicitly listed in the learning goals.

Table A.1 contains abbreviated versions of the Concepts, Practices, and Ideas. You can find longer descriptions within the NGSS, and we'll look at one as an illustrative example. Consider the full description of Crosscutting Concept #3:

> *Scale, proportion, and quantity.* In considering phenomena, it is critical to recognize what is relevant at different measures of size, time, and energy and to recognize how changes in scale, proportion, or quantity affect a system's structure or performance.

It seems likely that most Americans do not have a good and durable understanding of this concept, yet it has relevance to many aspects of their daily lives. The same could be said of most, if not all, of the remaining concepts on the list.

The Crosscutting Concepts are described in some detail in Appendix G of NGSS, and the Scientific and Engineering Practices are described in Appendix F.

Such understandings are almost certainly more important than knowing particular facts about geologic history or the nature of disease (two topics not given deep attention in the NGSS). Indeed, it's only possible to understand geologic history or the nature of disease if you also understand these concepts!

While your favorite topics may not be explicitly mentioned in NGSS, that doesn't necessarily preclude them from being taught. There's a tremendous amount of content in these *Teacher-Friendly Guides*™ that are not mentioned in NGSS, yet we believe that all of the contents of the Guides *support* teaching that is aligned with the NGSS. Different topics, such as glaciers or mineral resources, can serve as our pedagogical partners in building understandings of the Crosscutting Concepts, Scientific and Engineering Practices, and the Disciplinary Core Ideas that make up the NGSS. In other words, we can and should teach these topics, but understanding the particular topic isn't the primary goal. The primary goal is to use the teaching of these topics as a means to build an understanding of those bigger ideas.

It isn't clear if K–12 science curricula designed to bring the NGSS's vision to fruition will be more or less rigorous than today's common K–12 curricula, but rigor shouldn't be the goal of education. Education should develop citizens who can reason critically and use evidence to inform their actions. This isn't to say that schooling shouldn't be challenging, but rather that its challenges should be in the service of meeting other goals. Building deep and interconnected understandings of the three dimensions of NGSS will not be a simple task, but it has the potential to better prepare for students for citizenships, college, and careers.

Connecting "Why does this place look the way it does?" and Virtual Fieldwork to NGSS

This *Teacher-Friendly Guide™* is one part of a large project designed to help educators teach about Regional and Local (ReaL) Earth system science in an inquiry-based way. This ReaL Earth Inquiry Project, and all of its related resources, support educators and students in the investigation of the project's driving question: "*Why does this place look the way it does?*" The "place" of the question is anywhere you happen to be, but we hope and expect users of these materials will start by studying areas outside their backdoor or their classroom door. The *Fieldwork* chapter (Chapter 11) addresses both actual and Virtual Fieldwork, and we believe the coupling of virtual and actual fieldwork is an excellent way to teach and learn, and it's an approach that is fully three dimensional, in the NGSS's sense of that term.

Read through the Practices outlined in **Table A.1** with an eye towards engaging in and documenting fieldwork. See the graphic organizer and the question list in Chapter 11 and consider how these questions can be asked of any site, and how they can serve to inspire new questions that are site-specific. Then, consider the making of Virtual Fieldwork Experiences (VFEs) to document the site, allowing for continued investigation after leaving the field, and sharing findings with others in the community and beyond. This approach provides opportunities to engage *all* of the practices. To build rich explanations of the range of processes at play in a field site requires application of *all* of the Crosscutting Concepts. There are also opportunities for using field sites to build understandings of *all* of the DCIs, though selected ones from the Life and Earth & Space Sciences have the most direct correspondence. The use of virtual and actual fieldwork is scalable to fit the educational need, so a particular lesson or activity would be

Appendix

Scientific and Engineering Practices	Crosscutting Concepts
1. Asking Questions and Defining Problems 2. Developing and Using Models 3. Planning and Carrying Out Investigations 4. Analyzing and Interpreting Data 5. Using Mathematics and Computational Thinking 6. Constructing Explanations and Designing Solutions 7. Engaging in Argument from Evidence 8. Obtaining, Evaluating, and Communicating Information	1. Patterns 2. Cause and Effect 3. Scale, Proportion, and Quantity 4. Systems and System Models 5. Energy and Matter 6. Structure and Function 7. Stability and Change 8. Interdependence of Science, Engineering, and Technology 9. Influence of Engineering, Technology, and Science on Society and the Natural World

Disciplinary Core Ideas			
Physical Sciences	*Life Sciences*	*Earth and Space Sciences*	*Engineering, Technology, and the Applications of Science*
PS 1: Matter and its interactions **PS 2:** Motion and stability: Forces and interactions **PS 3:** Energy **PS 4:** Waves and their applications in technologies for information transfer	**LS 1:** From molecules to organisms: Structures and processes **LS 2:** Ecosystems: Interactions, energy, and dynamics **LS 3:** Heredity: Inheritance and variation of traits **LS 4:** Biological evolution: Unity and diversity	**ESS 1:** Earth's place in the universe **ESS 2:** Earth's systems **ESS 3:** Earth and human activity	**ETS 1:** Engineering design **ETS 2:** Links among engineering, technology, science, and society

Table A.1: Summary of NGSS's Three Dimensions. For more detailed descriptions, see the relevant appendices in The Next Generation Science Standards.

expected to target just one or two, but a program of fieldwork across a course would allow for the addressing of many of the Concepts, Practices, and Ideas.

Look again to the graphic organizers from Chapter 11: Fieldwork. It is easy to see how, especially in Earth science, biology, or environmental science courses, most of the units in these courses play out in some meaningful way outside the classroom door. As the DCIs are akin to umbrellas relative to a course's units, these too largely play out in meaningful ways outside the classroom door. The NGSS recognizes that in order to understand big ideas, years of coordinated study are required. The coordinated study of the local and regional environment provides an excellent opportunity for this. A field site can be studied using increasingly sophisticated approaches across the K–12 experience, and for the students, this does not entail repetition, but rather the opportunity to study a site from different disciplinary vantage points across all or part of the K–12 continuum. If such an approach is adopted broadly, kids who move during the course of their schooling can bring in new eyes, and information, to compare and contrast the environment in their new school with the environment where they used to live.

How to Read the NGSS

Each standard in the NGSS includes multiple interconnected parts. They have an architecture that can be seen in *Figure A.1*. This diagram is taken directly from the NGSS website's page, "How to Read the Next Generation Science Standards." This page includes a short written overview and an accompanying video as well as links to more detailed information. The standards are designed to be read online, with features like pop-ups, choices for highlighting different parts of the text (the different dimensions) in different colors, and links to related content elsewhere within the NGSS. If you're not familiar with how they work, you should follow the link above and then explore around the NGSS a bit before reading further.

Know that the appearance of the Standards can be a bit intimidating, with all the abbreviations, acronyms, codes, and different colors, but after a bit of time working with the text, its logic does become understandable.

Example of ReaL Connections to Performance Expectations

Earth and Space Science Disciplinary Core Idea #2 is "Earth's Systems," and it has five supporting concepts:

- ESS2.A: Earth Materials and Systems
- ESS2.B: Plate Tectonics and Large-Scale System Interactions
- ESS2.C: The Roles of Water in Earth's Surface Processes
- ESS2.D: Weather and Climate
- ESS2.E: Biogeology

In the middle school grade band of NGSS, there are six performance expectations associated with ESS2. All six are listed below, but not in their complete form. "Clarification Statements" and "Assessment Boundaries" are not included in the full list, but we'll look at one of the Performance Expectations in greater detail. See the full list (and the full standard) at http://nextgenscience.org/msess2-earth-systems.

Appendix

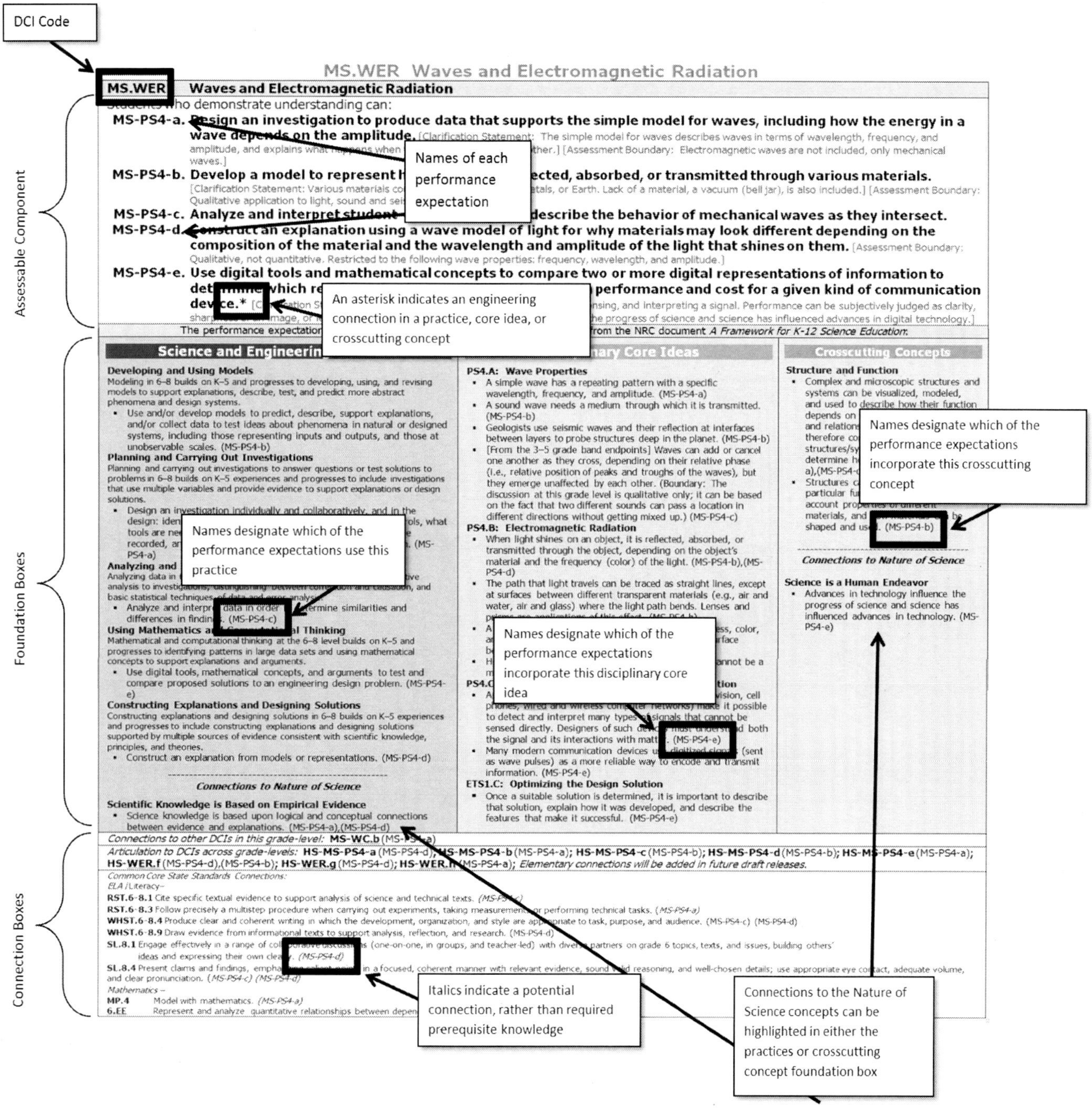

Figure A.1: The architecture of a standard. The NGSS is designed with the web in mind and features of its online architecture make it easier to understand than this diagram might indicate.

Appendix

MS-ESS2 Earth's Systems (Middle School-Earth System Science 2)
Students who demonstrate understanding can:

MS-ESS2-1. ***Develop a model to describe*** ***__the cycling of__*** **Earth's materials and the flow of energy that drives this process.**

MS-ESS2-2. ***Construct an explanation based on evidence for how*** **geoscience processes have changed Earth's surface** *__at varying time and spatial scales__.*

MS-ESS2-3. ***Analyze and interpret data*** *__on the distribution__* **of fossils and rocks, continental shapes, and seafloor structures** *__to provide evidence__* **of the past plate motions.**

MS-ESS2-4. ***Develop a model to describe*** **the cycling of water through Earth's systems** *__driven by energy__* **from the sun and the force of gravity.**

MS-ESS2-5. ***Collect data to provide evidence for how*** **the motions and complex interactions of air masses** *__results in changes__* **in weather conditions.**

MS-ESS2-6. ***Develop and use a*** *__model to describe how__* **unequal heating and rotation of the Earth cause patterns of** *__atmospheric and oceanic circulation that determine__* **regional climates.**

Each of the six above Performance Expectations (PEs) incorporates aspects of each of the three dimensions. The color-coding helps to reveal some of that. "Science and Engineering Practices" are shown in blue (***italics*** here) and Crosscutting Concepts are shown in green (*__underlined italics__* here). Disciplinary Core Ideas are in black. This is one of the color-coding options in the online presentation. Pop-ups (which can be disabled) appear when the different colored parts of the PE are scrolled over with the mouse. *Figure A.2* is a screen grab of the first three PEs for ESS2, with a pop-up showing the Crosscutting Concepts related to "MS-ESS2-2."

All of these Performance Expectations directly aligns with "*Why does this place look the way it does?*" We'll take a closer look at MS-ESS2-2, which addresses how geoscience processes have shaped the Earth's surface at varying time and spatial scales. This Guide coupled with the development of a VFE of a site local to your school, provides rich opportunities for addressing both this particular PE, along with all of the others within this standard. The Clarification Statements often provide helpful examples, and Assessment Boundaries indicate what will not be addressed in the assessments now under development. Importantly, this is not an indication that these topics are out of bounds. These standards represent minimum expectations—exceeding these expectations is often appropriate.

Appendix

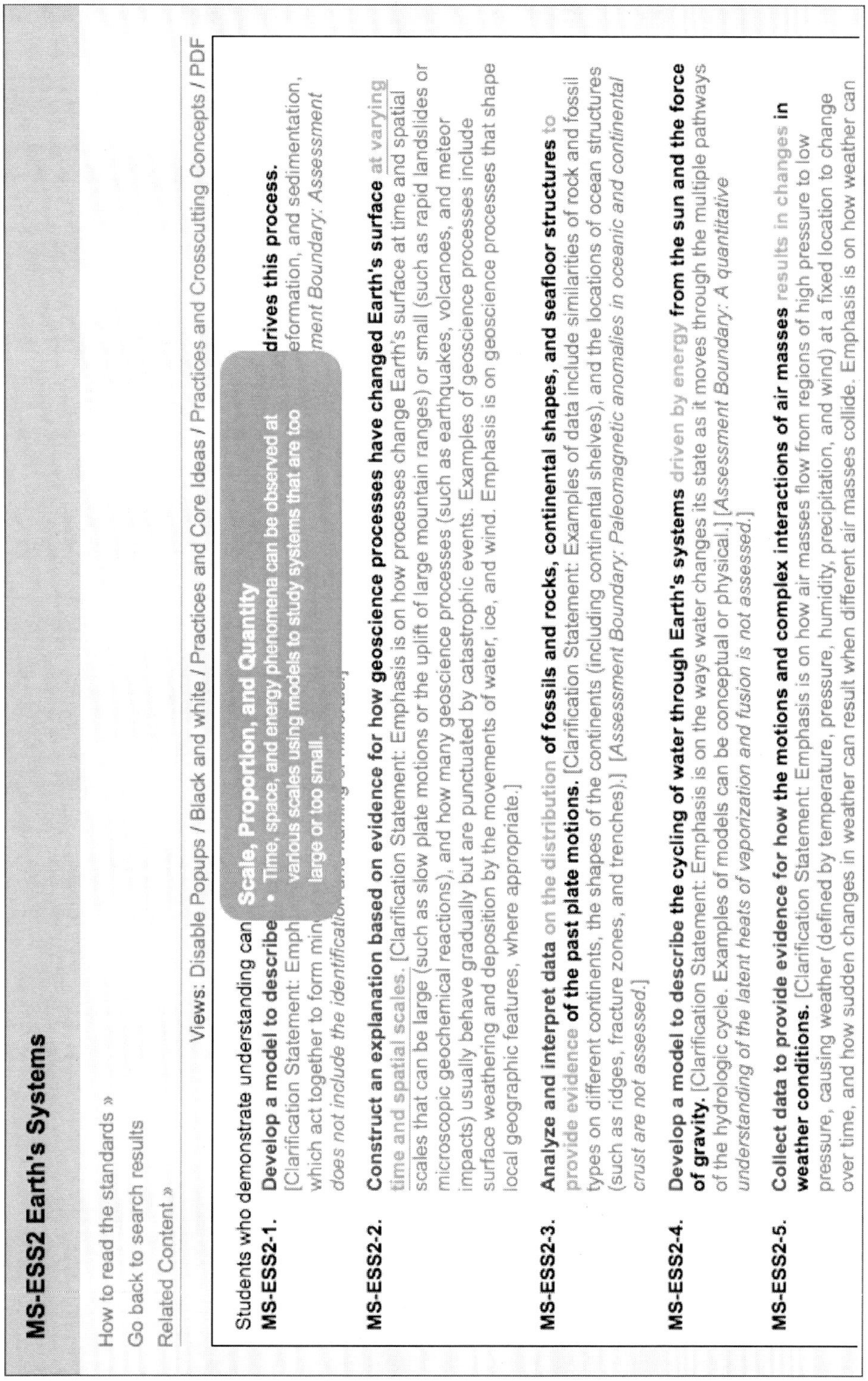

MS-ESS2 Earth's Systems

How to read the standards »
Go back to search results
Related Content »

Views: Disable Popups / Black and white / Practices and Core Ideas / Practices and Crosscutting Concepts / PDF

Students who demonstrate understanding can

Scale, Proportion, and Quantity

- Time, space, and energy phenomena can be observed at various scales using models to study systems that are too large or too small.

MS-ESS2-1. **Develop a model to describe** … **drives this process.** [Clarification Statement: Emph… …eformation, and sedimentation, which act together to form min… …*ment Boundary: Assessment does not include the identificatio…*

MS-ESS2-2. **Construct an explanation based on evidence for how geoscience processes have changed Earth's surface** at varying time and spatial scales. [Clarification Statement: Emphasis is on how processes change Earth's surface at time and spatial scales that can be large (such as slow plate motions or the uplift of large mountain ranges) or small (such as rapid landslides or microscopic geochemical reactions), and how many geoscience processes (such as earthquakes, volcanoes, and meteor impacts) usually behave gradually but are punctuated by catastrophic events. Examples of geoscience processes include surface weathering and deposition by the movements of water, ice, and wind. Emphasis is on geoscience processes that shape local geographic features, where appropriate.]

MS-ESS2-3. **Analyze and interpret data** on the distribution **of fossils and rocks, continental shapes, and seafloor structures** to provide evidence **of the past plate motions.** [Clarification Statement: Examples of data include similarities of rock and fossil types on different continents, the shapes of the continents (including continental shelves), and the locations of ocean structures (such as ridges, fracture zones, and trenches).] [*Assessment Boundary: Paleomagnetic anomalies in oceanic and continental crust are not assessed.*]

MS-ESS2-4. **Develop a model to describe the cycling of water through Earth's systems** driven by energy **from the sun and the force of gravity.** [Clarification Statement: Emphasis is on the ways water changes its state as it moves through the multiple pathways of the hydrologic cycle. Examples of models can be conceptual or physical.] [*Assessment Boundary: A quantitative understanding of the latent heats of vaporization and fusion is not assessed.*]

MS-ESS2-5. **Collect data to provide evidence for how the motions and complex interactions of air masses** results in changes **in weather conditions.** [Clarification Statement: Emphasis is on how air masses flow from regions of high pressure to low pressure, causing weather (defined by temperature, pressure, humidity, precipitation, and wind) at a fixed location to change over time, and how sudden changes in weather can result when different air masses collide. Emphasis is on how weather can

Figure A.2: A screen-grab of part of the middle school standard on Earth Systems: MS-ESS2. Shown here are the first three PEs, with the first partially obscured by a pop-up related to the CC in the second.

Figure A.2 only shows a piece of the standard—only the first few Performance Expectations. Like the example in the previous section, this PE also includes Foundation Boxes, which highlight what pieces of each of the three dimensions is addressed in the standard and Connection Boxes, which highlight connections to other disciplines and grade levels. Drawing these connections is important in helping fortify understandings of both the particular content and how that content is contextualized in broader human and natural systems.

Appendix

Resources

Following are some of the most commonly used and cited publications on science education standards and benchmarks.

AAAS, 1993, *Benchmarks for Science Literacy*, Oxford University Press. [AAAS is American Association for Advance of Science.] http://www/[rpkect2-61.org/publications/bsl/online/index.php.

Bransford, J. D., Brown, A. L., and Cocking, R. R., eds., 2000, *How People Learn: Brain, Mind, Experience, and School, expanded edition*, National Academies Press, Washington, DC, http://www.nap.edu/openbook.php?record_id=9853.

Common Core State Standards Initiative. [While not focused on science education directly, standards on math and non-fiction reading impact are importantly related.] http://www.corestandards.org.

NCSE, 2013, *Evolution and Climate Change in the NGSS*, National Center for Science Education, http://ncse.com/news/2013/04/evolution-climate-change-ngss-0014800.

NGSS@NSTA website, National Science Teacher Association, http://ngss.nsta.org/.

NRC, 1996, *National Science Education Standards,* National Academies Press, Washington, DC. [NRC is National Research Council, and body of the National Academy of Sciences.] http://www.nap.edu/openbook.php?record_id=4962.

NRC, 2011, *Successful K-12 STEM Education: Identifying Effective Approaches in Science, Technology, Engineering, and Mathematics*, National Academies Press, Washington, DC. http://www.nap.edu/openbook.php?record_id=13158.

NRC, 2012, *A Framework for K-12 Science Education: Practices, Crosscutting Concepts, and Core Ideas*, National Academies Press, Washington, DC. http://www.nap.edu/openbook.php?record_id=13165.

NRC, 2013, *Next Generation Science Standards: For States, By States*, National Academies Press, Washington, DC, http://www.nextgenscience.org/.

Wysession, M., 2013, The Next Generation Science Standards and the Earth and Space Sciences, *The Science Teacher*, April/May issue. [Duggan-Haas, author of this Appendix, worked with Wysession on NRC's Conceptual Framework for New Science Education Standards.] http://nstahosted.org/pdfs/ngss/resources/201304_NGSS-Wysession.pdf.

a Glossary

	Note: Words in bold font are also defined in this glossary.
Acadian Orogeny	a **Devonian** mountain-building event involving the collision of the east coast of North America and the **accreted terrane** of **Avalon**. The event caused **metaphorphism**, folding, and **faulting** in an area from New York to Newfoundland; sediments **eroded** from the mountains accumulated in thick strata, the Catskill Delta, in the **Appalachian Basin** of New York and Pennsylvania. See also: **orogeny**
accretion, accrete	the process by which a body of rock increases in size due to the addition of further **sedimentary** particles or of large chunks of land, such as **terranes**.
acritarch	a microscopic organic-walled fossil, often approximately spherical in shape, found in rocks from over 2 billion to about 500 million years ago. They peaked in abundance and diversity late in the **Precambrian**. Acritarchs are widely believed to be **eukaryotes**, probably **protists**. They may be related to ancestors of the modern photosynthetic group dinoflagellates.
active plate boundary, active plate margin	the boundary between two **plates** of the Earth's **crust** that are colliding, pulling apart, or moving past each other. See also: **plate tectonics**
aeolian	pertaining to, caused by, or carried by the **wind**. Aeolian sediments are often polished, giving them a "frosty" appearance. The name comes from Aeolus, the Greek god of wind.
aerosol	tiny solid or liquid particles in the air. Examples include dust, smoke, mist, and human-made substances such as particles emitted from factories and cars.
agate	a crystalline **silicate** rock with a colorful banded pattern. It is a variety of **chalcedony** (**quartz**). Agates usually occur as **nodules** in **volcanic** rock.
aggregate	crushed stone or naturally occurring un**lithified** sand and gravel, used for construction, agriculture, and industry. Aggregate properties depends on the properties of the component rock. Rock quarried for crushed stone includes, for example, **granite** and **limestone**.
Alfisols	a **soil order**; these are highly fertile and productive agricultural **soils** in which **clays** often accumulate below the surface. They are found in humid and subhumid **climates**.
alluvium, alluvial	a layer of river-deposited sediment.
aluminum	a metallic chemical element (Al), and the most abundant metal in the Earth's **crust**. Aluminium has a low **density** and an excellent ability to resist corrosion. Structural components made from the metal and its alloys are commonly used in the aerospace industry, transportation, and household goods.
ammonoid, ammonite	a group of **extinct cephalopods** belonging to the Phylum Mollusca, and possessing a spiraling, tightly-coiled shell characterized by ridges, or septa.
amphibole	a group of dark-colored **silicate minerals**, or either **igneous** or **metamorphic** origin.
andesite	a fine-grained **extrusive volcanic** rock, with a **silica** content intermediate between that of **basalt** and **dacite**.
Andisols	a **soil order**; these are highly productive **soils** often formed from **volcanic** materials. They possess very high water- and nutrient-holding capabilities, and are commonly found in cool areas with moderate to high levels of precipitation.

Glossary a

angiosperms	Flowering plants, including taxa from deciduous trees to shrubs and grasses. Angiosperms represent the highest diversity of plants today, and are dominant in most modern environments. They differ from other seed-bearing plants in possessing flowers, an endosperm (nutrition within the seed), and a fruit enclosing the seed. Angiosperms diversified in the **Cretaceous**, and probably originated in the **Jurassic**.
anthracite	a dense, shiny **coal** that has a high carbon content and little volatile matter. Anthracite is as much as 95% carbon. Found in deformed rocks, anthracite is the cleanest burning of the three types of coal, because it contains the highest amount of pure carbon.
anthropogenic	caused or created by human activity.
anticline	a layer of rock folded (bent) along an axis, concave side down (i.e., in an upside down "U" or "V" shape). Thus rocks at the center of the anticline, along the fold (crest), are lifted up relative to the rest of the layer.
Antler Orogeny	a period of mountain building that deformed rocks in a belt extending from the California–Nevada border northward into Idaho. The Antler Orogeny began began in the late **Devonian** and continued into the **Carboniferous**. See also: **orogeny**
Appalachian Basin	an **inland basin**, formed by the **Taconic** and **Acadian** mountain-building events. The **crust** was **downwarped** as a result of the colliding **plates**, and the basin was later filled with an **inland sea**.
aquifer	a water-bearing formation of **gravel**, **permeable** rock, or **sand** that is capable of providing water, in usable quantities, to springs or wells.
archaeocyathid	a vase-shaped organism with a **carbonate** skeleton, generally believed to be a **sponge**. Archaeocyathids were the first important animal **reef** builders, originating in the early **Cambrian**. They were very diverse, but went extinct by the end of the Cambrian. Archeocyathids are often easiest to recognize in **limestones**, by their distinctive cross-sections.
Archean	a **geologic time** interval that extends from 4 billion to 2.5 billion years ago. It is part of the **Precambrian**.
archosaur	a member of an evolutionary group of reptiles that includes non-avian **dinosaurs** and birds, **pterosaurs**, and crocodiles, among others. All archosaurs share common features of the skull (such as a hole in front of the eye socket and in the lower jaw) and limbs. They originated in the early **Triassic**, and, though they suffered a major extinction at the end of the **Cretaceous** (in dinosaurs and pterosaurs), are still abundant and diverse through birds.
arête	a thin ridge of rock with an almost knife-like edge, formed when two **glaciers erode** parallel valleys.
Aridisols	a **soil order**; these are formed in very dry (arid) **climates**. The lack of moisture restricts **weathering** and leaching, resulting in both the accumulation of **salts** and limited subsurface development. They are commonly found in deserts.
arkose	**sandstone** that is more than one quarter **feldspar**, reflecting its relative lack of transport and **weathering**.
arthropod	an invertebrate animal, belonding to the Phylum Arthropoda, and posessing an external skeleton (exoskeleton), body segments, and jointed appendages. Arthropods include crustaceans, arachnids, and insects, and there are over a million described arthropod species living today. **Trilobites** are a major group of extinct arthropods.
asbestos	a fibrous **silicate mineral** that is resistant to heat, flames, and chemical action. As a very slow conductor of heat, asbestos was once commonly used as a fireproofing material and electrical insulation. Concerns over its health effects on the lungs have led to its removal from most common uses.

a–b Glossary

asphalt	a black, sticky, semi-solid, and viscous form of **petroleum**.
asthenosphere	a thin, semifluid layer of the Earth, below the outer rigid **lithosphere**, forming much of the upper **mantle**. The **heat** and pressure created by the overlying lithosphere make the solid rock of the asthenosphere bend and move like metal when heated. The layer is thought to flow vertically and horizontally with circular **convection** currents, enabling sections of lithosphere to subside, rise, and undergo lateral movement.
atmosphere	a layer of gases surrounding a planet. Earth's atmosphere protects living organisms from damage by solar ultraviolet radiation, and it is mostly composed of nitrogen. Oxygen is used by most organisms for respiration. Carbon dioxide is used by plants, algae, and **cyanobacteria** for photosynthesis.
Avalon	an early **Paleozoic microcontinent** offshore of what is now the eastern coast of North America. Avalon collided with and became the eastern edge of North America during the **Acadian Orogeny**.
badlands	a type of **eroded topography** that forms in semi-arid areas experiencing occasional periods of heavy rainfall. Sloping ground composed of **sandstones** and calcareous sediments underlain by **clay** or other soft materials is eroded over time into an intricate series of gullies and ravines. Different layers of rock **weather** at different rates, resulting in a variety of sculpted spurs and buttresses, as well as tall pillars of softer rock with a hard **capstone**.
Baltica	a late-**Proterozoic**, early-**Paleozoic** continent that included ancient Europe (northern Europe without Ireland and Scotland). Baltica began moving toward North America in the **Ordovician**, starting the **Taconic Orogeny**. North America fully collided with Baltica in the **Devonian**, resulting in the **Acadian Orogeny** on the eastern edge of the continent.
basalt	an **extrusive igneous rock**, and the most common rock type on the surface of the Earth. It forms the upper surface of all oceanic **plates**, and is the principal rock of ocean/seafloor ridges, oceanic islands, and high-volume continental eruptions. Basalt is fine-grained and mostly dark-colored, although it often **weathers** to reds and browns because of its high **iron** content. Basaltic **magmas** are produced by partial melting of the upper **mantle**. Materials melt when we increase their temperature, but a second way to melt a solid is to decrease the pressure. In the interior of the Earth this second mechanism—decompression—is far more important. When pressure on the mantle is released as it is forced up through the crust, it becomes basaltic magma.
basaltic andesite	a dark, fine-grained rock that is intermediate between **basalt** and **andesite** in **silica** content. Basaltic andesite is produced when the **magmatic** source of eruption is in transition between a deeper source, which tends to produce basalt, and a shallower source, which tends to produce andesite.
basement rocks	the foundation that underlies the surface geology of an area, generally composed of **igneous** or **metamorphic** crystalline rock. In certain areas, basement rock is exposed at the surface because of **uplift** or **erosion**.
batholith	a large exposed structure of **intrusive igneous rock** that solidified at depth, and covers an area of over 100 square kilometers (40 square miles). While batholiths may appear uniform, they are actually composed of multiple **plutons** that converged to form one mass.
bauxite	a whitish, grayish, brown, yellow, or reddish-brown rock composed of hydrous **aluminum oxides** and aluminum hydroxides; the principal commercial source of aluminum.
bentonite	a **clay**, formed from decomposed volcanic ash, with a high content of the **mineral** montmorillonite.
beryl	a white, blue, yellow, green, or pink **mineral**, found in coarse **granites** and **igneous rocks**. It is a source of beryllium and used as a **gemstone**; the green variety is called emerald, the blue is known as aquamarine.
biodiversity	the number of kinds of organisms at any given time and place. Global changes in biodiversity through **geologic time** tells paleontologists that something is happening to the rate of **extinction** or the rate of origin of new species. Regional changes are influenced by migration, or the number of species supported by available food and space resources.

Glossary b

biofuel	carbon-based **fuel** produced from renewable sources of **biomass** such as plants and garbage. Energy is obtained through combustion, so **greenhouse gases** are still produced. Because plants get their carbon from the air, burning them for **energy** and re-releasing it into the air has less effect on **climate** than **fossil fuels**, whose carbon is otherwise sequestered away from the **atmosphere**.
biomass	organic material from one or more organisms.
biostratigraphy	the branch of geology that uses **fossils** to determine the relative age of **sedimentary** layers.
biota	the organisms living in a given region, including plants, animals, fungi, **protists**, and bacteria.
bitumen	any of various flammable mixtures of hydrocarbons and other substances, occurring naturally or obtained by distillation from **coal** or **petroleum**, that are a component of asphalt and tar and are used for surfacing roads and for waterproofing.
bituminous coal	a relatively soft **coal** containing a tar-like substance called **bitumen**, which is usually formed as a result of high pressure on **lignite**.
bivalve	a marine or freshwater invertebrate animal belonging to the Class Bivalvia (or Pelecypoda) in the Phylum Mollusca. Bivalves are generally called "clams," but they also include scallops, mussels, cockles, and oysters. Bivalves are characterized by right and left calcareous shells (valves) joined by a hinge. Most are **filter feeders**, collecting food particles from the water with their gills. During the **Paleozoic**, bivalves lived mostly on the surface of the ocean floor. In the **Mesozoic**, bivalves became extremely diverse and some evolved the ability to burrow into ocean floor sediments.
blastoid	an **extinct** form of stemmed **echinoderm**, similar to **crinoids**. Blastoids possessed a nut-shaped body covered with interlocking plates, which was covered with fine hairlike structures for use in **filter feeding**. The body was held above the sea floor by a stalk of stacked disc-shaped plates.
body fossils	**fossils** that consist of an actual part of an organism, such as a bone, shell, or leaf.
brachiopod	a marine invertebrate animal belonging to the Phylum Brachiopoda, and characterized by upper and lower calcareous shell valves joined by a hinge, and a crown of tentacles (lophophore) used for filter feeding and respiration. Brachiopods are the most common **fossil** in **Paleozoic sedimentary rocks**. Brachiopods look somewhat similar to the clams that you find at the beach today. Brachiopods and **bivalves** both have a pair of hinged shells (valves) to protect themselves while feeding. However, the soft parts of modern brachiopods tell us that they are completely unrelated to bivalves. Brachiopods have a special structure formed by tissue with thousands of tiny hair-like tentacles stretched along a coiled piece of internal shell material. These tentacles catch and move small particles toward the mouth. This body plan is very different from that of bivalves, which have a larger, fleshy body and collect particles with their gills. To tell the difference between a brachiopod and a bivalve, look for symmetry on the surface of the shell. Bivalve valves are of equal size and mirror image shapes. Brachiopods' bottom valves, however, are slightly bigger and often have a different shape.
braided stream	a stream consisting of multiple, small, shallow channels that divide and recombine numerous times, forming a pattern resembling strands of braided hair. A braided stream carries more sediment than a typical stream, causing the formation of sandbars and a network of crisscrossing streams.
breccia	a **pyroclastic** rock composed of **volcanic** fragments from an explosive eruption.
brine	See hydrothermal solution
British Thermal Unit (BTU or Btu)	the most commonly used unit for **heat energy**. One Btu is approximately the amount of heat required to raise one pound of water by one degree Fahrenheit. A Btu is also about the amount of energy released by burning a single wooden match.

b–c Glossary

bryozoan	a marine or freshwater, colonial invertebrate animal belonging to the Phylum Bryozoa, and characterized by an encrusting or branching calcareous skeleton from which multiple individuals (zooids) extend from small pores to **filter feed** using crowns of tentacles (lophophores). Bryozoans have a long and exemplary **fossil** record. One of the more common **Paleozoic** varieties looks like fine-mesh cloth with numerous tiny holes in which the individual animals in the colony lived. Although they function somewhat like coral, and are often found in similar environments, bryozoans are more closely related to **brachiopods**.
butte	an isolated hill with steep, often vertical sides and a small, relatively flat top.
calcite	a **carbonate mineral**, consisting of **calcium carbonate** ($CaCO_3$). Calcite is a common constituent of **sedimentary rocks**, particularly **limestone**.
calcium carbonate	a chemical compound with the formula $CaCO_3$, commonly found in rocks in the **mineral** forms **calcite** and aragonite, as well as the shells and skeletons of marine organisms.
caldera	a collapsed, cauldron-like **volcanic** crater formed by the collapse of land following a volcanic eruption.
caliche	a zone of cemented material within **soil**, formed when water infiltrates the soil, dissolves soluble materials, and evaporates, leaving behind precipitates (particularly **calcium carbonate**) in the pore space between soil grains. Layers of caliche accumulate to tens of feet in some locations. Caliche is commonly collected for use as an additive in cement.
calyx	the head of a **crinoid**.
Cambrian	a **geologic time** period lasting from 541 to 485 million years ago. During the Cambrian, multicellular marine organisms became increasingly diverse, as did their mineralized **fossils**. The Cambrian is part of the **Paleozoic** era.
Canadian Shield	the stable core of the North American continental landmass, containing some of the oldest rocks on Earth. The shield has experienced very little tectonic activity (**faulting** or folding) for millions of years. As the stable cores of all continents, shields are often covered by layers of younger material.
capstone, caprock	a harder, more resistant rock type that overlies a softer, less resistant rock. The harder rock typically helps to control the rate of **erosion**.
carbonate rocks	rocks formed by accumulation of **calcium carbonate**, often made of the skeletons of aquatic organisms such as corals, clams, **snails**, **bryozoans**, and **brachiopods**. These organisms thrive in warm, clear shallow waters common to tropical areas, therefore modern carbonate rocks are observed forming in places such as the Florida Keys and the Bahamas. They are also one of the dominant rock forms of the bottom of the ocean, where sediments form from the skeletons of planktonic organisms such as **foraminifera**. Carbonate rocks include **limestone**, **dolostone**, and **dolomite**.
Carboniferous	a **geologic time** period that extends from 359 to 299 million years ago. It is divided into two subperiods, the **Mississippian** and the **Pennsylvanian**. By the Carboniferous, terrestrial life had become well established. The name Carboniferous means "**coal**-bearing," and it is during this time that many of today's coal beds were formed. The Carboniferous is part of the **Paleozoic**.
cementation	the precipitation of **minerals**, such as **silica** and **calcite**, that bind together particles of rock, bones, etc., to form a solid mass of **sedimentary rock**.

Glossary C

Cenozoic	the **geologic time** period spanning from 66 million years ago to the present. The Cenozoic is also known as the age of mammals, since extinction of the large reptiles at the end of the **Mesozoic** allowed mammals to diversify. The Cenozoic includes the **Paleogene**, **Neogene**, and **Quaternary** periods.
cephalopod	a marine invertebrate animal belonging to the Class Cephalopoda in the Phylum Mollusca, and characterized by a prominent head, arms, and tentacles with suckers, and jet propulsion locomotion. Cephalopods are swimming predators with beak-shaped mouthparts. The shells of cephalopods range from long straight cones to spirals, but some have internal shells or no significant shell at all, such as the octopus.The group includes belemnites, **ammonoids**, nautilus, squid, and octopuses. A **mass extinction** between the **Cretaceous** and **Paleogene** eliminated many varieties of cephalopods.
chalcedony	a crystalline **silicate mineral** that is a microcrystalline variety of **quartz**.
chalcopyrite	a yellow **mineral** consisting of a **copper-iron** sulfide ($CuFeS_2$). Chalcopyrite is the most common and important source of copper, and can also be called copper **pyrite**.
chalk	a soft, fine-grained, easily pulverized, white-to-grayish variety of **limestone**, composed of the shells of minute planktonic single-celled algae.
chemical fossils	chemicals produced by an organism that leave behind an identifiable trace in the geologic record. Chemical **fossils** provide some of the oldest evidence for life on Earth.
chemical reaction	a process that involves changes in the structure and **energy** content of atoms, molecules, or ions but not their nuclei.
chert	a **sedimentary rock** composed of microcrystaline **quartz**. It is often found as **nodules** or **concretions** in **limestone** and other marine sedimentary rocks. As these rocks form, water moving through them transports small amounts of **silicon** dioxide that accumulate into clumps of microscopic crystals. The resulting rocks are extremely strong and have no planes of weakness. For thousands of years, humans exploited these qualities, breaking chert nodules into blades and other tools.
chordate	an animal that possesses the following five traits during at least one stage of its development: a notochord (the flexible rod that, in vertebrates, becomes the backbone), a hollow dorsal nerve cord, pharyngeal gill slits, an endostyle (precursor to the thyroid gland), and a post-anal tail.
cinder	a type of **pyroclastic** particle in the form of gas-rich **lava** droplets that cool as they fall.
cirque	a large bowl-shaped depression carved by **glacial erosion** and located in mountainous regions.
clay	the common name for a number of very fine-grained, earthy materials that become plastic (flow or change shape) when wet. Chemically, clays are hydrous **aluminum silicates**.
cleavage	a physical property of **minerals**. Cleavage occurs when a mineral breaks in a characteristic way along a specific plane of weakness. **Mica** and **graphite** have very strong cleavage, allowing them to easily break into thin sheets.
climate	a description of the average temperature, range of temperature, humidity, precipitation, and other **atmospheric**/hydrospheric conditions a region experiences over a period of many years (usually more than 30). These factors interact with and are influenced by other parts of the Earth **system**, including geology, geography, insolation, currents, and living things. The climate of a region represents the average **weather** over a long period of time.

C Glossary

climate change	See **global warming**
coal	a combustible, compact black or dark-brown carbonaceous rock formed by the compaction of layers of partially decomposed vegetation. By far the greatest abundance of coal is located in strata of **Carboniferous** age.
coal ball	masses of **calcium carbonate** that crystalize inside **coals** from **minerals** dissolved in groundwater, protecting the plants they contain from alteration.
coalification	the process by which **coal** is formed from plant materials through **compression** and **heating** over long periods of time.
cold front	the boundary between the warm air and the cold air moving into a region. At this boundary, **denser**, colder air moves in, making the less dense, warm air rise. This displaced warm air cools as it rises because air pressure decreases with increasing height in the **atmosphere**. As the air cools, it becomes saturated with water vapor, and condensation begins to occur, eventually leading to dramatic rainstorms.
color (mineral)	a physical property of **minerals**. Color is determined by the presence and intensity of certain elements within the mineral.
color (soil)	a physical property of **soils**. Soil color is influenced by **mineral** content, the amount of organic material, and the amount of water it routinely holds. These colors are identified by a standard soil color chart called the Munsell chart.
columnar joint	five- or six-sided columns that form as cooling **lava** contracts and cracks. Columnar joints are often found in **basalt** flows, but can also form in ashflow **tuffs** as well as shallow **intrusions**. The columns are generally vertical, but may also be slightly curved
commodity	a good for which there is demand, but which is treated as equivalent across all markets, no matter who produces it.
compression, compressional force	forces acting on an object from all or most directions, resulting in compression (flattening or squeezing). Compressional forces occur by pushing objects together.
concretion	a hard, compact mass, usually of spherical or oval shape, found in **sedimentary rock** or **soil**. Concretions form when **minerals** precipitate around a particulate nucleus within the sediment.
condylarth	a member of an informal (not evolutionary) group of ungulates (a large group of mammals that include most of the relatively large herbivorous mammals, such as horses, rhinos, deer, and others). Condylarths include a variety of hoofed animals, evolving convergently in two or more lineages, that diversified in the **Paleocene**; species ascribed to the group range from the late **Cretaceous** to **Oligocene**.
conglomerate	a **sedimentary rock** composed of multiple large and rounded fragments that have been **cemented** together in a fine-grained **matrix**. The fragments that make up a conglomerate must be larger than grains of **sand**.
conifer	a woody plant (**tree**) of the division Coniferophyta. Conifers bear cones that contain their seeds.
conodont	an extinct, eel-shaped animal classified in the class Conodonta and thought to be related to primitive **chordates**. Originally, conodonts were only known from small phosphatic tooth-like micro**fossils**, which have been widely used for **biostratigraphy**. Knowledge about their soft tissues still remains limited.
Conservation of Energy	a principle stating that **energy** is neither created nor destroyed, but can be altered from one form to another.

contact metamorphism	the process by which a **metamorphic rock** is formed through direct contact with **magma**. Changes that occur due to contact metamorphism are greatest at the point of contact. The farther away the rock is from the point of contact, the less pronounced the change.
convection	the rise of buoyant material and the sinking of denser material. In the **mantle**, variations in **density** are commonly caused by the melting of **subducting** materials.
convergent boundary	an **active plate boundary** where two tectonic **plates** are colliding with one another. **Subduction** occurs when an oceanic plate collides with a continental plate or another oceanic plate. If two continental plates collide, mountain building occurs. See also: **plate tectonics**
copper	a ductile, malleable, reddish-brown metallic element (Cu). Copper is used extensively as wiring in the electrical industry as well as in alloys such as brass and bronze.
Cordilleran Ice Sheet	one of two continental **glaciers** that covered Canada and parts of the Western US during the last major **Pleistocene ice age**.
corundum	an **aluminum oxide mineral** (Al_2O_3) that is, after **diamond**, the hardest known natural substance. Corundum is best known for its **gem** varieties, ruby (red) and sapphire (blue).
craton	the old, underlying portion of a continent that is geologically stable relative to surrounding areas. The portion of a craton exposed at the surface is termed a shield, while that overlain by younger layers is often referred to as a platform. A craton can be thought of as the heart of a continent—it is typically the oldest, thickest, and most stable part of the bedrock. It is also usually far from the margins of tectonic **plates**, where new rock is formed and old destroyed. This rock has usually been **metamorphosed** at some point during its history, making it resistant to **erosion**.
creodont	a member of an informal (not evolutionary) group of carnivorous mammals. Species ascribed to them arose in the **Paleocene**, were the most abundant and diverse terrestrial carnivores by the late Paleocene, and survived until the **Miocene**.
Cretaceous	a **geologic time** period spanning from 144 to 66 million years ago. It is the youngest period of the **Mesozoic**. The end of the Cretaceous bore witness to the **mass extinction** event that resulted in the demise of the **dinosaurs**. "Cretaceous" is derived from the Latin word *creta*, meaning "chalk." The white (**chalk**) cliffs of Dover on the southeastern coast of England are a famous example of Cretaceous chalk deposits.
crinoid	a marine invertebrate animal belonging to the Class Crinoidea of the Phylum Echinodermata, and characterized by a head (**calyx**) with a mouth on the top surface surrounded by feeding arms. Several groups of stemmed **echinoderms** appeared in the early **Paleozoic**, including crinoids, **blastoids**, and **cystoids**. Crinoids have five-fold symmetry and feathery arms (sometimes held off the sea floor on a stem) that collect organic particles from the water. The stems, the most often preserved part, are made of a series of stacked discs. Upon death, these stems often fall apart and the individual discs are preserved separately in the rock. The crinoid's feathery arms make it look something like a flower on a stem. Thus, crinoids are commonly called sea lilies, although they are animals, not plants.
cross-bedding	layering within a bed in a series of rock strata that does not run parallel to the plane of stratification. Cross-beds form as flowing water or wind pushes sediment downcurrent, creating thin beds that slope gently in the direction of the flow as migrating ripples. The downstream slope of the ripple may be preserved as a thin layer dipping in the direction of the current, across the natural flat-lying repose of the beds. Another migrating ripple will form an additional layer on top of the previous one.

c–d Glossary

crust	the uppermost, rigid outer layer of the Earth. Two types of crust make up the **lithosphere**, which is broken into tectonic **plates**. Oceanic crust is denser but significantly thinner than continental crust, while continental crust is much thicker but less dense (and therefore buoyant). When continental crust collides with oceanic crust, the denser oceanic crust will be dragged (**subducted**) under the buoyant continental crust. Although mountains are created by these oceanic/continental crust collisions due to the **compression** of the two plates, much taller ranges are produced by continental/continental collisions. When two buoyant continental crusts collide, there is nowhere for the crust to go but up! The modern Himalayas, at the collision site of the Asian and Indian plates, are a good example of very tall mountains formed by a collision between two continental crusts.
Cryogenian	a **geologic time** period lasting from 850 to 635 million years ago, during the **Precambrian**. During this period, the Earth was subject to a 200-million-year-long **ice age**.
crystal form	a physical property of **minerals**, describing the shape of the mineral's crystal structure (not to be confused with **cleavage**). A mineral might be cubic, rhomboidal, hexagonal, or polyhedral.
cyanobacteria	a group of bacteria, also called "blue-green algae," that obtain their **energy** through photosynthesis.
cycad	a palm-like, terrestrial seed plant (**tree**) belonging to the Class Cycadopsida, and characterized by a woody trunk, a crown of stiff evergreen leaves, seeds without protective coatings, and no flowers. Cycads were very common in the **Mesozoic**, but are much reduced in diversity today, restricted to the tropical and subtropical regions of the planet.
cystoid	**extinct**, stalked **echinoderms** related to **crinoids**, but with an ovoid body and triangular pore openings.
dacite	a fine-grained **extrusive igneous rock**, with a **silica** content intermediate between those of **andesite** and **rhyolite**.
debris flow	a dangerous mixture of water, mud, rocks, **trees**, and other debris that can move quickly down valleys. Such flows can result from sudden rainstorms or snowmelt that create flash floods. Areas that have experienced a recent wildfire are particularly vulnerable to debris flows, since there is no vegetation to hold the **soil**.
degrade (energy)	the transformation of **energy** into a form in which it is less available for doing work, such as **heat**.
delta, deltaic	a typically wedge-shaped deposit formed as sediment is eroded from mountains and transported by streams across lower elevations. The Mississippi Delta is a modern delta containing sediment being transferred from the Mississippi River into the Gulf of Mexico.
dendritic drainage	a drainage pattern where many smaller streams join and contribute to ever larger streams. The pattern looks somewhat like a **tree**, in which smaller branches connect to progressively larger branches.
density	a physical property of **minerals**, describing the mineral's mass per volume.
derrick	a lifting device in the form of a framework steel tower that is built over a deep drill hole, typically an **oil** well. An oil derrick is composed of machinery for hoisting and lowering tools required during the drilling process, and readying the well for extraction of **petroleum**.
Devonian	a **geologic time** period spanning from 419 to 359 million years ago. The Devonian is also called the "age of fishes" due to the diversity of fish that radiated during this time. On land, seed-bearing plants appeared and terrestrial **arthropods** became established. The Devonian is part of the **Paleozoic**.
diamond	a **mineral** form of carbon, with the highest **hardness** of any material. Most natural diamonds are formed at high temperature and pressure deep in the Earth's **mantle**.

diatom	a unicellular, microscopic, usually planktonic marine alga (plant) characterized by cell walls composed of **silica**.
diatreme	a rock-filled crack in the Earth's **crust** through which **magma** traveled upward before contacting groundwater and exploding due to expanding water vapor and **volcanic** gases. Such cracks are often known as diatreme pipes.
dike	a sheet of **intrusive igneous** or **sedimentary rock** that fills a crack cutting across a pre-existing rock body.
dimension stone	the commercial term applied to quarried blocks of rock cut to specific dimensions and used for buildings, monuments, facing, and curbing.
dinosaur	a member of a group of terrestrial reptiles with a common ancestor and thus certain anatomical similarities, including long ankle bones and erect limbs. All of the large reptile groups, including the dinosaurs, disappeared at or before the **mass extinction** at the end of the **Cretaceous**.
divergent plate boundary	an **active plate boundary** where two tectonic **plates** are pulling apart from one another, causing the **mantle** to well up at a **rift**. Mid-ocean ridges are the most common divergent boundary and are characterized by the eruption of bulbous pillow-shaped **basalt lavas** and **hydrothermal** fluids.
dolomite	a **carbonate mineral**, consisting of calcium magnesium carbonate ($CaMg(CO_3)_2$). Dolomite is an important reservoir rock for **petroleum**, and also commonly hosts large ore deposits.
dolostone	a rock (also known as dolomitic **limestone** and once called magnesian limestone) primarily composed of **dolomite**, a **carbonate mineral**. It is normally formed when magnesium bonds with **calcium carbonate** in limestone, forming dolomite.
double refraction	the result of light passing through a material that splits it into two polarized sets of rays, doubling images viewed through that material. For example, a single line on a sheet of paper will appear as two parallel lines when viewed through a clear **calcite** crystal.
downwarp	a segment of the Earth's **crust** that is broadly bent downward.
dynamic metamorphism	See **regional metamorphism**
earthquake	a sudden release of **energy** in the Earth's **crust** that creates **seismic waves**. Earthquakes are common at **active plate boundaries**.
echinoderm	a member of the Phylum Echinodermata, which includes starfish, sea urchins, and **crinoids**. Echinoderms have radial symmetry (which is usually five-fold), and a remarkable ability to regenerate lost body parts.
effervesce	to foam or fizz while releasing gas. **Carbonate minerals** will effervesce when exposed to hydrochloric acid.
efficiency	the use of a relatively small amount of **energy** for a given task, purpose, or service; achieving a specific output with less energy input.
energy	the **power** derived from the use of physical or chemical resources. Everything we do depends upon energy—without it there would be no civilization, no sunlight, no food, and no life. Energy moves people and goods, produces electricity, heats our homes and businesses, and is used in manufacturing and other industrial processes.
energy carrier	a source of **energy**, such as electricity, that has been subject to human-induced energy transfers or transformations.

Glossary

Entisols	a **soil order**; these are **soils** of relatively recent origin with little or no **horizon** development. They are commonly found in areas where erosion or deposition rates outstrip rates of soil development, such as **floodplains**, mountains, and **badland** areas.
Eocene	a **geologic time** period extending from 56 to 33 million years ago. The Eocene is an epoch of the **Paleogene** period.
epeirogenic	large-scale **crustal uplift** caused by hot or upwelling **mantle** underlying the surface, which buoys up the overlying crust but does not significantly fold or **fault** the rocks.
erg	an area of desert, greater than 125 square kilometers (48 square miles), covered by **wind**-blown **sand**. It is also known as a "sand sea" or "dune sea."
erosion	the transport of **weathered** materials. Rocks are worn down and broken apart into finer grains by **wind**, rivers, wave action, freezing and thawing, and chemical breakdown. Over millions of years, weathering and erosion can reduce a mighty mountain range to low rolling hills. Some rocks wear down relatively quickly, while others can withstand the power of erosion for much longer. Softer, weaker rocks, such as **shale** and poorly **cemented sandstone** and **limestone**, are much more easily worn than hard, crystalline **igneous** and **metamorphic rocks**, or well-cemented sandstone and limestone. Harder rocks are often left standing as ridges because the surrounding softer, less resistant rocks were more quickly worn away.
estuary	a place where freshwater and saltwater mix, created when sea level rises to flood a river valley.
eukaryotes	organisms with complex cells containing a nucleus and organelles. **Protists** and all multicellular organisms are eukaryotes.
evaporite	a **sedimentary rock** created by the precipitation of **minerals** directly from seawater, including **gypsum**, **calcite**, **dolomite**, and **halite**.
exsolve	to come out of solution and, in the case of a gas, form bubbles.
extinction	the end of species or other taxonomic groups, marked by death of the last living individual. Paleontologists estimate that over 99% of all species that have ever existed are now extinct. The species of modern animals that we study in biology today represent less than 1% of what has lived throughout **geologic time**.
extrusion, extrusive rock	an **igneous rock** formed by the cooling of **lava** after **magma** escapes onto the surface of the Earth through **volcanic** craters and cracks in the Earth's **crust**.
fault	a fracture in the Earth's **crust** in which the rock on one side of the fracture moves measurably in relation to the rock on the other side.
fault scarp	an escarpment directly beside a **fault** line, where the ground on one side of the fault has moved vertically with respect to the other side, creating step-like **topography**.
feldspar	an extremely common group of rock-forming **minerals** found in **igneous**, **metamorphic**, and **sedimentary rocks**. There are two groups of feldspar: alkali feldspar (which ranges from potassium-rich to sodium-rich) and plagioclase feldspar (which ranges from sodium-rich to calcium-rich). Potassium feldspars of the alkali group are commonly seen as pink crystals in igneous and metamorphic rocks, or pink grains in sedimentary rocks. Plagioclase feldspars are more abundant than the alkali feldspars, ranging in color from light to dark. Feldspars are commercially used in ceramics and scouring powders.
felsic	**igneous rocks** with high **silica** content and low **iron** and magnesium content. They are light in color and are typically found in continental **crust**.

Glossary f–g

filter feeder	an animal that feeds by passing water through a filtering structure that traps food. The water may then be expelled and the food digested. This strategy is employed by a wide range of animals today, from clams and krill to flamingos and whales.
flint	a hard, high-quality form of **chert** that occurs mainly as **nodules** and masses in **sedimentary rock**. Due to its strength and the fact that it splits into thin, sharp flakes, flint was often used to make tools during the Stone Age. Flint will also create sparks when struck against steel, and has been used to ignite gunpowder in more modern times.
floodplain	the land around a river that is prone to flooding. This area can be grassy, but the sediments under the surface are usually deposits from previous floods.
fluorite, fluorspar	the **mineral** form of calcium fluoride (CaF_2). Fluorite is used in a variety of commercial applications, including as lenses for microscopes, the production of some glass, and the chemical industry. Fluorite lent its name to the phenomenon of fluorescence, which occurs in some fluorites due to impurities in the crystal.
fluvial	See **outwash plain**
foliation	the arrangement of the constituents of a rock in leaflike layers, as in **schists**. During **metamorphism**, the weight of overlying rock can cause **minerals** to realign perpendicularly to the direction of pressure, layering them in a sheet-like pattern.
foraminifera	a class of aquatic **protists** that possess a calcareous or **siliceous** exoskeleton. Foraminifera have an extensive **fossil** record.
fossil	preserved evidence of ancient life, including, for example, preserved skeletal or tissue material, molds or casts, and traces of behavior. Fossilization may alter biological material in a variety of ways, including **permineralization**, **replacement**, and **compression**. Remains are often classified as fossils when they are older than 10,000 years, the traditional start of the **Holocene** (Recent) epoch. However, this date is only a practical guideline—scientists studying successions of plant or animal remains would not recognize any sudden change in the material at 10,000 years, and would typically refer to all material buried in sediments as fossil material. The word fossil is derived from the Latin word *fossilis*, meaning "dug up."
fossil fuels	**fuel** for human use that is made from the remains of ancient **biomass**, referring to any hydrocarbon fuel source formed by natural processes from anaerobically decomposed organisms, primarily **coal**, **petroleum**, **natural gas** (methane), and **peat**. Fossil fuels are non-renewable, meaning that because they take thousands to millions of years to form, the rate of use is far greater than the rate of formation, and eventually we will run out.
fracture (mineral)	a physical property of **minerals**, formed when a mineral crystal breaks; also a crack in rocks, sometimes known as a **joint**. This process is separate from **cleavage**, which occurs when a mineral breaks in a characteristic way along a specific plane of weakness.
fuel	a material substance possessing internal potential **energy** that can be transferred to the surroundings for specific uses—included are **petroleum**, **coal**, and **natural gas** (the **fossil fuels**), and other materials, such as uranium, hydrogen, and **biofuels**.
gabbro	a coarse-grained, **mafic**, and **intrusive igneous rock**. Most oceanic **crust** contains gabbro.
galena	an abundant **sulfide mineral** with cubic crystals. It is the most important **ore** of **lead**, as well as an important source of **silver**.

g Glossary

gastropod	a marine, freshwater, or terrestrial invertebrate animal belonging to the Class Gastropoda of the Phylum Mollusca, and characterized by a single, coiled, calcareous shell, a muscular foot for gliding, and internal asymmetry caused by an embryonic process (torsion). Gastropods include snails and slugs.
Gellisols	a **soil order**; these are weakly **weathered soils** formed in areas that contain **permafrost** within the soil profile.
gem, gemstone	a **mineral** that has aesthetic value and is often cut and polished for use as an ornament.
geologic time scale	a standard timeline used to describe the age of rocks and fossils, and the events that formed them. It spans Earth's entire history, and is often subdivided into four major time intervals: the **Precambrian**, **Paleozoic**, **Mesozoic**, and **Cenozoic**.
ginkgo	a terrestrial **tree** belonging to the plant division Ginkgophyta, and characterized by broad fan-shaped leaves, large seeds without protective coatings, and no flowers. Ginkgos were very common and diverse in the **Mesozoic**, but today only one species exists, *Ginkgo biloba.*
glacier	a body of dense ice on land that does not melt away annually and has sufficient mass to move under its own weight. Glaciers form when snow accumulates faster than it melts over many years. As long as melt does not exceed accumulation, the ice and snow pile up and become a self-sustaining system. As glaciers slowly flow, they abrade and erode the landscape around them to create grooves, scratches, **moraines**, and other distinguishing features. Glaciers form only on land, and are much thicker than ice that forms on the surface of water. 99% of Earth's glacial ice exists as vast polar **ice sheets**, but glaciers are also found high in the mountains of every continent except Australia.
glassy rock	a **volcanic** rock that cooled almost instantaneously, resulting in a rock with tiny crystals or no crystals at all. **Obsidian**, **basalt** glass, and **pumice** are examples of glassy rocks.
global warming	the current increase in the average temperature worldwide, caused by the build-up of **greenhouse gases** in the **atmosphere**. With the coming of the Industrial Age and exponential increases in human population, large amounts of gases have been released into the atmosphere (especially carbon dioxide) that give rise to global warming. The term "climate change" is preferred because warming contributes to other climatic changes such as precipitation and storm strength.
gneiss	a **metamorphic rock** that may form from **granite** or layered **sedimentary rock** such as **sandstone** or siltstone. Parallel bands of light and dark **minerals** give gneiss its banded texture.
gold	a soft, yellow, corrosion-resistant element (Au), which is the most malleable and ductile metal on Earth. Gold has an average abundance in the crust of only 0.004 parts per million. It can be profitably mined only where **hydrothermal solutions** have concentrated it.
Gondwana, Gondwanaland	the supercontinent of the Southern Hemisphere, composed of Africa, Australia, India, and South America. It combined with the North American continent to form **Pangaea** during the late **Paleozoic**.
granite	a common and widely occurring type of **igneous rock**. Granite usually has a medium- to coarse-grained texture, and is at least 20% **quartz** by volume.
granodiorite	a coarse-grained **plutonic rock** rich in the elements sodium and calcium, and in the **minerals** potassium **feldspar** and **quartz**.

Glossary g–h

graphite	a **mineral**, and the most stable form of carbon. Graphite means "writing stone," a reference to its use as pencil lead. Graphite occurs in **metamorphic rocks**, **igneous rocks**, and **meteorites**.
graptolite	an **extinct** colonial invertebrate animal belonging to the Class Graptolithina of the Phylum Hemichordata, and characterized by individuals housed within a tubular or cup-like structure. The soft parts of a graptolite's body have never been clearly identified.
gravel	unconsolidated, semi-rounded rock fragments larger than 2 millimeters (0.08 inches) and smaller than 75 millimeters (3 inches).
greenhouse gas	a gas in the **atmosphere** that absorbs and emits **heat**. The primary greenhouse gases in the Earth's atmosphere are water vapor, carbon dioxide, methane, nitrous oxide, and ozone.
gypsum	a soft, **sulfate mineral** that is widely mined for its use as fertilizer and as a constituent of plaster. Alabaster, a fine-grained light colored variety of gypsum, has been used for sculpture making by many cultures since ancient times.
halite	See **salt**
hardness	a physical property of **minerals**, specifying how hard the mineral is, and its resistance to scratching. Hardness helps us understand why some rocks are more or less resistant to **weathering** and **erosion**. See also: **Mohs Scale of Hardness**
heat	a form of **energy** transferred from one body to another as a result of a difference in temperature or a change in phase. Heat is transmitted through solids and fluids by conduction, through fluids by **convection**, and through empty space by radiation.
heat island effect	a phenomenon in which cities experience higher temperatures than do surrounding rural communities.
heat wave	a period of excessively hot **weather** that may also accompany high humidity. Temperatures of just 3°C (6°F) to 6°C (11°F) above normal are enough to reclassify a warm period as a heat wave. Under high humidity, the mechanism of sweating does little to cool people down because the humidity prevents sweat from evaporating and cooling off the skin.
hectare	a metric unit of area defined as 10,000 square meters.
helium	a gaseous chemical element (He), which is the second most abundant and second lightest element in the universe. Helium is used in cryogenics, as a coolant; it is also used in industrial applications including pressurizaton, welding, and leak detection. Balloons and blimps, although probably the most well-known and visible application of helium, take up less than an eighth of its total use.
hematite	a **mineral** form of **iron oxide** (Fe_2O_3). The name hematite has its origins in the Greek word *haimatos*, meaning "blood." It is very common in **Precambrian** banded iron formations. Iron from hematite is used in the manufacture of steel. The vivid red pigments that iron lends to the mineral also makes it valuable as a commercial pigment.
Histosols	a **soil order**; these are organic-rich **soils** found along lake coastal areas where poor drainage creates conditions of slow decomposition and **peat** (or muck) accumulates.
Holocene	the most recent portion of the **Quaternary**, beginning about 11,700 years ago and continuing to the present. It is the most recent (and current) **interglacial**, an interval of glacial retreat. The Holocene also encompasses the global growth and impact of the human species.

Glossary

horizon (soil)	a layer in the **soil**, usually parallel to the surface, which has physical characteristics (usually **color** and texture) that are different from the layers above and below it. Each type of soil usually contains three or four horizons.
hornblende	a dark **silicate mineral** that can occur in a variety of forms. Hornblende is a common constituent of many **igneous** and **metamorphic rocks**.
horsetail	See **sphenopsid**
hot spot	a **volcanic** region thought to be fed by underlying **mantle** that is anomalously hot compared with the mantle elsewhere. Hot spots form from plumes of hot material rising through the mantle. Magma from the hot spot pushes its way up through the **crust**, creating an **igneous intrusion** and sometimes a volcano. Although the hot spot remains fixed, the **plates** of the **lithosphere** continue to move above it. As a plate continues to move over the hot spot, the original volcano shifts off of the hot spot and a new intrusion or volcano is formed. This gradually produces a chain of **volcanic islands** such as the Hawaiian Islands. **Erosion** of volcanoes may eventually wear down the crust to reveal the igneous intrusions that formed the volcano's magma chamber.
humus	the organic component of soil; a major part of the **soil horizon** containing organic matter.
Huronian glaciation	a **glaciation** beginning about 2.4 billion years ago, that covered the entire surface of the Earth in ice for as long as 300 million years.
hurricane	a rapidly rotating storm system with heavy **winds**, a low-pressure center, and a spiral arrangement of thunderstorms. These storms tend to form over large, warm bodies of water. Once winds have reached 119 kilometers per hour (74 miles per hour), such a storm is classified as a hurricane. Hurricanes usually develop an eye, which is visible as a small, round, cloud-free area at the center of the storm. The eye is an area of relative calm and low **atmospheric** pressure. The strongest thunderstorms and winds circulate just outside the eye, in the eyewall.
hydrographic	the science of the physical aspects of water on the surface of the Earth. This includes the shoreline's shape, depth, and bottom **topography**; sediment of water bodies; tides, currents, and waves in water bodies; and the flow of streams. It is most often applied to navigation.
hydrothermal solution	hot, **mineral**-rich water moving through rocks. These solutions are often enriched in **salts** (such as sodium chloride, potassium chloride, and calcium chloride) and thus are called "brines." The brine is as salty or even saltier than seawater. Salty water can contain minute amounts of dissolved minerals such as **gold**, **lead**, **copper**, and **zinc**. The presence of salt in the water suppresses the precipitation of the metallic minerals from the brine because the chlorides in the salt preferentially bond with metals. Additionally, because the brine is hot, minerals are more easily dissolved, just as hot tea dissolves sugar more easily than cold tea.
hyperthermic	**soils** in which the mean annual temperature 50 centimeters (20 inches) below the surface is at least 22°C (71.6°F), with 5 or more degrees of variation between winter and summer. The term is often used as part of the phrase "hyperthermic temperature regime." Only the most southern parts of the US contain hyperthermic soils.
ice age	a period of global cooling of the Earth's surface and **atmosphere**, resulting in the presence or expansion of **ice sheets** and alpine **glaciers**. Throughout the Earth's history, it has been periodically plunged into **ice ages**, dependent upon the **climate** and position of the continents. Over the past 2.6 million years, North America has experienced about 50 glacial advances and retreats. The most recent ice age ended approximately 12,000 years ago.
ice cap	an **ice field** that lies over the tops of mountains.

Glossary

i

ice field	an extensive area of interconnected **glaciers** spanning less than 50,000 square kilometers (19,305 square miles). Ice fields are usually constrained by an area's **topography**. Ice fields that lie over the tops of mountains are called **ice caps**.
ice lobe	a broad, rounded section of a continental **glacier** that flows out near the glacier's terminus, often through a broad trough.
ice sheet	a mass of **glacial** ice that covers part of a continent and has an area greater than 50,000 square kilometers (19,000 square miles).
ichthyosaur	an **extinct Mesozoic** marine reptile that was probably similar in size and habitat to the toothed whales, dolphins, and large sharks of today.
igneous rocks	rocks derived from the cooling of **magma** underground or molten **lava** on the Earth's surface. Igneous rocks differ not only in their cooling rates and subsequent crystal sizes, but also in their chemical compositions. Rocks found in continental **crust**, such as **granite**, have high **silica** content and low **iron** and magnesium content. They are light in color and are called **felsic**. Rocks found in oceanic crust, such as **basalt**, are low in silica and high in iron and magnesium. They are dark in color and are called **mafic**. Although the composition of magma can be the same as lava, the texture of the rocks will be quite different due to different rates of cooling. It is because of this difference in genesis that geologists are able to make the distinction between **extrusive** and **intrusive** igneous rocks when encountered at an outcrop at the Earth's surface.
ilmenite	an **ore** of **titanium**, produced for use as a white pigment in paint.
Inceptisols	a **soil order**; these are **soils** that exhibit only moderate **weathering** and development. They are often found on steep (relatively young) **topography** and overlying **erosion**-resistant bedrock.
index fossil	a **fossil** used to determine the relative age of **sedimentary** deposits. An ideal index fossil lived during a short period of time, was geographically and environmentally widespread, and is easy to identify. Some of the most useful index fossils are hard-shelled organisms that were once part of the marine plankton.
inland basin	a depression located inland from the mountains, and formed by the buckling (**downwarping**) of the Earth's **crust**. Basins naturally preserve thick sediment layers because they accumulate eroded sediment and commonly continue to subside under the weight of the sediment.
inland sea	a shallow sea covering the central area of a continent during periods of high sea level. An inland sea is located on continental **crust**, while other seas are located on oceanic crust. An inland sea may or may not be connected to the ocean. For example, Hudson Bay is on the North American **plate** and connects to the Atlantic and Arctic oceans, while the Caspian Sea is on the European plate but does not drain into any ocean at all.
intensity (earthquake)	a subjective measurement that classifies the amount of shaking and damage done by an **earthquake** in a particular area.
interglacial	a period of geologic time between two successive **glacial** stages.
intermontane	between or among mountains.
intertidal	areas that are above water during low tide and below water during high tide.
intrusion, intrusive rock	a **plutonic igneous rock** formed when **magma** from within the Earth's **crust** escapes into overlying strata. As the magma rises, pushing through overlying layers of rock, it begins to cool. The cooling magma can crystallize and harden to become intrusive igneous rock, locked within layers of older rock.

Glossary

iron	a metallic chemical element (Fe). Iron is most often found in combination with other elements, such as oxygen and **sulfur**, to form ores like **hematite**, **magnetite**, siderite, and **pyrite**. The ready availability of iron at Earth's surface made it one of the earliest mined **mineral** resources in the US.
isostasy	an equilibrium between the weight of the **crust** and the buoyancy of the **mantle**.
jade	a word applied to two green **minerals** that look similar and have similar properties: jadeite (a kind of **pyroxene**) and nephrite (a kind of **amphibole**). Both minerals are formed during **metamorphism** and are found primarily near **subduction** zones, which explains why jade is abundant in a variety of locations along **active plate boundaries**.
jasper	a speckled or patterned **silicate** stone that appears in a wide range of **colors**. It is a variety of **chalcedony**. Jasper forms when silica precipitates in a fine particulate material such as soft sediment or **volcanic ash**. The particulates give the stone its color and patterns.
joint	a surface or plane of **fracture** within a rock.
joule (J)	the **energy** expended (or work done) to apply a force of one newton over a distance of one meter.
Jurassic	the **geologic time** period lasting from 201 to 145 million years ago. During the Jurassic, **dinosaurs** dominated the landscape and the first birds appeared. The Jurassic is the middle period of the **Mesozoic**.
kaolinite	a **silicate clay mineral**, also known as "china clay." Kaolinite is the main ingredient in fine china dishes such as Wedgwood.
karst topography	a kind of landscape defined by bedrock that has been **weathered** by dissolution in water, forming features like sinkholes, caves, and cliffs. Karst **topography** primarily forms in **limestone** bedrock.
kerogen	an immature, waxy, solid organic material, similar to **oil**, that must be artificially heated to convert it into synthetic oil (or a hydrocarbon gas).
kimberlite	a kind of rock formed of **iron-** and magnesium-rich **magma**, mixed with gases such as water vapor and carbon dioxide, from deep in the Earth's **mantle**. As kimberlite magma rises along vertical cracks in the **crust**, the gases expand and may erupt explosively. Cracks filled with kimberlite rock are often known as kimberlite pipes. Kimberlites are of special interest both because they are a source for many of the world's **diamonds** and because they provide information about deep geological processes.
kinetic energy	the **energy** of a body in motion (e.g., via friction).
Köppen system	a commonly used system of **climate** categorization developed by Russian climatologist Wladimir Köppen. It is based on the kinds of vegetation that areas sustain, and defines 12 climate types: rainforest, monsoon, tropical savanna, humid subtropical, humid continental, oceanic, Mediterranean, steppe, subarctic, tundra, polar **ice cap**, and desert. Updated by Rudolf Geiger, it has been refined to five groups each with two to four subgroups.
laccolith	an **intrusive igneous rock** body that forms from **magma** that has, through pressure, been forced between two **sedimentary** layers.
lacustrine	of or associated with lakes.

Glossary

Lagerstätte (pl. Lagerstätten)	**fossil** deposit containing animals or plants that are preserved unusually well, sometimes even including the soft organic tissues. Lagerstätten form in chemical environments that slow decay of organic tissues or enhance preservation through mineralization. Also, quick burial of the organism leaves no opportunity for disturbance of the fossils. Lagerstätten are important for the information they provide about soft-bodied organisms that we otherwise would know nothing about.
lahar	a **pyroclastic debris flow** or mudflow that typically flows down river valleys after a **volcanic** eruption. Lahars can be very destructive, as they can reach thicknesses of over 140 meters (460 feet) and travel at tens of meters (yards) per second.
landslide	the rapid slipping of a mass of earth or rock from a higher elevation to a lower level under the influence of gravity and water lubrication. Landslides include rock falls, avalanches, **debris flows**, mudflows, and the **slumping** of rock layers or sediment. See also: **mass wasting**
Laramide Orogeny	a period of mountain building that began in the late **Cretaceous**, and is responsible for the formation of the Rocky Mountains. See also: **orogeny**
lava	molten rock located on the Earth's surface. When **magma** rises to the surface, typically through a **volcano** or **rift**, it becomes lava. Lava cools much more quickly than magma because it is at the surface, exposed to the **atmosphere** or ocean water where temperatures are much cooler. Such rocks, with little time to crystallize, have small or no crystals.
lava tube	a natural tube formed by **lava** moving beneath the hardened surface of a lava flow.
Law of Superposition	the geologic principle that states that unless rock layers have been overturned or **intruded**, older rocks are found at the bottom and younger rocks are found at the top of a **sedimentary** sequence. See also: **stratigraphy**
lead	a metallic chemical element (Pb). Lead was one of the first metals mined in North America, where it was sought after especially for making shot. It is used in batteries, communication systems, and building construction.
lignite	a soft, brownish black **coal** in which the alteration of plant matter has proceeded farther than in **peat** but not as far as in **bituminous coal**.
lime	an inorganic white or grayish-white compound made by roasting **limestone** (**calcium carbonate**, $CaCO_3$) until all the carbon dioxide (CO_2) is driven off. Originating from limestone, **dolomite**, or **marble**, lime is very important to agriculture, in which it is regularly applied to make **soils** "sweeter" (less acidic).
limestone	a **sedimentary rock** composed of **calcium carbonate** ($CaCO_3$). Most limestones are formed by the deposition and consolidation of the skeletons of marine invertebrates; a few originate in chemical precipitation from solution. Limestone is ordinarily white but can be colored by impurities such as **iron oxide** (making it brown, yellow, or red), or organic carbon (making it blue, black, or gray). The rock's texture varies from coarse to fine.
liquefaction	a process by which water-saturated unconsolidated sediment temporarily loses strength and behaves as a fluid when vibrated.
lithification	the process of creating **sedimentary rock** through the compaction or **cementation** of soft sediment. The word comes from the Greek *lithos*, meaning "rock."

l–m Glossary

lithium	a metallic chemical element (Li) used in the manufacture of ceramics, glass, greases, and batteries.
lithosphere	the outermost layer of the Earth, comprising a rigid **crust** and upper **mantle** broken up into many **plates**. The plates of the lithosphere move with the underlying **asthenosphere**, on average approximately 5 centimeters (2 inches) per year and as much as 18 centimeters (7 inches) per year.
loam	a **soil** containing equal amounts of **clay**, **silt**, and **sand**.
loess	very fine-grained, **wind**-blown sediment, usually **rock flour** left behind by the grinding action of flowing **glaciers**.
luminescence	the emission of light.
luster	a physical property of **minerals**, describing the appearance of the mineral's surface in reflected light, and how brilliant or dull it is. Luster can range from metallic and reflective to opaque, vitreous like glass, translucent, or dull and earthy.
lycopod	an **extinct**, terrestrial **tree** belonging to the plant division Lycopodiophyta, and characterized by a tall, thick trunk covered with a pattern of diamond-shaped leaf scars, and a crown of branches with simple leaves. Lycopods, or "scale trees," grew up to 98 feet (30 meters) high in **Mississippian** and **Pennsylvanian** forests. The plant division Lycopodiophyta survives today but only as very small plants on the forest floor, sometimes called "ground pines."
maar	a shallow **volcanic** crater that forms in a place where trapped water vapor expands explosively when **magma**, rising near the surface, comes into contact with groundwater. Maars are often filled with groundwater, and the area around the maar may be covered by rock made of **volcanic ash** and broken surface rock.
mafic	**igneous rocks** that contain a group of dark-colored **minerals**, with relatively high concentrations of magnesium and **iron** compared to **felsic** igneous rocks.
magma	molten rock located below the surface of the Earth. Magma can cool beneath the surface to form **intrusive igneous rocks**. However, if magma rises to the surface without cooling enough to crystallize, it might break through the crust at the surface to form **lava**.
magnetic	affected by or capable of producing a magnetic field.
magnetite	a **mineral** form of **iron oxide** (Fe_3O_4). It is the most **magnetic** naturally occurring mineral. The molecules in magnetite align with the North and South poles when rocks containing magnetite **ore** are formed. By examining the alignment today, scientists can reconstruct how the rocks have moved since their formation, giving them clues about the previous arrangement of the continents. Magnetite lodestones were used as an early form of compass. Huge deposits of magnetite have been found in **Precambrian banded iron formations**.
magnitude (earthquake)	a logarithmic scale used to measure the seismic energy released by an **earthquake**. Magnitudes follow a numerical scale, with M1 earthquakes classified as micro, M2 earthquakes classed as minor, and earthquakes of M8 or greater being classified as great.
mammoth	an **extinct** terrestrial vertebrate animal belonging to the Order Proboscidea of the Class Mammalia. Mammoths are from the same line of proboscideans that gave rise to African and Asian elephants. They had tall bodies with a rather high "domed" skull, and teeth with numerous parallel rows of ridges. Mammoths are among the most common **Pleistocene** vertebrate **fossils** in North America, Europe, and Asia.
manganese	a metallic chemical element (Mn). Manganese is used in the production of steel.

Glossary m

mantle	the layer of the Earth between the **crust** and core. It consists of solid **silicate** rocks that, over long intervals of time, flow like a highly viscous liquid. Convection currents within the **mantle** drive the motion of **plate tectonics**.
marble	a **metamorphic rock** composed of recrystallized **carbonate minerals**, most commonly **calcite** or **dolomite**. Not everything commercially called a marble is "true marble," which lacks **fossils** and is recrystallized from **limestone** or **dolostone**.
marl	a fine-grained **sedimentary rock** consisting of **clay minerals**, **calcite** and/or aragonite, and **silt**.
mass extinction	the **extinction** of a large percentage of the Earth's species over a relatively short span of **geologic time**. Unfortunately, this is not just a phenomenon of the past: it is estimated that the extinction rate on Earth right now may be as much as 1000 times higher than normal, and that we are currently experiencing a mass extinction event.
mass wasting	a process in which **soil** and rock move down a slope in a large mass. This can occur both on land (such as a **landslide**) or underwater (such as a **turbidity current**).
mastodon	an **extinct** terrestrial vertebrate animal belonging to the Order Proboscidea of the Class Mammalia, and characterized by an elephant-like shape and size, and massive molar teeth with conical projections. Mastodons are among the most common **Pleistocene** vertebrate fossils in North America.
matrix	a fine-grained mass of material around and embedding larger grains or crystals. The term matrix can also describe sediment or rock in which a **fossil** is embedded.
Mesozoic	a **geologic time** period that spans from 252 to 66 million years ago. This period is also called the "age of reptiles" since **dinosaurs** and other reptiles dominated both marine and terrestrial ecosystems. During this time, the last of the Earth's major supercontinents, **Pangaea**, formed and later broke up, producing the Earth's current geography. The Mesozoic contains the **Triassic**, **Jurassic**, and **Cretaceous** periods.
metamorphism, metamorphic rocks	rocks formed by the recrystallization and realignment of **minerals** in pre-existing **sedimentary**, **igneous**, and metamorphic rocks when exposed to high enough temperature and/or pressure. This can be a result of **plate** movements, very deep burial, or contact with molten rock or superheated water. This process destroys many features in the rock that would have revealed its previous history, transforming it into an entirely new form. Tectonic forces can cause minerals to realign perpendicularly to the direction of pressure, layering them in a pattern called **foliation**, as exemplified in **gneiss** and **schist. Recrystallization**, as seen in **marble** and **quartzite**, results as rock is heated to high temperatures, and individual grains reform as interlocking crystals, making the resulting metamorphic rock harder than its parent rock.
meteorite	a stony or metallic mass of matter that has fallen to the Earth's surface from outer space.
mica	a large group of sheet-like **silicate minerals**.
microcontinent	a piece of continental **crust**, usually **rifted** away from a larger continent. Microcontinents and other smaller fragments of continental crust (**terranes**) each had their own, often complex, geologic history before they were tacked onto the margin of another continent.
Milankovitch Cycles	cyclical changes in the amount of **heat** received from the Sun, associated with how the Earth's orbit, tilt, and wobble alter its position with respect to the Sun. These changes affect the global **climate**, most notably alterations of **glacial** and **interglacial** intervals.

m–n Glossary

mineral	a naturally occurring inorganic solid with a specific chemical composition and a well-developed crystalline structure. Minerals are identified based on their physical properties, including **hardness**, **luster**, **color**, **crystal form**, **cleavage**, **density**, and **streak**. There are over 4900 identified minerals. However, the number of common rock-forming minerals is much smaller. The most common minerals that form **igneous**, **metamorphic**, and **sedimentary rocks** include **quartz**, **feldspars**, **micas**, **pyroxenes**, and **amphiboles**.
mineralogy	the branch of geology that includes study of the chemical and physical properties and formation of **minerals**.
minette	an unusual, **mafic**, fine-grained **igneous rock** made mostly of the potassium-containing **minerals** biotite **mica** and orthoclase (K-**feldspar**). It is usually found in narrow **dikes** and sheets associated with the injection of **lava**.
Miocene	a **geologic time** unit extending from 23 to 5 million years ago. During the Miocene, the Earth experienced a series of **ice ages**, and hominid species diversified. The Miocene is the first epoch of the **Neogene** period.
Mississippian	a subperiod of the **Carboniferous**, spanning from 359 to 323 million years ago.
Mohs Scale of Hardness	the scale of relative **hardness** of **minerals**, developed by the Austrian mineralogist Frederich Mohs in 1824. The scale is very useful as a means for identifying minerals or quickly determining hardness. A piece of glass has a hardness of approximately 5 on the scale; our fingernails are just over 2; a knife blade is just over 5. **Diamond** ranks at 10 as the hardest mineral.
Mollisols	a **soil order**; these are agricultural **soils** made highly productive due to a very fertile, organic-rich surface layer.
molybdenum	a metallic chemical element (Mo) which has the sixth-highest melting point of any element at 2623°C (4753°F). Molybdenum is mainly used in the creation of alloys, such as stainless steel and cast **iron**, and its strong ability to withstand **heat** makes it useful in applications that utilize extreme heat, such as the manufacture of motors and aircraft parts.
moraine	an accumulation of unconsolidated **glacial** debris (**soil** and rock) that can occur in currently glaciated and formerly glaciated regions, such as those areas acted upon by a past **ice age**. The debris is scraped from the ground and pushed forward by the glacier, to be left behind when the ice melts. Thus, many moraines mark the terminus or edge of a glacier. Lateral moraines can also occur in between and at the sides of glaciers or **ice lobes**.
mosasaur	an **extinct**, carnivorous, marine vertebrate reptile. Mosasaurs were characterized by a streamlined body for swimming, a powerful fluked tail, and reduced, paddle-like limbs. They were common in **Cretaceous** seas and were were powerful swimmers, reaching 12–18 meters (40–59 feet) in length.
multituberculate	a member of a group of rodent-like animals with a long evolutionary history, from the **Jurassic** period until at least the **Oligocene** epoch. They were the most dominant mammalian group in the **Cretaceous** and **Paleocene**.
natural gas	a hydrocarbon gas mixture composed primarily of methane (CH_4), but also small quantites of hydrocarbons such as ethane and propane. See also: **fossil fuel**
natural hazards	events that result from natural processes and that have significant impacts on human beings.

Neogene	the **geologic time** period extending from 23 to 2.6 million years ago. During the Neogene, global **climate** cooled, the continents moved close to their current positions, mammals and birds continued to evolve, and the first hominins appeared. The Neogene is a portion of the **Cenozoic**.
nickel	a ductile, silvery-white metallic element (Ni). Nickel in its pure form is rarely found on Earth's surface; large quantites of nickel are typically found in **meteorites**. On Earth, nickel is generally found in combination with **iron**. Nickel is resistant to corrosion and is commonly used to plate metals, coat chemistry equipment, and manufacture alloys such as electrum.
nodule	a small, irregular or rounded **mineral** deposit that has a different composition from the **sedimentary rock** that encloses it. Nodules typically form when minerals precipitate from a supersaturated solution within or around features such as **biotic** remains.
nuclear	pertaining to a reaction, as in fission, fusion, or **radioactive** decay, that alters the energy, composition, or structure of an atomic nucleus.
obsidian	a **glassy volcanic** rock, formed when **felsic lava** cools rapidly. Although obsidian is dark in color, it is composed mainly of **silicon** dioxide (SiO_2), and its dark color is a result of impurities such as **iron** and magnesium. Obsidian is extremely brittle and breaks with very sharp edges. It was valuable to Stone Age cultures for its use as cutting implements or arrowheads.
oil	See **petroleum**
Oligocene	a **geologic time** interval spanning from about 34 to 23 million years ago. It is an epoch of the **Paleogene**.
olivine	an **iron**-magnesium **silicate mineral** ($(Mg,Fe)_2SiO_4$) that is a common constituent of magnesium-rich, silica-poor **igneous rocks**.
opal	a **silicate gemstone** lacking a rigid crystalline structure (and therefore a "mineraloid" as opposed to a **mineral**). It forms when silica-rich water precipitates in fissures of almost any type of rock, as well as occasional organic matter.
Ordovician	a **geologic time** period spanning from 485 to 443 million years ago. During the Ordovician, invertebrates dominated the oceans and fish began to diversify. The Ordovician is part of the **Paleozoic**.
ore	a type of rock that contains **minerals** with valuable elements, including metals, that are economically viable to extract.
orogeny	a mountain-building event generally caused by colliding **plates** and **compression** of the edge of the continents. Orogeny is derived from the Greek word *oro*, meaning "mountain."
orographic precipitation	rainfall caused when **wind** pushes a mass of humid air up the side of an elevated land formation like a mountain. As the air rises, it cools, and the moisture precipitates out.
outwash plain	large **sandy** flats created by sediment-laden water deposited when a **glacier** melts. Outwash sediments are also called fluvial material.
oxidation, oxide	a **chemical reaction** involving the loss of at least one electron when two substances interact; most often used to describe the interaction between oxygen molecules and the substances they come into contact with. Oxidation causes effects such as rust and cut apples turning brown.

o–p Glossary

Oxisols	a **soil order**; these are very old, extremely leached and **weathered soils** with a subsurface accumulation of **iron** and **aluminum oxides**. They are commonly found in humid, tropical environments.
Paleocene	a **geologic time** interval spanning from about 66 to 56 million years ago. It is an epoch of the **Paleogene** period.
paleoecology	the study of the relationships of **fossil** organisms to one another and their environment.
Paleogene	the **geologic time** interval extending from 66 to 23 million years ago. During the Paleogene, mammals and birds diversified into many of the niches that had previously been held by **dinosaurs**. The Paleogene is the first part of the **Cenozoic**.
Paleozoic	a **geologic time** period that extends from 541 to 252 million years ago. **Fossil** evidence shows that during this time period, life evolved in the oceans and gradually colonized the land. The Paleozoic includes the **Cambrian**, **Ordovician**, **Silurian**, **Devonian**, **Carboniferous**, and **Permian** periods.
Pangaea	a supercontinent, meaning "all Earth," which formed over 300 million years ago and lasted for almost 150 million years, during which all of the Earth's continents were joined in a giant supercontinent. Pangaea eventually **rifted** apart and separated into the continents in their current configuration.
parent material	the original geologic material from which **soil** formed. This can be bedrock, pre-existing soils, or other transported sediment such as **till** or **loess**.
passive margin	a tectonically quiet continental edge, such as the eastern margin of North America, where **crustal** collision or **rifting** is not occurring.
peat	an accumulation of partially decayed plant matter. Under proper **heat** and pressure, it will turn into **lignite coal** over geologic periods of time. As much as 9 meters (30 feet) of peat might need to accumulate to produce an economically profitable coal seam. By the time that a peat bed has been turned into a layer of **anthracite**, the layer is one-tenth its original thickness.
peds	clumps of **soil**, identified by their shape, which may take the form of balls, blocks, columns, and plates. These structures are easiest to see in recently plowed fields, where the soil is often granular and loose or lumpy.
pegmatite	a very coarse-grained **igneous rock** that formed below the surface, usually rich in **quartz**, **feldspars**, and **micas**. Pegmatite **magmas** are very rich in water, carbon dioxide, **silicon**, **aluminum**, and potassium, and form as the last fluids to crystallize from magma or the first **minerals** to melt at high temperatures during **metamorphism**.
Pennsylvanian	a subperiod of the **Carboniferous**, spanning from 323 to 299 million years ago.
perennial	continuous; year-round or occurring on a yearly basis.
periglacial zone	a region directly next to an **ice sheet**, which, although it was never covered or **scoured** by ice, has its own distinctive landscape and features because it was next to the ice margin. The average annual air temperature in a periglacial area is between −12° and 3°C (10° and 37°F). Though the surface of the ground may melt in the summer, it refreezes in the winter.

permafrost	a layer of **soil** below the surface that remains frozen all year round. Its thickness can range from tens of centimeters (inches) to a few meters (yards). Permafrost is typically defined as any soil that has remained at a temperature below the freezing point of water for at least two years.
permeable, permeability	a capacity for fluids and gas (such as water, oil and natural gas) to move through **fractures** within a rock, or the spaces between its grains. **Sandstone**, **limestone**, and fractured rocks of any kind generally are permeable. **Shale**, on the other hand, is usually impermeable because the small, flat **clay** particles that make up the rock are tightly packed into a **dense** rock with very little space between particles. Poorly sorted **sedimentary rocks** can also be impermeable because smaller grains fill in the spaces between the bigger grains, restricting the movement of fluids.
Permian	the **geologic time** period lasting from 299 to 252 million years ago. During the Permian, the world's landmass was combined into the supercontinent **Pangaea**. The Permian is the last period of the **Paleozoic**. It ended with the largest **mass extinction** in Earth's history, which wiped out 70% of terrestrial animal species and 90% of all marine animal species.
permineralization	a **fossilization** method in which empty spaces (such as in a bone or shell) are filled by **minerals**.
petroleum	a naturally occurring, flammable liquid found in geologic formations beneath the Earth's surface and consisting primarily of hydrocarbons. Petroleum, also called oil, is a **fossil fuel**, formed when large masses of dead organisms (usually algae or plankton) are buried underneath sediments and subjected to intense **heat** and pressure. Today, petroleum is used to manufacture a wide variety of materials, and it is commonly refined into various types of fuels. It is estimated that 90 million barrels are consumed globally every day.
Phanerozoic	a generalized term used to describe the entirety of geological history after the **Precambrian**, from 541 million years ago to the present.
phenocryst	a large and generally conspicuous crystal which has been enclosed in a much finer-grained **igneous rock**. Phenocrysts may occur in all types of igneous rock, but are most common in **felsic** rocks.
phosphate	an inorganic **salt** of phosphoric acid, and a nutrient vital to biological life.
physiography	a subfield of geography that studies the Earth's physical processes and patterns, including consideration of the shape (not just the height) of land forms, as well as the bedrock, **soil**, water, vegetation, and **climate** of an area, and how they interacted in the past to form the landscape we see today.
phytosaur	an **extinct** reptile from the late **Triassic** period. Phytosaurs were semi-aquatic relatives of the crocodile with heavily armored bodies. Their **fossils** have been found in North America, Europe, and India.
pillow basalt	**basaltic lava** that forms in a characteristic "pillow" shape due to its **extrusion** underwater.
placer deposit	a **mineral** deposit occurring in rivers and streams where less **dense** sediment has been carried downstream but denser minerals such as **gold** have been left behind.
placoderms	an **extinct** class of heavily armored fishes. Placoderms lived from the **Silurian** to the **Devonian**.
plate tectonics	the process by which Earth's tectonic **plates** move and interact with one another at their boundaries. The Earth is dynamic, consisting of constantly moving plates that are made of rigid continental and oceanic **lithosphere** overlying a churning, plastically flowing **asthenosphere**. These plates are slowly pulling apart, colliding, or sliding past one another with great force, creating strings of **volcanic islands**, new ocean floor, **earthquakes**, and mountains.

p Glossary

plates	large, rigid pieces of the Earth's **crust** and upper **mantle**, which move and interact with one another at their boundaries. See also: **plate tectonics**
playa lakes	ephemeral or dry lakebeds that occasionally contain only a thin layer of quickly evaporating water. Soluble **minerals** such as **halite**, **gypsum**, and **calcite** precipitate from evaporating playa lakes, leaving behind rock **salt**, **gypsum**, and **limestone**.
Pleistocene	a subset of the **Quaternary**, lasting from 2.5 million to about 11,700 years ago. During the Pleistocene, continental **ice sheets** advanced south and retreated north several dozen times.
plesiosaur	a member of a group of **extinct**, long-necked, **Mesozoic** marine reptiles.
Pliocene	a **geologic time** interval extending from roughly 5 to 2.5 million years ago. The Pliocene epoch is a subdivision of the **Neogene** period, and is the time period directly preceding the onset of **Pleistocene glaciations**.
pluton, plutonic rock	a body of **intrusive igneous rock** that formed under the Earth's surface through the slow crystallization of **magma**. The term comes from the name of Pluto, Roman god of the underworld.
pluvial lake	a landlocked basin that fills with rainwater or meltwater during times of **glaciation**.
porosity	the percentage of openings in a body of rock such as pores, **joints**, channels, and other cavities, in which gases or liquids may be trapped or migrate through.
porphyry, porphyritic	an **igneous rock** consisting of large grained crystals, or **phenocrysts**, embedded in a fine-grained **matrix**.
potash	a name used for a variety of **salts** containing potassium, with mined potash being primarily potassium chloride (KCl). The majority of potash is used as fertilizer, but an increasing amount is being used in a variety of other ways: water softening, snow melting, a variety of industrial processes, as a medicine, and to produce potassium **carbonate** (K_2CO_3).
power (energy)	the rate at which **energy** is transferred, usually measured in **watts** or, less frequently, horsepower.
Precambrian	a **geologic time** interval that spans from the formation of Earth (4.6 billion years ago) to the beginning of the **Cambrian** (541 million years ago). Relatively little is known about this time period since very few **fossils** or unaltered rocks have survived. What few clues exist indicate that life first appeared on the planet as long as 3.9 billion years ago in the form of single-celled organisms. The Precambrian contains the Hadean, **Archean**, and **Proterozoic** eons.
primary energy source	a source of **energy** found in nature that has not been subject to any human-induced energy transfers or transformations (like conversion to electricity). Examples include **fossil fuels**, solar, **wind**, and hydropower.
Proterozoic	a **geologic time** interval that extends from 2.5 billion to 541 million years ago. It is part of the **Precambrian**. During this eon, the Earth transitioned to an oxygenated **atmosphere** and **eukaryotic** cells, including fungi, plants, and animals, originated.
protists	a diverse group of single-celled **eukaryotes**.
protolith	the original parent rock from which a **metamorphosed** rock is formed.

Glossary p–r

pteraspidomorph	a member of a group of jawless fish with extensive head armor that lived mostly in coastal marine, and possibly freshwater, bottom environments. Though originally considered a possible ancestor to modern fish, they are now considered more likely to share a common ancestor with (but not be an ancestor to) armored jawed fish and all other jawed vertebrates. Pteraspidomorphs range from the **Ordovician** to **Devonian** periods.
pterosaurs	**extinct** flying reptiles with wingspans of up to 15 meters (49 feet). They lived during the same time as the **dinosaurs**.
pumice	a **pyroclastic** rock that forms as frothing and sputtering **magmatic** foam cools and solidifies. It is so **vesicular** that it can float. Pumice is a common product of explosive eruptions. Today it is used in a variety of mediums, including construction materials and abrasives.
pyrite	an **iron** sulfide **mineral** (FeS_2). Pyrite's superficial resemblance to **gold** has led to the common nickname "fool's gold."
pyroclastic rocks	rocks that form during explosive **volcanic** eruptions, and are composed from a variety of different volcanic ejecta. The term comes from Greek, and means "broken fire." Pyroclastic debris of all types is known as **tephra**.
pyroxene	dark-colored, rock-forming **silicate minerals** containing **iron** and magnesium, found in many **igneous** and **metamorphic rocks**. They are often present in **volcanic** rocks.
quartz	the second most abundant **mineral** in the Earth's continental **crust** (after the **feldspars**), made up of silicon and oxygen (SiO_2). It makes up more than 10% of the **crust** by mass. There are a wide variety of types of quartz: onyx, **agate**, and petrified wood are fibrous, microcrystalline varieties collectively known as **chalcedony**. Although agate is naturally banded with layers of different **colors** and porosity, commercial varieties of agate are often artificially colored. **Flint**, **chert**, and **jasper** are granular microcrystalline varieties of quartz, with the bright red color of jasper due to the inclusion of small amounts of **iron** within the mineral structure. The most common, coarsely crystalline varieties include massive quartz veins, the distinct, well formed, crystals of "rock crystal," and an array of colored quartz, including amethyst (purple), rose quartz (pink), smoky quartz (gray), citrine (orange), and milky quartz (white).
quartzite	a hard **metamorphic rock** that was originally **sandstone**. Quartzite usually forms from sandstone that was metamorphosed through tectonic **compression** within **orogenic** belts. Quartzite is quarried for use as a building and decorative stone.
Quaternary	a **geologic time** period that extends from 2.6 million years ago to the present. This period is largely defined by the periodic advance and retreat of continental **glaciers**. The Quaternary is part of the **Cenozoic**.
radioactivity	the emission of radiation (**energy**) by an unstable atom.
radon	a naturally occurring **radioactive**, colorless, odorless gas. It is one of the products of decay from the breakdown of radioactive elements in **soil**, rock, and water, released by **weathering**.
rare earth elements	a set of 17 heavy, **lustrous** elements with similar properties, some of which have technological applications. Although they are relatively common in the **crust**, these metals are not usually found concentrated in economically viable **ore** deposits.

r Glossary

recrystallization	the change in structure of **mineral** crystals that make up rocks, or the formation of new mineral crystals within the rock. Recrystallization commonly occurs during **metamorphism**. When rocks are metamorphosed, individual grains that make up the original rock are melted slightly and recrystallize. The pressure allows crystals to grow into a tighter, interlocking arrangement than in an unmetamorphosed rock.
recurrence interval	the time elapsed between major events, such as floods.
reef	a feature lying beneath the surface of the water, which is a build-up of sediment or other material built by organisms, and which has positive **relief** from the sea floor. While some reefs result from abiotic processes such as deposition or wave action, the best-known reefs are built by corals and other marine organisms.
regional metamorphism	a **metamorphic rock** that has been altered due to deep burial and great pressure. This type of metamorphic rock tends to occur in long belts at the center of mountain ranges. Different types of metamorphic rock are created depending on the gradients of **heat** and pressure applied.
regression	a drop in sea level.
relief (topography)	the change in elevation over a distance.
renewable energy, renewable resource	**energy** obtained from sources that are virtually inexhaustible (defined in terms of comparison to the lifetime of the Sun) and replenish naturally over small time scales relative to human life spans.
replacement	a **fossilization** method by which the original material is chemically replaced by a more stable **mineral**.
residual weathering deposit	a **mineral** deposit formed through the concentration of a **weathering**-resistant mineral, in which the other minerals around it have been **eroded** away.
rhyolite, rhyolitic	a **felsic volcanic** rock high in abundance of **quartz** and **feldspar**.
rift	a break or crack in the **crust** that can be caused by tensional stress as a landmass breaks apart into separate **plates**.
rift basin	a **topographic** depression caused by **subsidence** within a **rift**; the basin, since it is at a relatively low evelation, usually contains freshwater bodies such as rivers and lakes.
ripple marks	surface features created when sediment deposits are agitated, typically by water currents or **wind**. The crests and troughs formed by this agitation are occasionally **lithified** and preserved, providing information about the flow of water or wind in the paleoenvironment.
rock flour	very fine sediments and **clay** resulting from the grinding action of **glaciers**.
Rodinia	a supercontinent that contained most or all of Earth's landmass, between 1.1 billion and 750 million years ago, during the **Precambrian**. Geologists are not sure of the exact size and shape of Rodinia. It was analagous to but not the same supercontinent as **Pangaea**, which formed was assembled several hundred million years later during the **Permian**.
rudist	an **extinct** group of box- or tube-shaped **bivalves** that arose during the **Jurassic**. They were major **reef** formers, but went extinct at the end of the **Cretaceous**.
rugose coral	an **extinct** group of corals that were prevalent from the **Ordovician** through the **Permian**. Solitary forms were most common; these were horn-shaped, leading to their common name, "horn corals."

Glossary S

salt	a **mineral** composed primarily of sodium chloride (NaCl). In its natural form, it is called "rock salt" or **halite**. Salt is essential for animal life, and is a necessary part of the diet. In addition, salt is used for de-icing roads in winter and is also an important part of the chemical industry.
salt dome	a largely subsurface geologic structure, consisting of a vertical cylinder of **salt** embedded in horizontal or inclined **sedimentary** strata. Salt buried under thousands of feet of overlying sediment often deforms plastically. Because it is less **dense** than the rocks above it, it flows upward toward areas of lower pressure, forming geological structures named for their shapes (e.g., domes, canopies, tables, and lenses).
sand	rock material in the form of loose, rounded, or angular grains, and formed as a result of the weathering and decomposition of rocks. Particles of sand are between 0.05 and 2 millimeters (0.00016 and 0.0065 inches) in diameter, and are most commonly composed of **quartz** and **feldspar**.
sandstone	**sedimentary rock** formed by cementing together grains of **sand**.
sauropod	a member of an evolutionary group of herbivorous **dinosaurs** characterized by very large body size, and long necks and tails. They appeared in the late **Triassic**, were among the dominant dinosaur herbivores of the **Jurassic**, and survived until the **extinction** of (non-avian) dinosaurs at the end of the **Cretaceous**. Genera well-known to the general public include *Brontosaurus/Apatosaurus*, *Brachiosaurus*, and *Diplodocus*, though many dozens of others existed. Sauropods are by far the largest terrestrial animals to have existed.
schist	a medium-grade **metamorphic rock** with sheet-like crystals flattened in one plane. The flattened crystals are often muscovite or biotite **mica**, but they can also be **talc**, **graphite**, or **hornblende**.
scleractinian coral	a modern "stony" coral; a colonial or solitary marine invertebrate animal belonging to the Order Scleractinia in the Class Anthozoa of the Phylum Cnidaria, and characterized by an encrusting calcareous skeleton from which multiple individuals (polyps) extend from small pores to capture prey with small tentacles equipped with stinging cells (nematocysts). Although scleractinians look somewhat similar to extinct **rugose** and **tabulate corals**, each group possesses distinctive features in the shape of the skeletal cup holding the individual polyps. Modern scleractinians host commensal algae (zooxanthellae) whose photosynthetic activities supply the coral with energy.
scour, scouring	**erosion** resulting from **glacial** abrasion on the landscape.
sedimentary rocks	rocks formed through the accumulation and consolidation of grains of broken rock, crystals, skeletal fragments, and organic matter. Sediment that forms from **weathering** is transported by **wind** or water to a depositional environment such as a lakebed or ocean floor; here they build up, burying and compacting lower layers. As water permeates the sediment, dissolved **minerals** may precipitate out, filling the spaces between particles and **cementing** them together. Sedimentary rocks may also accrete from fragments of the shells or skeletal material of marine organisms such as clams and coral. Sedimentary rocks are classified by their sediment size or their mineral content. Each one reveals the story of the depositional environment where its sediments accumulated and the history of its **lithification**.
seed fern	an **extinct** terrestrial plant belonging to the plant division Pteridospermatophyta, and characterized by a fern-like appearance, but bearing seeds instead of spores. Seed ferns lived from the **Mississippian** to the **Jurassic**.
seismic belt	a narrow geographic zone along which most **earthquakes** occur.

s

Glossary

seismic waves	the shock waves or vibrations radiating in all directions from the center of an **earthquake** or other tectonic event.
seismic zone	a regional zone that encompasses areas prone to seismic hazards, such as **earthquakes** or **landslides**.
sessile	unable to move, as in an organism that is permanently attached to its substrate.
Sevier Orogeny	a mountain-building event resulting from **subduction** along the western edge of North America, occurring mainly during the **Cretaceous**. During this **orogeny**, **compressive** forces and **heating** resulted in major **crustal** folding and thrust **faulting**.
shale	a dark, fine-grained, laminated **sedimentary rock** formed by the **compression** of successive layers of **silt**- and **clay**-rich sediment. Shale is weak and often breaks along thin layers. Shale that is especially rich in unoxidized carbon is dark grey or black. These organic-rich black shales are often source rocks for **petroleum** and **natural gas**.
shark	a large fish characterized by a cartilaginous skeleton and five to seven gill slits on the side of the head. Sharks first appeared 420 million years ago, and have since diversified to over 470 species.
shearing, shear	the process by which **compressive** stress causes the **fracturing** and **faulting** of brittle rocks.
shield	See **craton**
shield volcano	a **volcano** with a low profile and gradual slope, so named for its likeness to the profile of an ancient warrior's shield. Shield volcanoes erupt low-viscosity **magma** that is more fluid than the "sticky" **silica**-rich **lavas** that build up **stratovolcanoes**. Repeated eruptions of fluid lava build large, gently sloping mountains with an expansive size.
silica, silicon, silicate	a chemical compound also known as silicon dioxide (SiO_2). Silica is most commonly found as **quartz**, and is also secreted as skeletal material in various organisms. It is one of the most abundant materials in the **crust**.
silt	granular sediment most commonly composed of **quartz** and **feldspar** crystals. Particles of silt have diameters of less than 0.074 millimeters.
Silurian	a **geologic time** period spanning from 443 to 419 million years ago. During the Silurian, jawed and bony fish diversified, and life first began to appear on land. The Silurian is part of the **Paleozoic**.
silver	a metallic chemical element (Ag). Silver is used in photographic film emulsions, utensils and other tableware, and electronic equipment.
slate	a fine-grained, **foliated metamorphic rock** derived from a **shale** composed of **volcanic ash** or **clay**.
slump	a slow-moving **landslide** in which loosely consolidated rock or **soil** layers move a short distance down a slope. See also: **mass wasting**
snail	See **gastropod**

soapstone	a **metamorphic schistose rock** composed mostly of **talc**. Soapstone has a flaky texture and a greasy or soapy feel, and is an effective medium for carving.
soil	the collection of natural materials that collect on Earth's surface above the bedrock. Soil consists of layers (**horizons**) of two key ingredients: plant litter, such as dead grasses, leaves, and fallen debris, and sediment derived from the **weathering** of rock. Each of these components can influence the texture and consistency of the soil, as well as the **minerals** available for consumption by plants. The word is derived from the Latin *solum*, which means "floor" or "ground."
soil orders	the twelve major units of **soil taxonomy**, which are defined by diagnostic **horizons**, composition, **soil** structures, and other characteristics. Soil orders depend mainly on **climate** and the organisms within the soil. These orders are further broken down into 64 suborders based on properties that influence soil development and plant growth, with the most important property being how wet the soil is throughout the year.
soil taxonomy	the system used to classify **soils** based on their properties.
solution mining	the extraction of soluble **minerals** from subsurface strata by the injection of fluids, and the controlled removal of mineral-laden solutions.
speleothem	an often delicate **mineral** deposit in **limestone** or **dolostone** caves, formed through the dissolution of **carbonate** minerals.
sphenopsid	a terrestrial plant belonging to the Family Equisetaceae in the plant division Pteridophyta, and characterized by hollow, jointed stems with reduced, unbranched leaves at the nodes. Sphenopsids, or horsetails, reached over 33 feet (10 meters) high during the **Pennsylvanian**.
Spodosols	a **soil order**; these are acidic **soils** in which **aluminum** and **iron oxides** accumulate below the surface. They typically form under pine vegetation and sandy parent material.
sponge	a marine invertebrate belonging to the Phylum Porifera, and characterized by a soft shape with many pores and channels for water flow. Because they have no nervous, digestive, or circulatory systems, some consider them to be colonies of specialized single cells. Sponges come in a variety of shapes and body forms, and have been around at least since the **Cambrian**. Entire sponges are rarely preserved, but their tiny skeletal pieces (spicules) are common in **sedimentary rocks**. See also: **archaeocyathid**
stratiform deposit	a **mineral** deposit that forms when a **hydrothermal solution** moves through a **sedimentary** deposit such as an **aquifer**, and deposits minerals parallel to the sedimentary layers. Stratiform deposits may form in a variety of geological environments, for example, in association with **volcanism**.
stratigraphy, stratigraphic	the branch of geology specifically concerned with the arrangement and age of rock units. See also: **Law of Superposition**
stratovolcano	a conical **volcano** made up of many **lava** flows as well as layers of **volcanic ash** and **breccia** from explosive eruptions. Stratovolcanoes are often characterized by their periodic violent eruptions, which occur due to their presence at **subduction** zones. While young stratovolcanoes tend to have steep cone shapes, the symmetrical conical shape is readily disfigured by massive eruptions. Many older stratovolcanoes contain collapsed craters called **calderas**.
streak	a physical property of **minerals**, obtained by dragging the mineral across a porcelain plate and effectively powdering it. During identification, the color of the powder eliminates the conflating variables of external **weathering**, **crystal form**, or impurities.

Glossary

stromatolite	regularly banded accumulations of sediment created by the trapping and **cementation** of sediment grains in bacterial mats (especially photosynthetic **cyanobacteria**). Cyanobacteria emit a sticky substance that binds settling **clay** grains and creates a chemical environment leading to the precipitation of **calcium carbonate**. The calcium carbonate then hardens the underlying layers of bacterial mats, while the living bacteria move upward so that they are not buried. Over time, this cycle of growth combined with sediment capture creates a rounded structure filled with banded layers. Stromatolites peaked in abundance around 1.25 billion years ago, and likely declined due to the evolution of grazing organisms. Today, stromatolites exist in only a few locations worldwide, such as Shark Bay, Australia. Modern stromatolites form thick layers only in stressful environments, such as very **salty** water, that exclude animal grazers. Even though there are still modern stromatolites, the term is often used to refer specifically to **fossils**.
subduction	the process by which one **plate** moves under another, sinking into the **mantle**. This usually occurs at **convergent plate boundaries**. **Denser** plates are more likely to subduct under more buoyant plates, as when oceanic **crust** sinks beneath continental crust.
subsidence	the sinking of an area of the land surface.
subsoil	the layer of **soil** beneath the **topsoil**, composed of **sand**, **silt**, and/or **clay**. Subsoil lacks the organic matter and **humus** content of topsoil.
sulfur, sulfate	a bright yellow chemical element (S) that is essential to life. It acts as an **oxidizing** or reducing agent, and occurs commonly in raw form as well as in **minerals**.
supervolcano	an explosive **volcano** capable of producing more than 1000 cubic kilometers (240 cubic miles) of ejecta.
sustainable	able to be maintained at a steady level without exhausting natural resources or causing severe ecological damage, as in a behavior or practice.
suture	the area where two continental **plates** have joined together through continental collision. See also: **convergent boundary**, **plate tectonics**
syenite	a durable, coarse-grained **intrusive igneous rock**, which is similar to granite but contains less **quartz**. It can exhibit **columnar jointing**.
synapsid	a group of **tetrapod** vertebrates posessing one opening in the skull behind each orbit (eye hole), and a bony arch beneath. All mammals are synapsids.
system	a set of connected things or parts forming a complex whole; in particular, a set of things working together as parts of a mechanism or an interconnecting network.
tabulate coral	an **extinct** form of colonial coral that often formed honeycomb-shaped colonies of hexagonal cells.
Taconic Orogeny	a late **Ordovician** mountain-building event involving the collision and accretion of a **volcanic island** arc along the eastern coast of North America, from New England to eastern Canada. Sediments **eroded** from the resulting mountains accumulated in thick strata, the Queenston Delta, in the **Appalachian Basin** from New York to Quebec. See also: **orogeny**
talc	hydrated magnesium silicate, formed during **hydrothermal** alteration accompanying **metamorphism**. Talc can be formed from **calcite**, **dolomite**, **silica**, and some **ultramafic** rocks.
talus	debris fields found on the sides of steep slopes, common in **periglacial** environments.

Glossary t

tephra	fragmented material produced by a **volcanic** eruption. Airborne tephra fragments are called **pyroclastic**.
terrace	a flat or gently sloped embankment or ridge occurring on a hillside, and often along the margin of (or slightly above) a body of water, representing a previous water level.
terrane	a piece of **crustal** material that has broken off from its parent continent and become attached to another **plate**. Due to their disparate origins, terranes have distinctly different geologic characteristics than the surrounding rocks. Florida is a good example of an exotic terrane, originating as part of the supercontinent **Gondwana**. Parts of the western coast of North America (including Alaska and the Northeastern US) are also terranes that have been **sutured** onto the coast.
Tertiary	an unoffical but still commonly used term for the time period spanning from 66 million to 2.5 million years ago, including the **Paleogene**, **Neogene**, and part of the **Pleistocene**. Although the Tertiary period was officially phased out in 2008 by the International Commission on Stratigraphy, it can still be found in scientific literature. (In contrast, the **Carboniferous** and **Pennsylvanian** and **Mississippian** periods all enjoy official status, with the latter pair being more commonly used in the US.)
tetrapod	the first four-limbed animals (early land vertebrates) and all of their descendants, including all amphibians, reptile groups (including birds), and **synapsids** (including mammals). Although most tetrapods today have four limbs, some, such as snakes and whales, have secondarily lost limbs.
theropod	an evolutionary group of bipedal **dinosaurs**, including all of the carnivorous dinosaurs. Birds, as a lineage of small theropods, are considered to be theropod dinosaurs.
thorium	a **radioactive** rare earth element, with potential applications in next-generation **nuclear** reactors that could be safer and more environmentally friendly than current uranium reactors.
till	unconsolidated sediment that is **eroded** from the bedrock, then carried and eventually deposited by **glaciers** as they recede. Till may include a mixture of **clay**, **sand**, **gravel**, and even boulders. The term originated with farmers living in glaciated areas who were constantly removing rocks from their fields while breaking the **soil** for planting, a process known as tilling.
tillite	**glacial till** that has been compacted and **lithified** into solid rock.
titanium	a metallic chemical element (Ti). Titanium is important because of its lightweight nature, strength, and resistance to corrosion.
topographic inversion	a landscape with features that have reversed their elevation relative to other features, most often occurring when low areas become filled with **lava** or sediment that hardens into material more resistant to **erosion** than the material that surrounds it.
topography	the landscape of an area, including the presence or absence of hills and the slopes between high and low areas. These changes in elevation over a particular area are generally the result of a combination of deposition, **erosion**, **uplift**, and subsidence. These processes that can happen over an enormous range of timescales.
topsoil	the surface or upper layer of **soil**, as distinct from the subsoil, and usually containing organic matter.
tornado	a vertical funnel-shaped storm with a visible horizontal rotation. The word tornado has its roots in the Spanish word *tonar*, which means "to turn."
trace fossils	**fossils** that record the actions of organisms, such as footprints, trails, trackways, and burrows. Trace fossils cannot always be associated at least with a group of organisms or way of life. The first trace fossils appear a couple hundred million years before the first animal (**body**) fossils.

Glossary

trachyte	a fine-grained **extrusive igneous rock**, with a composition high in alkali **feldspar**.
trackway	a set of impressions in soft sediment, usually a set of footprints, left by an animal. Trackways preserved as **fossils** are known as **trace fossils**.
transform boundary	an **active plate boundary** in which the **lithospheric plates** move sideways past one another.
transgression	a relative rise in sea level in a particular area, through global sea level rise or **subsidence** of land.
tree	any woody **perennial** plant with a central trunk. Not all trees are closely related; different kinds of plants have evolved the tree form through geological time. The trees of the **Paleozoic** were more closely related to club mosses or ferns than they were to today's trees.
Triassic	a **geologic time** period that spans from 252 to 201 million years ago. During this period, **dinosaurs**, **pterosaurs**, and the first mammals appear and begin to diversify. The Triassic began directly after the **Permian**-Triassic **mass extinction** event, and is the first period of the **Mesozoic**.
trilobite	an **extinct** marine invertebrate animal belonging to the Class Trilobita of the Phylum Arthropoda, and characterized by a three-part body and a chitinous exoskeleton divided longitudinally into three lobes. Trilobites have been extinct since the end of the **Paleozoic**. Trilobites were primitive **arthropods** distantly related to horseshoe crabs. As bottom dwellers, they were present in a variety of environments. Like crabs and lobsters, trilobites molted their exoskeletons when they grew. Most **fossils** of trilobites are actually molts, broken as they were shed off the trilobite. Thus, it is common to find only parts of trilobites, such as the head, mid-section, or tail.
tuff	a **pyroclastic** rock made of consolidated **volcanic ash**. Tuff is the result of pyroclastic flows, in which the violent expansion of hot gas shreds the erupting **magma** into tiny particles that cool in the air to form dense clouds of volcanic ash. The tremendous explosions that are necessary to create ash-flow tuffs are caused by **rhyolitic** magma, which is **felsic**. High **silica** content makes the magma quite viscous, preventing gas bubbles from easily escaping, thus leading to pressure buildups that are released by explosive eruptions. The ash flows from these violent explosions tend to hug the ground, eventually solidifying into tuffs. Tuffs and other pyroclastic materials are **vesicular** (**porous**) due to gases expanding within the material as it cools.
turbidity current	a submarine sediment avalanche. These fast-moving currents of sediment are often caused by **earthquakes** or other geological disturbances that loosen sediment on a continental shelf. These massive sediment flows have extreme **erosive** potential, and often carve out underwater canyons. Turbidity currents deposit huge amounts of sediment during flow; such deposits are called turbidites. Because of the rate at which turbidity currents deposit **dense** sediments, they are often responsible for the effective preservation of many **fossil** organisms, which are swept up from shallow marine environments and buried in the deep sea.
Ultisols	a **soil order**; these are **soils** with subsurface **clay** accumulations that possess low native fertility and are often red hued (due to the presence of **iron oxides**). They are found in humid tropical and subtropical **climates**.
ultramafic rocks	**igneous rocks** with very low **silica** content (< 45%), which are composed of usually greater than 90% **mafic minerals**. The Earth's **mantle** is composed of **ultramafic** rocks, which are dark green to black in **color** due to their high magnesium and **iron** content.
unconformity	the relation between adjacent rock strata for which the time of deposition was separated by a period of nondeposition or **erosion**; a break in a **stratigraphic** sequence.

uplift	upward movement of the **crust** due to **compression**, **subduction**, or mountain building. Uplift can also occur as a rebounding effect after the removal of an **ice sheet** reduces the amount of weight pressing on the crust.
vanadium	a metallic element (V) that occurs naturally in **fossil fuel** deposits as well as in a variety of different **minerals**. Vanadium is mainly used to produce specialty steel alloys.
Vertisols	a **soil order**; these are **clayey soils** with a high moisture capacity. During dry periods, these soils shrink and develop wide cracks; during wet periods, they swell with moisture.
vesicular	porous or pitted with vesicles (cavities). Some **extrusive igneous rocks** have a vesicular texture.
volcanic ash	fine, unconsolidated **pyroclastic** grains under 2 millimeters (0.08 inches) in diameter. Consolidated ash becomes **tuff**.
volcanic islands	a string of islands created when molten rock rises upward through oceanic **crust**. Volcanic islands are common in several contexts, including at **subduction** zones between colliding oceanic **plates**, above oceanic **hot spots**, and along mid-ocean ridges. At subduction zones, the friction between the plates generates enough **heat** and pressure to melt some of the crust. In the case of hot spots, islands form as **magma** from the mantle breaks through the sea floor.
volcanic, volcanism	the eruption of molten rock onto the surface of the **crust**. Most volcanic eruptions occur along tectonic **plate** boundaries, but may also occur at **hot spots**. Rocks that form from molten rock on the surface are also called volcanic. Prior to eruption, **magma** ascends from the **mantle** to a relatively shallow (1–10 kilometers / 0.5–6 miles) magma chamber. Upward movement reduces the pressure on the magma until it is low enough to permit dissolved gas to exsolve (come out of solution and form bubbles). All eruptions are driven by the exsolution of dissolved gas. As the gas forms bubbles, it expands in volume and forces the magma out of the vent/chamber system onto the surface. The combination of magma viscosity and gas content can produce a range of eruptive styles, from gentle, effusive eruptions to violent explosions.
water table	the upper surface of groundwater, that is, the underground level at which groundwater is accessible.
watt	a unit of **power** measuring the rate of **energy** conversion or transfer designated by the International System of Units as one **joule** per second.
weather	the measure of short-term conditions of the **atmosphere** such as temperature, **wind** speed, and humidity. These conditions vary with the time of day, the season, and yearly or multi-year cycles.
weathering	the breakdown of rocks by physical or chemical means. Rocks are constantly being worn down and broken apart into finer and finer grains by **wind**, rivers, wave action, freezing and thawing, and chemical breakdown. Over millions of years, **weathering** and **erosion** can reduce a mighty mountain range to low rolling hills. Some rocks wear down relatively quickly, while others can withstand the power of erosion for much longer. Softer, weaker rocks such as **shale** and poorly **cemented sandstone** and **limestone** are much more easily worn away than hard, crystalline **igneous** and **metamorphic rocks**, or well-cemented sandstone and limestone. Harder rocks are often left standing alone as ridges because surrounding softer, less resistant rocks were more quickly worn away.
wind	the movement of air from areas of high pressure to areas of low pressure. The greater the temperature difference, the greater the air pressure difference and, consequently, the greater the speed at which the air will move.

w–z Glossary

wind shear	when **wind** speed and/or direction changes with increasing height in the **atmosphere**. Wind shear can happen when a **cold front** moves rapidly into an area with very warm air. There, the condensing water droplets mix with the cooler, drier air in the upper atmosphere to cause a downdraft.
windward	upwind; facing into the prevailing winds, and thus subject to **orographic precipitation**.
xenolith	a chunk of pre-existing rock incorporated into **magma** that becomes part of an **igneous rock** body. Xenoliths are generally pieces of rock from the rocky walls (often of quite different origin and composition than the magma) of subsurface spaces through which magma is flowing.
zeolites	**porous** aluminosilicate **minerals**, often formed some time after **sedimentary** layers have been deposited, or where **volcanic** rocks and ash react with alkaline groundwater. Zeolites are often used as catalysts and water softeners, and their microporous surface sturcture makes them useful in concentrating and condensing molecular substances.
zinc	a metallic chemical element (Zn). Zinc is typically used in metal alloys and galvanized steel.

General Resources

On the Earth System Science of North America

Books and Websites

Bally, A. W., and Palmer, A. R., eds., 1989, *The geology of North America—An Overview, vol. A of The Geology of North America*, Geological Society of America, Boulder, CO, 619 p.

Birdsall, S. S. & J. Florin, United States Geography, http://countrystudies.us/united-states/geography.htm.

National Park Geologic Resources, http://www.nature.nps.gov/geology/.

Maps (printed)

Thelin, G. P. and R. J. Pike, 1991, *Landforms of the Conterminous United States—A Digital Shaded-Relief Portrayal*, USGS Miscellaneous Investigations Series Map I-2206, http://pubs.usgs.gov/imap/i2206/.

Muehlberger, W. R., compiler, 1992, *Tectonic Map of North America, Scale 1:5,000,000*, American Association of Petroleum Geologists, Tulsa, OK.

Reed, J.C., and C.A. Bush, 2007, *Geology: the National Atlas of the United States* 32" x 28", http://pubs.usgs.gov/circ/1300/.

Reed, J.C., and C. A., Bush, 2007, About the geologic map in the National Atlas of the United States of America, *US Geological Survey Circular* 1300, 52 pp., http://pubs.usgs.gov/circ/1300/pdf/Cir1300_508.pdf.

USGS, 2005, *Resources for the Geologic Map of North America*, http://ngmdb.usgs.gov/gmna/.

Vigil, J.F., R. J. Pike, and D. G. Howell, 2000, *A Tapestry of Time and Terrain*, US Geological Survey Geologic Investigations Series 2720, 1 plate scale 1:2,500,000, 1 pamphlet, http://pubs.usgs.gov/imap/i2720/.

Maps (online)

American Geological Institute's *Earth Comm 2nd edition*, Map Resources. [A compilation of online map resources.] http://www.agiweb.org/education/earthcomm2/maps.html.

Geologic Maps of the 50 United States, by A. Alden, About.com, http://geology.about.com/od/maps/ig/stategeomaps/.

Geologic Provinces of the United States: Records of an Active Earth, US Geological Survey, http://geomaps.wr.usgs.gov/parks/province/.

Geology of the Conterminous United States, http://mrdata.usgs.gov/geology/us/.

Google Earth, http://www.google.com/earth/.

The National Atlas of the United States. [Custom-make maps.] http://nationalatlas.gov/mapmaker.

The National Map, http://nationalmap.gov.

The National Map: Historical Topographic Map Collection. [Online historic topographic maps.] http://nationalmap.gov/historical/index.html.

OneGeology. [A collaboration among many national geological surveys to create a dynamic digital geological map data for the world.] http://portal.onegeology.org/.

US Topo Quadrangles—Maps for America. [Online topographic maps.] http://nationalmap.gov/ustopo/index.html.

Vigil, J. F., R. J. Pike, & D. G. Howell, 2000, *A Tapestry of Time and Terrain*, US Geological Survey, Geologic Investigations Series 2720, http://pubs.usgs.gov/imap/i2720/.

Other General Resources on Earth System Science

Geologic Time Resources

Gradstein, F. M., J. G. Ogg, M. D. Schmitz, & G. M. Ogg, *The Geologic Time Scale 2012, 2 vols.*, Elsevier, New York, https://engineering.purdue.edu/Stratigraphy/charts/Stratigraphic_Chart_GTS2012.pdf.

International Commission on Stratigraphy, http://www.stratigraphy.org/.

Janke, P. R., 2013, *Correlated History of the Earth Chart (laminated), vol. 8*, Pan Terra, Hill City, SD.

The Paleontology Portal, http://paleoportal.org.
Walker, J. D., J. W. Geissman, S. A. Bowring, & L. E. Babcock, compilers, 2012, *Geologic Time Scale Poster*, Geological Society of America, Boulder, CO.

Dictionaries

Allaby, M., 2013, *A Dictionary of Geology and Earth Sciences*, Oxford University Press, Oxford, UK.
Bates, R. L., & J. A. Jackson, 1984, *Dictionary of Geological Terms, 3rd edition*, Anchor Press, Garden City, NY.
McGraw-Hill Education, 2003, *McGraw-Hill Dictionary of Geology and Mineralogy*, McGraw-Hill, New York.
Neuendorf, K. K. E., J. P. Mehl, Jr., & J. A. Jackson, 2011, *Glossary of Geology, 5th edition, revised*, American Geological Institute, Alexandria, VA, 800 pp.

Earth System Science Organizations

American Association of State Geologists, http://www.stategeologists.org/.
American Geological Institute (AGI is an umbrella organization representing over 40 other geological organizaitons), http://agiweb.org.
American Geophysical Union, http://agu.org.
Association for Women Geoscientists, http://awg.org.
Geological Society of America, http://geosociety.org.
Natural Resources Conservation Service, http://www.nrcs.usda.gov/wps/portal/nrcs/site/national/home/ (NRCS helps US farmers, ranchers and forest landowners conserve soil, water, air and other natural resources).
Paleontological Research Institution, http://priweb.org (publisher of this volume).
Paleontological Society, http://paleosoc.org.
US Geological Survey, http://usgs.gov.

General Earth Science Education Resources

Websites

Digital Library for Earth System Education (DLESE), http://dlese.org.
Earth Science World Image Bank, American Geological Institute, http://www.earthscienceworld.org/ imagebank/.
Resources for Earth Science and Geography Instruction, by Mike Francek, Central Michigan University, http://webs.cmich.edu/resgi/.
Science in Your Backyard, US Geological Survey. [State-by-state compilation of Earth science-related data, most of which will need to be adapted for education uses.] http://www.usgs.gov/state/.
SERC (The Science Education Resource Center) K-12 Resources. [Hundreds of classroom activities organized by grade level and topic as well as guidance on effective teaching.] http://serc.carleton.edu/k12/index.html.
SERC Earth Exploration Toolbook. [A collection of online Earth system science activities introducing scientific data sets and analysis tools.] http://serc.carleton.edu/eet/index.html.
Windows to the Universe, from the National Earth Science Teachers Association, http://www.windows2universe.org/.

Blogs

AGU Blogosphere, American Geophysical Union. [A collection of over a dozen blogs by geoscientists on recent events and perspectives on the geosciences.] http://blogs.agu.org/blogs/.
The Best of Geology and Earth Science on the Web: the Latest from the Blogosphere, http://all-geo.org/.
Geobulletin: News from the Geoblogosphere, stratigraphy.net, http://www.geobulletin.org/?action=news.
Geotripper: News and Views from the Geologic Realm, by Garry Hayes, Modesto Junior College. [Many, but not all, pertain to the western US.] http://geotripper.blogspot.com/.

Science education organizations

National Association of Geoscience Teachers. [Focused on undergraduate geoscience education, but includes active secondary school educators.] http://nagt.org.
National Earth Science Teacher Association. [Focused on secondary school Earth science education.] http://nestanet.org.
National Science Teacher Association, http://nsta.org.

General Resources by State

Geologic maps of individual US states. (Digital geologic maps of US states with consistent lithology, age, GIS database structure, and format.) http://mrdata.usgs.gov/geology/state.

Multistate Areas of the Southwest

Books

Abbott, L. & T. Cook, 2004, *Hiking the Grand Canyon's Geology*, The Mountaineers Books, Seattle, WA, 301 pp.

Baars, D. L., 2000, *The Colorado Plateau: a Geologic History, revised edition*, University of New Mexico Press, Albuquerque, 268 pp.

Baars, D. L., 2002, A *Traveler's Guide to the Geology of the Colorado Plateau*, The University of Utah Press, 294 pp.

Baldridge, W., 2004, *Geology of the Southwest: a Journey Through Two Billion Years of Plate-Tectonic History*, Cambridge University Press, New York, 314 pp.

Blakey, R. & W. Ranney, 2008, *Ancient Landscapes of the Colorado Plateau*, Grand Canyon Association, Grand Canyon, AZ, 156 pp.

Chronic, H. & L. Chronic, 2004, *Pages of Stone: Geology of the Grand Canyon and Plateau Country National Parks and Monuments, 2nd edition*, The Mountaineers Books, Seattle, WA, 176 pp.

Hopkins, R., 2002, *Hiking the Southwest's Geology: Four Corners Region*, The Mountaineers Books, Seattle, WA, 286 pp.

McPhee, J., 1981, *Basin and Range*, Farrar, Straus, Giroux, NY, 240 pp.

Maps

Southern Rocky Mountain Region Geological Highway Map, revised edition, 1990, AAPG, Tulsa, OK. [Includes Arizona, Colorado, New Mexico, and Utah.]

Websites

Encyclopedia of the Great Plains: Geology, University of Nebraska, Lincoln, http://plainshumanities.unl.edu/encyclopedia/doc/egp.pe.028.

List of National Parks within the Colorado Plateau, USGS Geology in the Parks, http://geomaps.wr.usgs.gov/parks/province/colo.html.

Arizona

Books and Articles

Abbott, L. & T. Cook, 2007, *Geology Underfoot in Northern Arizona*, Mountain Press Publishing Company, Missoula, MT, 330 pp.

Beus, S.S., & M. Morales, eds., 1990, **Grand Canyon Geology**, Oxford University Press and Museum of Northern Arizona Press, 518 pp.

Bezy, J. V., J. T. Gutmann, and G. B. Haxel, 2000, *A Guide to the Geology of Organ Pipe Cactus National Monument and the Pinacate Biosphere Reserve*, Arizona Geological Survey, 64 pp.

Bryan, T., 2013, *Arizona Rocks! A Guide to Geologic Sites in the Grand Canyon State*, Mountain Press Publishing Company, Missoula, MT, 112 pp.

Chronic, H., 1983, *Roadside Geology of Arizona*, Mountain Press Publishing Company, Missoula, MT, 322 pp.

Hanson, S. L., 2003, *Roadside Geology: Wupatki Crater and Sunset Crater Volcano National Monuments*, Arizona Geological Survey, 32 pp.

Lucchitta, I., 2001, *Hiking Arizona's Geology*, Mountaineers Books, Seattle, WA, 290 pp.

Nations, D. and E. Stump, 1996, *Geology of Arizona, 2nd edition*, Kendall/Hunt Publishing Company, Dubuque, IA, 272 pp.

Price, L., 1999, *An Introduction to Grand Canyon Geology*, Grand Canyon Association, Grand Canyon, AZ, 64 pp.

Ranney, W., 2012, *Carving Grand Canyon: Evidence, Theories, and Mystery*, Grand Canyon Association, Grand Canyon, AZ, 200 pp.

Maps

Kamilli, R. and S. Richard, 1998, *Geologic Highway Map of Arizona*, Arizona Geological Society, Arizona Geological Survey, Tucson.

Websites

Arizona Geological Society, http://www.arizonageologicalsoc.org/.

Arizona Geological Survey, http://www.azgs.az.gov/.

Arizona Geology, Blog of the State Geologist of Arizona, http://arizonageology.blogspot.com/.

The Geologic Origin of the Sonoran Desert, by R. Scarborough, Arizona-Sonora Desert Museum, Tucson, AZ, http://www.desertmuseum.org/books/nhsd_geologic_origin.php.
Grand Canyon Association, https://www.grandcanyon.org/.

Colorado

Books and Articles

Abbott, L. and T. Cook, 2012, *Geology Underfoot Along Colorado's Front Range*, Mountain Press Publishing Company, Missoula, MT, 352 pp.
Baars, D. L., 1992, *The American Alps: The San Juan Mountains of Southwest Colorado*, University of New Mexico Press, Albuquerque, 204 pp.,
Blair, R., 1996, *The Western San Juan Mountains: Their Geology, Ecology, and Human History*, University of Colorado Press, Boulder, 406 pp.
Chronic, H., 1984, *Time, Rocks, and the Rockies: a Geologic Guide to Rocks and Trails of Rocky Mountain National Park*, Mountain Publishing Company, Missoula, MT, 120 pp.
Chronic, H. and F. Williams, 2002, *The Roadside Geology of Colorado*, Mountain Press Publishing Company, Missoula, MT, 398 pp.
Fillmore, R., 2011, *Geological Evolution of the Colorado Plateau of Eastern Utah and Western Colorado*, University of Utah Press, Salt Lake City, 524 pp.
Foutz, D., 1994, *Geology of Colorado Illustrated*, Your Geologist Dell R. Foutz, Grand Junction, CO, 181 pp.
Griffitts, M. O., 1990, *Guide to the Geology of Mesa Verde National Park*, Mesa Verde Museum Association, Mesa Verde National Park, CO, 88 pp.
Hopkins, R. and L. B. Hopkins, 2000, *Hiking Colorado's Geology*, Mountaineers Books, Seattle, WA, 240 pp.
Matthews, V., K. KellerLynn, & B. Fox, 2003, *Messages in Stone: Colorado's Colorful Geology*, Colorado Department of Natural Resources, Colorado Geological Survey, Denver, 157 pp.
Milito, S., 2010, A survey of fossils and geology of Red Rock Canyon Open Space, Colorado Springs, Colorado, *The Mountain Geologist*, 47(1): 1–14.
Obee, R., 2012, *History in Stone: the Story of Red Rock Canyon*, Johnson Books, Boulder, CO, 265 pp.
Reed, J. and E. Ellis, 2009, *Rocks Above the Clouds: a Hiker's and Climber's Guide to Colorado Mountain Geology*, Colorado Mountain Club Press, Golden, CO, 128 pp.
Raup, O. B., 1996, *Geology Along Trail Ridge Road [of Rocky Mountain National Park]: a Self Guided Tour for Motorists*, Rocky Mountain Nature Association with Falcon Press Publishing Company, 75 pp.
Williams, F. and H. Chronic, 2014, *Roadside Geology of Colorado*, Mountain Press Publishing Company, Missoula, MT, 416 pp.

Maps

GTR Mapping, 2013, *Colorado Geologic Highway Map and Shaded Elevation Map with 14,000 Foot Peaks: Selected Mining Districts, & Dinosaur Localities*, GTR Mapping, Cañon City, CO.

Websites

Bob Weimer Mines Geology Trail, Geology & Geological Engineering, Colorado School of Mines, http://inside.mines.edu/Bob-Weimer-Trail.
Colorado Geological Survey, http://coloradogeologicalsurvey.org/.
Colorado Geology Photojournals: a Tribute to Colorado's Physical Past and Present, by Jeremy McCreary, http://www.cliffshade.com/colorado/index.htm.
Rock Talk, Colorado Geological Survey, http://coloradogeologicalsurvey.org/publications/rocktalk/.
The What, Where, When, How and Why of Colorado's Physical Past and Present, http://www.cliffshade.com/colorado/geo_overview.htm.

New Mexico

Books and Articles

Chronic, H., 1987, *Roadside Geology of New Mexico*, Mountain Press Publishing Company, Missoula, MT, 255 pp.
Goff, F., 2009, *Valles Caldera: A Geologic History*, University of New Mexico Press, Albuquerque, 128 pp.
Hart, D., 2013, *Old Forty-Four: a Historical and Geological Excursion Over New Mexico's Old Route 44 (now part of US-550),* Sunstone Press, Santa Fe, NM, 354 pp.
Mack, G. and K. Giles, 2004, *The Geology of New Mexico: a Geologic History*, New Mexico Geological Society, Socorro, NM, 474 pp. [Technical]
Price, L., 2010, *The Geology of Northern New Mexico's Parks, Monuments, and Public Lands*, New Mexico Bureau of Geology and Mineral Resources, Socorro, 372 pp.

Maps

Wilks, M. E., 2005, *New Mexico Geologic Highway Map*, New Mexico Geological Society, Albuquerque, NM

Websites
New Mexico Bureau of Geology and Mineral Resources, https://geoinfo.nmt.edu/.
Virtual Geologic Tour of New Mexico, New Mexico Bureau of Geology and Mineral Resources, https://geoinfo.nmt.edu/tour/home.html.

Utah

Books and Articles
Barnes, F. A., 1993, *Geology of the Moab Area*, Canyon Country Publications, Moab, UT.
Eldredge, S., 1996, *Canyon Country: a Geologic Guide to the Canyonlands Travel Region*, Utah Geological Survey, Public Information Series 34, 24 pp.
Fillmore, R., 2000, *The Geology of the Parks, Monuments, and Wildlands of Southern Utah*, University of Utah Press, Salt Lake City, 280 pp.
Fillmore, R., 2011, *Geological Evolution of the Colorado Plateau of Eastern Utah and Western Colorado*, University of Utah Press, Salt Lake City, 524 pp.
Hamilton, W. L., 1992, *The Sculpturing of Zion: a Road Guide to the Geology of Zion National Park*, Zion Natural History Association, 132 pp.
Hintze, L. F., & B. J. Kowallis, 2009, Geologic history of Utah, *Brigham Young University Geology Studies Special Publication* 9, 225 pp. [Somewhat technical]
Orndorff, R., R. Wieder, and D. Futey, 2006, *Geology Underfoot in Southern Utah*, Mountain Press Publishing Company, Missoula, MT, 286 pp.
Parry, W., 2016, *Geology of Utah's Mountains, Peaks, and Plateaus: Including Descriptions of Cliffs, Valleys, and Climate History*, FriesenPress, Victoria, BC, 244 pp.
Sprinkel, D., T. Chidsey, and P. Anderson, 2003, *Geology of Utah's Parks and Monuments, 3rd edition*, Bryce Canyon Natural History Association and Utah Geological Association, Salt Lake City, UT, 562 pp.
Stokes, W. L., 1986, *Geology of Utah*, Utah Museum of Natural History and Utah Geological and Mineralogical Survey, 280+ pp.
Williams, F., L. Chronic, and H. Chronic, 2014, *Roadside Geology of Utah*, Mountain Press Publishing Company, Missoula, MT, 416 pp.

Websites
Utah Geological Survey, http://geology.utah.gov/.
UtahGeology.com, compiled by L. Weaver, Utah Geological Survey, http://utahgeology.com/.

Acknowledgments

We are grateful to the following individuals, each of whom reviewed one or more chapters of *The Teacher-Friendly Guide™ to the Earth Science of the Southwestern US*: Jim Davis, Don Duggan-Haas, Rod Feldmann, Karl Flessa, Bryan Isacks, Vince Matthews, Thomas McGuire, Cori Myers, David Lindsey, Judith Parrish, Donald Rasmussen, David Tabet, and Alex Wall. For assistance with the text and figures in Chapter 3 (Fossils), we are grateful to: Loren Babcock, Wade Greenberg-Brand, Kirk Johnson, Alana McGillis, Corinne Myers, Rene O'Connell, Susannah Porter, Samantha Richards, Chris Schur, Leslie Skibinski, Christi Sobel, and Peter Stifel.

Richard Kissel managed early content development of the Guide, and was aided in content research by Sara Auer Perry. The glossary was developed by Paula Mikkelsen and Andrielle Swaby.

Funding for this Guide came from National Science Foundation DR K-12 grant DRL-0733303 to the Paleontological Research Institution. Funding to start *The Teacher-Friendly Guide™* series was provided by the Arthur Vining Davis Foundations. Jane (Ansley) Picconi did page layout for the first Guide in the series, *The Teacher-Friendly Guide™ to the Geology of the Northeastern US* (Paleontological Research Institution special publication 24, 2000), many features of which have been adopted for this Guide.

Figure Credits

Creative Commons enables the sharing and reuse of media through a variety of free copyright licenses. The Creative Commons licenses used by images in this publication can be found at the following URLs:

CC-BY-2.0: https://creativecommons.org/licenses/by/2.0/
CC-BY-2.5: https://creativecommons.org/licenses/by/2.5/
CC-BY-3.0: https://creativecommons.org/licenses/by/3.0/
CC-BY-SA-2.0: https://creativecommons.org/licenses/by-sa/2.0/
CC-BY-SA-2.5: https://creativecommons.org/licenses/by-sa/2.5/
CC-BY-SA-3.0: https://creativecommons.org/licenses/by-sa/3.0/
CC-BY-SA-4.0: https://creativecommons.org/licenses/by-sa/4.0/
CC-BY-NC-2.0: https://creativecommons.org/licenses/by-nc/2.0/
CC-BY-NC-4.0: http://creativecommons.org/licenses/by-nc/4.0/
CC-BY-NC-ND-2.0: https://creativecommons.org/licenses/by-nc-nd/2.0/
CC-BY-ND-2.0: https://creativecommons.org/licenses/by-nd/2.0/

Chapter 1: Geologic History

1.1: Jim Houghton
1.2: Jane Picconi
1.3: Wade Greenberg-Brand
1.4: Wade Greenberg-Brand, adapted from image by USGS
1.5: Wade Greenberg-Brand, adapted from image by Karlstrom et al. (2012), Vishnu basement rocks of the Upper Granite Gorge: Continent formation 1.84 to 1.66 billion years ago, *Geological Society of America Special Papers* 489, pp 7–24
1.6: Adapted from image by John Goodge, USGS
1.7: Adapted from image by Ron Blakey, NAU Geology
1.8: Wade Greenberg-Brand
1.9: Adapted from image by Ron Blakey, NAU Geology
1.10: Jim Houghton
1.11-1.12: Adapted from image by Ron Blakey, NAU Geology
1.13: US Department of Energy
1.14: Adapted from image by Ron Blakey, NAU Geology
1.15: NOAA
1.16: Wade Greenberg-Brand, adapted from image by New Mexico Institute of Mining and Technology
1.17: Wade Greenberg-Brand, adapted from image by William A. Cobban and Kevin C. McKinney, USGS
1.18: "Kbh3rd" [CC-BY-SA-3.0] via Wikimedia Commons
1.19: Wade Greenberg-Brand
1.20: Wade Greenberg-Brand, adapted from image by Illinois State Geological Survey
1.21: Wade Greenberg-Brand
Crust Box: Jim Houghton
Pangaea Box: Wade Greenberg-Brand, adapted from image by USGS
Plate Boundaries Box: Jose F. Vigil, USGS

Chapter 2: Rocks

2.1: Jane Picconi
2.2: National Park Service
2.3: Wade Greenberg-Brand
2.4: National Park Service
2.5: James St. John [CC-BY-2.0] via Flickr
2.6: National Park Service
2.7: USGS
2.8: Kevin Walsh [CC-BY-2.0] via Flickr
2.9: Wade Greenberg-Brand, adapted from image by Doelling et al. (2000), *Geology of Grand Staircase-Escalante National Monument, Utah*, Utah Geological Association Publication 28
2.10: James St. John [CC-BY-2.0] via Flickr
2.11: Keith Yahl [CC-BY-2.0] via Flickr
2.12: Dirk Van de Velde [CC-BY-2.0] via Flickr
2.13: Wade Greenberg-Brand, adapted from image by Chan and Archer (2000), *Cyclic Eolian Stratification on the Jurassic Navajo Sandstone, Zion National Park: Periodicities and Implications for Paleoclimate*, Utah Geological Association Publication 28
2.14: Wade Greenberg-Brand, adapted from image by William A. Cobban and Kevin C. McKinney, USGS
2.15: James St. John [CC-BY-2.0] via Flickr

- 2.16: Ken Lund [CC-BY-SA-2.0] via Flickr
- 2.17: Luca Galuzzi - www.galuzzi.it [CC-BY-SA-2.5] via Wikimedia Commons
- 2.18: Wade Greenberg-Brand, adapted from image by Goff and Janik (2002), Gas geochemistry of the Valles Caldera Region, New Mexico and comparisons with gases at Yellowstone, Long Valley and other geothermal systems, *Journal of Volcanology and Geothermal Research*, 116(3–4): 299–323
- 2.19: Wade Greenberg-Brand, adapted from image by USGS
- 2.20: Graeme Churchard [CC-BY-2.0] via Flickr
- 2.21: Bowie Snodgrass [CC-BY-2.0] via Wikimedia Commons
- 2.22: Doc Searls [CC-BY-SA-2.0] via Flickr
- 2.23: Wade Greenberg-Brand, adapted from image by USGS
- 2.24: Bryant Olsen [CC-BY-NC-2.0] via Flickr
- 2.25: Daniel Mayer [CC-BY-SA-3.0] via Wikimedia Commons
- 2.26: Jim Houghton
- 2.27: Bureau of Land Management
- 2.28: Tristan Higbee [CC-BY-SA-2.0] via Flickr
- 2.29: "summitcheese" [CC-BY-SA-2.0] via Flickr
- 2.30–2.31: James St. John [CC-BY-2.0] via Flickr
- 2.32: "Loco Steve" [CC-BY-2.0] via Flickr
- 2.33: Sam Howzit [CC-BY-2.0] via Flickr
- 2.34: James St. John [CC-BY-2.0] via Flickr
- 2.35: Brittany Woiderski [CC-BY-2.0] via Flickr
- 2.36: Carol M. Highsmith
- 2.37: James St. John [CC-BY-2.0] via Flickr
- 2.38: Bryce Bradford [CC-BY-ND-2.0] via Flickr
- 2.39: James St. John [CC-BY-2.0] via Flickr
- Surface Rocks Box: Jim Houghton
- Dune Box: Wade Greenberg-Brand
- Columnar Jointing Box: Wendy Van Norden
- Metamorphism Box: Jim Houghton
- Sedimentary Environments Box: Jim Houghton

Chapter 3: Fossils

- 3.1: Andrielle Swaby
- 3.2: Wade Greenberg-Brand
- 3.3: Carl Bowman, National Park Service [CC-BY-2.0] via Wikimedia Commons
- 3.4: Susannah Porter, reproduced with permission.
- 3.5: Walcott, C. D. (1886), Second Contribution to the Studies on the Cambrian Fauna of North America, *Bulletin of the US Geological Survey* No. 30, Government Printing Office, Washngton
- 3.6: A) and B) Chris Schur, reproduced with permission; C) © Christi Sobel
- 3.7: A) Nicholson, H.A. (1896), *The Ancient Life-History of the Earth*, Appleton, New York, 407 pp; B) Eastman, C.R. (ed.) (1937), *Text-book of Paleontology, Vol. I*, MacMillan, London, 839 pp; C) Shimer, H. W., and R. R. Schrock (1944), *Index Fossils of North America: A New Work Based on the Complete Revision and Reillustration of Grabau and Shimer's 'North American Index Fossils,'* MIT Press, Cambridge, MA
- 3.8: Michael Quinn, National Park Service
- 3.9: © Christi Sobel
- 3.10: McKee, E.D. (1982), The Supai Group of Grand Canyon, *US Geological Survey Professional Paper* 1173, 504 p.
- 3.11: A) Michael Quinn, National Park Service; B) © Christi Sobel
- 3.12: © Christi Sobel
- 3.13: A) Photo by Wade Greenberg-Brand, specimen TU1061 in the collection of the Paleontological Research Institution; B), C), D), E), F), and G) McKee, E. D. (1938), The environment and history of the Toroweap Formation and Kaibab formations of northern Arizona and southern Utah, *Carnegie Institution of Washington Publication* No. 492, Washington, DC, 268 pp; H) and I) © Christi Sobel
- 3.14: A) D. M. S. Watson; B) Nobu Tamura [CC-BY-3.0] via Wikimedia Commons
- 3.15: Wade Greenberg-Brand
- 3.16: Will Sillin, reproduced with permission
- 3.17: A) and D) © Christi Sobel; B) Alana McGillis; C) Nobu Tamura [CC-BY-3.0] via Wikimedia Commons; E) David Starr Jordan
- 3.18: James St. John [CC-BY-2.0] via Flickr and Jeff Martz, National Park Service
- 3.19: A) and B) Frank Kovalchek [CC-BY-2.0] via Wikimedia Commons and Jeff Martz, National Park Service; C) Mark Byzewski [CC-BY-2.0] via Wikimedia Commons and Jeff Martz, National Park Service; D) © Christi Sobel and Jeff Martz, National Park Service
- 3.20: "Moondigger" [CC-BY-SA-2.5] via Wikimedia Commons
- 3.21: © Christi Sobel
- 3.22: © Robert S Oakes & Joseph D Lavenburg/National Geographic Creative
- 3.23: © Christi Sobel
- 3.24: Wade Greenberg-Brand

3.25: © Christi Sobel
3.26: Kyle Greenberg [CC-BY-2.0] via Flickr
3.27: Alana McGillis, adapted from data by Hunt et al. (2005), Vertebrate trace fossils from Arizona with special reference to tracks preserved in National Park Service units and notes on the Phanerozoic distribution of fossil footprints, in Heckert, A.B., and Lucas, S.G., eds., *Vertebrate Paleontology in Arizona*, New Mexico Museum Of Natural History And Science Bulletin 29: 158–166; Lockley et al. (1986), North America's largest dinosaur trackway site: implications for Morrison Formation paleoecology, *Geological Society of America Bulletin*, 97(10): 1163–1176; Lockley et al. (2000), Dinosaur tracksites in New Mexico: A review, in Lucas, S.G., and A.B. Heckert, eds, *Dinosaurs of New Mexico*, New Mexico Museum of Natural History and Science Bulletin, 17: 9–16; Lockley and Lucas (2013), *Evazoum gatewayensis*, a new late Triassic archosaurian ichnospecies from Colorado: implications for footprints in the ichnofamily Otozoidae, in *The Triassic System*, New Mexico Museum of Natural History and Science, Bulletin 61: 345–353; Lockley and Lucas (eds.) (2014), *Fossil footprints of Western North America*, New Mexico Museum of Natural History and Science Bulletin 62; Milan et al. (2008), Crouching theropod and *Navahopus* sauropodomorph tracks from the Early Jurassic Navajo Sandstone of USA, Acta Palaeontologica Polonica, 53(2): 197–205;
3.28: Alana McGillis, adapted from images by Tidwell, W.D. (1998) *Common Fossil Plants of Western North America*, Smithsonian Institution Press, Washington, DC, 299 pp, and Foster, J. (2007), *Jurassic West: The Dinosaurs of the Morrison Formation and Their World*, Indiana University Press, Bloomington, 416 pp
3.29: Mary Beth Ihnken
3.30: Sampson et al. (2010), New horned dinosaurs from Utah provide evidence for intracontinental dinosaur endemism, *PLoS ONE*, 5(9): e12292, [CC-BY-2.5] via Wikimedia Commons
3.31: © Christi Sobel
3.32: A) Wade Greenberg-Brand; B) Warren Allmon
3.33: Zach Tirrell [CC-BY-SA-2.0] via Wikimedia Commons and © Christi Sobel
3.34: Alana McGillis
3.35: Wade Greenberg-Brand
3.36: A) and B) © Christi Sobel; C) Stokes, W. L. (1986), *Geology of Utah*, Utah Museum of Natural History and Utah Geological and Mineralogical Survey, 280+ pp.
3.37: Wade Greenberg-Brand
3.38: A) and C) © Christi Sobel; B) Birmingham Paleontological Society, reproduced with permission
3.39: A), B), C), F), and G) Cobban, W. (1977), Characteristic marine mollusan fossils from the Dakota Sandstone and intertongued Mancos Shale, west-central New Mexico, *US Geological Survey Professional Paper* 1009, 30 pp, 21 pl.; D) Shimer, H. W., and R. R. Schrock (1944), *Index Fossils of North America: A New Work Based on the Complete Revision and Reillustration of Grabau and Shimer's 'North American Index Fossils,'* MIT Press, Cambridge, MA; E), H), and I) © Christi Sobel
3.40: A) Nobu Tamura [CC-BY-3.0] via Wikimedia Commons and Alana McGillis; B) Matthew, W. D. (1937), Paleocene faunas of the San Juan Basin, New Mexico, *Transactions of the American Philosophical Society*, 30: 1–510, and Robert Bruce Horsfall; C) and D) Matthew, W. D. (1937), Paleocene faunas of the San Juan Basin, New Mexico, *Transactions of the American Philosophical Society*, 30: 1–510, and © Christi Sobel; E) Scott, W. B. (1913), *A History of Land Mammals in the Western Hemisphere*, The Macmillan Company, New York, NY
3.41: Tidwell et al. (1981), Cretaceous and Tertiary floras of the San Juan Basin, in S. G. Lucas, J. K. Rigby, Jr., and B. S. Kues (eds.), *Advances in San Juan Basin Paleontology*, University of New Mexico Press, Albuquerque, pp. 307–332, reproduced with permission
3.42: A) National Park Service; B) and C) Stock, C. (1925), *Cenozoic Gravigrade Edentates of Western North America, with Special Reference to the Pleistocene Megalonychinae and Mylodontidae of Rancho Le Brea*, Carnegie Institution of Washington, Washington
3.43: A), C), D), E), G), and H) Robison, R. A., L. E. Babcock, and V. G. Gunther (2015), *Exceptional Cambrian Fossils from Utah: a Window into the Age of Trilobites*, Utah Geological Survey Miscellaneous Publication 15-1, 97 pp; B) Parent Géry [CC-BY-SA-3.0] via Wikimedia Commons; F) Micha L. Rieser [CC-BY-SA-3.0] via Wikimedia Commons
3.44: A), B), and D) Robison, R. A., L. E. Babcock, and V. G. Gunther (2015), *Exceptional Cambrian Fossils from Utah: a Window into the Age of Trilobites*, Utah Geological Survey Miscellaneous Publication 15-1, 97 pp; C) Virtual Fossil Museum [CC-BY-NC-4.0] via http://www.fossilmuseum.net and Matteo De Stefano/MUSE [CC-BY-SA-3.0] via Wikimedia Commons; E) Robison, R. A., L. E. Babcock, and V. G. Gunther (2015), *Exceptional Cambrian Fossils from Utah: a Window into the Age of Trilobites*, Utah Geological Survey Miscellaneous Publication 15-1, 97 pp and © Christi Sobel
3.45: A), C), G), H), and I) Hintze, L. F., and F. D. Davis (2003), *Geology of Millard County, Utah*, Utah Geological Survey Bulletin 133, 305 pp; B), D), and E) © Christi Sobel; F) Guensburg, T. E., and J. Sprinkle (2003), The oldest known crinoids (Early Ordovician, Utah) and a new crinoid plate homology system, *Bulletins of American Paleontology*, no. 364, 43 pp
3.46: Hintze, L. F., and F. D. Davis (2003), *Geology of Millard County, Utah*, Utah Geological Survey Bulletin 133, 305 pp
3.47: Stokes, W. L. (1986), *Geology of Utah*, Utah Museum of Natural History and Utah Geological

and Mineralogical Survey, 280+ pp

3.48: A) and B) © Christi Sobel; C) Teichert, C. (1965), Devonian rocks and paleogeography of central Arizona, *US Geological Survey Professional Paper* 464, 225 pp; D) Stokes, W. L. (1986), *Geology of Utah*, Utah Museum of Natural History and Utah Geological and Mineralogical Survey, 280+ pp

3.49: A) through H) Stokes, W. L. (1986), *Geology of Utah*, Utah Museum of Natural History and Utah Geological and Mineralogical Survey, 280+ pp; I) Teichert, C. (1965), Devonian rocks and paleogeography of central Arizona, *US Geological Survey Professional Paper* 464, 225 pp

3.50: A), H), I), and M) Hintze, L. F., and F. D. Davis (2003), *Geology of Millard County, Utah*, Utah Geological Survey Bulletin 133, 305 pp; B) through G), K), and L) Stokes, W. L. (1986), *Geology of Utah*, Utah Museum of Natural History and Utah Geological and Mineralogical Survey, 280+ pp; J) Scott, W. B. (1916), *An introduction to Geology*, MacMillan, New York, 816 pp

3.51: A) © Christi Sobel; B) Mark A. Wilson

3.52: A) © 2009 Birmingham Paleontological Society, All Rights Reserved, reproduced with permission; B) Ray Troll, reproduced with permission

3.53: A), B), C), D), and F) Stokes, W. L. (1986), *Geology of Utah*, Utah Museum of Natural History and Utah Geological and Mineralogical Survey, 280+ pp; E) Hintze, L. F., and F. D. Davis (2003), *Geology of Millard County, Utah*, Utah Geological Survey Bulletin 133, 305 pp

3.54: "Medtrails" [CC-BY-SA-3.0] via Wikimedia Commons

3.55: A) through I) Stokes, W. L. (1986), *Geology of Utah*, Utah Museum of Natural History and Utah Geological and Mineralogical Survey, 280+ pp; J) Hook, S. C., and W. A. Cobban (2013), Middle Turonian (Late Cretaceous) rudistids from the lower tongue of the Mancos Shale, Lincoln County, New Mexico, *New Mexico Geology*, 35(1): 13–19

3.56: Stokes, W. L. (1986), *Geology of Utah*, Utah Museum of Natural History and Utah Geological and Mineralogical Survey, 280+ pp

3.57: A) Scott, W. B. (1913), *A History of Land Mammals in the Western Hemisphere*, The Macmillan Company, New York, NY and © Christi Sobel; B) Scott, W. B. (1913), *A History of Land Mammals in the Western Hemisphere*, The Macmillan Company, New York, NY; C) © Christi Sobel

3.58: A) H. Zell [CC-BY-SA-3.0] via Wikimedia Commons; B) Michelle Pemberton [CC-BY-SA-3.0] via Wikimedia Commons

3.59: © Christi Sobel

3.60: A) Photo by Wade Greenberg-Brand, specimen PRI 50287 in the collection of the Paleontological Research Institution; B) Alana McGillis

3.61: © Jan Vriesen, Denver Museum of Nature and Science

3.62: Ian Miller, Denver Museum of Nature and Science, reproduced with permission

3.63: Wade Greenberg-Brand

3.64: A) Photo by Wade Greenberg-Brand, specimen K22294 in the collection of the Paleontological Research Institution; B) through H) Virtual Fossil Museum [CC-BY-NC-4.0] via http://www.fossilmuseum.net

3.65: A), C), and D) © Christi Sobel; B) Scott, W. B. (1913), *A History of Land Mammals in the Western Hemisphere*, The Macmillan Company, New York, NY

3.66: A), B), and C) National Park Service; D) Chris Light [CC-BY-SA-4.0] via Wikimedia Commons; E) Photo by Wade Greenberg-Brand, specimen 42415 in the collection of the Paleontological Research Institution

3.67: Wade Greenberg-Brand

3.68: A) Pirsson, L.V. and C. Schuchert (1915), *A Textbook of Geology, Part II - Historical Geology*, John Wiley & Sons, New York; B) through G) Cooper, G. A. and R. E. Grant (1974), Permian Brachiopods of West Texas, Vols II and III, *Smithsonian Contributions to Paleobiology* No. 15, Smithsonian Institution Press, Washington

3.69: "Valugi" [CC-BY-SA-3.0] via Wikimedia Commons

3.70: © Christi Sobel

Stromatolite Box: James St. John [CC-BY-2.0] via Wikimedia Commons

Conodont Box: © Christi Sobel

Mammal Teeth Box: Scott, W. B. (1913), *A History of Land Mammals in the Western Hemisphere*, Macmillan Company, New York

Crinoid Box: © Christi Sobel

Graptolite Box: © Christi Sobel

Brachiopod Box: Wade Greenberg-Brand

Ammonoid Box: Wade Greenberg-Brand, adapted from image by http://www.renmanart.com

Mammoths and Mastodons Box: © Christi Sobel

Chapter 4: Topography

4.1: Wendy Van Norden

4.2: Jim Houghton

4.3: Robin Lacassin [CC-BY-SA-3.0] via Wikimedia Commons

4.4: Adapted from image by USGS

4.5: Bryan Isacks

4.6: Andrielle Swaby

4.7: Rob Lee [CC-BY-ND-2.0] via Flickr
4.8: Wade Greenberg-Brand
4.9: "Palacemusic" [CC-BY-SA-3.0] via Wikimedia Commons
4.10: "Moondigger" [CC-BY-SA-2.5] via Wikimedia Commons
4.11: USGS
4.12: The Greater Southwestern Exploration Company [CC-BY-2.0] via Flickr
4.13: Pearson Scott Foresman
4.14: "Chensiyuan" [CC-BY-SA-4.0] via Wikimedia Commons
4.15-4.16: John Fowler [CC-BY-2.0] via Wikimedia Commons
4.17: Tyler Finvold
4.18: Paul L. Sealey [CC-BY-3.0] via Wikimedia Commons
4.19: Shane Torgerson [CC-BY-3.0] via Wikimedia Commons
4.20: Andrielle Swaby
4.21: Wade Greenberg-Brand
4.22: Karl Musser [CC-BY-SA-3.0] via Wikimedia Commons
4.23: USDA Forest Service
4.24: Donna Sutton [CC-BY-ND-2.0] via Flickr
4.25: Andrielle Swaby
4.26: New Mexico Bureau of Land Management [CC-BY-NC-ND-2.0] via Flickr
4.27: "iris" [CC-BY-ND-2.0] via Flickr
4.28: Andrielle Swaby
4.29: Eric Ward [CC-BY-SA-2.0] via Flickr
4.30: Jacob W. Frank [CC-BY-ND-2.0] via Flickr
4.31: Zach Dischner [CC-BY-2.0] via Flickr
4.32: Phil Armitage
4.33: Andrielle Swaby
4.34: Colin Jaskiewicz [CC-BY-2.0] via Flickr
4.35: R. D. Miller, USGS
4.36: Dave Bunnell [CC-BY-SA-2.5] via Wikimedia Commons
Colorado River Box: “Shannon” [CC-BY-SA-4.0] via Wikimedia Commons
Topographic Inversion Box: Jim Houghton
Karst Box: Wade Greenberg-Brand, adpated from image by USGS
Elevation Map: Andrielle Swaby

Chapter 5: Mineral Resources

5.1: Andrielle Swaby, adapted from USGS 2010–2011 and 2012–2013 State Minerals Yearbook
5.2: Andrielle Swaby, adapted from image by US Energy Information Administration
5.3: Wade Greenberg-Brand
5.4: Wade Greenberg-Brand, adapted from image by Swinsto101 [CC-BY-SA-3.0] via Wikimedia Commons
5.5: Wade Greenberg-Brand, adapted from image by the Salt Association
5.6: Doc Searls [CC-BY-2.0] via Wikimedia Commons
5.7: Dagny Gromer [CC-BY-NC-ND-2.0] via Flickr
5.8: Andrielle Swaby, adapted from USGS 2010–2011 and 2012–2013 State Minerals Yearbook
5.9: James St. John [CC-BY-2.0] via Flickr
5.10: Doc Searls [CC-BY-2.0] via Wikimedia Commons
5.11: "Kbh3rd" [CC-BY-SA-3.0] via Wikimedia Commons
5.12: James St. John [CC-BY-2.0] via Flickr
5.13: John Phelan [CC-BY-SA-3.0] via Wikimedia Commons
5.14: Rob Lavinsky, iRocks.com [CC-BY-SA-3.0] via Wikimedia Commons
5.15: "Triceratops56" [CC-BY-ND-2.0] via Flickr
5.16: Andrielle Swaby, adapted from USGS 2010–2011 and 2012–2013 State Minerals Yearbook
5.17: Andrielle Swaby, adapted from image by Colorado Geologic Survey
5.18: W. G. Dayton [CC-BY-ND-2.0] via Flickr
5.19-5.21: James St. John [CC-BY-2.0] via Wikimedia Commons
5.22: Andrielle Swaby, adapted from image by Colorado Geologic Survey
Elements Box: Jane Picconi
Hydrothermal Solutions Box: Jim Houghton

Chapter 6: Energy

6.1: Jim Houghton
6.2: Wade Greenberg-Brand, adapted from image by US Energy Information Administration
6.3: Andrielle Swaby, adapted from image by USGS
6.4: Andrielle Swaby, adapted from image by Peter Nester
6.5: Andrielle Swaby, adapted from image by US Energy Information Administration
6.6: Wade Greenberg-Brand
6.7: Wade Greenberg-Brand, adapted from image by Britannica Online for Kids
6.8: Wade Greenberg-Brand
6.9: US Department of Energy
6.10: Andrielle Swaby, adapted from image by US Energy Information Administration

- 6.11: Paul R. Kucher [CC-BY-2.0] via Wikimedia Commons
- 6.12: Andrielle Swaby, adapted from images by US Energy Information Administration and the National Renewable Energy Laboratory
- 6.13–6.14: US Department of Energy
- 6.15: Adapted from image by National Renewable Energy Laboratory
- 6.16: US Department of Energy
- 6.17: J. N. Stuart [CC-BY-NC-ND-2.0] via Flickr
- 6.18: Bureau of Land Management
- 6.19: Andy Pernick, Bureau of Reclamation
- 6.20: WildEarth Guardians [CC-BY-NC-ND-2.0] via Flickr
- 6.21: Jim Houghton
- 6.22: "Plazak" [CC-BY-SA-4.0] via Wikimedia Commons
- 6.23: Andrielle Swaby, adapted from image by the National Renewable Energy Laboratory
- Oil and Gas Box: Jim Houghton
- Coal Box: Jim Houghton
- Geothermal Box: Wade Greenberg-Brand

Chapter 7: Soils

- 7.1–7.2: Wade Greenberg-Brand, adapted from images by USDA NRCS
- 7.3: Wade Greenberg-Brand
- 7.4: USDA NRCS
- 7.5–7.10: Andrielle Swaby, adapted from images by USDA
- 7.11: Wade Greenberg-Brand, adapted from image by Brady, Nyle (1984), *The Nature and Properties of Soils (9th edition)*, Macmillan, New York, NY
- 7.12: Matt Lavin [CC-BY-SA-2.0] via Flickr
- 7.13: US Forest Service
- 7.14: Jessica Reeder [CC-BY-2.0] via Flickr
- 7.15: Jimmy Thomas [CC-BY-SA-2.0] via Flickr
- 7.16: "Cosmic Kitty" [CC-BY-2.0] via Flickr
- 7.17: Eric Ward [CC-BY-SA-2.0] via Flickr
- 7.18: Brandon Fick [CC-BY-2.0] via Flickr
- 7.19: Kimon Berlin [CC-BY-SA-2.0] via Flickr
- 7.20: Beatrice Murch [CC-BY-2.0] via Flickr

Chapter 8: Climate

- 8.1: Jim Houghton
- 8.2: Robert Rohde [CC-BY-SA-3.0] via Wikimedia Commons
- 8.3: Adapted from Wikipedia
- 8.4-8.5: Adapted from image by Ron Blakey, NAU Geology
- 8.6: Wade Greenberg-Brand, adapted from image by USGS
- 8.7: Wade Greenberg-Brand, adapted from image by William A. Cobban and Kevin C. McKinney, USGS
- 8.8: Wade Greenberg-Brand, adapted from images by Utah Geological Society and Utah Geological Survey
- 8.9–8.10: Adapted from images by Scenarios for Climate Assessment and Adaptation
- 8.11: Adapted from image by Alex Matus [CC-BY-SA-3.0] via Wikimedia Commons
- 8.12: NOAA
- 8.13-8.15: National Climate Assessment
- 8.16: US Bureau of Land Reclamation
- Köppen Climate Box: Wade Greenberg-Brand

Chapter 9: Earth Hazards

- 9.1: USGS
- 9.2: Adapted from image by USGS
- 9.3: R. L. Schuster, US Geological Survey
- 9.4: Jon White, Colorado Geological Survey
- 9.5: USGS
- 9.6: Colorado Department of Transportation
- 9.7: Utah Department of Natural Resources
- 9.8: Wade Greenberg-Brand
- 9.9: Wade Greenberg-Brand, adapted from image by USGS
- 9.10: Arizona Geological Survey
- 9.11: US Fish and Wildlife Service
- 9.12: Wade Greenberg-Brand
- 9.13: USGS
- 9.14-9.15: Utah Geological Survey
- 9.16: Adapted from image by USGS
- 9.17: Lee Ruk [CC-BY-SA-2.0] via Flickr
- 9.18: Gretchen Baker, reproduced with permission

9.19: Adapted from image by Tobin and Weary, USGS
9.20: Louis J. Maher, Jr.
9.21: USGS
9.22: National Cave and Karst Research Institute, Carlsbad, New Mexico, reproduced with permission
9.23: Wade Greenberg-Brand, adapted from image by USGS
9.24: EPA
9.25: Wade Greenberg-Brand, adapted from image by EPA
9.26: "Shannon" [CC-BY-SA-4.0] via Wikimedia Commons
9.27: NOAA
9.28: Utah National Guard, 2-222nd Field Artillery
9.29: NASA
9.30: Adapted from image by Alex Matus [CC-BY-SA-3.0] via Wikimedia Commons
9.31: City of Roswell, New Mexico
9.32: "Junebug172"
9.33: NOAA
9.34: Sgt. William Smith, Fort Carson [CC-BY-2.0] via Flickr
9.35: USGS
9.36: Union of Concerned Scientists

Chapter 10: Fieldwork
10.1–10.2: Paleontological Research Institution
10.3: Don Duggan-Haas

Appendix
A.1–A.3: Next Generation Science Standards